Chicken Health Handbook

Chicken Health Handbook

Edited by Caroline Anderson

MURPHY & MOORE
www.murphy-moorepublishing.com

Published by Murphy & Moore Publishing,
1 Rockefeller Plaza,
New York City, NY 10020, USA
www.murphy-moorepublishing.com

Chicken Health Handbook
Edited by Caroline Anderson

International Standard Book Number: 978-1-63987-103-2 (Hardback)

Cataloging-in-Publication Data

Chicken health handbook / edited by Caroline Anderson.
p. cm.
Includes bibliographical references and index.
ISBN 978-1-63987-103-2
1. Chickens--Health. 2. Poultry--Health. I. Anderson, Caroline.
SF487 .C45 2022
636.5--dc23

TABLE OF CONTENTS

PREFACE

The world is advancing at a fast pace like never before. Therefore, the need is to keep up with the latest developments. This book was an idea that came to fruition when the specialists in the area realized the need to coordinate together and document essential themes in the subject. That's when I was requested to be the editor. Editing this book has been an honour as it brings together diverse authors researching on different streams of the field. The book collates essential materials contributed by veterans in the area which can be utilized by students and researchers alike.

Chicken is one of the common domesticated animal. They are reared for meat and eggs, and are kept as pets. The susceptibility of chickens to various parasites such as lice, fleas, ticks and intestinal worms is high. They are also affected by a number of diseases. These are bird flu, fowl cholera, red mite, blackhead disease, yeast infection, mushy chicken diseases, and various other infection based diseases. These diseases can be prevented by taking precautionary measures such as routine vaccines and use of appropriate antibiotics. Proper feeding techniques, farm hygiene and veterinary care play a crucial role in disease management. This book contains some path-breaking studies in the field of chicken health. It consists of contributions made by international experts. A number of latest researches have been included to keep the readers up-to-date with the global concepts in this area of study.

Each chapter is a sole-standing publication that reflects each author′s interpretation. Thus, the book displays a multi-facetted picture of our current understanding of application, resources and aspects of the field. I would like to thank the contributors of this book and my family for their endless support.

Editor

1

Diversity of *Escherichia coli* strains involved in vertebral osteomyelitis and arthritis in broilers in Brazil

Juliana Fortes Vilarinho Braga[1,2†], Nathalie Katy Chanteloup[2], Angélina Trotereau[2], Sylvie Baucheron[2], Rodrigo Guabiraba[2], Roselene Ecco[1†] and Catherine Schouler[2*]

Abstract

Background: Locomotor disorders and infections by *Escherichia coli* represent major concerns to the poultry industry worldwide. Avian pathogenic *E. coli* (APEC) is associated with extraintestinal infections leading to respiratory or systemic disease known as colibacillosis. The most common lesions seen in cases of colibacillosis are perihepatitis, airsacculitis, pericarditis, peritonitis/salpingitis and arthritis. These diseases are responsible for significant economic losses in the poultry industry worldwide. *E. coli* has been recently isolated from vertebral osteomyelitis cases in Brazil and there are no data on molecular and phenotypic characteristics of *E. coli* strains isolated from lesions in the locomotor system of broilers. This raised the question whether specific *E. coli* strains could be responsible for bone lesions in broilers. The aim of this study was to assess these characteristics of *E. coli* strains isolated from broilers presenting vertebral osteomyelitis and arthritis in Brazil.

Results: Fifteen *E. coli* strains from bone lesions were submitted to APEC diagnosis and setting of ECOR phylogenic group, O serogroup, flagella type, virulence genes content, genetic patterns by Pulsed Field Gel Electrophoresis (PFGE) and Multilocus Sequence Typing (MLST). In addition, bacterial isolates were further characterized through a lethality test, serum resistance test and antibiotic resistance profile. *E. coli* strains harbored different genetic pattern as assessed by PFGE, regardless of flock origin and lesion site. The strains belonged to seven sequence types (STs) previously described (ST117, ST101, ST131, ST 371 and ST3107) or newly described in this study (ST5766 and ST5856). ECOR group D (66.7 %) was the most frequently detected. The strains belonged to diverse serogroups (O88, O25, O12, and O45), some of worldwide importance. The antibiotic resistance profile confirmed strains' diversity and revealed a high proportion of multidrug-resistant strains (73 %), mainly to quinolones and beta-lactams, including third generation cephalosporin. The percentage of resistance to tetracycline was moderate (33 %) but always associated with multidrug resistance.

Conclusions: Our results demonstrated that vertebral osteomyelitis and arthritis in broilers can be associated with highly diverse *E. coli* based on molecular and phenotypic characteristics. There was no specific virulence patterns of the *E. coli* strains associated with vertebral osteomyelitis or arthritis. Also, *E. coli* strains were frequently multidrug resistant and belonged to STs commonly shared by APEC and human ExPEC strains.

Keywords: Broilers, Bacterial infections, APEC, Virulence genes, Pathology, Multidrug-resistant *E. coli*

* Correspondence: catherine.schouler@tours.inra.fr
†Equal contributors
[2]ISP, INRA, Université François Rabelais de Tours, 37380 Nouzilly, France
Full list of author information is available at the end of the article

Background

Escherichia coli is a genetically diverse bacteria comprising non-pathogenic intestinal strains and pathogenic strains responsible for intestinal and extra-intestinal disease [1]. The strains able to cause disease in chickens are known as Avian pathogenic *E. coli* (APEC). APEC is associated with extraintestinal infections leading to respiratory or systemic disease known as colibacillosis. These diseases are responsible for significant economic losses in the poultry industry worldwide, which may occur by decreased hatching rates, mortality, lowered production, carcass condemnation at processing and treatment costs [2]. The most common lesions associated with colibacillosis are perihepatitis, airsacculitis and pericarditis, although other syndromes such as osteomyelitis, arthritis, yolk peritonitis, peritonitis/salpingitis (SPS syndrome), coligranuloma, omphalitis and cellulitis can also be found [3].

Another challenge to modern poultry industry is locomotor disorders, which represent a major economic and welfare problem. Although these disorders may be classified according to underlying pathology as infectious, developmental and degenerative, this classification is difficult since these categories are not mutually exclusive [4]. Infectious conditions such as osteomyelitis, arthritis (or osteoarthritis) and synovitis can be associated with different etiologic agents [3]. Among bacteria, *Staphylococcus* sp. (mainly, *S. aureus*) was isolated from these diseases, although an increase in the incidence of musculoskeletal infection associated with *E. coli* has been reported [5].

Brazil, which is currently the largest exporter and the second largest producer of poultry meat in the world, faces challenges with colibacillosis and locomotor disorders the same form as other countries with relevant poultry industry. There are no data on molecular and phenotypic characteristics of *E. coli* strains isolated from lesions in the locomotor system of broilers, although *E. coli* has been recently isolated from vertebral osteomyelitis cases in Brazil [6]. This raises the question whether specific *E. coli* strain could be responsible for bone lesions in broilers. The aim of our work is to provide data on the phenotypic and molecular characteristics of *E. coli* strains isolated from vertebral osteomyelitis and arthritis cases in broilers from Brazil.

Methods

Samples

Fifteen *E. coli* strains isolated between 2012 and 2014 from broilers presenting vertebral osteomyelitis or arthritis at commercial poultry farms in the state of Minas Gerais, Southeast of Brazil, were studied. The broilers were from eight different flocks, which represent seven different farms. They had variable ages and gender. All experimental procedures were approved by the Universidade Federal de Minas Gerais (UFMG), Committee for Ethics in Animal Experimentation (CETEA) under protocol 205/2011.

Clinical signs and pathology

For clinical examination, broilers presenting locomotor disorders were placed in station and encouraged to move for change in gait and posture assessment. Then, broilers were euthanized by cervical dislocation for necropsy and gross evaluation. The locomotor system was analyzed for size, shape, color, flexibility and breaking strength. The vertebral column of all broilers was sectioned along the longitudinal midline for vertebral body and spinal cord analysis. The free thoracic vertebra was considered as T4. Articulations were analyzed for size and aspects of the synovial fluid in the articular space. Samples for bacterial isolation were collected aseptically from broilers presenting osteomyelitis or arthritis. Tissue sections were fixed in 10 % neutral buffered formalin for 48 to 56 hours. Then, formalin-fixed-vertebral column, intertarsal and femorotibial articulations with lesions were decalcified in 24 % formic acid. For slide preparation, tissues were dehydrated in increasing ethanol concentrations, diaphoanised in xylene, embedded in paraffin to obtain 4-μm thick serial sections and then stained with hematoxylin-eosin (HE) and Goodpasture for further analysis under a light microscope.

Bacterial isolation and identification

Swabs of the lesions were inoculated onto two blood agar (BA) plates and one MacConkey agar (MCK) plate. One BA plate was incubated in microaerophilic conditions at 37 °C for 24 to 72 h, while the others were incubated at the same temperature and time under aerobic conditions. After the initial growth, morphology of isolated colonies was characterized and these same colonies were subcultured, Gram stained and submitted to catalase and oxidase tests. Bacterial isolates were subjected to automatic bacterial identification through VITEK 2 system (bioMérieux, Inc. Hazelwood, MO, USA) using commercially available identification cards for Gram-negative bacteria in accordance to the manufacturer's recommendations. After bacterial identification, the colonies were inoculated into microtubes containing Brain-Heart Infusion (BHI) broth with 30 % glycerol and stored at - 80 °C until subsequent molecular and phenotypic tests described below.

APEC diagnosis tests

The diagnosis of APEC strains was performed by different methods previously described. The ability of *E. coli* strains to induce lethality in 1-day-old specific-pathogen-free (SPF) chicks (detailed on section **Lethality test**) was considered gold standard test to assess strain pathogenicity. In addition, two molecular methods based on genetic profiles were used: 1) detection of minimal predictors described by Johnson et al. [7], which classify an *E. coli* strain as pathogenic based on the minimum detection of four out of five virulence genes (*iroN, ompT, hlyF, iss* and *iutA*); and 2) genotyping method developed by Schouler et al. [8], which is based on the identification of different associations of virulence genes (*iutA*, *sitA*, *aec26*, P (F11) fimbriae, O78, frz_{orf4}) that allow the APEC strains classification in four genetic patterns of virulence (A, B, C and D).

Serogrouping

Determination of O antigens was carried out by agglutination using antisera O1, O2, O5, O8, O15, O18, O25, O45, O78, O88, O111 and O120, according to the method described by Blanco et al. [9]. The O antisera were produced in the Laboratorio de Referencia de *Escherichia coli* (Lugo, Spain). Furthermore, PCR was performed to detect O1, O2, O4, O6, O7, O8, O12, O16, O18, O25a, O45a, O45b, O75, O78, O88 and O104 antigens, as previously described (Additional file 1: Table S1).

Flagellar type

The strains were submitted to PCR to determine flagella type H4, H7, H8, H21 and H25 (Additional file 1: Table S1). Those strains negative for all flagellar types tested by PCR were submitted to motility test. Briefly, bacteria were grown on LB broth overnight. Then, the strains were deeply inoculated in LB plates 0.3 % agar using a Pasteur pipette and then incubated at 37 °C overnight for motility evaluation the following day [10].

ECOR phylogenetic grouping

E. coli strains were classified into the four main ECOR phylogenetic groups by triplex PCR as described by Clermont et al. [11]. Strains were assigned to phylogenetic groups A, B1, B2, or D according to the amplification of the *chuA* and *yjaA* genes and the TspE4C2 fragment. Strains MG1655, ECOR26, ECOR62, and ECOR50 were used as controls for phylogenetic groups A, B1, B2, and D, respectively.

Virulence genotyping

Total DNA extracts were prepared by a rapid boiling method [12]. The presence of genes encoding virulence factors were determined using primers and PCR amplification programs previously described, together with positive control strains (Additional file 1: Table S2).

Single PCR assays were used to detect *sfaS, focG, tsh, ibeA, aatA, neuC, irp2, ireA, sat, vat, astA, fyuA, hlyA, traT, cva/cvi, iucD, hra, iha, pic, csgA, tia, malX* (=rpai), *KpsMTII, cnf 1* and *cnf 2*. Furthermore, some multiplex assays were performed to detect simultaneously *clbB* and *clbN*, and *fimA*, fim_{avMT78} and *fimH*. DNA fragments were amplified in a 25-μL PCR mix containing 1 U of GoTaq®G2 Flexi DNA polymerase (Promega), 12.5 pmol of the forward and reverse primers, and 5 nmol of deoxynucleotide triphosphate mix (Eurogentec) in 1x GoTaq®G2 Flexi buffer. The PCR conditions were as follows: initial denaturation at 94 °C for 4 to 5 min, followed by 30 cycles of 94 °C for 30 s, annealing temperature according to GC-content of primers for at least 30 s, 72 °C for 30 s to 45 s according to the size of the amplified fragment (1 min/kbp), and then a final extension at 72 °C for 7 min.

Pulsed-field gel eletrophoresis (PFGE)

Pulsed-field gel electrophoresis was conducted as previously described [13]. Bacterial cells (equivalent to an OD_{600} of 1.0) grown in BHI broth were harvested by centrifugation. The cellular pellet was resuspended in 500 μL of buffer TE 100 (10 mM Tris/HCl, pH 9, 100 mM EDTA) and incubated for 30 min at 37 °C. The bacterial suspension was mixed with an equal volume of 2.0 % low-melting-point agarose and dispensed into plug molds (Biorad). Agarose plugs were incubated in a lysozyme solution (10 mM Tris/HCl, pH 9, 100 mM EDTA, 5 mg lysozyme ml^{-1}, 0.05 % sarkosyl) for 2 h at 37 °C, and then incubated overnight at 55 °C in a lysis solution (10 mM Tris/HCl, pH 9, 100 mM EDTA, 1 mg proteinase K ml^{-1}, 1 % SDS). After lysis, agarose plugs were washed three times in a TE buffer (10 mM Tris/HCl, pH8, 1 mM EDTA) for 1 h at room temperature, where the first washing buffer was supplemented with 100 mM PMSF (Phenylmethylsulfonyl fluoride). For digestion, plugs were equilibrated in incubation buffer containing *XbaI* restriction enzyme (Takara) overnight. Pulsed-field gel electrophoresis was conducted in a CHEF-DRIII apparatus (Bio-Rad). Gels (1 % agarose) were run at 14 °C for 24 h in TBE buffer (4 mM Tris, 4 mM borate, 1 mM EDTA, pH 8.3) at 6 V cm^{-1}. Pulse times were increased from 10 to 30 s. *XbaI* restriction fragments of *Salmonella enterica* serovar Braenderup H9812 were used as size markers. Cluster analysis using Dice similarity indices was done in BioNumerics 6.6 software (at 0.5 % tolerance and 0.5 % optimization) (Applied Maths, Ghent, Belgium) to generate a dendrogram describing the relationships among PFGE profiles.

Multilocus Sequence Typing (MLST)

The phylogenetic relationships between strains were studied using MLST method initially described by Maiden et al. [14] and *E. coli* Achtman's scheme (http://mlst.warwick.ac.uk/mlst/dbs/Ecoli/). *E. coli* MLST scheme used internal fragments of seven house-keeping genes: *adk* (adenylate kinase), *fumC* (fumarate hydratase), *gyrB* (DNA gyrase), *icd* (isocitrate/isopropylmalate dehydrogenase), *mdh* (malate dehydrogenase), *purA* (adenylosuccinate dehydrogenase) and *recA* (ATP/GTP binding motif). They were amplified in a total volume of 20 μL containing 4 μL of DNA crude extract as a template, 2.5 U of GoTaq®G2 Flexi DNA polymerase (Promega), 10 pmol of each primer, 5 nmol of deoxynucleoside triphosphate 30 mM $MgCl_2$ in 1x buffer. PCR conditions were as follows: 94 °C for 5 min; 30 cycles of 94 °C for 40 s, variable annealing temperature (54 °C, 60 °C, 64 °C, 58 °C, 62 °C, 62 °C or 58 °C, respectively) for 45 s, and 72 °C for 45 s; and a final extension at 72 °C for 5 min. The amplicons were sequenced on both strands and sequence type (ST) of each allele was attributed according to Achtman's scheme. Novel STs described in this work were submitted to the *E. coli* MLST database and identified as ST5856 and ST5766.

Lethality test

Strain virulence was evaluated by a lethality test using 1-day-old chicks as previously described [15]. Lethality test was carried out in the experimental infection unit PFIE (Plateforme d'Infectiologie Expérimentale, INRA Val de Loire). For each strain, groups of five 1-day-old SPF chicks were inoculated subcutaneously with 0.5 mL of an overnight culture in LB-Miller broth without agitation (inoculum in stationary phase was ~10^8 CFU). Mortality was recorded at 4 days post inoculation and the strains were classified as pathogenic when at least one chick died [16]. Avian *E. coli* strains BEN2908 and BEN2269 (a non-pathogenic avian *E. coli* isolate of serogroup O2) were used as positive (5 chicks died) and negative control (no chicks died), respectively. The housing, husbandry and slaughtering conditions conformed to European Guidelines for care and use of laboratory animals. French regional ethics committee number 19 (Comité d'Ethique en Expérimentation Animale Val de Loire) approved this protocol under the reference 2012-11-5.

Serum bactericidal test

The serum bactericidal assay was performed as previously described by Dozois et al. [17] with some modifications. Briefly, bacteria were grown overnight in LB broth at 41 °C with agitation (180 rpm). Then, bacterial cultures were resuspended in fresh medium (OD_{600} = 0.02), incubated at 41 °C with agitation (180 rpm), and harvested during the logarithmic growth phase (DO_{600} = 0.35). Bacteria were washed at room temperature with dulbeco's phosphate-buffered saline (pH 7.0 ~ 7.3) and then resuspended to a concentration of $2x10^6$ CFU/mL. A volume of 500 μL of bacterial suspension was added to 500 μL of complement or inactivated (56 °C, 30 min) SPF chicken serum, which were incubated at 41 °C without agitation. Viable cell counts were counted at 0 h and 3 h by plating 10-fold dilutions in sterile saline solution on LB agar plates. Compared to the bacterial count in inactivated serum, a strain was considered resistant when the bacterial count increased or did not change, intermediate when the bacterial count decreased up to one order of magnitude, and sensitive when bacterial count decreased more than one order of magnitude. Serum resistant (BEN2908) and serum sensitive (BEN4134) *E. coli* strains were used as positive and negative controls.

Antibiotic susceptibility testing

Susceptibility testing was performed by the disk diffusion method according to the guidelines of the Antibiogram Committee of the French Society of Microbiology (http://www.sfm-microbiologie.org). The antibiotics tested belong to seven different classes: aminoglycosides (gentamicin, Gen; neomycin, Neo; apramycin, Apr), beta-lactams (amoxicillin, Amx; amoxicillin + clavulanic acid, Amc), cephalosporins (cephalotin, Cef; cefoxitin, Fox; ceftiofur, Xnl), phenicols (florfenicol, Ffc), polypeptides (colistin, Cst), quinolones (nalidixic acid, Nal; flumequine, UB; enrofloxacin, Enr), sulfonamides (trimethoprim, Tmp; Tmp + sulfamethoxazole, TmpStx), and tetracyclines (tetracycline, Tet). The presence of extended spectrum β-lactamases (ESBL) was detected by double-disk synergy method [18]. *E. coli* ATCC 25922 strain was used as quality control.

Results

Epidemiological features of *E. coli* strains and PFGE

E. coli strains were isolated from eight flocks in the municipalities of Belo Horizonte, Bom Jesus de Amparo, Igarapé, Igaratinga, Itabira and São Sebastião do Oeste, all located in the state of Minas Gerais, Brazil. Management and biosecurity practices varied among the farms, with broilers number per flock ranging from 20,000 to 40,000. Broiler farms usually raise broilers up to approximately 42 to 45 days before processing them. Broilers studied were 40 to 56 days-old (average of 46 days-old). Antibiotics usage in sampled farms included enrofloxacin, fosfomycin, amoxicillin, and trimethoprim sulfa, which were most commonly used to treat respiratory or enteric diseases. Antibiotics such as zinc bacitracin and colistin were confirmed to be frequently used as growth promoters, although information on its use was not available for all farms.

The fifteen *E. coli* strains presented different genetic profiles and revealed to be highly diverse, even for same flock isolates (Fig. 1).

Clinic and pathological findings of the diseases

Vertebral osteomyelitis

The clinic and pathological findings and the total of broilers examined were previously described in details by Braga et al. [6]. Clinically, affected broilers presented partial or total gait impairment according to the degree

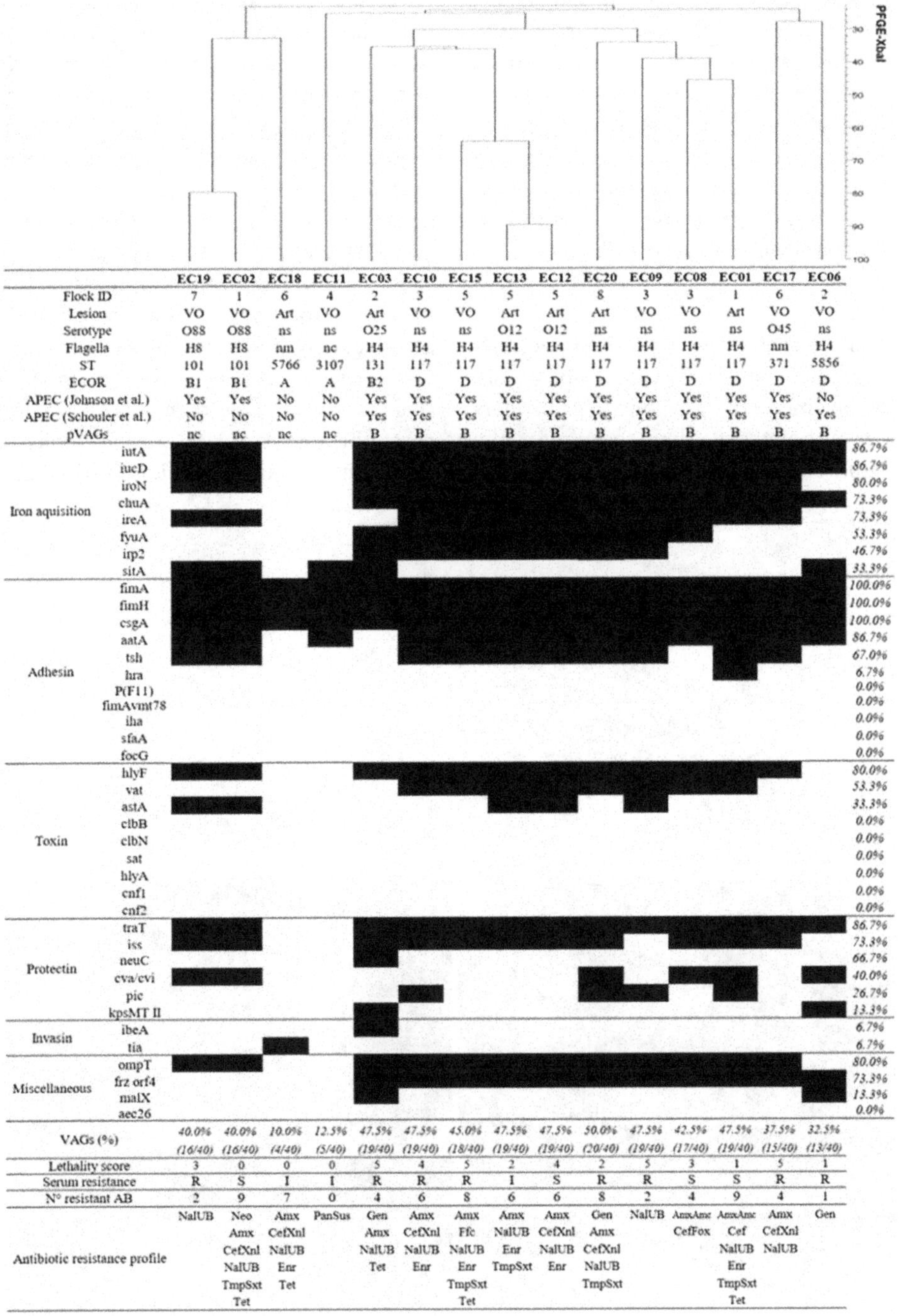

Fig. 1 Molecular and phenotypic characterization of 15 *Escherichia coli* strains isolated from broilers with osteomyelitis and arthritis. Black and white boxes represent positive and negative results, respectively. *Flock ID*, number of the flock of origin; *Lesion, VO*: vertebral osteomyelitis, *Art*: arthritis; *Serotype, ns*: non-serotyped; *Flagella, nm*: non-motile, *nc*: non-correspondent to any flagellar type tested; *ST*, Sequence type; *ECOR*: ECOR phylogenetic group; *APEC (Johnson* et al.*)*: APEC diagnosis according to Johnson et al. (2008); *APEC (Schouler* et al.*)*; APEC diagnosis according to Schouler et al. (2012); *Yes*: APEC strain, *No*: non-APEC strain; *pVAGs*, pattern of virulence genes described by Schouler et al. (2012), *nc*: non-correspondent to the described patterns; *Iron acquisition*, genes encoding iron acquisition system; *Adhesin*, genes encoding adhesins; *Toxin*, genes encoding toxins; *Protectin*, genes encoding protectins; *Invasin*, genes encoding invasins; *Miscellaneous*, genes encoding different kinds of virulence; *VAGs (%)*, percentage of APEC-associated virulence genes; *Lethality score*, number of chicks that died at the fourth day post-infection with *E. coli*; *Serum resistance*, *R*: serum resistant strain, *I*: intermediate resistant strain, *S*: serum sensitive strain; *N° resistant AB*: number of antibiotics to which the strain was resistant; *Antibiotic resistance profile*: gentamicin, Gen; neomycin, Neo; apramycin, Apr; amoxicillin, Amx; amoxicillin + clavulanic acid, Amc; cephalotin, Cef; cefoxitin, Fox; ceftiofur, Xnl; florfenicol, Ffc; colistin, Cst; nalidixic acid, Nal; flumequine, UB; enrofloxacin, Enr; trimethoprim, Tmp; Tmp + sulfamethoxazole, TmpStx; tetracycline, Tet; pansusceptible, PanSus

of vertebral body enlargement (T4 vertebra) and consequent spinal cord compression, which varied from mild to severe (Fig. 2a). Gross lesions were characterized by caseonecrotic osteomyelitis with protrusion of affected vertebral body and spinal cord compression (Fig. 2b, c). Histopathological evaluation of affected vertebral body included caseonecrotic osteomyelitis frequently associated with intralesional Gram-negative bacteria, besides degeneration and necrosis of overlying spinal cord (Fig. 3a).

Arthritis

Broilers presented different degrees of limited mobility depending on the joint lesion site (unilateral or bilateral). When there was involvement of only one leg, broilers could stay in station, although avoiding to support the affected limb on the floor. In bilateral cases, birds often remained in ventral recumbency, with retracted members and supporting their weight on the pectoral muscles (Fig. 2d). Gross evaluation showed swollen of affected joints and, on cut surface, the aspect of lesions varied according to the course of disease. In acute lesions, there was mild to moderate suppurative exudate within synovial fluid and involving articular capsule, occasionally extending to adjacent tendon sheaths, musculature, and subcutaneous tissue (Fig. 2e). In one case, the inflammatory process extended to adjacent proximal tibiotarsus leading to tibial osteomyelitis characterized by indistinct growth plate and metaphysis, which were replaced by necrosuppurative exudate (Fig. 2f). In chronic cases, there was moderate to severe caseofibrinous arthritis. Occasionally, acute arthritis in an antimere and chronic arthritis in the contralateral

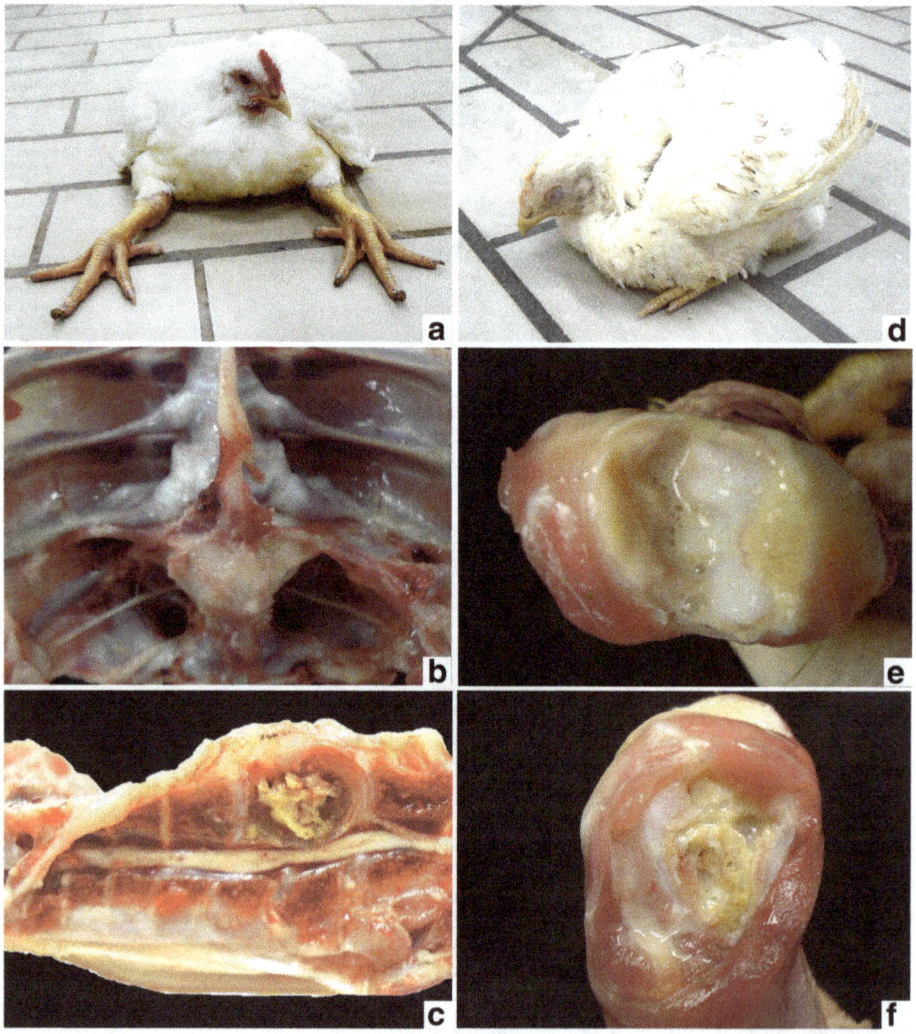

Fig. 2 Clinical signs and gross pathology of vertebral osteomyelitis (**a**, **b**, **c**) and arthritis (**d**, **e**, **f**) in broilers. **a** Broiler showing the classical clinical sign of severe cases of vertebral osteomyelitis. **b** Note the enlargement of affected vertebral body (T4), **c** which revels caseonecrotic material and spinal cord compression on longitudinal section. **d** Broiler with bilateral arthritis showing ventral recumbency and retracted legs. **e** Suppurative exudate in articular cavity in acute arthritis, **f** which extended to proximal tibiotarsus causing tibial osteomyelitis

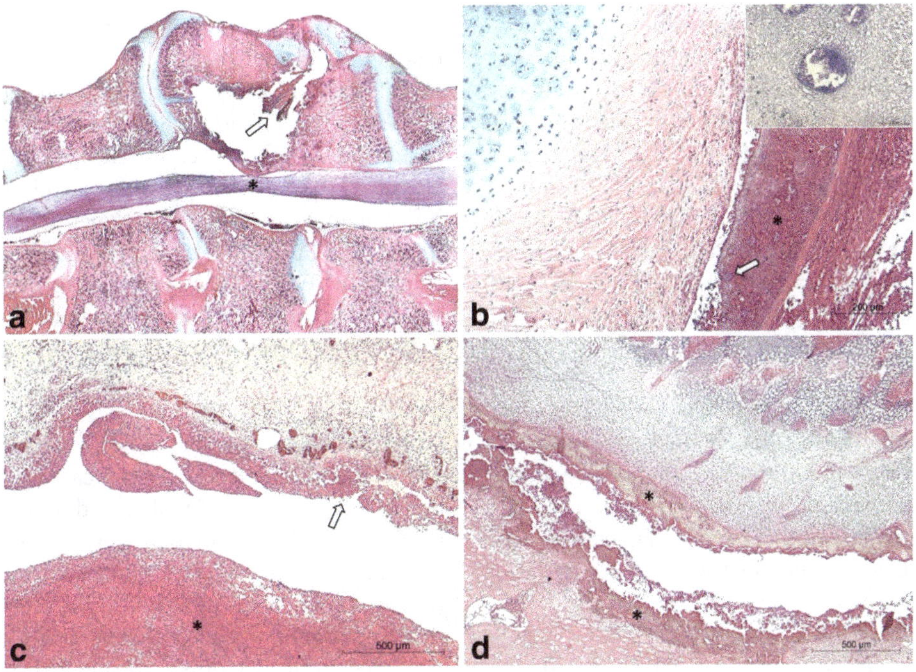

Fig. 3 Histopathology of osteomyelitis and arthritis in broilers. **a** Vertebral osteomyelitis showing enlargment of vertebral body (T4) by caseonecrotic material (remanescent, arrow), which compresses spinal cord (*); HE. **b** Caseonecrotic hererophilic and histiocytic exudate (*) in the articular space with intralesional bacterial colonies (arrow); HE. *Inset*: Gram-negative bacteria stained by Goodpasture. **c** Necrotic synovitis (arrow) associated with caseonecrotic exudate within the articular space (*); HE. **d** Proximal growth plate (physis) of tibiotarsus showing extensive necrosis (*) with heterophilic exudate in a case of tibial osteomyelitis; HE

antimere were observed in the same broiler. Histopathological analysis of acute lesions revealed moderate to intense fibrinoheterophilic and histiocytic arthritis. In chronic cases, there was caseonecrotic heterophilic and histiocytic arthritis, often associated with myriads of intralesional bacterial colonies (Fig. 3b). Furthermore, necrotic synovitis and synovial hyperplasia were occasionally observed (Fig. 3c). In lesions with greater intensity and extension, involvement of adjacent periarticular structures was characterized by degeneration and necrosis of skeletal muscles or osseous tissue (Fig. 3d) associated with hyperemia, infiltration of heterophils and macrophages and fibrin. In addition, proliferation of fibrous tissue in the articular capsule and adjacent tissue was found in more advanced cases.

APEC diagnosis

According to lethality test, 12 *E. coli* strains were considered as APEC, since four strains killed 5 out 5 chicks, two strains killed 4 out 5 chicks, two killed 3 out 5 chicks, two killed 2 out 5 chicks, and two killed 1 out 5 chicks. One strain (EC02) did not kill any chick, but was considered as APEC according to Johnson et al. [7], which classify an *E. coli* strain as pathogenic based on the presence of minimum four out of five virulence genes carried by plasmids associated with highly pathogenic APEC. The results of molecular tests previously described to diagnose APEC showed an agreement of 80.0 % (12/15) (Fig. 1). According to Johnson et al. [7] and Schouler et al. [8], 10 *E. coli* strains were considered APEC and two were considered as non-pathogenic strains. Although there were three discrepancies between both tests, these three APEC strains were diagnosed alternatively by one of the tests and the final criteria for APEC diagnosis was the lethality test.

Group O serotyping and flagella

Different serogroups were detected among the strains, mainly O12 (13.3 %), O88 (13.3 %), O25 (6.7 %), and O45 (6.7 %) (Fig. 1). High percentage (60.0 %) of strains did not correspond to any of the O somatic antigen surveyed in this study, and were classified as non-serotyped (NS). The most prevalent flagellar types were H4 (66.7 %) and H8 (13.3 %), and only for one motile strain it was not possible to identify the corresponding flagella (Fig. 1). H4 was detected in O12, O25 and non-serotyped strains, while O88 strains were H8.

MLST and ECOR phylogroups

The strains were assigned to seven different sequence types (STs) (Fig. 1). Most strains (86.7 %, 13/15) were grouped in known STs, while 13.3 % (2/15) were new STs described in this work. The most frequent ST was ST117 and represented 53.3 % (8/15) of *E. coli* strains, followed by ST101 (13.3 %, 2/15). ST131, ST371 and ST 3107 were identified once each. When classified into ECOR phylogenic groups, most strains were D (66.7 %, 10/15), followed by A (13.3 % 2/15), B1 (13.3 %, 2/15) and B2 (6.7 %, 1/15) (Fig. 1).

Virulence genes profile

APEC strains showed highly variable content of virulence genes, although those responsible for iron acquisition and adhesion were detected more frequently (Fig. 1). The non-pathogenic strains showed marked lack of virulence genes when compared to APEC strains, with higher content of adhesin encoding genes.

Bactericidal effect of serum

High percentage of *E. coli* strains, 53.3 % (8/15), was serum resistant, while 33.4 % (5/15) was characterized as serum sensitive and 13.3 % (2/15) as intermediate strains (Additional file 1: Figure S1).

Antibiotic resistance profile

The *E. coli* strains studied presented a large diversity of antibiotic resistance profiles (Fig. 1). One *E. coli* strain was pansusceptible, but high percentage (73.0 %) of strains were resistant to more than three classes of antibiotics and defined as multidrug-resistant *E. coli*. These eleven multidrug-resistance strains were mainly characterized by resistance to amoxicillin (100.0 %), enrofloxacin (54.5 %), ceftiofur (54.5 %), and tetracycline (45.4 %). The *E. coli* strains were more resistant to nalidixic acid (quinolone class) and amoxicillin (beta-lactam class), 80.0 % and 73.3 % respectively (Fig. 4). Susceptibility or low resistance to polypeptides (0.0 %,) and phenicols (6.7 %,) were observed. One non-pathogenic *E. coli* strain (EC18) was suspected to produce an ESBL by the synergy observed between amoxicillin/clavulanic acid and ceftiofur using the disk diffusion method.

Discussion

Our results showed that *E. coli* strains involved in vertebral osteomyelitis and arthritis cases in broilers in Brazil are highly diverse. We observed that the same disease (i.e., vertebral osteomyelitis) was caused by genetically diverse *E. coli* strains with different pathogenicity traits in the same flock (flock 3). Furthermore, genetically diverse strains were recovered from different diseases (i.e., vertebral osteomyelitis or arthritis) in the same flock (flocks 2, 5 and 6). These findings show that both diseases are not caused by a unique *E. coli* strain. Other authors also report genetic diverse populations of *E. coli* in field cases of colibacillosis in a single flock [19] or in different flocks [20].

E. coli is one of the bacteria described as etiological agent of vertebral osteomyelitis [5]. Recent data on etiological agents of this disease in broilers described involvement of single or mixed bacteria including

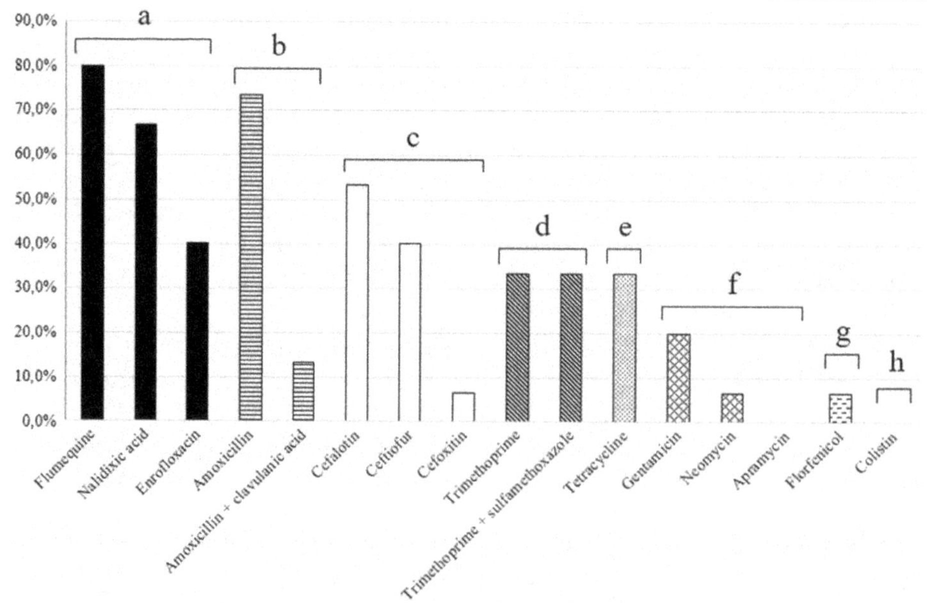

Fig. 4 Percentages of antibiotic resistance of *E. coli* strains isolated from vertebral osteomyelitis and arthritis in broilers by antibiotic class: (**a**) quinolones; (**b**) beta-lactams; (**c**) cephalosporins; (**d**) sulfonamides; (**e**) tetracyclines; (**f**) aminoglycosides; (**g**) phenicols; and (**h**) polypeptides

Enterococcus spp., *E. faecalis*, *E. hirae* and *Staphylococcus aureus*, besides *E. coli* [6]. This feature can be similar to what have been previously proposed on turkey osteomyelitis complex (TOC), in which bacterial arthritis and osteomyelitis are associated to involvement of many different opportunistic microorganisms, suggesting that is likely to be influenced by factors such as immunosuppression rather than by the pathogenicity intensity of these bacteria [21].

Diversity in serogroups among *E. coli* strains was also remarkable, as exemplified by the detection of serogroup O12, which up to now was not described in Brazilian *E. coli* strains from human or animal origin. Strains belonging to this serogroup exhibited profile O12:H4-ST117 and were isolated from two broilers from the same flock presenting only arthritis. Previous studies on serogroup determination of *E. coli* isolated from septicemic and healthy broilers revealed that O12 was involved in only 1 % of colisepticemia cases, but was one of the serogroups predominantly identified among septicemic *E. coli* [22]. O12 *E. coli* strain was also reported in human, isolated from an immunocompetent woman with a history of repeated amnion infections and spontaneous abortion [23].

An *E. coli* strain O45:HNM-D-ST371 was also detected in this study. This type of strain has been described previously in 16.4 % (9/55) of O45 *E. coli* strains isolated from avian colibacillosis cases in Europe and it was identified only in O45 *E. coli* strains of avian origin, different from O45:K1:H7-B2-ST95 identified in avian and human *E. coli* isolates [24]. However, this last one was not detected in Brazilian APEC strains described here and in previous studies [25].

ST 117, which represents more than half of our *E. coli* strains, and ST 131 were involved in osteomyelitis and arthritis cases. These STs are commonly shared by APEC and human ExPEC strains [25]. Close genetic relations have been detected in ST 117 *E. coli* strains of animal and human origin, which have been identified in large poultry producers such as Brazil [25], USA [26], and also Egypt [27], Denmark [28], Sri Lanka [29], and South Korea [30].

We also identified two ST101 APEC strains, which belonged to phylogroup B1, serotype O88:H8 and were non-ESBL as evaluated by the disk diffusion method. This ST was not related to infections caused by APEC until recently, when one O15:H10-B1-ST101 APEC strain was isolated from colibacillosis associated lesions in Spain [20].

ST131 is a globally disseminated multidrug resistance clone, responsible for urinary tract and bloodstream infections in humans. Its rapid emergence and successful spread is strongly associated with antibiotic resistance [31–33]. One O25:H4-B2-ST131 *E. coli* strain was detected in a broiler joint with arthritis in this study. In Brazil, this clone was previously detected in APEC strains recovered from broilers with different visceral lesions [34] and from APEC and human ExPEC collections [25]. Although few data are available on this clonal group from poultry, Mora et al. [35] reported an increasing presence of clonal group O25b:H4-ST131 in retail chickens. Interestingly, a retail chicken sample revealed macrorestriction profile indistinguishable from an *E. coli* strain isolated from a human with urinary tract infection [36].

High percentage of multidrug resistant *E. coli* was detected in this study. It is known that *E. coli* strains isolated from poultry frequently show resistance to more than one antimicrobial drug [37], which represents a global concern. It has been shown that poultry workers may have increased risk of carrying multidrug-resistant *E. coli*, which demonstrates that occupational exposure to antimicrobial-resistant *E. coli* from live-animal contact in the broiler industry may be an important route of entry for antimicrobial-resistant *E. coli* into the community [38].

Most *E. coli* strains analyzed in this study exhibited resistance to at least one antibiotic from different main classes: beta-lactams, cephalosporins and quinolones. Resistance to these antibiotic classes is a chronic problem described for avian *E. coli* strains isolated in Brazil [34, 39]. A concern is the increasing resistance to ceftiofur, which was evident when we compared our *E. coli* strains to those isolated from broilers in previous years in Brazil [34]. This finding is probably the result of increasing usage of this drug in poultry and highlights the need for responsible and controlled use of antibiotics in animals. A major public health concern is that the use of third-generation cephalosporins, such as ceftiofur, in food animals is leading to resistance to other extended-spectrum cephalosporin molecules, which are used in the treatment of many different human infections [40].

Tetracycline resistance level of the *E. coli* strains studied was lower than that described in other regions of Brazil [34, 39] and in other countries, such as China, where resistance to tetracycline can reach about 90 % [41]. For many years, tetracycline was used as prevention and as growth promoter in poultry, but the use of antibiotics with these purposes was banned since 2009 in Brazil. In the state of Minas Gerais, where samples were collected, tetracycline use has no longer being recommended by poultry veterinarians due to its prohibition and bacterial resistance (personal information). These data suggest that the discontinued use of tetracycline in poultry in the region may have provided an increase in the number of *E. coli* strains susceptible to this drug, as described for *Salmonella* strains in USA, where it was observed significant reduction of humans and swine strains resistant to tetracycline after its prohibition as prophylactic drug in animal feed [42].

All 15 *E. coli* strains studied were isolated from the exudate of osteomyelitis or arthritis lesions. The strains EC11 and EC18 classified as non-pathogenic were also isolated from broilers with vertebral osteomyelitis and arthritis, respectively. In these cases, necrotic and inflammatory lesions were associated with bacterial colonies constituted by Gram negative rods, including strains classified as non-APEC. The single or double colonies picked up from the pure culture of *E. coli* probably resulted in the selection of a non-APEC clone, once it is known that in the same lesion it is possible to find distinct *E. coli* clone populations. In order to avoid the selection of a non-representative bacterial clone, it is recommended to select and mix several colonies from the pure culture isolated from the lesion for further evaluation [43]. This procedure provides more efficient results, especially regarding to antimicrobial susceptibility, since it can reduce a possible variation in the susceptibility of isolated clones and improve the selection of antibiotics for treatment.

The broilers had no additional gross lesions in other sites at necropsy, except in two: one exhibited vertebral osteomyelitis and intertarsal arthritis and another broiler had intertarsal arthritis with osteomyelitis in proximal tibiotarsus of the same antimere. Although the information on previous respiratory disease was not available for all flocks studied, it is known that colibacillosis is frequent in broilers of the region (laboratory and field observations). Localization of *E. coli* in bones and synovial tissues is a common sequel of colisepticemia [3]. Some studies with turkeys demonstrated that often multiple sites are involved and the bones most often affected are tibiotarsus, femur, thoracolumbar vertebra, and humerus [44]. Bacteria colonizing the vascular sprouts that invade the physis of a growing bone, provoke an inflammatory response that results in osteomyelitis. Transphyseal blood vessels in birds serve as conduits for the process to spread bacteria into the joint and surrounding soft tissues [45].

Conclusions

Our results showed that highly diverse *E. coli* strains can be recovered from osteomyelitis and arthritis in broilers, even in the same flock. Based on molecular and phenotypic characteristics, there is no specific virulence pattern of the *E. coli* associated with vertebral osteomyelitis or arthritis. Some of the strains involved in these diseases are belonged to STs commonly shared by animals and humans, similar to others previously isolated from different lesions in broilers. Most of these strains are multidrug resistant, with increasing rates of ceftiofur resistance, which is a public and animal health concern. These findings highlight the importance of appropriate management practices, which are valuable in preventing and controlling colibacillosis, thus reducing the need for antibiotics use in animals.

Acknowledgments

The authors thank Nathalie Lallier for her valuable technical assistance on PFGE and lethality tests, and the personnel from the experimental unit PFIE (Plateforme d'Infectiologie Expérimentale) at the INRA- Centre Val de Loire, Nouzilly (France). We also thank Jorge Blanco and Ghizlane Dahbi from Laboratorio de Referencia de *E. coli* (LREC) of the University of Santiago de Compostela (Spain) for kindly providing the antisera.

Funding

This study was supported by INRA (France), Conselho Nacional de Desenvolvimento Científico e Tecnológico (CNPq) project 14/2010 and Coordenação de Aperfeiçoamento de Pessoal de Nível Superior (CAPES), Brazil.

Authors' contributions

RE, JFVB and CS designed of the study. JFVB and RE performed sampling and pathological analysis. JFVB, NKC, AT, RG and CS performed the tests for molecular and phenotypic characteristics of *E. coli*, except antibiotic resistance test, which was performed by JFVB and SB. All the authors reviewed the literature, read and approved the final version of the manuscript.

Competing interests

The authors declare that they have no competing interests.

Author details

[1]Departamento de Clínica e Cirurgia Veterinárias, Escola de Veterinárias, Universidade Federal de Minas Gerais, Av. Antônio Carlos, 6627, Campus Pampulha, 30161-970 Minas Gerais, Brazil. [2]ISP, INRA, Université François Rabelais de Tours, 37380 Nouzilly, France.

References

1. Dale AP, Woodford N. Extra-intestinal pathogenic *Escherichia coli* (ExPEC): Disease, carriage and clones. J Infect. 2015;71:615–26. doi:10.1016/j.jinf.2015.09.009.
2. La Ragione RM, Woodward MJ. Virulence factors of *Escherichia coli* serotypes associated with avian colisepticaemia. Res Vet Sci. 2002;73:27–35. doi:10.1016/S0034-5288(02)00075-9.
3. Barnes HJ, Nolan LK, Vaillancourt JP. Colibacillosis. In: Saif YM, Fadly AM, Glisson JR, McDougald LR, Nolan LK, Swayne DE, editors. Diseases of Poultry. Ames (Iowa): Blackwell Publishing; 2008. p. 691–732.
4. Bradshaw RH, Kirkden RD, Broom DM. A review of the aetiology and pathology of leg weakness in broilers in relation to welfare. Avian Poult Biol Rev. 2002;13:45–103. doi:10.3184/147020602783698421.
5. Riddell C. Leg problems still important. Poult Diag. 1997;56:28–31.
6. Braga JFV, Silva CC, Teixeira MPF, Martins NRS, Ecco R. Vertebral osteomyelitis associated with single and mixed bacterial infection in

broilers. Avian Path. Published on line first July 2016. doi: 10.1080/03079457.2016.1193843.

7. Johnson TJ, Wannemuehler Y, Doetkott C, Johnson SJ, Rosenberger SC, Nolan LK. Identification of minimal predictors of Avian Pathogenic *Escherichia coli* virulence for use as a rapid diagnostic tool. J Clin Microbiol. 2008;46:3987–96. doi:10.1128/JCM.00816-08.
8. Schouler C, Schaeffer B, Brée A, Mora A, Dahbi G, Biet F, Oswald E, Mainil J, Blanco J, Moulin-Schouleur M. Diagnostic strategy for identifying avian pathogenic *Escherichia coli* based on four patterns of virulence genes. J Clin Microbiol. 2012;50:1673–78. doi:10.1128/JCM.05057-11.
9. Blanco M, Blanco JE, Blanco J, González EA, Mora A, Prado C, Fernández L, Rio M, Ramos J, Alonso MP. Prevalence and characteristics of *Escherichia coli* serotype O157:H7 and other verotoxin-producing *E. coli* in healthy cattle. Epidemiol Infect. 1996;117:251–7. doi:10.1017/S0950268800001424.
10. Clarke MB, Sperandio V. Transcriptional regulation of *flhDC* by QseBC and σ^{28} (FliA) in enterohaemorrhagic *Escherichia coli*. Mol Microbiol. 2005;57:1734–49. doi:10.1111/j.1365-2958.2005.04792.
11. Clermont O, Bonacorsi S, Bingen E. Rapid and simple determination of the *Escherichia coli* phylogenetic group. Appl Environ Microbiol. 2000;66:4555–8. doi:10.1128/AEM.66.10.4555-4558.2000.
12. Sambrook J, Fritsch EF, Maniatis T. Molecular cloning: a laboratory manual. New York: Cold Spring Harbor Laboratory Press; 1989.
13. Moulin-Schouleur M, Schouler C, Tailliez P, Kao MR, Brée A, Germon P, Oswald E, Mainil J, Blanco M, Blanco J. Common virulence factors and genetic relationships between O18:K1:H7 *Escherichia coli* isolates of human and avian origin. J Clin Microbiol. 2006;44:3484–2. doi:10.1128/JCM.00548-06.
14. Maiden MC, Bygraves JA, Feil E, Morelli G, Russell JE, Urwin R, Zhang Q, Zhou J, Zurth K, Caugant DA, Feayers IM, Achtman M, Spratt BG. Multilocus sequence typing: a portable approach to the identification of clones within populations of pathogenic microorganisms. Proc Natl Acad Sci. 1998;95:3140–5. doi:10.1073/pnas.95.6.3140.
15. Dho M, Lafont JP. Adhesive properties and iron uptake ability in *Escherichia coli* lethal and nonlethal for chicks. Avian Dis. 1984;28:1016–25. doi:10.2307/1590278.
16. Dozois CM, Dho-Moulin M, Brée A, Fairbrother JM, Desautels C, Curtiss R. Relationship between the Tsh autotransporter and pathogenicity of avian *Escherichia coli* and localization and analysis of the *tsh* genetic region. Infect Immun. 2000;68:4145–54. doi:10.1128/IAI.68.7.4145-4154.2000.
17. Dozois CM, Fairbrother JM, Harel J, Bossé M. Pap-and pil-related DNA sequences and other virulence determinants associated with *Escherichia coli* isolated from septicemic chickens and turkeys. Infect Immun. 1992;60:2648–56. DOI http://dx.doi.org/0019-9567/92/072648-09.00/0.
18. Jarlier V, Nicolas M, Fournier G, Philippon A, Jarlier V, Nicolas M, Fournier G, Philippon A. Extended broad-spectrum, 3-lactamases conferring transferable in resistance to newer 13-lactam hospital agents Enterobacteriaceae: prevalence and susceptibility patterns P-lactamase of nosocomial infections. Rev Infect Dis. 1998;10:867–78. doi:10.1093/clinids/10.4.867.
19. Ozaki H, Murase T. Multiple routes of entry for *Escherichia coli* causing colibacillosis in commercial layer chickens. J Vet Med Sci. 2009;71:1685–9. doi:10.1292/jvms.001685.
20. Solà-Ginés M, Cameron-Veas K, Badiola I, Dolz R, Majó N, Dahbi G, Viso S, Mora A, Blanco J, Piedra-Carrasco N, González-López JJ, Migura-Garcia L. Diversity of multi-drug resistant Avian Pathogenic *Escherichia coli* (APEC) causing outbreaks of colibacillosis in broilers during 2012 in Spain. PLoS One. 2015;10:e0143191. doi:10.1371/journal.pone.0143191.
21. Huff GR, Huff WE, Rath NC, Balog JM. Turkey osteomyelitis complex. Poult Sci. 2000;79:1050–6. doi:10.1093/ps/79.7.1050.
22. Blanco JE, Blanco M, Mora A, Jansen WH, García V, Vázquez ML, Blanco J. Serotypes of *Escherichia coli* isolated from septicaemic chickens in Galicia (Northwest Spain). Vet Microbiol. 1998;61:229–35. doi:10.1016/S0378-1135(98)00182-5.
23. Blum-Oehler G, Heesemann J, Kranzfelder D, Scheutz F, Hacker J. Characterization of *Escherichia coli* serotype O12:K1:H7 isolates from an immunocompetent carrier with a history of spontaneous abortion and septicemia. Eur J Clin Microbiol Infect Dis. 1997;16:153–5. doi:10.1007/BF01709475.
24. Mora A, Viso S, López C, Alonso MP, García-Garrote F, Dabhi G, Maman R, Herrera A, Marzoa J, Blanco M, Blanco JE, Moulin-Schouleur M, Schouler C, Blanco J. Poultry as reservoir for extraintestinal pathogenic *Escherichia coli* O45:K1:H7-B2-ST95 in humans. Vet Microbiol. 2013;167:506–2. doi:10.1016/j.vetmic.2013.08.007.
25. Maluta RP, Logue CM, Casas MRT, Meng T, Guastalli EAL, Rojas TCG, Montelli AC, Sadatsune T, Ramos MDC, Nolan LK, Silveira WD. Overlapped sequence types (STs) and serogroups of avian pathogenic (APEC) and human extra-intestinal pathogenic (ExPEC) *Escherichia coli* isolated in Brazil. PLoS One. 2014;9:1–9. doi:10.1371/journal.pone.0105016.
26. Danzeisen JL, Wannemuehler Y, Nolan LK, Johnson TJ. Comparison of multilocus sequence analysis and virulence genotyping of *Escherichia coli* from live birds, retail poultry meat, and human extraintestinal infection. Avian Dis. 2013;57:104–8.
27. Hussein AH, Ghanem IA, Eid AA, Ali MA, Sherwood JS, Li G, Nolan LK, Logue CM. Molecular and phenotypic characterization of *Escherichia coli* isolated from broiler chicken flocks in Egypt. Avian Dis. 2013;57:602–11.
28. Pires-dos-Santos T, Bisgaard M, Christensen H. Genetic diversity and virulence profiles of *Escherichia coli* causing salpingitis and peritonitis in broiler breeders. Vet Microbiol. 2013;162:873–80.
29. Dissanayake DRA, Octavia S, Lan R. Population structure and virulence content of avian pathogenic *Escherichia coli* isolated from outbreaks in Sri Lanka. Vet Microbiol. 2014;168:403–12.
30. Lim JS, Choi DS, Kim YJ, Chon JW, Kim HS, Park HJ, Moon JS, Wee SH, Seo KH. Characterization of Escherichia coli-Producing Extended-Spectrum β-Lactamase (ESBL) Isolated from Chicken Slaughterhouses in South Korea. Foodborne Pathogens and Dis. 2015;12:741–8. doi:10.1089/fpd.2014.1921.
31. Nicolas-Chanoine MH, Blanco J, Leflon-Guibout V, Demarty R, Alonso MP, Caniça MM, Park YJ, Lavigne JP, Pitout J, Johnson JR. Intercontinental emergence of *Escherichia coli* clone O25:H4-ST131 producing CTX-M-15. J Antimicrob Chemother. 2008;61:273–81.
32. Peirano G, Pitout JDD. Molecular epidemiology of Escherichia coli producing CTX-M β-lactamases: the worldwide emergence of clone ST131 O25:H4. Int J Antimicrob Agents. 2010;35:316–21.
33. Rogers BA, Sidjabat HE, Paterson DL. *Escherichia coli* O25b-ST131:pandemic, multiresistant, community-associated strain. J Antimicrob Chemother. 2011; 66:1–14. doi:10.1093/jac/dkq415.
34. Barbieri NL, Oliveira AL, de Tejkowski TM, Pavanelo DB, Matter LB, Pinheiro SRS, Vaz TM, Nolan LK, Logue, CM, Brito BG, de Horn F. Molecular characterization and clonal relationships among *Escherichia coli* strains isolated from broiler chickens with colisepticemia. Foodborne Pathog Dis. 2015;12:74–83. doi:10.1089/fpd.2014.1815.
35. Mora A, Herrera A, Mamani R, López C, Alonso MP, Blanco JE, Blanco M, Dahbi G, García-Garrote F, Pita JM, Coira A, Bernárdez MI, Blanco J. Recent emergence of clonal group O25b:K1:H4-B2-ST131 ibeA strains among *Escherichia coli* poultry isolates, including CTX-M-9-producing strains, and comparison with clinical human isolates. Appl Environ Microbiol. 2010;76: 6991–7. doi:10.1128/AEM.01112-10.
36. Vincent C, Boerlin P, Daignault D, Dozois CM, Dutil L, Galanakis C, Reid-Smith RJ, Tellier PP, Tellis PA, Ziebell K, Manges A R. Food reservoir for *Escherichia coli* causing urinary tract infections. Emerg Infect Dis. 2010;16:88–95.
37. Zanatta GF, Kanashiro AMI, Castro AGM, Cardoso ALSP, Tessari ENC, Pulici SCP. Suscetibilidade de amostras de *Escherichia coli* de origem aviária a antimicrobianos. Arq Inst Biol. 2004;71:283–6.
38. Price LB, Graham JP, Lackey LG, Roess A, Vailes R, Silbergeld E. Elevated risk of carrying gentamicin-resistant *Escherichia coli* among U.S. poultry workers. Environ Health Perspect. 2007;115:1738–42. doi:10.1289/ehp.10191.
39. Korb A, Nazareno ER, Costa LD, Nogueira KS, Dalsenter PR, Tuon FFB, Pomba MC. Tipagem molecular e resistência aos antimicrobianos em isolados de *Escherichia coli* de frangos de corte e de tratadores na Região Metropolitana de Curitiba, Paraná. Pesq Vet Bras. 2015;35:258–64. doi:10.1590/S0100-736X2015000300008.
40. Zhao S, White DG, Mcdermott PF, Friedman S, English L, Ayers S, Maurer JJ, Holland R, Walker RD, Meng J. Identification and expression of cephamycinase bla_{CMY} genes in *Escherichia coli* and *Salmonella* isolates from food animals and ground meat. Antimicrob Agents Ch. 2001;45:3647–50. doi:10.1128/AAC.45.12.3647.
41. Zhang T, Wang CG, Lv JC, Wang RS, Zhong XH. Survey on tetracycline resistance and antibiotic-resistant genotype of avian *Escherichia coli* in North China. Poult Sci. 2012;91:2774–7. doi:10.3382/ps.2012-02453.

42. Manie T, Khan S, Brözel VS, Veith WJ, Gouws PA. Antimicrobial resistance of bacteria isolated from slaughtered and retail chickens in South Africa. Lett Appl Microbiol. 1998;26:253–8. doi:10.1046/j.1472-765X.1998.00312.x.
43. Clermont O, Glodt J, Burdet C, Pognard D, Lefort A, Branger C, Denamur E. Complexity of *Escherichia coli* bacteremia pathophysiology evidenced by comparison of isolates from blood and portal of entry within single patients. Int J Med Microbiol. 2013;303:529–32. doi:10.1016/j.ijmm.2013.07.002.
44. Mutalib A, Miguel B, Brown T, Maslin W. Distribution of Arthritis and Osteomyelitis in Turkeys with Green Liver Discoloration. Avian Dis. 1996;40: 661–4. doi:10.2307/1592278.
45. Bayyari GR, Huff WE, Rath NC, Balog JM, Newberry LA, Villine JD, Skeeles JK. Immune and physiological responses of turkeys with green-liver osteomyelitis complex. Poult Sci. 1997;76:280–8.

2

A multiplex xTAG assay for the simultaneous detection of five chicken immunosuppressive viruses

Feng Cong[1†], Yujun Zhu[1†], Jing Wang[1], Yuexiao Lian[1,2], Xiangnan Liu[1,2], Li Xiao[1], Ren Huang[1], Yu Zhang[1], Meili Chen[1*] and Pengju Guo[1*]

Abstract

Background: Chicken anemia virus (CAV), avian reovirus (ARV), infectious bursal disease virus (IBDV), Marek's disease virus (MDV) and reticuloendotheliosis virus (REV) all cause immunosuppressive disease in birds through vertical or horizontal transmission. Mixed infections with these immunosuppressive pathogens lead to atypical clinical signs and obstruct accurate diagnoses and epidemiological investigations. Therefore, it is essential to develop a high-throughput assay for the simultaneous detection of these immunosuppressive viruses with high specificity and sensitivity. The aim of this study was to establish a novel method using a RT-PCR assay combined with fluorescence labeled polystyrene bead microarray (multiplex xTAG assay) to detect single or mixed viral infections.

Results: The results showed that the established xTAG assay had no nonspecific reactions with avian influenza virus (AIV), infectious bronchitis virus (IBV), newcastle disease virus (NDV), infectious laryngotracheitis virus (ILTV), *Mycoplasma gallisepticum* (MG) and *Mycoplasma synoviae* (MS). The limit of detection was 1.0×10^3 copies/μL for IBDV and 1.0×10^2copies/μL for the other four viruses. Ninety field samples were tested and the results were confirmed using conventional RT-PCR methods. The detection results of these two methods were 100% consistent. The established multiplex xTAG assay allows a high throughput and simultaneous detection of five chicken immunosuppressive viruses.

Conclusion: The multiplex xTAG assay has been showed to be an additional tool for molecular epidemiology studies of five chicken immunosuppressive viruses in the poultry industry.

Keywords: Simultaneous detection, Multiplex xTAG assay, Chicken anemia virus, Avian reovirus, Chicken immunosuppressive viruses

Background

Immunosuppression caused by viral infections is a common syndrome in the poultry industry. The result is temporary or permanent immune suppression, leaving the flock highly susceptible to other pathogenic agents. Young bird flocks are susceptible to five main immunosuppressive viruses, such as avian reovirus (ARV), chicken anemia virus (CAV), infectious bursal disease virus (IBDV), Marek's disease virus (MDV) and reticuloendotheliosis virus (REV) [1–11]. Immunosuppression often leads to considerable losses in poultry production including a negative impact on vaccinations with low protection rates. Immunosuppression is especially problematic because primary disease control strategies frequently rely on efficient vaccines. In addition, several reports showed that co-infection with two or more immunosuppressive viruses either experimentally or through natural exposure enhanced the pathogenicity [5, 7, 8, 12–16]. Therefore, a rapid and cost-effective early diagnostic method would help prevent transmission within the farm.

Virus isolation and electron microscopy is the current standard for virus identification. However, these processes are time and labor intensive and require technical proficiency in these methods. Serological tests are

* Correspondence: chml@gdlami.com; vetbio2016@hotmail.com
[†]Feng Cong and Yujun Zhu contributed equally to this work.
[1]Guangdong Laboratory Animals Monitoring Institute and Guangdong Provincial Key Laboratory of Laboratory Animals, Guangzhou 510633, China

sufficient to separate infected birds and are widely used on poultry production farms although immunosuppressed animals may test as falsely negative. Molecular biology techniques are now in use for characterization of viral pathogens. Conventional PCR and RT-PCR tests detect specific genes from target viral genomes. Agarose gel electrophoresis-based detection is insensitive compared with real-time and multiplex PCR/RT-PCR assays [17–19]. These latter approaches provide a fast and effective alternative with good sensitivity and specificity but the number of fluorophores is limited for multiplex real-time PCR instruments.

An alternative to these procedures is the commercial MagPlex-TAG system that enables high-throughput detection of multiple analytes in a single reaction. This allows simultaneous detection of viral target in a mixed infection [20, 21]. The method involves specific primers designed for target sequence amplification by PCR. The amplicons are detected by incubations with beads and streptavidin-R-phycoerythrin. In the assay reader, lasers identify the bead's color set and the phycoerythrin reporter dye. In this study, a multiplex xTAG assay was established to distinguish five avian pathogens.

Results

Specificity, sensitivity and reproducibility tests

The specificity of each pair of target primers for viral sequence amplification was examined using DNA and RNA templates by both the multiplex xTAG assay and conventional PCR. There were no cross-reactions with avian influenza virus (AIV), infectious bronchitis virus (IBV), newcastle disease virus (NDV), infectious laryngotracheitis virus (ILTV), *Mycoplasma gallisepticum* (MG) and *Mycoplasma synoviae* (MS) (see Fig. 1 and Additional file 1: Table S1).

The minimum requirement for detection of target sequences was determined as the median fluorescence intensity (MFI) of the replicates above the cut-off value from the highest plasmid or RNA dilution. The limit of detection was 1.0×10^3copies/μL RNA for IBDV and 1×10^2copies/μL for ARV, CAV, MDV and REV. The sensitivity of multiplex xTAG assay for CAV was ten-fold higher than conventional PCR and had similar sensitivity for the other four viruses (see Fig. 2 and Table 1).

DNA or RNA standards at the range of 1×10^5 and 1×10^7 copies/μL were prepared for reproducibility test. The inter-assay variation fell between 1.4 to 4.6%. The intra-assay variations were blind tested in duplicates by another user using three parallel reactions at 1×10^5 and 1×10^7 copies/μL. These variations ranged from 1.2 to 3.3% (Table 2).

Application of field samples

Among 90 field samples tested, six samples were identified as ARV positive and one specimen as IBDV positive. Eight samples were MDV positive only, fourteen samples were CAV/MDV co-infections, six were REV positive only, one was CAV/REV co-infections and one case was a CAV/MDV/REV mixed infection. The rest of samples were CAV positive with an occurrence rate of 67.8%. DNA sequence analysis verified that the amplified target fragments were identical to specific viral sequences (Table 3 and Additional file 2: Table S2).

Artificial mixed infection analysis

To evaluate potential interference of the xTAG assay in multiplex infection, artificial mixed samples were obtained by spiking. Three samples were prepared as double infections and two samples were made as triple infection. Meanwhile one sample was prepared as CAV/

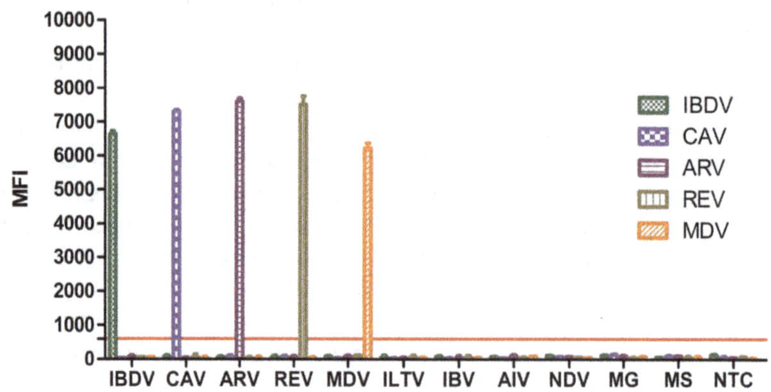

Fig. 1 Result of the specificity testing. Viral target specificity was tested in a multiplex mode using target-specific primers. The targets of IBDV, CAV, ARV, REV and MDV were tested against each other and the non-target viral pathogens: avian influenza virus (AIV), infectious bronchitis virus (IBV), newcastle disease virus (NDV), infectious laryngotracheitis virus (ILTV), Mycoplasma gallisepticum (MG) and Mycoplasma synoviae (MS). The cut-off value was 600, defined as the mean of the net MFI from negative PCR controls with three times of standard deviation (MFI +3SD). All the samples were performed in triplicate

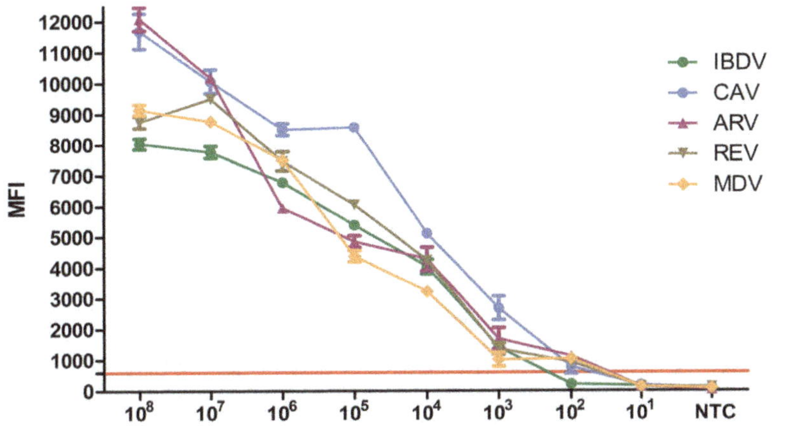

Fig. 2 Result of Sensitivity testing. The concentration of DNA or RNA standards ranged from 1×10^8 copies/ul to 1×10^1 copies/ul. The cut-off value was 600. the limit of detection was 1×10^3 copies/ul for IBDV and 1×10^2 copies/ul for CAV, ARV, REV, MDV. All the samples were performed in triplicate

MDV/ARV/REV mixed infection and one was CAV/MDV/ARV/REV/IBDV co-infection (Fig. 3). The results showed that the developed multiplex xTAG assay is no potential interference when multiple viral targets were present.

Discussion

Immunosuppression among chicken flocks causes significant economic losses in the poultry industry. These immunosuppressed birds are highly susceptibility to secondary infections with other viruses or bacteria. Mixed infections with two or more pathogens would obscured typical clinical signs and increase the difficulty for accurate diagnosis [1, 22]. Mixed viral infections are common in bird flocks and infections with CAV, IBDV, REV, ARV and MDV all lead to a decline of humoral and cell-mediated immunity [1, 3, 4, 7, 9]. An assay that can detect multiple viruses simultaneously would greatly enhance clinical diagnoses. In comparison with conventional PCR and ELISA, the multiplex PCR assay allows multiplexing and reduces reagent consumption and material preparation time. However, multiple primer pair interactions in the multiplex PCR assay can adversely affect accuracy [23].

The multiplex xTAG assay established in this study is available to support high-throughput and simultaneous molecular diagnosis of five immunosuppressive viral pathogens in a cost-effective manner with high specificity and sensitivity. Target selection is important to provide a successful diagnostic method [20]. For example, although CAV sero-positive distribution in chickens is worldwide, there is no significant genetic difference for isolates from various regions and only one serotype has been classified [24]. Based on sequence alignments, the

Table 1 The average of the MFI values with SD of sensitivity test by the xTAG assay and conventional PCR/RT-PCR

Sample (copies/ul)	IBDV		CAV		ARV		REV		MDV	
	xTAG assay		xTAG assay		xTAG assay		xTAG assay		xTAG assay	
	MFI values ±SD	RT-PCRP/N	MFI values ±SD	RT-PCRP/N	MFI values ±SD	RT-PCRP/N	MFI values ±SD	RT-PCRP/N	MFI values ±SD	RT-PCRP/N
10^8	8035 ± 178	+	11,687.5 ± 574	+	12,380 ± 382	+	8742.5 ± 191	+	9143.5 ± 172	+
10^7	7764 ± 84	+	10,063.5 ± 279	+	10,208 ± 465	+	9302 ± 170	+	8702 ± 152	+
10^6	6782 ± 138	+	8503 ± 191	+	5944 ± 119	+	7470.5 ± 310	+	7496 ± 169	+
10^5	5380.5 ± 110	+	8567 ± 185	+	4803 ± 135	+	5938 ± 97	+	4350 ± 198	+
10^4	4034 ± 230	+	5107.5 ± 72	+	4277 ± 382	+	4233 ± 134	+	3227 ± 157	+
10^3	1402 ± 38	+	2680.5 ± 122	+	1683 ± 347	+	1358.5 ± 182	+	1806.5 ± 220	+
10^2	215 ± 184	–	767.5 ± 206	+	1121 ± 26	+	924 ± 37	+	1039 ± 128	+
10^1	167 ± 138	–	191.5 ± 139	–	137 ± 38	–	141.5 ± 46	–	138.5 ± 59	–
blank	78 ± 45	–	89 ± 17	–	48 ± 35	–	128 ± 83	–	85 ± 21	–

MFI median fluorescent intensity
+: positive (P)
-: negative (N)

Table 2 CV% of reproducibility tests

	Copies/uL	Intra-assay MFI			CV(%)	Inter-assay MFI			CV(%)
		1	2	3		1	2	3	
CAV	1×10^5	8684	8353	8664	2.1	8567	8216	8317	2.2
	1×10^7	10,177.5	9745	10,268	2.7	10,063.5	9602	10,057	2.6
MDV	1×10^5	4374.5	4536	4142	4.6	4350	4637	4575	3.3
	1×10^7	8761	8816	8529	1.8	8702	8923	8627	1.8
IBDV	1×10^5	5420.5	5465.5	5255.5	2.1	5380.5	5268	5528	2.4
	1×10^7	7764	7848	7680	1.4	7764	7901	7658	1.6
ARV	1×10^5	4947	4678	4784	2.8	4803	4917	4639	2.9
	1×10^7	10,082	10,723.5	9819	4.6	10,208	9957	10,084	1.2
REV	1×10^5	6051	5874	5890	1.6	5938	5987	5703	2.5
	1×10^7	9499	9213.5	9196	1.8	9302	9248.5	9645	2.3

VP1 gene was selected for CAV detection due to its function as main structural protein for capsid assembly. On the other hand, the IBDV strains have multiple presentations of antigenicity and virulence. A very virulent IBDV emerged three decades ago and had a high mortality rate. Therefore, the *VP2* gene was the better choice for the multiplex xTAG assay because it is able to induce neutralizing antibody and possesses a highly conserved region that can differentiate between vaccine and wild type strains [25]. For MDV, the *MEQ* gene is a major latency antigen expressed in the oncogenic serotype 1 but not serotypes 2 and 3 [9, 26]. MEQ is ideal to distinguish between these serotypes in the PCR assay. The *ENV* gene in REV encodes a glycoprotein that comprises the binding pocket for neutralizing antibody. The σC protein expressed by the *S1* gene in ARV mediates viral attachment and antibody-specific neutralization [5, 27–29]. Thus the conserved regions in these two genes were chosen for primer design.

The PH value of the hybridization buffer is also important to MFI value. When the PH value is 8.0, the MFI value of the negative sample is the lowest, the MFI value of the positive sample is the highest. When the PH value is not optimal, the cut-off value is higher, which lead to the false negative results.

The specificity and sensitivity of xTAG-multiplex PCR assay were determined in this study and it was found no cross-reactions with other common chicken viral pathogens, indicating that xTAG-multiplex PCR assay had a high specificity. The limit of detection of xTAG-Multiplex PCR assay for IBDV was 1.0×10^3copies/μL while the limit of detection for the other four viruses was $> 1.0 \times 10^2$copies/μL. Compared to conventional PCR, the sensitivity of xTAG multiplex PCR for CAV was ten-fold higher while it remains similar sensitivity for the other four viruses.

Among 90 field samples, the detection results using Multiplex xTAG assay were 100% consistent with those tested by conventional PCR assay, demonstrated that the multiplex xTAG assay is high-throughput and suitable for molecular epidemiology studies of five immunosuppression viruses. More than 60% CAV positive rate revealed that the viral infection control strategies on the farm were not fully successful. Five cases of mixed infections of CAV and MDV were found based on the result of xTAG assay, which may induce a synergistic effect on infection and cause severe immunosuppression. A high positive infection rate of CAV also suggests vaccination failure. An annual veterinary evaluation would do much to draw attention to potential problems of immunosuppression and vaccine failure.

Conclusion

The multiplex xTAG assay can perform the simultaneous detection of five avian immunosuppressive viruses in one reaction. This will enhance clinical diagnoses and epidemiological investigations.

Methods

Viruses and field samples

The virus strains IBDV B87, REV C15 and the vaccine strain MDV 814 were purchased from Guangdong Animal Epidemic Prevention and Material Reserve Center. Cell culture adapted strains CAV GD-2014, ARV GD-2, Avian influenza virus (AIV) H7N2 subtype, newcastle disease virus (NDV) F48E9 strain, infectious bronchitis virus (IBV) Massachusetts 41, infectious laryngotracheitis virus (ILTV) N-71851 strain (ATCC VR-783) were maintained in our laboratories. *Mycoplasma gallisepticum* (MG) and *Mycoplasma synoviae* (MS) were purchased from the China Veterinary Culture Collection.

90 chicken samples (cloacal swabs, livers, spleens and kidneys) from a poultry farm in Guangdong province were submitted to our laboratory at Guangdong Laboratory Animals Monitoring Institute for routine animal health surveillance.

Table 3 Screening results for 90 clinical samples using the xTAG assay

Sample	CAV	MDV	IBDV	ARV	REV	Sample	CAV	MDV	IBDV	ARV	REV
T1	+++	–	–	–	–	T46	++	++	–	–	–
T2	+++	–	–	–	–	T47	++	+	–	–	–
T3	+++	–	–	–	–	T48	++	++	–	–	–
T4	+++	–	–	–	–	T49	++	++	–	–	–
T5	+++	–	–	–	–	T50	++	+	–	–	–
T6	+++	+	–	–	–	T51	++	–	–	–	–
T7	+++	–	–	–	–	T52	++	+	–	–	–
T8	+++	–	–	–	–	T53	++	++	–	–	–
T9	+++	–	–	–	–	T54	++	++	–	–	–
T10	+++	–	–	–	–	T55	++	++	–	–	–
T11	+++	–	–	–	–	T56	–	–	–	–	–
T12	+++	–	–	–	–	T57	–	–	–	–	–
T13	–	++	–	–	–	T58	–	–	–	–	–
T14	+++	+	–	–	–	T59	–	–	–	–	–
T15	+++	–	–	–	–	T60	–	–	–	–	–
T16	+	++	–	–	–	T61	–	–	–	–	–
T17	+++	–	–	–	–	T62	–	–	–	++	–
T18	–	+	–	–	–	T63	–	–	–	–	++
T19	+++	–	–	–	–	T64	–	–	–	–	++
T20	++	–	–	–	–	T65	–	–	–	–	++
T21	+++	–	–	–	–	T66	–	–	–	–	++
T22	+++	–	–	–	–	T67	–	–	–	–	++
T23	+++	–	–	–	–	T68	–	–	–	–	++
T24	+	–	–	–	–	T69	–	–	–	–	–
T25	+++	–	–	–	–	T70	–	–	–	–	–
T26	+++	–	–	–	–	CS1	+++	–	–	–	–
T27	+++	–	–	–	–	CS2	+++	+	–	–	–
T28	–	+++	–	–	–	CS3	+++	–	–	–	–
T29	+++	–	–	–	–	CS4	–	+	–	–	–
T30	+++	–	–	–	–	CS5	++	–	–	–	–
T31	++	–	–	–	–	CS6	++	–	–	–	–
T32	+++	–	–	–	–	CS7	–	++	–	–	–
T33	++	–	–	–	–	CS8	+++	++	–	–	–
T34	++	–	–	–	+	CS9	++	–	–	–	–
T35	+++	–	–	–	–	CS10	+++	–	–	–	–
T36	+++	–	–	–	–	CS11	+++	–	–	–	–
T37	+++	++	–	–	+	CS12	–	–	–	+	–
T38	+++	–	–	–	–	CS13	–	–	–	++	–
T39	+++	–	–	–	–	CS14	–	++	–	–	–
T40	–	–	–	++	–	CS15	–	–	++	–	–
T41	–	–	–	+	–	CS16	+	–	–	–	–
T42	–	–	–	++	–	CS17	++	–	–	–	–
T43	–	+	–	–	–	CS18	++	–	–	–	–
T44	+++	–	–	–	–	CS19	++	–	–	–	–

Table 3 Screening results for 90 clinical samples using the xTAG assay *(Continued)*

Sample	CAV	MDV	IBDV	ARV	REV	Sample	CAV	MDV	IBDV	ARV	REV
T45	–	++	–	–	–	CS20	++	–	–	–	–

T tissue specimens, *CS* cloacal swabs;
+++: strong positive (MFI > 5 × cutoff);
++: positive (3 × cutoff <MFI < 5 × cutoff);
+: weak positive cutoff (MFI < 3 × cutoff);
-: negative (MFI < cutoff)

Nucleic acid extraction

Frozen chicken organ samples were homogenized for 15–30s before use. Cultured cell harvests were frozen and thawed several times followed by low speed centrifugation. The supernatants were stored at – 80 °C until use. Lyophilized vaccine samples were suspended in phosphate buffered saline following the manufacturer's protocol. Viral genomic DNA and RNA were extracted from homogenates and lysed cell culture supernatants using TIANamp Virus DNA/RNA Kit following the company's protocol (DP202, Tiangen Biotech, BeiJing). The samples after extraction were eluted in 70 μL RNase-free ddH_2O and stored at – 80 °C.

Primer design and selection

Full-length viral genome sequences were retrieved from the National Center for Biotechnology Information (NCBI) GenBank database (https://www.ncbi.nlm.nih.gov/) and each primer pair was designed based on multiple sequence alignment from Clustal W implemented in Geneious version 5.4.3 (Biomatters, Auckland, NZ). Primer oligomers were evaluated using Primer Premier 5.0 software (www.premierbiosoft.com).

xTAG assay uses target-specific primers which have linker overhangs (TAG sequences) on one primer which allow the PCR products to attach to a differently colored bead sets (xTAG beads, Luminex, Austin, TX, USA) and biotin on the other primer so that the PCR products contain biotin. Attached to each differently colored bead is an anti-TAG sequence which only binds to the complementary TAG sequence on the primer. After PCR amplification, color-coded beads were hybridized with the target PCR products via the TAG and anti-TAG complementarity. And then streptavidin-R-phycoertyhrin was added to bind to the biotin on the PCR amplicons. The intensity of fluorescence of the beads and R-phycoerythrin was determined by Luminex® 200 instrument. All primer sets were synthesized and purified by polyacrylamide gel electrophoresis (PAGE) (Invitrogen, Guangzhou, China). The gene targets were CAV (*VP1* gene), MDV (*MEQ* gene), REV (*ENV* gene), ARV (*S1* gene) and IBDV (*VP2* gene) (Table 4).

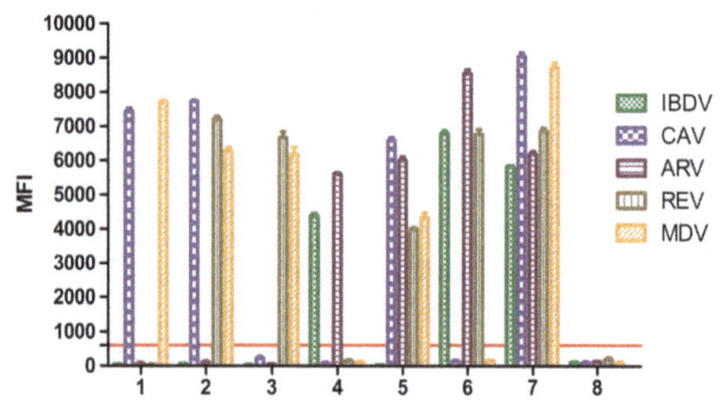

Fig. 3 Result of artificial mixed infection testing. Each bar represents the average median fluorescence intensity (MFI) of triplicate samples with standard deviation. The cut-off value was 600. 1: CAV and MDV positive sample; 2: CAV, MDV and REV positive sample; 3: MDV and REV positive sample; 4: IBDV and ARV positive sample; 5: CAV, MDV, ARV and REV positive sample; 6: IBDV, ARV and REV positive sample; 7: IBDV, ARV, REV, CAV, MDV positive sample; 8: negative control

Multiplex nucleic acid amplification and hybridization assay

The nucleic acid amplifications were performed using the OneStep RT-PCR Kit (Qiagen, Valencia, CA, USA; Cat. no. 210212). Each 20 μL reaction contained 1 μL nucleic acid template, 0.8 μL OneStep RT-PCR Enzyme Mix, 250 nM of each primer, 0.8 μL dNTP mixes and 4 μL OneStep RT-PCR buffer. Cycling conditions for the multiplex PCR were 50 °C for 30 min, 95 °C for 15 min, 40 cycles of 94 °C for 25 s, 55 °C for 25 s, and 72 °C for 20 s and 72 °C for 10 min. After reaction completion, 5 μl of the reaction mix was added to 75 μl of streptavidin-R-phycoerythrin (SAPE, Life Technologies GmbH) and 20 μl beads (approximately 2500 beads). The hybridization process was carried out at 37 °C for 30 min in a thermal cycler. After the hybridization reaction, the products were analyzed on the Luminex® 200 instrument.

Data analysis and cut-off value determination

The median fluorescent intensity (MFI) was calculated based on conditions suggested by the manufacturer and at least 50 beads were counted for each bead set in a single experiment. The data was analyzed using the Luminex xPONENT software 3.1. Negative controls containing all the hybridization components except target nucleic acid were used in each experiment to reduce background interference. The cut-off value was defined as the mean of the net MFI from negative PCR controls plus three times of standard deviation (MFI + 3SD).

Table 4 Primer pairs used in this study

Targeted virus	Forward primers (5′-3′)	Reverse primers (5′-3′)	Amplicon size (bp)	Bead region	Source
CAV	ATACTTTACAAACAAATAACACAC-spacer18-GCTCTCCAAGAAGATACTCCA	GCATTCGCGCAGCCACAC	149	19	this study
MDV	TACTTAAACATACAAACTTACTCA-spacer18-CCCATTCCCTCTTCTGCC	GCTGAGCGTAAACCGTC	114	65	this study
IBDV	TACTTCTTTACTACAATTTACAAC-spacer18-ATGCGGAGCCTTCTGA	ATTAGCCCTGACCCTGTG	128	15	this study
ARV	ACTTATTTCTTCACTACTATATCA-spacer18-GGATTCCGTCTCCATTCT	GAGTTTCCGTCAACCGTA	102	34	this study
REV	CTTAAACTCTACTTACTTCTAATT-spacer18-GACTGCCTTGTGACTGCT	ACTCCCACTGTTGTCTAAATC	153	56	this study
CAV	GACTGTAAGATGGCAAGACGAGCTC	GGCTGAAGGATCCCTCATTC	675	/	Caterina et al., 2004 [17]
IBDV	AGCCTTCTGATGCCAACAAC	ATCTGTCAGTTCACTCAGGC	365	/	Caterina et al., 2004 [17]
ARV	GGTGCGACTGCTGTATTTGGTAAC	AATGGAACGATAGCGTGTGGG	532	/	Caterina et al., 2004 [17]
MDV	GTGATGGGAAGGCGATAGAA	TCCGCATATGTTCCTCCTTC	225	/	Cao et al., 2013 [30]
REV	AATGGTTGTAAAGGGCAGAT	CTCCTCTCACTGCCAATCT	200	/	Cao et al., 2013 [30]

Assay sensitivity and specificity tests

Sensitivity testing was determined by using 10-fold serial dilutions of DNA or RNA standards as assay templates. PCR products of the five viral PCR products were cloned in plasmid vector pGEM T Easy (Promega, Madison, WI. USA). The constructs of IBDV, REV and ARV were in vitro transcribed using a RiboMax Large Scale RNA production system T7 (Promega) according to the manufacturer's instructions. RNase-free DNase (40 U) (Promega) was used for processing. Trizol LS reagent (Invitrogen, Carlsbad, CA) was used for RNA isolation to eliminate DNA interference. RNA concentrations were estimated by UV spectrophotometry from the average of five repeats. The copy numbers of DNA or RNA standards were calculated based on the nucleic acid concentration and molecular weight. Assay sensitivity was compared to conventional PCR using 10-fold serial dilutions of DNA or RNA standards. All the samples were performed in triplicate.

Assay specificity was tested using samples of non-target viruses in the reactions, such as AIV, IBV, NDV, ILTV, MG and MS. PCR products from specificity experiments were excised from 1.0% agarose gels using an AxyPrep DNA gel extraction kit (Axygen Scientific, Inc., CA) and sequenced (Invitrogen, Guangzhou, China).

Sample tests

Homogenates of organ and swab samples were used for RNA extraction and simultaneously tested for the presence of CAV, MDV, IBDV, REV and ARV using the multiplex xTAG assay and conventional PCR (Table 4) [30]. PCR products were sequenced. To further evaluate the xTAG assay, artificial mixed samples with combinations of the different viruses were tested.

Abbreviations

AIV: Avian influenza virus; ARV: Avian reovirus; CAV: Chicken anemia virus; IBDV: Infectious bursal disease virus; IBV: Infectious bronchitis virus; ILTV: Infectious laryngotracheitis virus; MDV: Marek's disease virus; MG: Mycoplasma gallisepticum and; MS: Mycoplasma synoviae; NDV: Newcastle disease virus; REV: Reticuloendotheliosis virus

Acknowledgements

We highly appreciate the help from Dr. Rajeev Dhawan (Senior Director of BioAssay Services at Charles River Labortories, USA) for the manuscript revision.

Funding

The study was supported by the Science and Technology Research Program (2017B030314171, 2015A030313693, 2016201604030059, 2017A070702001 and 2016A030303024) from Department of Science and Technology of Guangdong Province and Guangzhou Municipal. This study was also supported by the National Key Technology R&D Program (2015BAI07B01).

Authors' contributions

RH and PJG conceived and designed the study; YJZ and FC carried out experiments and wrote the manuscript; JW and LX collected samples; YZ and MLC were responsed for sample preparation and reproducibility test; YXL and XNL carried out statistical analysis; all authors read and approved the final manuscript.

Competing interests

The authors declare that they have no competing interests.

Author details
[1]Guangdong Laboratory Animals Monitoring Institute and Guangdong Provincial Key Laboratory of Laboratory Animals, Guangzhou 510633, China. [2]Guangdong Provincial Key Laboratory of Zoonosis Prevention and Control, College of Veterinary Medicine, South China Agricultural University, Guangzhou 510640, China.

References

1. Adair B. Immunopathogenesis of chicken anemia virus infection. Dev Comp Immunol. 2000;24(2):247–55.
2. Benavente J, Martínez-Costas J. Avian reovirus: structure and biology. Virus Res. 2007;123(2):105–19.
3. Brandt M, Yao K, Liu M, Heckert RA, Vakharia VN. Molecular determinants of virulence, cell tropism, and pathogenic phenotype of infectious bursal disease virus. J Virol. 2001;75(24):11974–82.
4. Li Y, Cui S, Cui Z, Chang S, Zhao P. Genome analysis and pathogenicity of reticuloendotheliosis virus isolated from a contaminated vaccine seed against infectious bursal disease virus: first report in China. J Gen Virol. 2016; 97(11):2809–15.
5. Lu H, Tang Y, Dunn PA, Wallner-Pendleton EA, Lin L, Knoll EA: Isolation and molecular characterization of newly emerging avian reovirus variants and novel strains in Pennsylvania, USA, 2011–2014. Sci Rep 2015, 5.
6. Lütticken D. Viral diseases of the immune system and strategies to control infectious bursal disease by vaccination. Acta Vet Hung. 1997;45(3):239–49.
7. Markowski-Grimsrud CJ, Schat KA. Infection with chicken anaemia virus impairs the generation of pathogen-specific cytotoxic T lymphocytes. Immunology. 2003;109(2):283–94.
8. McNulty M. Chicken anaemia agent: a review. Avian Pathol. 1991;20(2):187–203.
9. Shamblin CE, Greene N, Arumugaswami V, Dienglewicz RL, Parcells MS. Comparative analysis of Marek's disease virus (MDV) glycoprotein-, lytic antigen pp38-and transformation antigen Meq-encoding genes: association of meq mutations with MDVs of high virulence. Vet Microbiol. 2004;102(3): 147–67.
10. Sharma JM, Kim I-J, Rautenschlein S, Yeh H-Y. Infectious bursal disease virus of chickens: pathogenesis and immunosuppression. Dev Comp Immunol. 2000;24(2):223–35.
11. Zhao K, He W, Xie S, Song D, Lu H, Pan W, Zhou P, Liu W, Lu R, Zhou J. Highly pathogenic fowlpox virus in cutaneously infected chickens, China. Emerg Infect Dis. 2014;20(7):1200.
12. K-y B, Y-p Z, H-w Z, Lv H-c, Y-l G, J-f W, H-l G, H-y C, Y-q W, X-g R. Isolation and full-genome sequence of two reticuloendotheliosis virus strains from mixed infections with Marek's disease virus in China. Virus Genes. 2015;50(3):418–24.
13. Cloud S, Rosenberger J, Lillehoj H. Immune dysfunction following infection with chicken anemia agent and infectious bursal disease virus. II. Alterations of in vitro lymphoproliferation and in vivo immune responses. Vet Immunol Immunopathol. 1992;34(3–4):353–66.
14. De Boer G, Van Roozelaar D, Moormann R, Jeurissen S, Van Den Wijngaard J, Hilbink F, Koch G. Interaction between chicken anaemia virus and live Newcastle disease vaccine. Avian Pathol. 1994;23(2): 263–75.
15. Gallardo RA, van Santen VL, Toro H. Effects of chicken anaemia virus and infectious bursal disease virus-induced immunodeficiency on infectious bronchitis virus replication and genotypic drift. Avian Pathol. 2012;41(5):451–8.
16. Otaaki Y, Nunoya T, Tajima M, Kato A, Nomura Y. Depression of vaccinal immunity to Marek's disease by infection with chicken anaemia agent. Avian Pathol. 1988;17(2):333–47.
17. Caterina KM, Frasca S, Girshick T, Khan MI. Development of a multiplex PCR for detection of avian adenovirus, avian reovirus, infectious bursal disease virus, and chicken anemia virus. Mol Cell Probes. 2004;18(5):293–8.
18. Davidson I, Raibshtein I, Al-Touri A. Quantitation of Marek's disease and chicken anemia viruses in organs of experimentally infected chickens and commercial chickens by multiplex real-time PCR. Avian Dis. 2013; 57(2s1):532–8.
19. Gopal S, Manoharan P, Kathaperumal K, Chidambaram B, Divya KC. Differential detection of avian oncogenic viruses in poultry layer farms and turkeys by use of multiplex PCR. J Clin Microbiol. 2012;50(8):2668–73.
20. Cong F, Zhu Y, Liu X, Li X, Chen M, Huang R, Guo P. Development of an xTAG-multiplex PCR array for the detection of four avian respiratory viruses. Mol Cell Probes. 2017.
21. Navidad JF, Griswold DJ, Gradus MS, Bhattacharyya S. Evaluation of Luminex xTAG gastrointestinal pathogen analyte-specific reagents for high-throughput, simultaneous detection of bacteria, viruses, and parasites of clinical and public health importance. J Clin Microbiol. 2013;51(9):3018–24.
22. Cui Z, Sun S, Zhang Z, Meng S. Simultaneous endemic infections with subgroup J avian leukosis virus and reticuloendotheliosis virus in commercial and local breeds of chickens. Avian Pathol. 2009;38(6):443–8.
23. Xu W, Zhai Z, Huang K, Zhang N, Yuan Y, Shang Y, Luo Y. A novel universal primer-multiplex-PCR method with sequencing gel electrophoresis analysis. PLoS One. 2012;7(1):e22900.
24. Islam M, Johne R, Raue R, Todd D, Müller H. Sequence analysis of the full-length cloned DNA of a chicken Anaemia virus (CAV) strain from Bangladesh: evidence for genetic grouping of CAV strains based on the deduced VP1 amino acid sequences. Zoonoses Public Health. 2002; 49(7):332–7.
25. Mundt E. Tissue culture infectivity of different strains of infectious bursal disease virus is determined by distinct amino acids in VP2. J Gen Virol. 1999; 80(8):2067–76.
26. Lupiani B, Lee LF, Cui X, Gimeno I, Anderson A, Morgan RW, Silva RF, Witter RL, Kung H-J, Reddy SM. Marek's disease virus-encoded Meq gene is involved in transformation of lymphocytes but is dispensable for replication. Proc Natl Acad Sci U S A. 2004;101(32):11815–20.
27. Bodelón G, La L, Martínez-Costas J, Benavente J. The avian reovirus genome segment S1 is a functionally tricistronic gene that expresses one structural and two nonstructural proteins in infected cells. Virology. 2001;290(2):181–91.
28. Shih WL, Hsu HW, Liao MH, Lee LH, Liu HJ. Avian reovirus σC protein induces apoptosis in cultured cells. Virology. 2004;321(1):65–74.
29. Sinha A, Johnson WE. Retroviruses of the RDR superinfection interference group: ancient origins and broad host distribution of a promiscuous Env gene. Curr Opin Virol. 2017;25:105–12.
30. Cao W, Mays J, Dunn J, Fulton R, Silva R, Fadly A. Use of polymerase chain reaction in detection of Marek's disease and reticuloendotheliosis viruses in formalin-fixed, paraffin-embedded tumorous tissues. Avian Dis. 2013;57(4):785–9.

3

Effect of housing arrangement on fecal-oral transmission of avian hepatitis E virus in chicken flocks

Baoyuan Liu[1,2†], Yani Sun[1,2†], Yiyang Chen[1,2], Taofeng Du[1,2], Yuchen Nan[1,2], Xinjie Wang[1,2], Huixia Li[1,2], Baicheng Huang[1,2], Gaiping Zhang[3], En-Min Zhou[1,2*] and Qin Zhao[1,2*]

Abstract

Background: Avian hepatitis E virus (HEV) infection is common in chicken flocks in China, as currently no measures exist to prevent the spread of the disease. In this study, we analyzed the effect of caged versus cage-free housing arrangements on avian HEV transmission. First, 127 serum and 110 clinical fecal samples were collected from 4 chicken flocks including the two arrangements in Shaanxi Province, China and tested for HEV antibodies and/or virus. Concurrently, 36 specific-pathogen-free chickens were divided equally into four experimental living arrangement groups, designated cage-free (Inoculated), caged (Inoculated), cage-free (Negative) and caged (Negative) groups. In caged groups, three cages contained 3 chickens each. Three chickens each from cage-free (Inoculated) and caged (Inoculated) groups (one chicken of each cage) were inoculated by cutaneous ulnar vein with the same dose of avian HEV, respectively. The cage-free (Negative) and caged (Negative) groups served as negative control. Serum and fecal samples were collected at 1 to 7 weeks post-inoculation (wpi) and liver lesions were scored at 7 wpi.

Results: The results of serology showed that the avian HEV infection rate (54.10%) of the cage-free chickens was significantly higher than the one (12.12%) for caged chickens ($P < 0.05$). Also, the rate of detection of avian HEV RNA in the clinical fecal samples was significantly higher in the cage-free (22.80%, 13/57) than caged birds (5.66%, 3/53). Moreover, under experimental conditions, the infected number of uninoculated cage-free chickens (6) was significantly higher than the one for the uninoculated caged birds (2), as evidenced by seroconversion, fecal virus shedding, viremia and gross and microscopic liver lesions.

Conclusions: These results suggest that reduction of contact with feces as seen in the caged arrangement of housing chickens can reduce avian HEV transmission. This study provides insights for prevention and control of avian HEV infection in chicken flocks.

Keywords: Avian HEV, Housing arrangement, Virus transmission, Prevention

Background

Avian hepatitis E virus (HEV) is the causative agent of big liver and spleen disease and hepatitis-splenomegaly syndrome in chickens [1, 2]. The disease is characterized by increased mortality (1%–4%), a decrease in egg production (10%–40%) and enlarged livers and spleens in broiler breeder and laying hens aged 30–72 weeks [3, 4]. The virus can also cause mild clinical disease in chicken flocks, characterized by a decrease in egg production when the external environment is altered (such as with changes in climate and feed) and/or if infection occurs concurrently with other pathogen infections [5]. Although vertical transmission of the virus has been reported recently [6], avian HEV is believed to be transmitted mainly by the fecal-oral route in flocks [1, 7, 8]. To date the virus has been reported to be present in many countries [1, 2, 8–13].

* Correspondence: zhouem@nwsuaf.edu.cn; qinzhao_2004@nwsuaf.edu.cn
†Equal contributors
[1]Department of Preventive Veterinary Medicine, College of Veterinary Medicine, Northwest A&F University, 22 Xinong Road, Yangling, Shaanxi, China

Avian HEV is a non-enveloped single-stranded, positive sense RNA virus that is classified as an *Orthohepevirus* B species within the *Hepeviridae* family [14]. Sharing only 48% identity with human and swine HEVs, the avian HEV genome is approximately 6.6 kb in size and consists of three open-reading frames (ORFs) and two noncoding regions [15]. The ORF2 gene encodes a capsid protein containing the major viral epitopes; this capsid protein thus serves as the target for serological diagnosis and vaccine design [8, 16–18]. Although avian HEV strains have been divided into 4 major genotypes [19], they all belong to a single serotype [20].

In China in 2010, an avian HEV strain infecting a broiler breeder chicken flock exhibiting hepatitis-splenomegaly syndrome was isolated and characterized [11]. Subsequently, several serological surveys have indicated that avian HEV infection is widespread in chicken flocks in China [16, 21]. However, due to the lack of effective vaccines and drugs, no practical measures yet exist to prevent and control the disease, which hampers healthy development of poultry. Ultimately, blocking fecal-oral transmission should prevent the spread of virus infection [22], especially since this route has been shown to be the main avian HEV transmission route in chicken flocks [1, 7, 8]. Therefore, we evaluated the efficacy of disease control through inhibition of chicken contact with feces.

Results

Detection of avian HEV antibodies and RNA in clinical samples

The overall anti-avian HEV seropositivity rate was 32.28% (41/127), while the seropositivity rates for flocks A, B, C and D were 60% (18/30), 11.76% (4/34), 48.39% (15/31) and 12.5% (4/32), respectively (Table 1). The OD_{450nm} value distributions of the serum samples tested for antibody detection using indirect ELISA are shown in Fig. 1. For the two types of living arrangements, the positive rates of cage-free and caged chickens were 54.10% (33/61) and 12.12% (8/66), respectively. Statistical analyses showed that the difference in positive rates based on the type of living arrangements was significant ($P < 0.05$).

Consequently, agarose gel analysis showed that 16 of the samples exhibited bands of the expected size of 571 bp (Fig. 2). The 16 positive PCR products were sequenced and the sequences were analyzed using BLAST searches, nucleotide sequence alignments as well as phylogenetic analyses with known avian HEV sequences in the GenBank. The results showed that the 16 sequences shared high identity (82%–99%) with other known avian HEV strains from GenBank, indicating the 16 fecal samples contained avian HEV RNA. The total HEV positive rate of these fecal samples was 14.55% (16/110) and the positive rates of A, B, C and D flocks were 25% (7/28), 4% (1/25), 20.69% (6/29) and 7.14% (2/28), respectively (Table 1). For the two living arrangements, the positive infection rates of the cage-free and caged groups were 22.80% (13/57) and 5.66% (3/53), respectively. Statistical analyses also showed that the difference in positive rates between cage-free and caged chickens for avian HEV RNA in the fecal samples was significant ($P < 0.05$).

Avian HEV antibody seroconversion in experimental chickens

Prior to inoculation, all SPF chickens were shown to be seronegative for avian HEV (Fig. 3). Based on the cut-off value (0.368) of the indirect ELISA, anti-avian HEV IgG antibodies were detected in the 6 inoculated chickens of both cage-free (Inoculated) and caged (Inoculated) groups (Nos. 2, 5, 8, 11, 14 and 18) at 3 wpi (Fig. 3). At the end of the experiment (7 wpi), all 6 chickens were still positive for anti-avian HEV antibodies (Fig. 3). The remaining 6 uninoculated chickens in cage-free (Inoculated) group seroconverted to IgG anti-avian HEV antibodies by 5 wpi and were still positive at 7 wpi (Table 2, Fig. 3). However, in caged (Inoculated) group, all remaining chickens from the three cages remained seronegative throughout the study (Table 2, Fig. 3). In the cage-free (Negative) and caged (Negative) groups, all the chickens were seronegative throughout the study (Fig. 3).

Detection of avian HEV RNA in fecal and serum samples from experimental chickens

The fecal and serum samples from all chickens were negative for avian HEV RNA at pre-inoculation (Table 3). Fecal virus shedding was detected in the 5 of 6 inoculated chickens from cage-free (Inoculated) and caged (Inoculated) groups at 1 wpi (Table 3) while all 6 inoculated chickens were positive at 2 wpi (Table 3). At 5 wpi,

Table 1 Detection of avian HEV antibodies and RNA in sera and feces from 4 layer flocks

Chicken flock	Scale	Breed	Housing arrangements	Serum samples			Fecal samples		
				Detected number	Sum	Positive rate	Detected number	Sum	Positive rate
A	2000	Hy-Line Brown	Cage-free	18	30	60%	7	28	25%
B	10,000	Hy-Line Brown	Caged	4	34	11.76%	1	25	4%
C	3000	Hy-Line Brown	Cage-free	15	31	48.39%	6	29	20.69%
D	8000	Hy-Line Brown	Caged	4	32	12.5%	2	28	7.14%

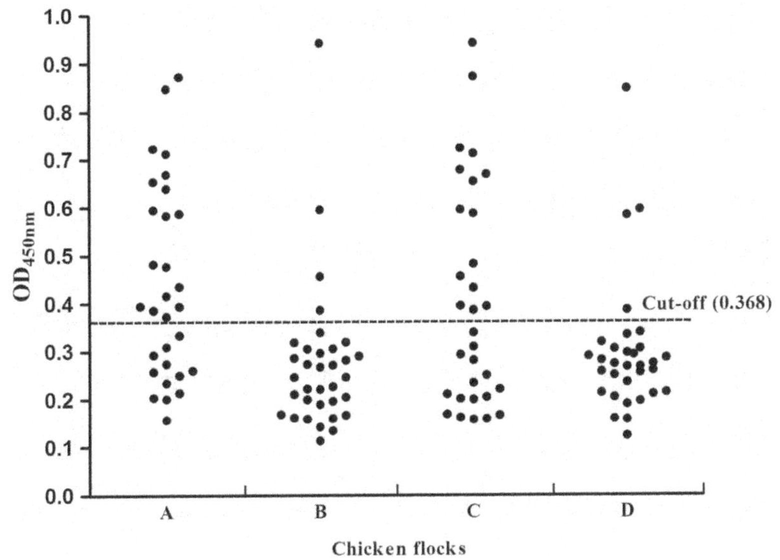

Fig. 1 Distribution of OD_{450nm} values of sera from the caged and cage-free arrangements using indirect ELISA. The dotted line represents the cut-off value (0.368)

there was no more shedding of virus in the feces of these chickens (Table 3). For the uninoculated chickens, fecal virus shedding was first detected in 2 of 6 chickens in cage-free (Inoculated) group at 3 wpi; at 6 wpi all 6 chickens were positive (Tables 2 and 3). However, in caged (Inoculated) group, only two uninoculated chickens housed in the two cages containing the inoculated chickens exhibited fecal virus shedding at 6 wpi (Tables 2 and 3). The remaining four uninoculated chickens in the cages were all negative (Table 3). Viremia was also first observed in inoculated chickens at 1 wpi (Table 3). All 6 inoculated chickens of cage-free (Inoculated) and caged (Inoculated) groups exhibited viremia at 3 wpi (Table 3). For the uninoculated chickens in cage-free (Inoculated) group, viremia was observed in 2 of 6 chickens at 3 wpi, 4/6 chickens at 4 wpi, 3/6 chickens at 5 wpi and 3/6 chickens at 6 wpi (Tables 2 and 3). However, in caged (Inoculated) group, viremia was only observed in 1/6 uninoculated chickens at 6 wpi (Tables 2 and 3). In cage-free (Negative) and caged (Negative) groups, all chickens remained negative throughout the experiment (Data not shown).

The positive PCR products from serum samples and fecal swabs of inoculated and uninoculated chickens of cage-free (Inoculated) and caged (Inoculated) groups were sequenced and BLAST results showed that the viruses recovered from the infected chickens originated from the original virus inoculum.

Gross lesions

Gross lesions were observed clearly in the 6 inoculated chickens from cage-free (Inoculated) and caged (Inoculated) groups at 7 wpi. Subcapsular hemorrhages were observed in the livers of the 6 chickens and several hemorrhagic spots and regions were also observed in livers (Fig. 4). In cage-free (Inoculated) group, hemorrhagic spots and regions were observed in 5 of the 6 remaining chickens (Table 2). However, in caged

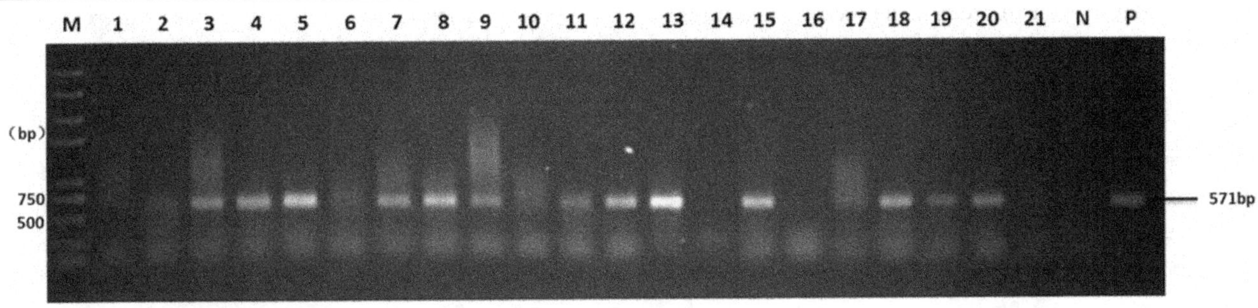

Fig. 2 Agarose gel analysis of PCR products for the avian HEV partial ORF1 gene from faeces. M:Trans2K® Plus II DNA marker;1–8:A flock;9–10:B flock;11–18:C flock;19–21:D flock;N:Ultrapure water;P:A bile sample containing avian HEV

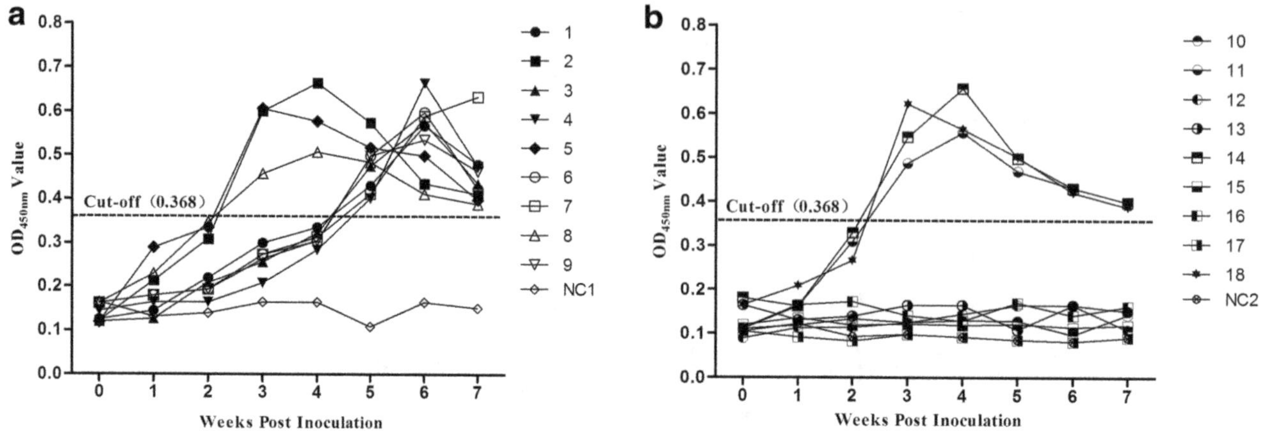

Fig. 3 Seroconversion to avian HEV in the experimental chickens from the cage-free and caged arrangements. **a**: Cage-free (Inoculated) groups;(**b**): Caged (Inoculated) groups. The dotted line represents the cut-off values of the indirect ELISA used for detection of the avian HEV antibodies in the serum samples. NC1: Cage-free (Negative) group;NC2: Caged (Negative) group. The week post-inoculation on 0 the X axis represents the period prior to inoculation

(Inoculated) group, gross lesions in the 6 remaining chickens were not evident (Table 2). In addition, compared with the uninfected chickens, the liver/body weight ratios from the 6 inoculated chickens and 2 contact-infected chickens of cage-free (Inoculated) were significantly higher (Table 4). In cage-free (Negative) and caged (Negative) groups, all chickens had no evident liver lesions.

Microscopic lesions

Microscopically, hepatic lesions were all observed in the livers of the 6 inoculated chickens from cage-free (Inoculated) and caged (Inoculated) groups, showing severe lymphocytic portal phlebitis and periphlebitis (Fig. 4). For the uninoculated chickens in cage-free (Inoculated) group, five contact-infected chickens exhibited mild lymphocytic phlebitis in their livers (Fig. 4). However, in caged (Inoculated) group, microscopic lesions were only evident in 1 of 6 uninoculated chickens. The data on microscopic lesions in livers from all the chickens are summarized (Table 4) and show that the lesion scores of the infected chickens kept in cage-free conditions were significantly higher than that of the caged chickens (Table 4). In cage-free (Negative) and caged (Negative) groups, all chickens had no evident liver microscopic lesions.

Discussion

Avian HEV causes not only big liver and spleen disease, but also subclinical infections in chickens [5]. The virus infection is endemic in many countries and has resulted in serious economic losses in the poultry industries in some countries [1, 2, 8–13]. Avian HEV was first characterized in 2010 in China and subsequent serological and molecular epidemiological investigations indicated that it is now common in Chinese chicken flocks [11, 16, 21]. In this study, we detected anti-avian HEV antibodies and avian HEV RNA in four seemingly healthy chicken flocks of Shaanxi Province, China. The positive rates for antibodies (32.28%) and viral RNA (14.55%) suggest that avian HEV infection is common in these 4 flocks from Shaanxi Province, which aligns with HEV detection rates in other provinces in China [16].

In China, two major chicken housing arrangements are used in the poultry industry. One is the cage-free arrangement in which the birds are housed and fed in a shared region and can mingle and move freely about. Another is the caged condition, in which the birds are housed and fed in different cages and cannot mingle freely. Previous studies have documented that resistance to diseases varies for animals kept under different living conditions [23, 24]. In this study the positive rates for

Table 2 Number of the uninoculated chickens infected by avian HEV in the cage-free and caged groups

Housing arrangements	Seroconversion to avian HEV	Fecal virus shedding	Viremia	Gross lesions	Microscopic lesions
Cage-free (Inoculated)	6/6	6/6	6/6	5/6	5/6
Caged (Inoculated)	0/6	2/6	1/6	0/6	1/6
Cage-free (Negative)	0/6	0/6	0/6	0/6	0/6
Caged (Negative)	0/6	0/6	0/6	0/6	0/6

The chickens were infected as evidenced by seroconversion, fecal virus shedding, viremia and gross and microscopic lesions of livers. The number was shown as positive number/total number

Table 3 Fecal virus shedding and viremia of all the chickens in the cage-free (Inoculated) and caged (Inoculated) groups

Housing arrangements	Chicken No.	Detection of fecal virus shedding /viremia from different weeks post inoculation							
		0	1	2	3	4	5	6	7
Cage-free (Inoculated)	1	−/−	−/−	−/−	+/+	−/+	+/−	+/	−/−
	2	**−/−**	**+/+**	**+/−**	**−/+**	**+/−**	**−/−**	**−/−**	**−/−**
	3	−/−	−/−	−/−	−/−	−/−	+/+	+/+	−/−
	4	−/−	−/−	−/−	+/+	+/+	−/−	+/+	−/−
	5	**−/−**	**+/−**	**+/+**	**−/+**	**+/−**	**−/−**	**−/−**	**−/−**
	6	−/−	−/−	−/−	−/−	−/−	−/+	+/−	−/−
	7	−/−	−/−	−/−	−/−	−/+	+/+	+/−	−/−
	8	**−/−**	**+/+**	**+/−**	**+/+**	**+/−**	**−/−**	**−/−**	**−/−**
	9	−/−	−/−	−/−	−/−	+/+	+/−	+/+	−/−
Caged (Inoculated)	10	−/−	−/−	−/−	−/−	−/−	−/−	−/−	−/−
	11	**−/−**	**+/+**	**+/+**	**−/+**	**+/−**	**−/−**	**−/−**	**−/−**
	12	−/−	−/−	−/−	−/−	−/−	−/−	−/−	−/−
	13	−/−	−/−	−/−	−/−	−/−	−/−	+/−	−/−
	14	**−/−**	**+/−**	**+/+**	**−/+**	**+/−**	**−/−**	**−/−**	**−/−**
	15	−/−	−/−	−/−	−/−	−/−	−/−	−/−	−/−
	16	−/−	−/−	−/−	−/−	−/−	−/−	+/+	−/−
	17	−/−	−/−	−/−	−/−	−/−	−/−	−/−	−/−
	18	**−/−**	**−/+**	**+/−**	**+/+**	**+/−**	**−/−**	**−/−**	**−/−**

The bold numbers represent the inoculated chickens. For the caged group, the Nos 10–12 were in the same cage, the Nos 13–15 in the same cage and the Nos 16–18 in the same cage. "+" represents the chickens' positive for fecal virus shedding and viremia

avian HEV antibodies and RNA under both housing arrangements showed that the avian HEV infection rate was higher for cage-free compared to caged animals. Then, the animal experimental results also showed that the number of infected chickens present in the cage-free arrangement was greater than that for the caged group. The key point of difference between cage-free and caged chickens lies in whether or not chickens come into contact with feces of other chickens. Based on this difference and the known primary route of avian HEV transmission, we speculate that the reason for the higher rates in the cage-free arrangement was due to uninhibited contact between chickens and feces of their housing companions.

By 4 wpi, viremia was undetected in all 6 inoculated chickens of cage-free and caged groups while antibodies

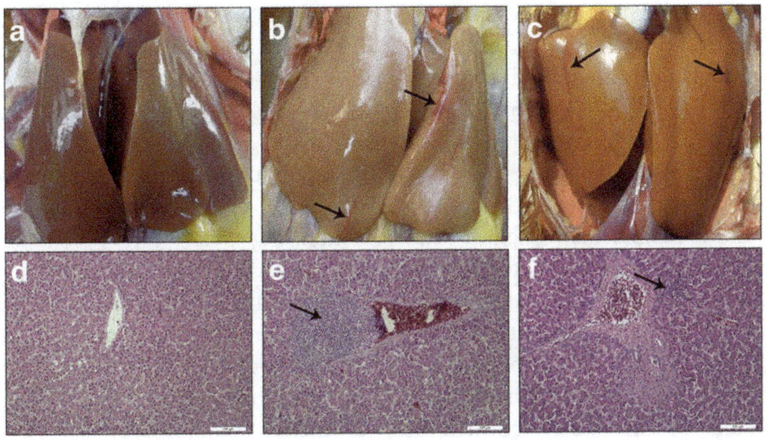

Fig. 4 Gross lesions and microscopic lesions of the livers from experimental chickens. Subcapsular hemorrhages (**a**, **b** and **c**) and lymphocytic periphlebitis (**d**, **e** and **f**) are indicated by arrows. **a** and **d**: uninfected chickens in cage-free (Negative) group; (**b**) and (**e**): inoculated chickens from cage-free (Inoculated) group; (**c** and **f**): chickens infected by contact with feces from cage-free (Inoculated) group. Tissues were stained with hematoxylin and eosin

Table 4 Scoring of microscopic liver lesions and liver/body weight ratios in all the chickens. For caged group, the Nos 10–12, 13–15, 16–18, 28–30, 31–33 and 34–36 chickens were kept separately in the same cage

Housing arrangements	No. of chickens (score [a], liver/body weight ratio [b])								
Cage-free (Inoculated)	1(1, 23.16)	**2(4, 40.92)**	3(2, 23.44)	4(2, 32.19)	**5(4, 37.98)**	6(3, 33.25)	7(2, 24.01)	**8(4, 36.78)**	9(2, 23.48)
Caged (Inoculated)	10(0, 24.84)	**11(4, 36.47)**	12(1, 23.39)	13(0, 22.08)	**14(4, 35.88)**	15(0, 24.12)	16(2, 23.98)	17 (1, 22.68)	**18(4, 38.11)**
Cage-free (Negative)	19(0, 23.98)	20(0, 24.12)	21(1, 22.65)	22(0, 23.65)	23(1, 24.09)	24(0, 23.51)	25(0, 22.98)	26(1, 23.65)	27(0, 24.01)
Caged (Negative)	28(1, 22.99)	29(0, 25.01)	30(0, 23.98)	31(0, 24.19)	32(1, 24.67)	33(2, 24.88)	34(0, 22.87)	35(1, 23.13)	36(0, 24..35)

[a] Liver lesion scores ranged from 0 to 4 (0, no lesions; 1, <5 foci; 2, 5 to 8 foci; 3,9 to 15 foci; 4,>15 foci)

[b] Liver to body weight ratio was calculated by (liver weight)/(body weight) × 100

The liver of each chicken was weighed and the liver/body weight ratios were calculated. Compared with the uninfected chickens, the mean ratios from the 6 inoculated chickens (bold) and 2 contact-infected chickens (Nos.4 and 6) were significantly higher($P < 0.05$)

against avian HEV peaked at 3 to 4 wpi, coinciding with the disappearance of virus from the blood stream. After 1 week pi, fecal virus shedding also became undetectable. Some of the uninoculated chickens were observed to be infected and this could have occurred through contact with feces from the inoculated chickens. However, compared to the latter, they experienced a delay of 2 or more weeks before seroconversion, viremia and fecal virus shedding occurred. This observation can be explained by the fact that in the inoculated chickens which were infected intravenously, the virus would have invaded the bloodstream directly before getting to the liver where massive replication takes place. Subsequently, it would pass through the bile into the gastrointestinal tract before being excreted from the body [25]. However, for the uninoculated chickens infected by fecal-oral routes, a small quantity of virus replicates first in the intestinal tract before getting into the blood and liver. This procedure takes a long time before virus replication [25].

For the uninoculated animals in the caged group, two chickens in the two cages were infected. Since the 1 inoculated and 2 uninoculated chickens were housed in the same cage for 7 weeks, it is possible that some avian HEV-contaminated feces from the inoculated chicken may have adhered to the edges of the cage, and this could have been ingested by the uninoculated chickens through the fecal-oral route. In addition, compared with cage-free chickens infected after feces contact, the two HEV RNA positive caged chickens showed no avian anti-HEV antibody seroconversion and no evident gross liver lesions. Moreover, these results also suggest that the two infected chickens may have been infected during the late phase of the study or were infected with a very small amount of virus after contact with contaminated feces. To date, there are few reports about prevention and control measures for avian HEV infection in chicken flocks. Based on the results of this study, control of chicken excrement pollution could be reduced of minimize avian HEV infection. Moreover, the caged living arrangement exhibited greater prevention of HEV transmission than did the cage-free arrangement.

Conclusions

Overall, we report that the rate of avian HEV infection in chicken flocks living in cage-free conditions was higher than that observed for caged birds. Next, using animal experiments, we confirmed that differences in rate of contact with feces can explain why the caged living arrangement was more effective than the cage-free arrangement for prevention of avian HEV infection. Therefore, this study provides some insights into the prevention and control of avian HEV infection in chicken flocks.

Methods

Clinical sample collection and processing

The profiles of four healthy chicken flocks in Shaanxi Province, China (A, B, C and D) that were used to provide 127 serum and 110 fecal samples for HEV testing are shown in Table 1.

Fecal samples were homogenized in 10% (*w*/*v*) sterile phosphate-buffered saline (PBS, pH 7.2). The fecal homogenates were clarified by centrifugation at 1000 *g* for 10 min at 4 °C and 200 μL supernatants were used for the detection of avian HEV RNA using reverse transcription-nested PCR (RT-nPCR). The serum samples were used for the detection of anti-avian HEV antibodies by indirect ELISA.

Virus

An avian HEV infectious stock was produced by intravenously inoculating four 8-week-old specific-pathogen-free (SPF) chickens with 200 μL of a clinical bile sample containing avian HEV isolated from a chicken aged 35 weeks (CaHEV, GenBank accession no. GU954430). This avian HEV stock contained 10^4 genomic equivalents (GE)/mL or 500 median chicken infective dose (CID_{50})/mL of the virus.

Chickens

Thirty-six 8-week-old, SPF female chickens were purchased from Beijing Merial Vital Laboratory Animal Technology Company. All birds were negative for avian HEV antibodies and RNA.

Animal experimental design

The 36 SPF chickens were randomly divided into 4 experimental groups, with 9 chickens per group. The chickens (Nos. 1 through 9) in cage-free (Inoculated) group were housed in a room with a floor space of 6 square meters and could regularly contact each other and their companions' feces. The 9 chickens (Nos. 10 through 18) in the caged (Inoculated) group were divided into 3 cages and located in a room. Each cage had a footprint of 2 square meters per cage and housed 3 chickens. The 3 cages were placed closely spaced and side-by-side such that chickens in different cages could not contact each other freely, although chickens within each cage could contact one another. All 9 chickens in caged (Inoculated) group were not expected to contact feces because the feces dropped from the cages and were cleared quickly, although some feces may have adhered to the edges of the cage. The 9 chickens (Nos. 19 through 27) in cage-free (Negative) group were housed same as the cage-free (Inoculated) group and the chickens (Nos. 28 through 36) in the caged (Negative) group were arranged same as the caged (Inoculated) group. The temperatures in the four rooms ranged from 20 to 21 °C and 10/14 h (light/dark) cycle was given for chickens to get enough sleep. They had free access to water and a commercial starter diet without supplementation of antibiotics.

In cage-free (Inoculated) and caged (Inoculated) groups, six chickens (Nos. 2, 5, 8, 11, 14, and 18) were randomly selected for inoculation by cutaneous ulnar vein with 800 μL of avian HEV stock. In cage-free (Negative) and caged (Negative) groups, chickens served as uninoculated controls under cage-free and caged conditions, respectively. All chickens were monitored for avian HEV infection for 7 weeks and then were necropsied. The animal experiments were approved by the Animal Care and Use Committee of Northwest Agricultural & Forestry University (NWSUAF, Permit Number: AE189135) with adherence to NWSUAF guidelines during handling of all experimental animals.

Animal experimental samples collection

Serum and fecal swab samples were collected prior to inoculation and weekly thereafter as described above. During each round of blood collection, the gloves, needles and gowns were changed to avoid introduction of cross-infection among the chickens by the procedures used. Serum samples were tested by indirect ELISA for anti-avian HEV IgG antibodies while serum and fecal samples were tested by RT-nPCR for avian HEV RNA. The indirect ELISA and RT-nPCR procedures for testing experimental samples were same methods used for clinical samples, as described below.

Gross and microscopic hepatic lesions

During necropsies, gross pathological lesions in the liver of each chicken were evaluated and recorded. In addition, the liver of each chicken was weighed and the liver/body weight ratios were calculated. The liver tissues were also fixed in 10% neutral buffered formalin and processed for routine histological examination. Histopathological lesions in the liver were evaluated and scored according to a standard scoring system documented previously by Billam et al. [25]. Liver lesion scores ranged from 0 to 4 (0, no lesions; 1, <5 foci; 2, 5 to 8 foci; 3, 9 to 15 foci; 4, >15 foci).

RT-nPCR to detect avian HEV RNA

The partial ORF1 gene of the avian HEV genome was amplified using RT-nPCR from both clinical and experimental fecal and serum samples [26]. Two nested sets of primers were used: external primer set DHF1, 5′-TGAGGGTTCGAGCTGACAG-3′ and DHR1, 5′-CAT-ACGCCTCGTCCACAAT-3′; and internal primer set DHF2, 5′-CAGCAGCCATCCGCAAAC-3′ and DHR2, 5′-GGACGCCTGATGAACAACG-.3′, which were described previously by Dong et al. [26]. The sizes of the expected PCR products for the first and second round PCRs were 921 bp and 571 bp, respectively.

Total RNA was extracted from each 200 μL serum sample or 10% fecal suspension with TRIzol reagent (ThermoFisher Scientific, USA) according to the manufacturer's instructions and was resuspended in 20 μL DNase-, RNase-, and proteinase-free water. Superscript® II Reverse Transcriptase (Invitrogen, USA) was used to perform reverse transcription using the DHR1 primer. The reaction was incubated at 50 °C for 30 min and 85 °C for 5 min. Next, 5 μL cDNA was used as the template for the first PCR and 2 μL first PCR product as the template for the second PCR with TransTaq® High Fidelity DNA polymerase (Beijing TransGen Biotech, Ltd., China) according to the manufacturer's instructions. The first and second PCR reaction parameters both included an initial incubation at 95 °C for 9 min, followed by 39 cycles of 94 °C for 45 s, 42 °C for 45 s and 72 °C for 50 s, with a final incubation step at 72 °C for 10 min. Final PCR products were electrophoresed on a 1% agarose gel.

Sequence analysis

All positive PCR products from clinical fecal samples and from experimental serum and fecal samples were purified using the QIAquick PCR Purification Kit (QIAGEN, USA) and sequenced using an ABI 3130 Genetic Analyzer automated sequencing system (Applied Biosystems, USA). Sequences were analyzed using BLAST searches (http://blast.ncbi.nlm.nih.gov/Blast.cgi).

ELISA for avian HEV antibodies

Anti-avian HEV IgG antibodies were detected in both clinical and experimental serum samples using an indirect ELISA described previously by Zhao et al. [16]. Briefly, a purified truncated recombinant CaHEV capsid protein expressed in *Escherichia coli* was used as the coating antigen for the indirect ELISA. After the coated plates were blocked and washed, the serum samples (100 μL/well) were added into the wells and incubated for 1 h at room temperature (RT). After three washes, a horseradish peroxidase-goat anti-chicken IgG diluted 1:4000 (100 μL/well) was added to the wells and incubated for 1 h again at room temperature. After a final three washes, 3,3′,5,5′-tetramethylbenzidine (TMB) was added to each well and the plates were incubated in the dark for 15 min at RT. The colorimetric reaction was stopped by adding 3 M H2SO4 (50 μL/well) and the optical density (OD) values were read at 450 nm using an automated ELISA plate reader (Bio-Rad, USA). All sera were tested in at least duplicate wells.

Statistical analyses

Data collection and analyses were performed using IBM SPSS Statistics 20 (IBM, USA). The student's t-test was used to estimate the differences in avian HEV infection rate between caged and cage-free chickens. For determination of the liver weight to body weight ratio and histologic lesion scores in the experimental inoculated chickens, statistical analysis was performed as described previously [27]. $P \leq 0.05$ was considered significant.

Abbreviations

ELISA: Enzyme-linked immunosorbent assay; HEV: Hepatitis E virus; NC: Negative Control; RT-nPCR: Reverse Transcription-nested Polymerase Chain Reaction; Wpi: Week post inoculation

Acknowledgements

We would like to thank our funding sources.

Funding

Funding: The study was funded by grants from National Natural Science Foundation of China (No.31372464) to EMZ and (No. 31402233 and No. 31672583) to QZ, Technology Research and innovation of Shaanxi province (2015NY170) to YNS, the Fundamental Research Funds for the Central Universities (2452015035) to YNS, (2014YB013) to TFD and (2452016048) to QZ.

Authors' contributions

BYL and YNS performed the experiment, arranged the data and drafted the manuscript. YYC, TFD, YCN, XJW and analyzed the data. BYL, QZ and HXL collected the clinical samples. BCH revised the manuscript. GPZ, QZ and EMZ analyzed the data and revised the manuscript. QZ and EMZ designed the study, analyzed the data and revised the manuscript. All of the authors read and approved the final manuscript.

Competing interests

The authors declare that they have no competing interests.

Author details

[1]Department of Preventive Veterinary Medicine, College of Veterinary Medicine, Northwest A&F University, 22 Xinong Road, Yangling, Shaanxi, China. [2]Scientific Observing and Experimental Station of Veterinary Pharmacology and Veterinary Biotechnology, Ministry of Agriculture, Yangling, Shaanxi, China. [3]Department of Preventive Veterinary Medicine, College of Veterinary Medicine, Henan Agriculture University, Zhengzhou, Henan 450002, China.

References

1. Haqshenas G, Shivaprasad HL, Woolcock PR, Read DH, Meng XJ. Genetic identification and characterization of a novel virus related to human hepatitis E virus from chickens with hepatitis-splenomegaly syndrome in the United States. J Gen Virol. 2001;82:2449–62.
2. Payne CJ, Ellis TM, Plant SL, Gregory AR, Wilcox GE. Sequence data suggests big liver and spleen disease virus (BLSV) is genetically related to hepatitis E virus. Vet Microbiol. 1999;68:119–25.
3. Clarke JK, Allan GM, Bryson DG, Williams W, Todd D, Mackie DP, McFerran JB. Big liver and spleen disease of broiler breeders. Avian Pathol. 1990;19:41–50.
4. Handlinger JH, Williams W. An egg drop associated with splenomegaly in broiler breeders. Avian Dis. 1988;32:773–8.
5. Gerber PF, Trampel DW, Willinghan EM, Billam P, Meng XJ, Opriessnig T. Subclinical avian hepatitis E virus infection in layer flocks in the United States. Vet J. 2015;206:304–11.
6. Troxler S, Pac K, Prokofieva I, Liebhart D, Chodakowska B, Furmanek D, Hess M. Subclinical circulation of avian hepatitis E virus within a multiple-age rearing and broiler breeder farm indicates persistence and vertical transmission of the virus. Avian Pathol. 2014;43:310–8.
7. Haqshenas G, Huang FF, Fenaux M, Guenette DK, Pierson FW, Larsen CT, Shivaprasad HL, Toth TE, Meng XJ. The putative capsid protein of the newly identified avian hepatitis E virus shares antigenic epitopes with that of swine and human hepatitis E viruses and chicken big liver and spleen disease virus. J Gen Virol. 2002;83:2201–9.
8. Huang FF, Haqshenas G, Shivaprasad HL, Guenette DK, Woolcock PR, Larsen CT, Pierson FW, Elvinger F, Toth TE, Meng XJ. Heterogeneity and seroprevalence of a newly identified avian hepatitis e virus from chickens in the United States. J Clin Microbiol. 2002;40:4197–202.
9. Agunos AC, Yoo D, Youssef SA, Ran D, Binnington B, Hunter DB. Avian hepatitis E virus in an outbreak of hepatitis–splenomegaly syndrome and fatty liver haemorrhage syndrome in two flaxseed-fed layer flocks in Ontario. Avian Pathol. 2006;35:404–12.
10. Morrow CJ, Samu G, Matrai E, Klausz A, Wood AM, Richter S, Jaskulska B, Hess M. Avian hepatitis E virus infection and possible associated clinical disease in broiler breeder flocks in Hungary. Avian Pathol. 2008;37:527–35.
11. Zhao Q, Zhou EM, Dong SW, Qiu HK, Zhang L, Hu SB, Zhao FF, Jiang SJ, Sun YN. Analysis of avian hepatitis E virus from chickens. China Emerg Infect Dis. 2010;16:1469–72.
12. Sprygin AV, Nikonova ZB, Zinyakov NG. Avian hepatitis E virus identified in Russian chicken flocks exhibits high genetic divergence based on the ORF2 capsid gene. Avian Pathol. 2012;41:459–63.
13. Hsu IW, Tsai HJ. Avian hepatitis E virus in chickens, Taiwan, 2013. Emerg Infect Dis. 2014;20:149–51.
14. Smith DB, Simmonds P, International Committee on Taxonomy of Viruses Hepeviridae Study G, Jameel S, Emerson SU, Harrison TJ, Meng XJ, Okamoto H, Van der Poel WH, Purdy MA. Consensus proposals for classification of the family Hepeviridae. J Gen Virol. 2014;95:2223–32.
15. Huang FF, Sun ZF, Emerson SU, Purcell RH, Shivaprasad HL, Pierson FW, Toth TE, Meng XJ. Determination and analysis of the complete genomic sequence of avian hepatitis E virus (avian HEV) and attempts to infect rhesus monkeys with avian HEV. J Gen Virol. 2004;85:1609–18.
16. Zhao Q, Sun Y, Zhao J, Hu S, Zhao F, Chen F, Clavijo A, Zhou EM, Xiao Y. Development and application of an indirect ELISA for detection of antibodies against avian hepatitis E virus. J Virol Methods. 2013;187:32–6.
17. Wang L, Sun Y, Du T, Wang C, Xiao S, Mu Y, Zhang G, Liu L, Widen F, Hsu WH, Zhao Q, Zhou EM. Identification of an antigenic domain in the N-terminal region of avian hepatitis E virus (HEV) capsid protein that is not common to swine and human HEVs. J Gen Virol. 2014;95:2710–5.
18. Guo H, Zhou EM, Sun ZF, Meng XJ. Protection of chickens against avian hepatitis E virus (avian HEV) infection by immunization with recombinant avian HEV capsid protein. Vaccine. 2007;25:2892–9.

19. Zhao Q, Syed SF, Zhou EM. Antigenic properties of avian hepatitis E virus capsid protein. Vet Microbiol. 2015;180:10–4.
20. Meng XJ. Hepatitis E virus: animal reservoirs and zoonotic risk. Vet Microbiol. 2010;140:256–65.
21. Liu B, Zhao Q, Sun Y, Wang X, Zhao J, Du T, Wang C, Xiao S, Mu Y, Zhang G, Luo J, Hsu WH, Zhou EM. Development of a blocking ELISA for detection of antibodies against avian hepatitis E virus. J Virol Methods. 2014;204:1–5.
22. Molaei G, Andreadis TG, Armstrong PM, Anderson JF, Vossbrinck CR. Host feeding patterns of Culex mosquitoes and West Nile virus transmission, northeastern United States. Emerg Infect Dis. 2006;12:468–74.
23. Hietala SK, Hullinger PJ, Crossley BM, Kinde H, Ardans AA. Environmental air sampling to detect exotic Newcastle disease virus in two California commercial poultry flocks. J Vet Diagn Investig. 2005;17:198–200.
24. Shields S, Greger M. Animal welfare and food safety aspects of confining broiler chickens to cages. Animals (Basel). 2013;3:386–400.
25. Billam P, Huang FF, Sun ZF, Pierson FW, Duncan RB, Elvinger F, Guenette DK, Toth TE, Meng XJ. Systematic pathogenesis and replication of avian hepatitis E virus in specific-pathogen-free adult chickens. J Virol. 2005;79:3429–37.
26. Dong S, Zhao Q, Lu M, Sun P, Qiu H, Zhang L, Lv J, Zhou EM. Analysis of epitopes in the capsid protein of avian hepatitis E virus by using monoclonal antibodies. J Virol Methods. 2011;171:374–80.
27. Kwon HM, LeRoith T, Pudupakam RS, Pierson FW, Huang YW, Dryman BA, Meng XJ. Construction of an infectious cDNA clone of avian hepatitis E virus (avian HEV) recovered from a clinically healthy chicken in the United States and characterization of its pathogenicity in specific-pathogen-free chickens. Vet Microbiol. 2011;147:310–9.

4

Crossed beaks in a local Swiss chicken breed

Sara Joller[1†], Flurina Bertschinger[2†], Erwin Kump[3], Astrid Spiri[4], Alois von Rotz[5], Daniela Schweizer-Gorgas[6], Cord Drögemüller[1*†] and Christine Flury[2†]

Abstract

Background: Crossed beaks have been reported to occur in Appenzeller Barthuhn, a local Swiss chicken breed. The assumed causes for this beak deformity which are also seen in other bird species including domestic chickens, range from environmental influences to genetic factors. The aim of this project was to characterize the prevalence, the phenotype, and the underlying genetics of crossed beaks in Appenzeller Barthuhn chickens.

Results: The estimated prevalence of 7% crossed beaks in Appenzeller Barthuhn was significantly higher compared to two other local Swiss chicken breeds. A breeding trial showed significantly higher prevalence of offspring with deformed beaks from mating of affected parents compared to mating of non-affected parents. Examination of 77 Appenzeller Barthuhn chickens with crossed beaks showed a variable phenotype presentation. The deviation of the beak from the median plane through the head ranged from 1° to 61°. In more than 60% of the cases, the upper and lower beak were bent in the same direction, whereas the remaining cases showed different forms of crossed beaks. Computed tomographic scans and bone maceration of the head of two chickens with crossed beaks revealed that the maxilla and the mandibula were affected, while other parts of the skull appeared to be normal. The gene *LOC426217*, a member of the keratin family, was postulated as a candidate gene for beak deformity in domestic chickens. Sequencing of the coding region revealed two significantly associated synonymous variants for crossed beaks in Appenzeller Barthuhn chickens. A genome-wide association study and a comparative analysis of runs of homozygosity based on high-density SNP array genotyping data of 53 cases and 102 controls showed no evidence of association.

Conclusions: The findings suggest a hereditary cause of crossed beaks in Appenzeller Barthuhn chickens. However, the observed variation in the phenotype, together with the inconclusive molecular genetic results indicates the need for additional research to unravel the genetic architecture of this beak deformity.

Keywords: Beak deformity, Indigenous breed, Congenital anomaly, *LOC426217*, GWAS

Background

A bird's beak is a defining feature of its head. It is important for the intake of food and water [1, 2], for grooming purposes and parasite removal [3–5] and as a tactile sensory organ [6]. The form of a beak is very well adapted to the trophic niche of its species [7, 8]. It is evident that divergence from a normally shaped beak poses a great problem for the bird. A well-known example is the red crossbill (*Loxia curvirostra*) showing distinctive mandibles, crossed at the tips, which enable them to extract seeds from conifer cones and other fruits [9]. A bird's beak consists of the osseous maxilla and mandibula, which are covered by a horny sheath of keratin (rhamphotheca). The rhamphotheca is essentially a modified epidermis, where cells of the stratum corneum contain tricalcium phosphate, hydroxyapatite and keratin, giving the beak its rigidity [10]. The rhamphotheca continues growing throughout life and maintains its intended form through continuous wear and tear [6, 11]. In chickens, two nares (nostrils are found laterally at the base of the upper beak which are partially covered by the operculum nasale [12]).

In domestic chickens, beak deformities affect welfare and production. The occurrence of different types of crossed beaks was already reported almost a century ago

* Correspondence: cord.droegemueller@vetsuisse.unibe.ch
†Equal contributors
[1]Institute of Genetics, Vetsuisse Faculty, University of Bern, Bern, Switzerland
Full list of author information is available at the end of the article

in various breeds [13, 14]. The crossed beak is a beak deformity that can be defined as misalignment of the upper and lower beak [11]. Until today, there are few reports on prevalence of crossed beaks in chicken [11, 14]. It was estimated that 0.32% of all wild birds have crossed beaks [11]. In a recent study, prevalence ranging from 1 to 3% was documented for local Chinese chicken breeds [15, 16].

Different reasons for the occurrence of beak deformities in birds have been proposed: they can result from the underlying bony structure, from a malocclusion of the mandibles or disturbances in the stratum germinativum of the epidermis [10, 17]. The occurrence of crossed beaks was reported due to breeding conditions [18], accidents or traumata [19], abnormal abrasion of the rhamphotheca [11], housing conditions [8], environmental influences such as toxins [20–24] or nutritional deficiencies [25, 26], infections [27, 28], or genetic causes [14, 23, 29–32]. Already in 1934 a simple recessive inheritance for crossed beaks was postulated in the budgerigar [33]. In 1938 Landauer [14] failed to establish a pure line of domestic chicken with crossed beaks in spite of massive inbreeding. He concluded that the phenomenon is caused by several genes with incomplete penetrance. Several possible candidate genes for deformed beaks in a Beijing-You chicken were proposed from digital gene expression profiling studies [15]. Most of the identified genes belonged to the keratin family, as well as genes coding for proteins important in the biosynthesis of unsaturated fatty acids and the metabolism of glycerolipids. A follow-up study on the most promising gene *LOC426217* found five variants that differed significantly between chickens with crossed beaks and normal controls [16].

In recent years, breeders have occasionally reported the occurrence of crossed beaks in Appenzeller Barthuhn chickens (Fig. 1). Along with the Appenzeller Spitzhaubenhuhn and the Schweizerhuhn, this breed represents one of three local chicken breeds found in Switzerland [34]. The aim of this study is to describe the crossed beak phenotype based on Appenzeller Barthuhn cases and to derive estimates of prevalence using data from a survey and a breeding trial. First attempts to unravel the genetic architecture of the crossed beak phenomenon are presented.

Methods

Breeding analysis

A total of 165 breeding reports, provided by breeders, were analyzed to derive the prevalence of crossed beaks for the breeding seasons 2012–2017. Breeders reported their annual breeding success and the number of hatched animals showing crossed beaks. Estimated prevalence were compared by Pearson's chi-squared test with Yates' continuity correction using the R software [35].

Breeding experiment

Animals from the 2016 breeding season were selected based on the shape of their beak at 12 weeks of age, and included in a targeted breeding experiment as the parental generation. This comprised in total of six breeding units of comparable size: three units consisting of affected animals and three units consisting of control animals (Additional file 1). In the three affected units, three to four affected hens and one affected rooster were grouped together in each unit, resulting in 11 affected hens and three affected roosters (Additional file 1). The three control units consisted of five hens and one rooster each, totaling 15 hens and three roosters with a normal beak. The six parental breeding units were grouped in August 2016 and were kept constantly under the same feeding regime and environmental conditions until the 2017 breeding season.

The F1-generation eggs of the six parental groups were collected during three periods of 10 days in spring 2017,

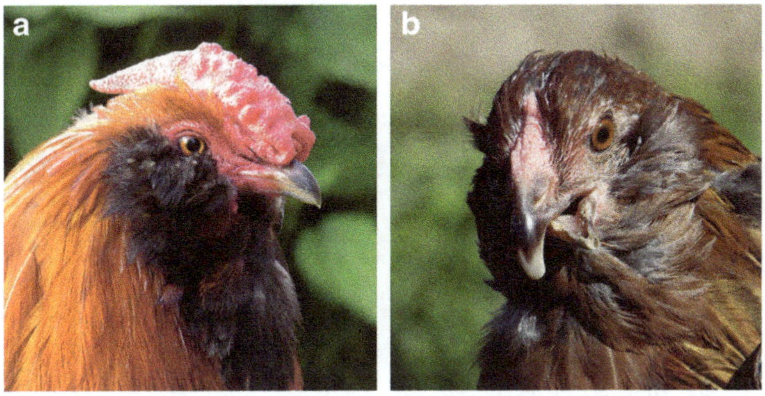

Fig. 1 Normal and crossed beak: (**a**) Appenzeller Barthuhn rooster with a normal beak and (**b**) Appenzeller Barthuhn hen with a crossed beak

and incubated under standardized conditions in a conventional hatchery. After hatching, chicks were individually tagged and reared until 12 weeks of age. The occurrence of beak deformities was visually controlled, and the final status was assessed between 10 and 12 weeks of age. Proportions of affected and unaffected offspring were compared using Fisher's exact test in the RStudio software [35].

Phenotype description

Between the years 2012 and 2017, a total of 77 chickens with complete, undamaged heads, displaying supposed beak deformities were presented by breeders of Swiss Appenzeller Barthuhn chickens and collected from the breeding trial. All heads were photographed and macroscopically examined for 11 morphological features (Additional file 2). In addition, the angles of the upper and lower beak were measured. The head was fixed with the beak in a horizontal position between two lateral and one caudal screw and a photo was taken from directly above. Based on the photograph, the 0° axis was defined as perpendicular to the two screws and through the median plane of the head (Additional file 3). The axis of the beak was defined by drawing a line along the base and middle part of the upper or lower beak, while ignoring all additional bending of the tip. The angle was then measured at the intersection of 0° axis and the axis of the beak (Additional file 2). Statistical analyses were conducted with the software RStudio software [35]. Angles of the upper beak, lower beak and in between beaks were tested for normal distribution with Shapiro-Wilk normality test. One sample t-test for normal distributed data and Wilcoxon signed rank test for non-normal distributed data were used respectively, to test whether the beaks were bent more to one side than the other. The Wilcoxon signed rank test was used to compare angles of upper and lower beaks. The binomial test was used to examine association between bending of the beak and bending of the tip. A detailed anatomical examination of two cases with crossed beaks and one normal Appenzeller Barthuhn control were carried out by computed tomography and maceration of the head bones (10% aqueous Biozyme SE-Solution, 60 °C, 20 h).

Candidate gene analysis

The annotated exon of the candidate gene *LOC426217* on chromosome 25 was sequenced in 53 affected Appenzeller Barthuhn chickens (specified in Additional file 4) and 114 breed controls. The DNA was isolated either from EDTA-blood or tissue samples using the Nucleon Bacc2 kit (GE Healthcare). In addition, DNA samples of 42 Appenzeller Spitzhaubenhuhn and 42 Schweizerhuhn were analyzed for comparison. PCR amplification was done as previously described [16] and subsequently sequenced by direct Sanger sequencing on the 3730 DNA-Analyzer (Thermofisher). The obtained sequence data was analyzed with Sequencher 5.1 software (Gene Codes). For all variants allele and genotype frequencies were compared between cases and controls using chi-squared and Fisher's exact tests. The Benjamini and Hochberg method [36] was used for multiple testing correction of p-values. Phased haplotypes were considered for further analysis.

Genome-wide analysis

Genomic DNA samples from 53 cases and 102 control Appenzeller Barthuhn chickens were genotyped with the Axiom™ Genome-Wide Chicken Array Kit (Affymetrix) for 580,961 SNP markers. PLINK v1.07 [37] was applied for quality control by removing: SNPs with a call rate below 99%, SNPs with a minor allele frequency below 1% and significant ($p \leq 0.0001$) deviation from the Hardy-Weinberg equilibrium. Additionally, genotypes of individuals with a call rate below 90% were excluded. In total, 155 individual genotypes consisting of 341,115 SNPs remained for the genome-wide association study (GWAS). A mixed linear model was fitted with the GenABEL package [38], including a random polygenic effect based on an autosomal genomic relationship matrix, in order to account for population structure and relatedness between animals. SNPs were considered to be genome-wide significantly associated if their p-values were below the 5% Bonferroni-corrected threshold for 341,115 independent tests (pBONF 1.47×10^{-7}). Runs of homozygosity (ROH) were detected using PLINK v1.90 [37] with a density of at least 1 SNP per 85 kb. The percentages of animals having a SNP in a ROH were compared between cases and controls at all the different loci on all chromosomes.

Results

The prevalence of crossed beaks differs between breeds

Across the three Swiss local chicken breeds, a total of 78 out of 2698 chicken (2.9%) were reported to show a crossed beak within the past six years (Table 1). The anomaly occurred in all three breeds, but the prevalence

Table 1 Occurrence of crossed beaks in three local chicken breeds based on breeder reports between 2012 and 2017

Breed	Number of reports	Total number of birds	Normal	Crossed beak
Appenzeller Barthuhn	55	797	738 (92.6%)	59 (7.4%)
Appenzeller Spitzhaubenhuhn	35	454	447 (98.5%)	7 (1.5%)
Schweizerhuhn	75	1447	1435 (99.2%)	12 (0.8%)
Total	165	2698	2620 (97.1%)	78 (2.9%)

differs. The Appenzeller Barthuhn had a significantly higher prevalence (7.4%) compared to Appenzeller Spitzhaubenhuhn (1.5%, $\chi2 = 18.726$, p-value < 0.001) and Schweizerhuhn (0.8%, $\chi2 = 70.351$, p-value < 0.001) (Table 1). No significant difference was observed in the prevalence between Appenzeller Spitzhaubenhuhn and Schweizerhuhn ($\chi2 = 1.1262$, p-value = 0.2886).

In the F1-generation of the breeding trial, 83 offspring of parents with crossed beaks and 102 offspring of parents with normal beaks were examined between 10 and 12 weeks of age (Additional file 1).

A significant difference in prevalence (p-value = 0.003) of crossed beak offspring was observed when comparing these two groups: 2.9% of the offspring from normal parents showed crossed beaks, whereas 15.7% affected animals were observed in the offspring of parents with crossed beaks (Table 2). Additionally, seven animals (3.8%) displayed deformed but not crossed beaks. This fraction did not differ significantly between the two parental groups (Table 2).

The crossed beak anomaly in Appenzeller Barthuhn chicken

The crossed beak is characterized by one or both beaks deviating laterally from the longitudinal axis of the head. Out of 77 cases collected from breeders and the breeding trial, 48 (62.3%) showed an upper and lower beak that deviated clearly either left or right (Fig. 2), while in 5 cases (6.5%) the upper and lower beak deviated on different sides. Of the 48 cases where both beaks deviated in the same direction, 44 (91.7%) showed a stronger bending of one of the two beaks. In 23 of 77 cases (29.9%) the upper beak was straight while the lower beak was bent. A bent upper beak without a bent lower beak occurred only in one case (1.3%).

Upper and lower beaks which deviated from the median plane showed a wide range of angles from only 1° up to 50° in the upper beak and from 2° up to 61° in the lower beak. About 50% of all deviation angles of beaks were between 1° and 9° for the upper beak, and 2° and 12° for the lower beak, respectively. The upper beak was more often bent to the right side (p-value = 0.029), while no significant difference in the direction of the lateral bend of the lower beak (p-value = 0.905) could be found. There was no significant difference between the angles of the upper and the lower beak (p-value = 0.5822). The angle between upper and lower beak varied between 0° (cases where both upper and lower beak deviated at exactly the same angle) and 61°.

In 51 out of 77 cases (66.2%), one or both tips of the beak were turned in the direction of the other tip, as if trying to find its counterpart (Fig. 2b, c). This was independent of the original bending of the beak. Out of 77 beaks available for examination, 47 (61.0%) showed a rotation of the beak (Fig. 3c). The direction of the rotation was associated with the direction of the bending of the beak; it usually rotated in the same direction as the bending (p-value = 0.002).

A total of 64 cases (83.1%) showed an excessively vertically downward bent upper beak. In 21 cases (27.3%), the lower beak was bent upwards and in 21 cases (27.3%) bent downwards (Fig. 3a, b). In 23 cases (29.9%) the upper beak of an affected animal was also bent upwards at the base, usually accompanied by a protruding of the palate (Fig. 3d). Out of 77 cases examined, 47 (61.0%) showed one or both beaks elongated (Fig. 3a, b). Both beaks were elongated in 31 cases (40.3%), only the upper beak in nine cases (11.7%) and in seven cases (9.1%) only the lower beak. In nine cases (11.7%) the upper beak was too short compared to a normal beak and in three (3.9%) of these cases, the short upper beak was paired with an elongated lower beak.

The rhamphotheca in animals with crossed beak showed no overall abnormality. Most commonly observed were brittle tips on overlong beaks, sometimes with pieces of tip missing (Fig. 4a). The normal beak of a domestic chicken has a sharp-cut tomium, however in 11 birds (14.3%), the tomium was rolled inwards (Fig. 2b). In 51 cases (66.2%) food and soil got stuck in one or both beaks, mostly in a part of the beak that was not covered by its counterpart. This was promoted by an inward rotation of the tomium (Fig. 2b).

The remainder of the head including the cranium, eyes, ears, feathers and wattles did not show any outward abnormality. The most rostral part of the rose comb was often bent with the upper beak. The nares, situated laterally and at the base of the upper beak were in some cases compressed on one side and stretched on the other side depending on the bending of the upper beak (Fig. 4b, c). The operculum on the convex side of the bending was stretched and entirely covered the naris (Fig. 4b), while the operculum on the concave side was folded and left the nostril wide open (Fig. 4c).

Table 2 Number of offspring with and without crossed beaks from affected and unaffected parents observed during the breeding experiment

	Number of offspring with normal beak (fraction in %)	Number of offspring with crossed beak (fraction in %)	Number of offspring with otherwise deformed beak (fraction in %)
Parents with crossed beaks	67 (80.7%)	13 (15.7%)	3 (3.6%)
Parents with normal beaks	95 (93.1%)	3 (2.9%)	4 (3.9%)
Total	162 (87.6%)	16 (8.6%)	7 (3.8%)

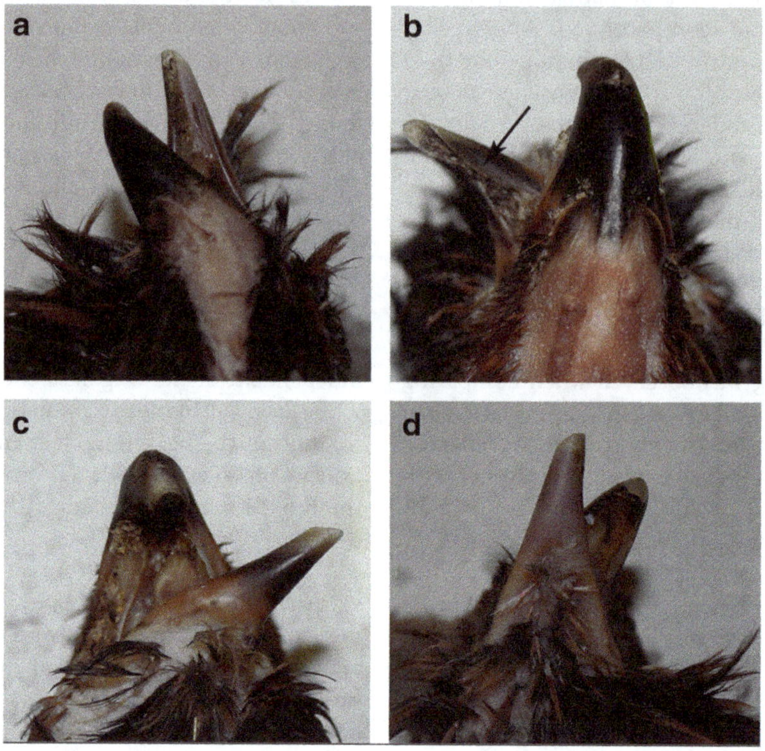

Fig. 2 Examples of typical crossed beak in Appenzeller Barthuhn chickens. Note that the beaks are bent in different directions and at a different location. **a** Dorsal view on an upper beak bent to the left at the base of the beak. **b** Dorsal view of an upper beak bent only at the tip, while the lower beak is bent at the base with an inwardly-rolled tomium. **c** Ventral view of a lower beak bent at the base. **d** Ventral view of a lower beak bent at the tip and an upper beak bent at the base

A detailed view on the crossed beak phenotype in Appenzeller Barthuhn chickens

Maceration and computed tomographic images showed that the shape of several bones was affected by the crossed beak (Fig. 5c–h) compared to normally formed beaks (Fig. 5a–f). In both examined cases, the rostral part of the skull was asymmetrical. The os praemaxillare and the os nasale were bent to the right in both cases and the beak was rotated around the longitudinal axis. The beak's tip was bent to the left. In both animals the mandibula also showed a flexion (Fig. 5d). The convex side of the pars intermedia was longer than the concave side (Fig. 5d, h). The pars caudalis of the mandibula was symmetrical (Fig. 5h). Computed tomographic transverse and dorsal reformatted images of affected heads showed the location of the bending. In the upper beak, the bending occurred at the height of the aboral end of the processus frontalis of the os praemaxillare and the nostrils (Fig. 5g). The more rostral part of the beak seemed straight and without inflection, except the tip, which was bent in the opposite direction. The mandibula was bent on the whole length of the pars intermedia, while the base of the mandibula, on the height of the pars caudalis, was still symmetrical (Fig. 5h).

Associated *LOC426217* variants in crossed beaks

Sequencing of *LOC426217* in 53 cases (of upper and lower beak deformity) and 114 controls revealed a total of 12 variants (Table 3). A single variant was located in the 5′-untranslated region, whereas the remaining variants were in the coding region, including 4 missense variants and 7 synonymous variants. There was a significant genotype frequency difference between cases and controls at two synonymous coding variants (Table 3). Subsequently, a haplotype estimation using the genotypes of these 12 variants in all animals was performed, which revealed three predominantly occurring *LOC426217* haplotypes (Table 4). In addition, further rare haplotypes occurring in seven animals were observed. Based on the diplotypes for these 160 animals carrying the three most frequent haplotypes, a significant (p-value < 0.001) difference in haplotype distribution between cases and controls was observed (Table 4). However, the most abundant haplotype A observed in Appenzeller Barthuhn chickens with crossed beaks was also by far the most common haplotype in the genotyped cohorts of animals belonging to the two other local Swiss breeds (Table 4, Additional file 5).

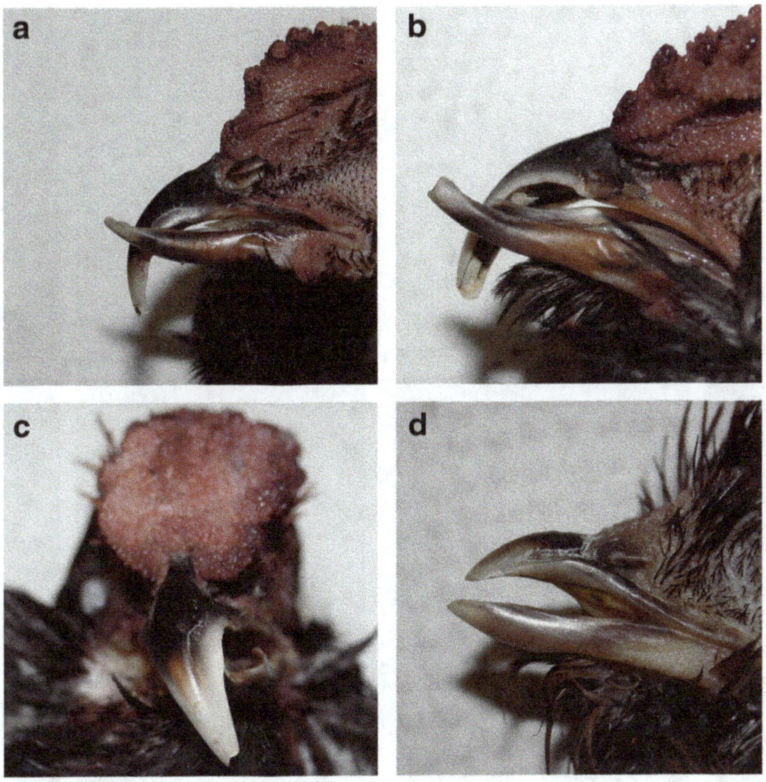

Fig. 3 Occasionally observed additional features in Appenzeller Barthuhn chickens with crossed beaks. **a + b** The upper beak bent excessively downwards, while the lower beak is bent upwards. Note the overgrowth of both beaks. **c** The beak is not only bent horizontally, but is also rotated along its longitudinal axis. **d** The upper beak is bent upwards at its base

Crossed beak phenotype shows no evidence of genetic association at the genome-wide level

The GWAS based on high-density SNP array genotyping data of 53 cases and 102 controls from the Appenzeller Barthuhn breed showed no evidence for association (Additional file 6). Additional GWAS testing of 20 cases showing mostly upper or 31 cases showing mostly lower beak deviation against all controls also revealed no evidence for association. A comparative analysis of runs of homozygosity was performed using the SNP genotyping data of all 53 cases and 102 controls. No major difference was observed between groups (Additional file 7).

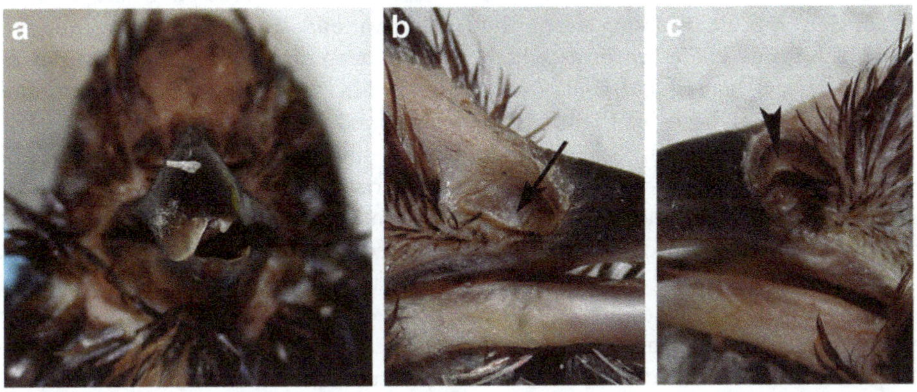

Fig. 4 Occasionally observed anomalies of the rhamphotheca and nares in Appenzeller Barthuhn chickens with crossed beaks. **a** Overgrowth of the beak resulting in an instable brittle tip. **b + c** Deformed nares occur due to the bending of the upper beak at the height of the nares. Note that the nostril on the convex side of the beak was nearly entirely closed due to a stretched operculum, while the operculum on the concave side was folded

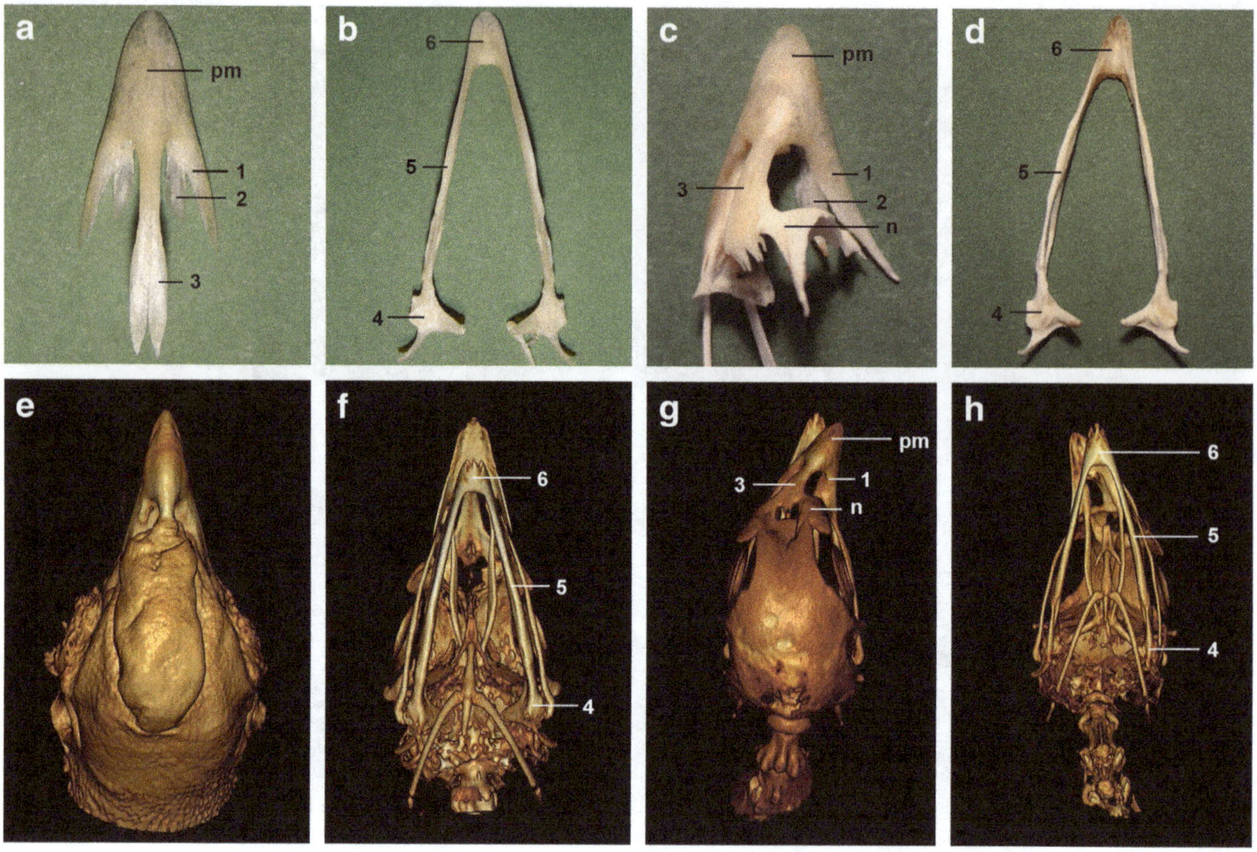

Fig. 5 Comparative view of the head of a normal Appenzeller Barthuhn chicken with an animal with crossed beaks. Macerated bones of the beak of a control chicken (**a** + **b**) and a chicken with crossed beak (**c** + **d**). Volume rendered three-dimensional computed tomographic model of the normal head of a control (**e** + **f**) and of a crossed beak (**g** + **h**): view from dorsal (**e** + **g**) and ventral (**f** + **h**). Note that both upper and lower beak of the crossed beak are bent to the right. Os praemaxillare (pm) of the maxilla with processus maxillaris (1), processus palatinus (2), processus frontalis (3), and the os nasale (n). Mandibula with pars caudalis (4), pars intermedia (5), and pars symphysialis (6)

Discussion

Within the last century, the sporadic occurrence of the crossed beak phenotype has been described by various authors in wildlife as well as in domestic birds including chickens [13, 15, 39, 40]. Pomeroy [11] defined a crossed beak as the upper or lower beak deviating from the median plane through the head. There are birds where this naturally occurs, such as the red crossbill (*Loxia curvirostra*). The upper and lower beaks of the crossbill are aligned when they hatch and cross with growth [14]. This development has also been reported in domestic chicks with a crossed beak [14]. The same holds for the offspring produced by the breeding experiment conducted here: the beak deformities were not present at hatching. Nonetheless, in some of the cases presented by breeders, chicks already had a crossed beak at a few days old. The onset of crossed beaks in the affected animals therefore varies. Based on the 185 offspring evaluated in the breeding experiment, it can be concluded that chickens should not be labelled free of crossed beaks before the age of at least 12 weeks.

This study reports for the first time a detailed phenotype characterization of crossed beak in chickens based on morphological measurements and anatomical examination. Nearly 80 years ago Landauer [14] developed a classification of crossed beaks based upon descriptive observations. The first two categories included crossed beaks accompanied by abnormalities of the eyes or the skull. This was not observed in our study, as none of the examined affected Appenzeller Barthuhn chickens had other obvious anomalies of the head. The third category comprised of crossed beaks that developed after 1 to 2 months of age, while the fourth category consisted of crossed beak present at hatching but later grow into a normally developed beak. As indicated above, the crossed beak phenotype in the Appenzeller Barthuhn chickens unambiguously manifested after two months of age and we therefore conclude that the third category is of major concern in this local breed. Landauer [14] also described that the affected beak was usually the maxilla and that the crossed beak had its origin in a malformed bone, not only a malformed rhamphotheca. The cases

Table 3 Genotype frequencies for *LOC426217* variants in normal Appenzeller Barthuhn chickens compared to animals with crossed beaks

Position in the *LOC426217* cDNA[a] and on chromosome 25 (Gallus_gallus-5.0)[b]		Genotype[c, d] AA	AB	BB	χ^2-value[e]	*p*-value	corrected *p*-value
c.-62G > T; 5'UTR	control	7	26	81	6.960	0.029	0.086
g.2013308C > A	case	0	6	46			
c.16C > G; p.Leu6Val	control	113	1	0	none	1	1
g.2013231G > C	case	52	0	0			
c.24 T > C; synonymous	control	17	56	41	3.087	0.214	0.320
g.2013223A > G	case	5	21	26			
c.36G > C; synonymous	control	35	59	20	16.849	< 0.001	0.003
g.2013211C > G	case	10	17	25			
c.48A > G; synonymous	control	82	26	7	5.538	0.063	0.103
g.2013199 T > C	case	46	5	1			
c.49G > A; p.Ala17Thrr	control	111	3	0	none	0.553	0.603
g.2013198C > T	case	52	0	0			
c.96C > T; synonymous	control	82	25	7	6.762	0.034	0.078
g.2013151G > A	case	47	6	0			
c.222 T > C; synonymous	control	17	56	41	2.819	0.244	0.326
g.2013025A > G	case	5	22	26			
c.252A > G; synonymous	control	82	25	7	8.013	0.018	0.068
g.2012995 T > C	case	48	5	0			
c.256_258delCTGinsTAT; p.Leu86Ile	control	82	25	7	6.762	0.034	0.078
g.2012991_2012989delGACinsATA	case	47	6	0			
c.289G > T; p.Gly97Cys	control	114	0	0	none	0.317	0.381
g.2012958C > A	case	52	1	0			
c.363 T > C; synonymous	control	21	58	35	15.011	< 0.001	0.003
g.2012884A > G	case	25	18	10			

[a]NCBI accession no. XM_423880
[b]NCBI accession no. NC_006112
[c]For some variants individual genotypes are missing (Additional file 5)
[d]A = reference allele, B = variant allele
[e]None = for these variants fisher test was used and no χ^2-value exists

described in this current study demonstrated both upper and lower beaks were affected.This partially confirms the observations of Landauer [23], as the crossed beak phenotype in Appenzeller Barthuhn clearly affects the bones of the beak. Contrary to this in wild birds, crossed beaks often seem to consist of an abnormally formed rhamphotheca over normally formed bones e.g. avian keratin disorder [11, 40, 41]. Nevertheless, there are reports of cases of affected wild birds, where the bone is clearly malformed as well [31]. One feature in the presented cases of crossed beaks in Appenzeller Barthuhn chickens is very similar to the avian keratin disorder, i.e. the overgrowth of the rhamphotheca. This might be due to the absence of abrasion because of the different use of the beak in domestic chicken and the missing conjunction of upper and lower beak. Previously it has been suggested that upper and lower beak influence each other's growth during development [11]. It may be due to such influences that traits like tips bent in the direction of the other tip, up and downwards bent tips or rotation of the beak develop. It could be speculated that this may be the result of physiological mechanisms trying to revert the missing conjunction of upper and lower beak in growth [42].

Table 4 Distribution of diplotypes at *LOC426217* in chickens with and without crossed beaks, and in two other local Swiss chicken breeds

Diplotypes for estimated *LOC426217 haplotypes*	Appenzeller Barthuhn Control	Case	Appenzeller Spitzhaubenhuhn	Schweizerhuhn
AA	20	24	40	41
AB	43	17	1	
BB	17	5		
BC	11	5		
CC	7	0		
AC	11	0		
other	5	2	1	1
Total	114	53	42	42

As hypothesized earlier a genetic cause is responsible for the development of crossed beaks in chickens. A significantly higher prevalence of affected animals was found in the Appenzeller Barthuhn breed compared with two other local chicken breeds. This information combined with the known differences in the breeds [43] suggests a major variation in the genetic background for this phenomenon in the Appenzeller Barthuhn chicken. In addition, the results of the targeted mating experiments further support the inheritance of this anomaly. By comparing status of offspring from normal parents with those from affected parents, a significantly higher number of affected offspring resulted from parental units with affected animals. Nonetheless, the occasional occurrence of affected offspring in the control units and the quite high number of non-affected offspring from affected parental units does not allow elucidation of the mode of inheritance.

The gene *LOC426217* located on chromosome 25 was recently proposed as a functional candidate for crossed beak [15, 16], therefore, allele- and haplotype frequencies based on 166 samples (53 affected and 114 controls) of the Appenzeller Barthuhn breed were investigated. From the five variants proposed by Bai et al. [16], three were not polymorphic in Appenzeller Barthuhn chickens. However, seven additional variants, mostly predicted to have no effect on the encoded amino acid sequence are presented. A comparison of the genotype frequencies revealed two variants with significant differences between cases and controls. These two variants seem to confirm the findings of Bai et al. [16] for *LOC426217* as a possible candidate gene for crossed beak, although the phenotype of that study included only lower beak deviation. As both associated variants do not cause a change in the amino acid sequence, their functional impact on the phenotype expression is not obvious. Furthermore, the subsequent haplotype association analysis also showed a significant difference between cases and controls in Appenzeller Barthuhn chickens. As the associated *LOC426217* haplotype also occurs almost exclusively in the genotyped control cohorts of Appenzeller Spitzhaubenhuhn and Schweizerhuhn, it is unlikely that these variants explain the crossed beak phenotype. It could be speculated that other DNA variants, e.g. structural variants affecting larger segments of DNA, situated in the flanking region containing regulatory important elements of *LOC426217* might be present on the associated haplotype in Appenzeller Barthuhn only which could lead to the development of crossed beaks. On the other hand, the identification of *LOC426217* as a candidate gene for beak deformities is based on transcript analysis from lower mandibles [16]. This gene belongs to the keratin family, and it could be concluded that the initially observed difference in expression represents a secondary effect due to the existence of the deviated lower beak.

Finally, the GWAS and the analysis of runs of homozygosity did not support the association with any specific region of the chicken genome. However, the breeding analysis suggests a genetic background. The previously identified single locus association for the tested *LOC426217* variants on chromosome 25 was not confirmed by the genome-wide study. The SNP array marker density was most likely sufficient so it could be assumed, that the previous result might be false positive. The current findings suggest that the crossed beak anomaly in Appenzeller Barthuhn chickens cannot be explained by simple Mendelian inheritance as was previously assumed [33]. The collection of 53 cases used for GWAS shows an obvious variable phenotype expression supporting a complex genetic background. Therefore it is concluded, that a case-control design and the available sample sizes are not sufficient to unravel the genetic architecture of the crossed beak. Additional studies using more genotypes of affected and unaffected animals with a clear description of the phenotype are therefore required.

Conclusion

Phenotypic variability in chickens with crossed beaks exists. Genetic predisposition in the Appenzeller Barthuhn chicken breed is evident. The targeted mating experiments confirmed a genetic influence on the occurrence of crossed beak chickens. Genetic analyses were inconclusive, indicating a complex mode of inheritance.

Additional files

Additional file 1: Scheme of the breeding trial. In total six breeding units of comparable size, made up the parental generation: three units consisting of affected animals (orange units, three to four affected hens and one affected rooster) and three units consisting of control animals (blue units, five hens and one rooster). For the F1-generation, eggs of the six parental groups were collected and incubated. After hatching, chicks were reared until the age of 12 weeks.

Additional file 2: Phenotype details of Appenzeller Barthuhn chickens with crossed beak and all cases used in the study.

Additional file 3: Phenotyping tool for the measurement of beak angles. The head is fixed between two screws, which sit in the orbita of the skull, and one caudal screw, which supports the caudal part of the head. The photograph is taken from directly above. Based on the photograph, the 0° axis is defined as perpendicular to the two screws and through the median plane of the head (central vertical line of the grid). The axis of the beak is defined by drawing a line along the base and middle part of the upper or lower beak (yellow line).

Additional file 4: Table with all cases for phenotype description and genetic analyses.

Additional file 5: Genotypes and diplotypes of Appenzeller Barthuhn and two other Swiss Chicken breeds. Genotypes of 12 variants and their diplotypes in *LOC426217* of Appenzeller Barthuhn, Appenzeller Spitzhaubenhuhn and Schweizerhuhn.

Additional file 6: GWAS of 53 cases and 102 controls. (A) Manhattan plot. The red line marks the 5% Bonferroni-corrected threshold for 341,115 independent tests (pBONF 1.47×10^{-7}). (B) MDS plot. (C) QQ plot.

Additional file 7: Comparison of runs of homozygosity between cases and controls on the autosomes. Chromosome-wise plots show on the x-axis the mega base (Mb) position on the chromosome (CHR), and on the y-axis the proportions of cases (yellow) and controls (black) being homozygous.

Abbreviations
Bp: Base pairs; GWAS: Genome-wide association study; Kb: Kilo base pairs; Mb: Mega base pairs; PCR: Polymerase chain reaction; ROH: Run of homozygosity; SNP: Single nucleotide polymorphism

Acknowledgements
We thank Nathalie Besuchet-Schmutz, Johannes Joller, Stefan Gfeller and Danya Wiederkehr for expert technical assistance.

Funding
The project was funded by the Swiss Federal Office for Agriculture (BLW; ARAMIS-Nr. 625000438 und REF-1602-41303). SJ was supported by the H. Wilhelm Schaumann Stiftung, Hamburg.

Authors' contributions
SJ and FB performed the phenotyping and the genetic investigations. AR and DSG analyzed anatomical details of selected cases. AS and EK assisted in data and sample collection. SJ and CD drafted the manuscript. CF and CD supervised the project and edited the manuscript. All authors approved the final version of the manuscript.

Competing interests
The corresponding author of this manuscript (CD) is Associate Editor for BMC Veterinary Research. The authors declare to have no financial or non-financial competing interests.

Author details
[1]Institute of Genetics, Vetsuisse Faculty, University of Bern, Bern, Switzerland. [2]School of Agricultural, Forest and Food Sciences, Bern University of Applied Sciences, Zollikofen, Switzerland. [3]ProSpecieRara, Basel, Switzerland. [4]Züchterverein für ursprüngliches Nutzgeflügel, Neukirch an der Thur, Switzerland. [5]Divison of Veterinary Anatomy, Vetsuisse Faculty, University of Bern, Bern, Switzerland. [6]Division of Clinical Radiology, Vetsuisse Faculty, University of Bern, Bern, Switzerland.

References
1. Benkman CW, Lindholm AK. The advantages and evolution of a morphological novelty. Nature. 1991;349:519–20. https://doi.org/10.1038/349519a0.
2. Temeles EJ, Kress WJ. Adaptation in a plant-hummingbird association. Science (80-). 2003;300:630–3. https://doi.org/10.1126/science.1080003.
3. Chen BL, Haith KL, Mullens BA. Beak condition drives abundance and grooming-mediated competitive asymmetry in a poultry ectoparasite community. Parasitology. 2011;138:748–57. https://doi.org/10.1017/S0031182011000229.
4. Clayton DH, Moyer BR, Bush SE, Jones TG, Gardiner DW, Rhodes BB, et al. Adaptive significance of avian beak morphology for ectoparasite control. Proc R Soc B Biol Sci. 2005;272:811–7. https://doi.org/10.1098/rspb.2004.3036
5. Vezzoli G, Mullens BA, Mench JA. Relationships between beak condition, preening behavior and ectoparasite infestation levels in laying hens. Poult Sci. 2015;94:1997. https://doi.org/10.3382/ps/pev171.
6. Salomon F-V, Krautwald-Junghans M-E. Anatomie der Vögel. In: Salomon F-V, Geyer H, Gille U, editors. Anatomie für die Tiermedizin. 2. Stuttgart: Enke; 2008. p. 754–814.
7. Boag PT, Grant PR. Intense natural selection in a population of Darwin's finches (Geospizinae) in the Galapagos. Science. 1981;214:82–5. https://doi.org/10.1126/science.214.4516.82.
8. Clabaut C, Herrel A, Sanger TJ, Smith TB, Abzhanov A. Development of beak polymorphism in the African seedcracker, Pyrenestes ostrinus. Evol Dev. 2009;11:636–46.
9. Red Crossbill. https://en.wikipedia.org/wiki/Red_crossbill. Accessed 20 Dec 2017.
10. Macwhirter P. Basic anatomy, physiology and nutrition. In: Handbook of avian medicine. Second Edi. Elsevier; 2009. p. 25–55. https://doi.org/10.1016/B978-0-7020-2874-8.00002-X.
11. Pomeroy DE. Birds with abnormal bills. Br Birds. 1962;55:233–8. https://britishbirds.co.uk/article/birds-with-abnormal-bills/643
12. Nickel R, Schummer A, Seiferle E. Lehrbuch der Anatomie der Haustiere, Band 5, Anatomie der Vögel. Third Edit. Stuttgart: Parey; 2004.
13. Lindner F. Kreuzschnabelmissbildung. Orn Mschr. 1902;27:59–62.
14. Landauer W. Notes on cross-beak in fowl. J Genet. 1938;37:51–68. https://doi.org/10.1007/BF02982143.
15. Bai H, Zhu J, Sun Y, Liu R, Liu N, Li D, et al. Identification of genes related to beak deformity of chickens using digital gene expression profiling. PLoS One. 2014;9:e107050. https://doi.org/10.1371/journal.pone.0107050.
16. Bai H, Sun Y, Zhu J, Liu N, Li D, Xue F, et al. Study on LOC426217 as a candidate gene for beak deformity in chicken. BMC Genet. 2016;17:44. https://doi.org/10.1186/s12863-016-0353-x.
17. Stettenheim P. The integument of birds. In: Farner DS, King JR, Parkes KC, editors. Avian biology, Vol. 2. New York: Academic Press; 1972. p. 1–63.
18. Robertson I. The influence of turning on the hatchability of hens eggs. J Agric Sci. 1961;57:57. https://doi.org/10.1017/S0021859600050000.
19. Doneley B. Avian medicine and surgery in practice. Second ed. Boca Raton: CRC Press; 2010. https://doi.org/10.1201/b15176.
20. Al-Qudsi F, Al-Jahdali A. Effect of monosodium glutamate on Chick embryo development. J Am Sci. 2012;8:499–509. https://doi.org/10.7537/marsjas081012.72.
21. Blankenship AL, Hilscherova K, Nie M, Coady KK, Villalobos SA, Kannan K, et al. Mechanisms of TCDD-induced abnormalities and embryo lethality in white leghorn chickens. Comp Biochem Physiol C Toxicol Pharmacol. 2003;136:47–62. https://doi.org/10.1016/S1532-0456(03)00166-2.
22. Handel CM, Van Hemert C. Environmental contaminants and chromosomal damage associated with beak deformities in a resident north American passerine. Environ Toxicol Chem. 2015;34:314–27. https://doi.org/10.1002/etc.2799.
23. Landauer W. Hereditary and induced cross-beak of fowl. Experiments with ethyl carbamate. J Exp Zool. 1956;132:25–38. https://doi.org/10.1002/jez.1401320103.
24. Smith F, Hu D, Young NM, Lainoff J, Jamniczky H, Maltepe E, et al. The effect of hypoxia on facial shape variation and disease phenotypes in chicken embryos. Dis Model Mech. 2013;6:915–24. https://doi.org/10.1242/dmm.011064.
25. Cooper RG, Horbanczuk JO. Crooked beak in a 14-month-old ostrich (Struthio camelus) hen : to the editor. J S Afr Vet Assoc. 2006;77:170. https://doi.org/10.4102/jsava.v77i4.371.
26. Tangredi BP. Environmental factors associated with nutritional secondary hyperparathyroidism in wild birds. Avian Poult Biol Rev. 2007;18:47–56. https://doi.org/10.3184/147020607X251212.
27. Galligan TH, Kleindorfer S. Naris and beak malformation caused by the parasitic fly, Philornis downsi (Diptera: Muscidae), in Darwin's small ground finch, Geospiza fuliginosa (Passeriformes: Emberizidae). Biol J Linn Soc. 2009;98:577–85. https://doi.org/10.1111/j.1095-8312.2009.01309.x.
28. Gartrell B, Alley M, Kelly T. Bacterial sinusitis as a cause of beak deformity in an Antipodes Island parakeet (Cyanoramphus unicolor). N Z Vet J. 2003;51:196–8. https://doi.org/10.1080/00480169.2003.36365.
29. Landauer WA. Semi-lethal mutation in fowl affecting length of the upper beak and of the long bones. Genetics. 1941;26:426–39.
30. Pourlis AF. Developmental malformations in avian species. Manifestations of unknown or genetic etiology-a review. Asian J Anim Vet Adv. 2011;6:401–15. https://doi.org/10.3923/ajava.2011.401.415.
31. Threlfall WA. Herring Gull chick (Larus argentatus) with an abnormal bill. Auk. 1967;85:506–8.
32. Tsudzuki M, Nakane Y, Wada A. Short beak: a new autosomal recessive semilethal mutation in Japanese quail. J Hered. 1998;89:175–8. https://doi.org/10.1093/jhered/89.2.175.
33. Schräpel J. Kreuzschnabelbildung. Dtsch Tierarztl Wochenschr. 1934;42:709–10.
34. ProSpecieRara - Tiere. 2017. https://www.prospecierara.ch/de/tiere. Accessed 20 Dec 2017.
35. R Core Team. R: A language and environment for statistical computing. 2017. https://www.R-project.org/.
36. Benjamini Y, Hochberg Y. Controlling the false discovery rate: a practical and powerful approach to multiple testing. J R Stat Soc. 1995;57:289–300. http://www.jstor.org/stable/2346101

37. Purcell S, Neale B, Todd-Brown K, Thomas L, Ferreira MAR, Bender D, et al. PLINK: a tool set for whole-genome association and population-based linkage analyses. Am J Hum Genet. 2007;81:559–75. https://doi.org/10.1086/519795.
38. Aulchenko YS, Ripke S, Isaacs A, van Duijn CM. GenABEL: an R library for genome-wide association analysis. Bioinformatics. 2007;23:1294–6. https://doi.org/10.1093/bioinformatics/btm108.
39. Craves JA. Passerines with deformed bills. North Am Bird Bander. 1994;19: 14–9.
40. Handel CM, Pajot LM, Matsuoka SM, Van Hemert C, Terenzi J, Talbot SL, et al. Epizootic of beak deformities among wild birds in Alaska: an emerging disease in North America? Auk. 2010;127:882–98. https://doi.org/10.1525/auk.2010.10111.
41. Van Hemert C, Handel CM. Beak deformities in northwestern crows: evidence of a multispecies epizootic. Auk. 2010;127:746–51. https://doi.org/10.1525/auk.2010.10132.
42. Wouterlood FG. The mutual relations of structures in the growing bill of Chick (Gallus Domesticus L.) and duck (Anas Platyrhynchos L.) embryos. Netherlands J Zool. 1975;25:353–68. https://doi.org/10.1163/002829675X00326.
43. Gfeller S, Joller S, Burren A, Neuditschko M, Spiri A, Kump E, et al. Genetic diversity of three local chicken breeds using high-density SNP-data. In: 67th annual meeting of the European Association for Animal production. Belfast; 2016.

5

Distribution and dissemination of antimicrobial-resistant *Salmonella* in broiler farms with or without enrofloxacin use

Ke Shang[†], Bai Wei[†] and Min Kang[*]

Abstract

Background: *Salmonella* is a major zoonotic food-borne pathogen that persists on poultry farms, and animals undergo reinfection with endemic strains. The present study aimed to investigate the characteristics and dissemination of antimicrobial-resistant *Salmonella* within and between broiler farms that used enrofloxacin and those that did not.

Results: Cloacal and environmental (litter, feed, and water) samples from two selected flocks in each of 12 farms owned by the same company were collected three times over a 30-day period of two production cycles during 2015–2016. The rate of *Salmonella* isolation was 7.8% (123/1584). Nine *Salmonella* serotypes (116 isolates) and seven untypable isolates were identified, and *Salmonella* Montevideo was the most prevalent serotype. Azithromycin-resistant (17.9%) and colistin-resistant (3.3%) isolates were detected, and multidrug-resistant isolates (43.1%) were also observed. No isolate was resistant to enrofloxacin or ciprofloxacin; however, intermediate resistance to enrofloxacin was significantly higher ($P < 0.05$) in farms that used enrofloxacin than in those that did not. The rate of multi-drug resistance among litter isolates (25/44, 56.8%) was significantly higher ($P < 0.05$) than that among cloacal swab (24/67, 35.8%) and feed (4/12, 33.3%) isolates. Pulsed-field gel electrophoresis (PFGE) analysis of strains of the same serotype was conducted to determine their epidemiological relationship. The PFGE types were classified into 31 groups with a 100% correlation cutoff in dendrograms for *Salmonella* Montevideo isolates, which showed 100% genomic identity based on age, sample type, flock, and production cycle within and between farms.

Conclusion: The present study highlights the occurrence of horizontal transmission and cyclic contamination with antimicrobial-resistant *Salmonella* in broiler farms owned by the same company. Litter may be a good indicator of indoor environmental contamination with antimicrobial-resistant *Salmonella* on farms. Additionally, enrofloxacin use may be one of the factors promoting resistance towards it in *Salmonella*.

Keywords: *Salmonella*, Antimicrobial resistance, Circulating contamination, Enrofloxacin, Broiler farm, Litter, PFGE

Background

Salmonella is the leading cause of food-borne illness worldwide, and dissemination of antimicrobial-resistant *Salmonella* through the food chain, especially through chicken, has important implications for the failure of salmonellosis treatment. There is increasing risk of chicken-mediated spread of antimicrobial-resistant *Salmonella* to public health [1]. The prevalence of *Salmonella* on farms is linked to the prevalence of *Salmonella* in the derived meat products [2]. Prevention of *Salmonella* contamination in chicken products requires detailed knowledge of the major sources of contamination. Although measures for eliminating *Salmonella* on breeder farms through vaccination, use of all-in/all-out replacement systems on broiler farms, and "antimicrobial-free" strategies have been implemented, high *Salmonella* prevalence rates and antimicrobial-resistance rates are still observed in broiler farms [3, 4]. Interventions performed at poultry farms, including enhanced biosecurity, rodent control, routine surveillance of the farm environment, feed decontamination, and use of

* Correspondence: vet.minkang@gmail.com
[†]Ke Shang and Bai Wei contributed equally to this work.
Department of Veterinary Infectious Diseases and Avian Diseases, College of Veterinary Medicine and Center for Poultry Diseases Control, Chonbuk National University, 79 Gobong-ro, Iksan 54596, South Korea

autogenous poultry vaccine, can reduce, but cannot eliminate *Salmonella* from live poultry [5]. Environmental samples, especially poultry litter, have been reported to be a good indicator for the presence of *Salmonella* in poultry farms [6, 7]. We speculated that the transmission of antimicrobial-resistant *Salmonella* among farms might occur through indoor environmental contamination in farms.

Salmonella strains can develop antimicrobial resistance (AMR), initially to the traditional first-line drugs chloramphenicol, ampicillin, and trimethoprim/sulfamethoxazole. Because of such resistance, fluoroquinolones (FQs), third-generation cephalosporins, and macrolides (azithromycin) have become critically important for treating salmonellosis in humans [8]. Prophylactic treatment of poultry with enrofloxacin (ENR), a fluoroquinolone antibiotic used to treat animal infections, has been implicated in the increasing resistance to ciprofloxacin, posing a risk to human health [9–11]. In Korea, ENR was licensed for veterinary use in 1987, and the quantity of ENR used has increased since its introduction [12]. In fact, the quantity of ENR sold was the highest of all antimicrobials used to treat chicken in Korea [13]. Concerns over the entry of antimicrobial-resistant zoonotic bacteria into the food chain and the consequent human infections led the Food and Drug Administration (FDA) to ban the use of ENR in poultry in the USA in September 2005 [14]. The use of antimicrobials in poultry farming and the accompanying selection pressure for resistant *Salmonella* have been the subjects of numerous studies [15, 16]. Antimicrobial use in farming has led to the widespread dissemination of antimicrobial-resistant *Salmonella* in broiler farms [4]. Therefore, to institute effective measures for reducing the infection of chicken with antimicrobial-resistant *Salmonella*, *Salmonella* contamination on broiler farms needs to be well-understood.

To explore production cycle-contamination with antimicrobial-resistant *Salmonella* in broiler farms that use antimicrobials and in those that do not, samples were collected from 12 farms (which either used or did not use ENR) owned by a single company during two production cycles. Trace-back investigations of the relationship between antimicrobial-resistant *Salmonella* and ENR use in broiler farms have not been previously conducted. The present study was aimed at determining the distribution and spread of indoor contamination with antimicrobial-resistant *Salmonella* through a comprehensive investigation of its prevalence within and between broiler farms. We also determined the epidemiological relationships among strains of the same serotype using electrophoresis-based DNA fingerprinting.

Methods

Sampling

The samples used in the present study were collected specifically for the purposes of the study. In total, 1584 fresh samples including 660 cloacal samples and 924 environmental samples (396 litter samples, 264 feed samples, and 264 water samples), were collected during 2015 and 2016 from 12 farms belonging to one of the largest integrated poultry companies in South Korea. In seven farms, ENR (50 mg/L) has been administered to young chicks via addition to drinking water for three consecutive days, while on the other five farms; no ENR was used [17]. The same ENR administration protocol was followed in all seven farms. The sampling farms contained an average of 70–100 thousands broilers, and 3–5 separate flocks. Two flocks per farm were sampled three times over a thirty-day period (chickens at 1, 15, and 30 days of age) during one production cycle. This sampling was then repeated during a second, separate production cycle. The cloacal samples were randomly collected from 25 broilers in entire area of the flock, and samples from five broilers were pooled into one test sample. The environmental samples, including litter ($n = 15$), feed ($n = 10$), and water ($n = 10$) were uniformly collected from equal areas of the flock, and five samples obtained from same area were pooled into one test sample. Finally cloacal swabs (S, $n = 5$), litter (L, $n = 3$), feed (F, $n = 2$) and water (W, $n = 2$) were collected from each flock.

Salmonella isolation

Samples were collected in sterile plastic conical tubes (50 mL; SPL Life Sciences Co., Ltd., Seoul, Korea) and were stored under refrigeration in the laboratory until analysis, which was performed within 48 h of arrival. Fresh samples [1 g (or mL)] were separately mixed with 9 mL (1:9 dilutions) of buffered peptone water (BPW; BD Difco, Sparks, MD, USA) and incubated at 37 °C for 24 h for enrichment. An aliquot of the enriched BPW culture (100 μL) was transferred to 10 mL of Rappaport Vassiliadis (RV) broth (Thermo Fisher Scientific, Oxoid Ltd., Basingstoke, UK), and incubated at 42 °C for 24 h for selective enrichment [18]. A loopful of each RV culture was streaked onto a xylose-lysine-deoxycholate (XLD) agar plate (BD Difco™ XLD agar, USA), and the plate was incubated overnight at 37 °C. Presumptive *Salmonella* colonies were then tested with a *Salmonella* latex test kit (Thermo Fisher Scientific, Oxoid Ltd., Basingstoke, UK).

Salmonella serotyping

The serogroup and serovar of each *Salmonella* isolate were identified following Edwards and Ewing's procedure for the identification of *Enterobacteriaceae* using an antisera kit (BD Difco, Sparks, MD, USA) [19]. Somatic O antigen (BD Difco) was identified using the slide agglutination test with a commercially available antiserum. Flagellar (H) antigens (phases 1 and 2) were identified via successive inoculation onto 0.3% brain heart infusion (BHI) agar (BD) to activate flagella, followed by inoculation into BHI broth. The broth was cultured overnight, fixed with 0.6% formalin, and then analyzed using a tube agglutination test [20].

Antimicrobial susceptibility

The minimum inhibitory concentrations (MICs) of the test antimicrobials nalidixic acid (NAL), ciprofloxacin (CIP), neomycin (NEO), gentamicin (GEN), streptomycin (STR), tetracycline (TET), azithromycin (AZM), amoxicillin/clavulanic acid (AMC), cefoxitin (FOX), ceftiofur (XNL), ampicillin (AMP), trimethoprim/sulfamethoxazole (SXT), colistin (COL), florfenicol (FFN), and chloramphenicol (CHL) were determined using the Sensititre panel KRNV4F (TREK Diagnostic Systems, Korea), while the MICs of enrofloxacin (ENR), tigecycline (TIG), and fosfomycin (FOS) were determined using the agar dilution method. *Escherichia coli* (ATCC 25922) were used as the quality-control strain. The susceptibility breakpoints of most tested antimicrobials were interpreted based on CLSI guidelines [21], while those of XNL, ENR, and FFN were interpreted based on CLSI BM31-A3 standards [22]. No CLSI interpretation criteria were available for STR, TIG, COL, and NEO; therefore, the following MIC values were considered to indicate resistance: STR, ≥ 64 μg/mL [23]; TIG, ≥ 8 μg/mL [24]; COL, ≥ 4 μg/mL [25]; and NEO, ≥ 16 μg/mL [26] (Table 1). *Salmonella* isolates resistant to at least three antimicrobial classes were designated multidrug resistant (MDR).

PFGE and BioNumerics analysis

Isolates of *Salmonella* Montevideo ($n = 75$) and *Salmonella* Senftenberg ($n = 16$) were genotyped using PFGE following protocols of the Centers for Disease Control and Prevention available on PulseNet, with some modifications. In brief, the *Salmonella* isolates were streaked onto MacConkey agar plates and incubated overnight at 37 °C. Then, the bacteria were suspended in PBS at an OD value of 0.6–0.8. Genomic DNA (extraction using 1% SDS and 1 mg/mL proteinase K, Biosesang, Seoul, Korea) samples were digested with 50 U of *Xba*I (Thermo Fisher Scientific, Inchon, Korea) at 37 °C for 3 h. The digested DNA was separated by electrophoresis in 0.5 x TBE buffer at 14 °C for 18 h using a CHEF-DR@ electrophoresis system (Bio-Rad, Hercules, CA, USA). The pulse time was ramped from 2.16 to 63.8 s.

Table 1 Antimicrobials used in the study and the tested concentration ranges

Antimicrobials[a]	Abbreviation	Breakpoints (μg/mL)			Concentration ranges (μg/mL)
		S[b]	I	R	
Nalidixic acid	NAL	16	-[c]	32	2–128
Ciprofloxacin	CIP	0.06	0.12–0.5	1	0.12–16
Enrofloxacin	ENR	0.25	0.5–1	2	0.12–64
Neomycin	NEO	–	–	16	2–32
Gentamicin	GEN	4	8	16	1–64
Streptomycin	STR	32	–	64	2–128
Tetracycline	TET	4	8	16	2–128
Azithromycin	AZM	16	–	32	0.25–64
Amoxicillin/clavulanic acid	AMC	8/4	16/8	32/16	2/1–64/32
Cefoxitin	FOX	8	16	32	1–32
Ceftiofur	XNL	2	4	8	0.5–8
Ampicillin	AMP	8	16	32	2–32
Trimethoprim/sulfamethoxazole	SXT	2/38	–	4/76	0.12/2.38–4/76
Colistin	COL	2	–	4	2–32
Florfenicol	FFN	–	–	32	2–64
Chloramphenicol	CHL	8	16	32	2–64
Tigecycline	TIG	2	4	8	0.12–32
Fosfomycin	FOS	64	128	256	0.25–256

[a]Sorted based on category of antimicrobials
[b]S, sensitivity; I, intermediate resistance; R, resistance
[c]-, no standard breakpoint value in related references

Salmonella Braenderup H9812, which was included as a molecular weight standard, was processed with each batch of isolate. The gels were stained with ethidium bromide, and the DNA patterns were visualized on a UV transilluminator (Bio Doc-It Imaging System, Upland, CA, USA). The DNA fingerprints obtained with PFGE were analyzed using BioNumerics (version 5.10 for Windows). Dice similarity coefficients were calculated based on pairwise comparison of the PFGE types of the isolates. The isolates were considered to have closely related band patterns based on molecular typing when their PFGE types had dice similarity coefficients of 100%, and were clustered at the 90% similarity level. Band-matching settings, with an optimization of 1.0% and a position tolerance of 1.0%, were applied.

Statistical analysis

The chi-square test was used to test for significant differences in the rates of *Salmonella* isolation, MDR prevalence, and AMR rates between farms that used ENR and those that did not. *P* values less than 0.05 were considered statistically significant. Analyses were performed using SPSS version 19.0 (IBM Co., Armonk, NY, USA).

Results

Prevalence and serovars of *Salmonella*

The cloacal and environmental samples ($n = 1584$) collected from the broiler farms were analyzed for the presence of *Salmonella*. In total, 123 (7.8%) samples were positive for *Salmonella*. There were no significant differences in isolation rates between the litter samples (44/396, 11.1%) and the cloacal swabs (67/660, 10.2%); however, significantly higher isolation rates were found in litter samples and cloacal swabs compared to feed (12/264, 4.5%), and water samples (0/264, 0.0%) (Fig. 1a). The *Salmonella* isolation rate from all types of samples was significantly lower ($P < 0.05$) in farms that used ENR (55/924, 6.0%) than on farms that did not (68/660, 10.3%; Table 2).

Isolates were assigned to nine serovars, most of which belonged to serogroups C1 (65.0%) and E1 (17.1%). *S.* Montevideo (75 isolates, 61.0%) was the dominant serovar, and *S.* Senftenberg (21 isolates, 17.1%) was a distant second, followed by *S.* Emek (9 isolates, 7.3%), *S.* Macclesfield (4 isolate, 3.3%), and *S.* Virchow (3 isolates, 2.4%). Only one isolate was detected for the four serotypes, *S.* Infantis, *S.* Edinburg, *S.* Hato, and *S.* Vellore, and seven isolates (5.7%) marked as *S.* spp. could not be assigned to specific serotypes (Table 3).

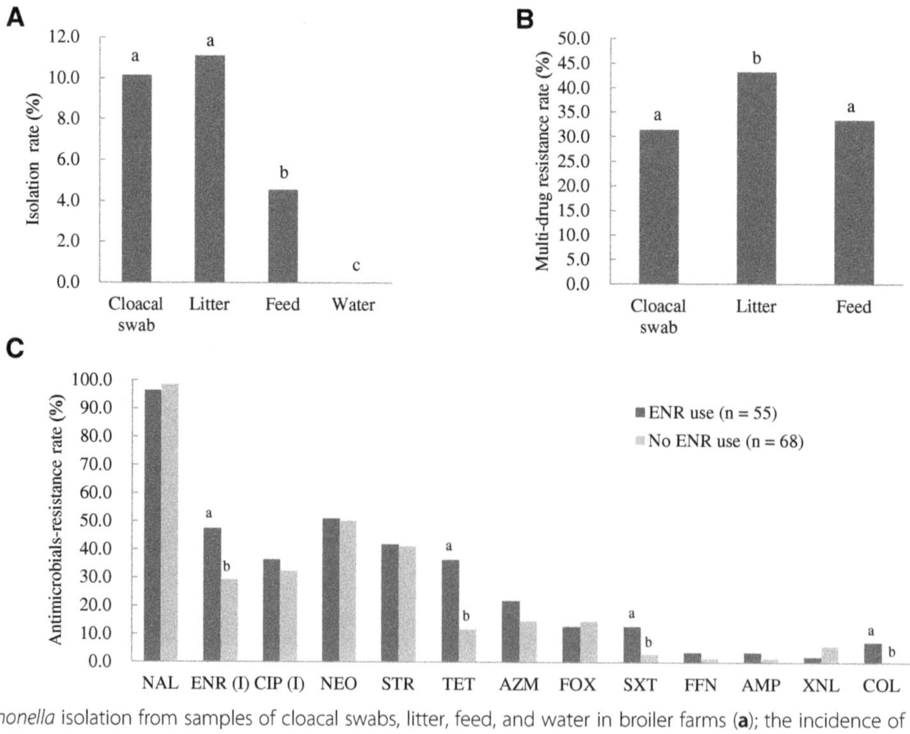

Fig. 1 Rates of *Salmonella* isolation from samples of cloacal swabs, litter, feed, and water in broiler farms (**a**); the incidence of multi-drug resistance in isolates from cloacal swabs, litter, feed, and water in broiler farms (**b**); rates of antimicrobial-resistance in isolates from farms that used ENR and in isolates from farms that did not (**c**). The chi-square test was used to assess the significance of differences. *P* values less than 0.05 were considered statistically significant, and were marked with lowercase letters (a/b/c)

Table 2 *Salmonella* isolation from different sample types in broiler farms with or without enrofloxacin (ENR) use *

ENR use	Sample no.	% (no. positive/total no. samples)				
		Cloacal swab	Litter	Feed	Water	Total
Yes	924	7.5 (29/385) [a]	8.2 (19/231) [a]	4.5 (7/154)	0.0 (0/154)	6.0 (55/924) [a]
No	660	13.8 (38/275) [b]	15.2 (25/165) [b]	4.5 (5/110)	0.0 (0/110)	10.3 (68/660) [b]
Total	1584	10.2 (67/660)	11.1 (44/396)	4.5 (12/264)	0.0 (0/264)	7.8 (123/1584)

*Lowercase (a/b) were used to indicate significant difference in isolation rates between farms with enrofloxacin use and those without; different letters indicate significant differences ($P < 0.05$)

Antimicrobial susceptibility analysis

All the isolates analyzed in this study showed resistance to at least one tested antimicrobial (Table 3). Resistance to NAL was the most prevalent (120/123, 97.6%), followed by that to NEO (62/123, 50.4%), STR (51/123, 41.5%), TET (28/123, 22.8%), AZM (22/123, 17.9%), FOX (17/123, 13.8%), XNL (5/123, 4.1%), AMP (3/123, 2.4%), SXT (9/123, 7.3%), COL (4/123, 3.3%), and FFN (3/123, 2.4%). several isolates showed intermediate resistance (IR) to ENR (46/123, 37.4%) and CIP (42/123, 31.1%). All isolates were sensitive to the five antimicrobials CHL, GEN, TIG, FOS, and AMC.

Thirty-one AMR phenotypes were observed among the *Salmonella* isolates from the 12 broiler farms; the AMR profile NAL + NEO + STR + TET was the most prevalent in this study (10 isolates, 8.1%) (Table 4). Further, seven MDR isolates comprising four *S.* Montevideo, two *S.* spp., and one *S.* Infantis isolates, were resistant to ≥5 antimicrobial classes, all of which were isolated from the farms that used ENR.

The MDR rate among litter isolates (25/44, 56.8%) was significantly higher ($P < 0.05$) than that among the cloacal swab (24/67, 35.8%) and feed (4/12, 33.3%) isolates (Fig. 1b). The ratio of IR to ENR was significantly higher ($P < 0.05$) in isolates from farms that used ENR (26/55, 47.3%) than in those from farms that did not (20/68, 29.4%; Fig. 1c).

Serotype distribution and genetic analysis of isolates from broiler farms

Serotype diversity was observed in the broiler farms sampled in the present study with 75% (9/12) of them showing contamination with at least two *Salmonella* serovars (Table 5). *S.* Montevideo and *S.* Senftenberg were isolated from 11 and 8 of the 12 farms, respectively. In farms B, H, I, and J, there were at least four different serotypes observed in a single flock. In one farm (farm H), isolates from the same flock at different ages had the same PFGE type (type 13), with a similarity index of 100% (isolates A15-CF-002-1S-3 and A15-CF-003-1 L-1); isolates from different flocks were of the same type (type 13; isolates A15-CF-003-1 L-1 and A15-CF-003-2S-2); and two isolates, A15-CF-002-2 L-2 and A15-CF-063-1 L-2, which were of the same PFGE type (type 11) and same AMR profile (including intermediate ENR resistance), were obtained from litter during different production cycles. Isolates of the same PFGE type were also observed in different farms (type 5 in farms H, F, I, K, and E; type 8 in farms B, E, D, J, C, F, and A; type 11 in farms A, H, and J; type 13 in farms B, E, H, and J; type 17 in farms G and J; type 22 in farms K and E; and type 26 in farms B and K; Fig. 2). In *S.* Senftenberg isolates, the same PFGE type was observed in two different farms in different cities, that neither of which used ENR (type 4 in farms H and J; type 6 in farms J and K; Additional file 1).

Discussion

In the present study, to explore the distribution and dissemination of antimicrobial-resistant *Salmonella* in broiler farms, we obtained samples during two production cycles from farms that used antimicrobials and those that did not (Table 5).

S. Montevideo, one of the serotypes prevalent in human infections [27], showed high resistance to ≥3 antimicrobials (49.3%; Table 3). In recent years, there has been an increase in *S.* Montevideo human infections and outbreaks worldwide, including in the USA, Europe, Australia, and Asian nations, such as South Korea [5, 28–30]. In the USA, the number of human illnesses caused by *S.* Montevideo increased from 728 cases in 2002 to 1203 cases in 2012 [31]. In South Korea, *S.* Montevideo is one of the most common serotypes affecting the poultry industry [32, 33]. Additionally, the most serious case of resistance found in the present study was an *S.* Montevideo isolate with resistance to eight antimicrobials (NAL + NEO + STR + TET + FOX + XNL + AMP + AZM; Table 4). In addition, there was an MDR *S.* Infantis isolate resistant to ≥5 antimicrobial classes (Table 3); and *S.* Infantis is one of the most prevalent serotypes in South Korea and second in frequency among serovars [34]. In countries including Germany [35], Hungary [36], Italy [37], and Japan [38], emergence and clonal dissemination of MDR *S.* Infantis strains in chicken has been recent reported, and has been related to an increased incidence of human infections [31]. Given these findings, further investigations of the antimicrobial resistance and distribution of *S.* Infantis in slaughterhouses and retail chicken meat are required.

Table 3 Antimicrobial resistance in *Salmonella* isolates from broiler farms [a]

Serovar (Serogroup)	N (%)	NAL		NEO		STR		TET		AZM		XNL		FOX		AMP		SXT		COL		FFN		> 1 agent	≥3 agents	≥5 classes
		$MIC_{50/90}$	*n* (%)	MIC 50/90	*n* (%)	MIC 50/90	*n* (%)	MIC 50/90	*n* (%)	MIC 50/90	*n* (%)	MIC 50/90	*n* (%)	MIC 50/90	*n* (%)	MIC 50/90	*n* (%)	MIC 50/90	*n* (%)	MIC 50/90	*n* (%)	MIC 50/90	*n* (%)	*n* (%)	*n* (%)	*n* (%)
Hato (B)	1 (0.8)	–	1 (100)	–	1 (100)	–	1 (100)	–	0	–	0	–	0	–	0	–	0	–	0	–	0	–	0	1 (100)	1 (100)	0
Vellore (B)	1 (0.8)	–	0	–	1 (100)	–	0	–	0	–	0	–	0	–	0	–	0	–	0	–	0	–	0	1 (100)	0	0
Montevideo (C1)	75 (61.0)	128/128	75 (100)	4/32	35 (46.7)	32/128	37 (49.3)	4/16	24 (32)	16/32	12 (16)	1/1	3 (4.0)	4/16	9 (12.0)	4/8	2 (2.7)	0.5/1	8 (10.7)	2/2	2 (2.7)	8/8	3 (4.0)	75 (100)	37 (49.3)	4 (5.3)
Virchow (C1)	3 (2.4)	128/128	3 (100)	16/32	3 (100)	64/64	2 (66.7)	8/8	0	64/64	2 (66.7)	1/1	0	16/32	1 (33.3)	8/8	0	0.25/0.25	0	2/2	0	8/8	0	3 (100)	2 (66.7)	0
Infantis (C1)	1 (0.8)	–	1 (100)	–	1 (100)	–	1 (100)	–	1 (100)	–	1 (100)	–	1 (100)	–	0	–	0	–	0	–	1 (100)	–	0	1 (100)	1 (100)	1 (100)
Edinburg (C1)	1 (0.8)	–	1 (100)	–	0	–	0	–	0	–	0	–	0	–	1 (100)	–	0	–	0	–	0	–	0	1 (100)	0	0
Emek (C2-C3)	9 (7.3)	128/128	9 (100)	4/16	3 (33.3)	8/32	1 (11.1)	1/4	0	16/32	3 (33.3)	0.5/1	0	4/16	0	2/4	0	0.5/0.5	0	2/2	0	4/8	0	9 (100)	3 (33.3)	0
Macclesfield (D2)	4 (3.3)	64/128	4 (100)	2/16	1 (25)	8/128	1 (25)	1/16	1 (6.3)	16/64	1 (25)	0.5/1	0	4/32	0	2/8	0	0.5/4	1 (25)	2/2	0	2/8	0	4 (100)	1 (25)	0
Senftenberg (E4)	21 (17.1)	128/128	21 (100)	4/32	10 (47.6)	32/128	4 (19)	1/4	0	16/16	2 (9.5)	0.5/1	0	4/16	2 (9.5)	2/4	0	0.5/0.5	0	2/2	0	4/8	0	21 (100)	2 (9.5)	0
spp. (C2-C3, B, D1)	7 (5.7)	–	6 (71.4)	–	7 (100)	–	4 (57.1)	–	2 (28.6)	–	1 (14.3)	–	1 (14.3)	–	4 (57.1)	–	1 (14.3)	–	0	–	1 (14.3)	–	0	7 (100)	4 (57.1)	2 (28.6)
Total	123	128/128	120 (97.6)	8/32	62 (50.4)	8/32	51 (41.5)	1/16	28 (22.8)	16/32	22 (17.9)	1/1	5 (4.1)	4/32	17 (13.8)	4/8	3 (2.4)	0.5/0.5	9 (7.3)	2/2	4 (3.3)	8/8	3 (2.4)	123 (100)	53 (43.1)	7 (5.7)

[a]-, not analyzed; the unit of $MIC_{50/90}$ was µg/mL

Table 4 Antimicrobial resistance profiles of *Salmonella* isolates from broiler farms ($n = 123$)

No.	Antimicrobial resistance profile	n^a (%)	Serovars (n)
–	Susceptible	0	–
1	NAL	27 (22.0)	Montevideo (13), Senftenberg (8), Emek (5), Macclesfield (1)
2	NEO	2 (1.6)	Vellore (1), spp. (1)
3	NAL + NEO	18 (14.6)	Montevideo (9), Senftenberg (6), Virchow (1), spp. (2)
4	NAL + STR	6 (4.9)	Montevideo (5), Senftenberg (1)
5	NAL + AZM	4 (3.3)	Montevideo (1), Senftenberg (1), Emek (1), Macclesfield (1)
6	NAL + FOX	3 (2.4)	Montevideo (1), Senftenberg (1), Edinburg (1)
7	NAL + SXT	9 (7.3)	Montevideo (8), Macclesfield (1)
8	NAL + COL	1 (0.8)	Montevideo (1)
9	NAL + NEO + STR	7 (5.7)	Montevideo (3), Senftenberg (2), Emek (1), Hato (1)
10	NAL + NEO + FFN	2 (1.6)	Montevideo (2)
11	NAL + NEO + AZM	4 (3.3)	Montevideo (1), Senftenberg (1), Emek (2)
12	NAL + NEO + COL	1 (0.8)	Montevideo (1)
13	NEO + STR + FOX	3 (2.4)	Montevideo (3)
14	NEO + STR + TET	4 (3.3)	Montevideo (4)
15	NAL + STR + AZM	2 (1.6)	Montevideo (2)
16	NEO + STR + TET + FOX	1 (0.8)	spp. (1)
17	NAL + STR + FOX + AZM	1 (0.8)	Montevideo (1)
18	NAL + NEO + STR + TET	10 (8.1)	Montevideo (10)
19	NAL + NEO + STR + FOX	2 (1.6)	Senftenberg (1), spp. (1)
20	NAL + NEO + STR + AZM	1 (0.8)	Virchow (1)
21	NAL + NEO + FFN + AZM	1 (0.8)	Montevideo (1)
22	NAL + NEO + STR + TET + XNL	2 (1.6)	Montevideo (2)
23	NAL + NEO + STR + TET + FOX	2 (1.6)	Montevideo (1), Macclesfield (1)
24	NAL + NEO + STR + TET + AZM	2 (1.6)	Montevideo (2)
25	NAL + NEO + STR + FOX + AZM	1 (0.8)	Virchow (1)
26	NAL + NEO + STR + FOX + COL + AZM	1 (0.8)	spp. (1)
27	NAL + NEO + STR + TET + FOX + AZM	2 (1.6)	Montevideo (2)
28	NAL + STR + TET + FFN + AMP + AZM	1 (0.8)	Montevideo (1)
29	NAL + NEO + STR + TET + FOX + XNL + AMP	1 (0.8)	spp. (1)
30	NAL + NEO + STR + TET + XNL + COL + AZM	1 (0.8)	Infantis (1)
31	NAL + NEO + STR + TET + FOX + XNL + AMP + AZM	1 (0.8)	Montevideo (1)

[a]*n*, number of isolates

All isolates showed resistance to at least one antimicrobial, suggesting that AMR *Salmonella* is widespread in broiler farms, as was reported previously in South Korea [33]. Contrary to the increasing prevalence of FQ-resistant *Salmonella* reported in other countries [39, 40], we did not observe any FQ-resistant isolates in this study; however, 31.1% of the isolates showed reduced susceptibility to CIP (with a MIC of 0.12 to 0.25 or 0.5 μg/mL), and 37.4% of the isolates showed IR to ENR [41, 42]. The rate of IR to CIP was higher in farms that used ENR than in farms that did not, although the difference was not statistically significant ($p > 0.05$). The ratio of IR to ENR was significantly higher ($P < 0.05$) in farms that used ENR than in those that did not (Fig. 1c). Resistance to SXT, TET, and COL was significantly higher in farms that used ENR than in those that did not ($P < 0.05$; Fig. 1c). Resistance to the three antimicrobials ENR, TET, and SXT was most commonly observed in isolates from chicken, in a previous study [43]. One plausible explanation is that the selective pressure exerted by the antimicrobials contributed to the co-selection of this antimicrobial resistance pattern. The co-selection of resistance to more than one antimicrobial, owing to the genetic linkage of resistance genes, is a common feature of resistance acquired by horizontal gene transfer [44, 45].

Table 5 Distribution of *Salmonella* serotypes and genotypes in broiler farms

Farm	ENR use [a]	Flock	Production cycle 1			Production cycle 2		
			1 day	15 days	30 days	1 day	15 days	30 days
A	Y	1	*S.* Montevideo (11)[b] *S.* Senftenberg (5)	-[c]	–	*S.* Montevideo (10) *S.* Macclesfield (NC[d])	*S.* Montevideo (8)	–
		2	–	–	–	–	–	*S.* spp. (NC) *S.* Vellore (NC)
B		1	*S.* Montevideo (7, 8, 15)	*S.* Montevideo (13)	*S.* Montevideo (26)	*S.* Senftenberg (7)	–	–
		2	*S.* Montevideo (2, 4) *S.* Senftenberg (10)	*S.* Montevideo (20) *S.* Emek (3)	*S.* Macclesfield (NC)	–	–	–
C		1	–	–	*S.* Montevideo (8)	*S.* Montevideo (28) *S.* Senftenberg (9) *S.* Emek (4)	–	–
		2	–	–	–	*S.* spp. (NC)	–	–
D		1	*S.* Senftenberg (2)	*S.* Montevideo (8)	*S.* Montevideo (8, 25)	–	–	–
		2	*S.* Macclesfield (NC)	*S.* Montevideo (8)	*S.* Montevideo (12)	–	–	*S.* Montevideo (8)
E		1	–	*S.* Montevideo (22)	*S.* Montevideo (5)	–	–	–
		2	*S.* Montevideo (8)	*S.* Montevideo (24)	*S.* Montevideo (3, 13)	–	–	–
F		1	–	*S.* Virchow (NC)	*S.* Montevideo (30) *S.* spp. (NC)	*S.* Montevideo (8, 9)	–	–
		2	–	*S.* Montevideo (5) *S.* Virchow (NC)	*S.* Infantis (NC)	–	–	–
G		1	–	*S.* Montevideo (17)	–	N[e]	N	N
		2	–	–	–	N	N	N
H	N	1	*S.* Senftenberg (11) *S.* Virchow (NC)	*S.* Montevideo (13, 16) *S.* Senftenberg (4)	*S.* Montevideo (13)	*S.* Macclesfield (NC)	*S.* Montevideo (13)	*S.* Montevideo (3, 11)
		2	*S.* Senftenberg (8)	*S.* Montevideo (5, 11)	*S.* Montevideo (13)	–	–	–
I		1	–	*S.* Montevideo (19)	–	*S.* Montevideo (4) *S.* Senftenberg (8) *S.* Emek (2)	*S.* Montevideo (5) *S.* Edinburg (NC)	–
		2	*S.* Montevideo (29)	*S.* Montevideo (28) *S.* Hato (NC)	–	*S.* Montevideo (28) *S.* Senftenberg (2)	*S.* Montevideo (16, 28) *S.* Senftenberg (2)	*S.* Montevideo (28)
J		1	*S.* Montevideo (13) *S.* Emek (1,2)	*S.* spp. (NC)	*S.* Montevideo (24)	*S.* Montevideo (14, 17, 18) *S.* Senftenberg (4)	*S.* Montevideo (8)	–
		2	*S.* Montevideo (11)	*S.* Montevideo (23)	*S.* Montevideo (23)	*S.* Montevideo (13) *S.* Senftenberg (4)	*S.* Montevideo (14) *S.* Senftenberg (6)	–

Table 5 Distribution of *Salmonella* serotypes and genotypes in broiler farms *(Continued)*

Farm	ENR use [a]	Flock	Production cycle 1			Production cycle 2		
			1 day	15 days	30 days	1 day	15 days	30 days
K		1	*S.* Senftenberg (6)	*S.* Montevideo (26)	*S.* Montevideo (22)	–	–	–
		2	*S.* Montevideo (5, 6, 21, 31)	*S.* Montevideo (27)	*S.* Montevideo (22)	–	–	*S.* spp. (NC)
L		1	*S.* Senftenberg (NC)	*S.* Senftenberg (NC)	–	N	N	N
		2	–	–	–	N	N	N

[a]ENR use, Y implies yes; N implies no.
[b]PFGE types are showed within parenthesis
[c]The dash (–) indicates that the farm was negative for *Salmonella*
[d]NC implies that PFGE was not conducted
[e]N denotes no sampling

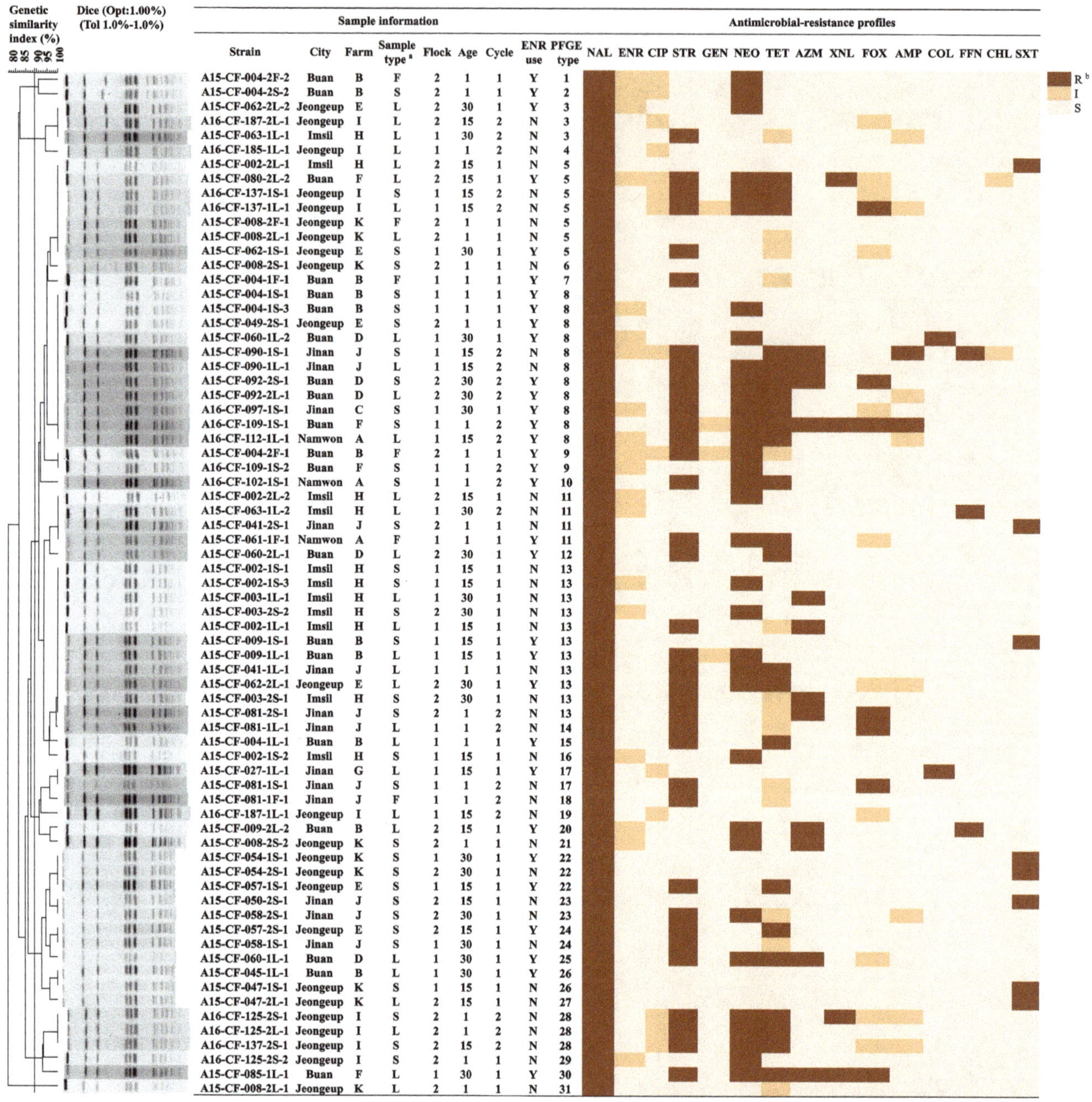

Fig. 2 Dendrograms showing pattern analysis on the basis of *Xba* I-PFGE of the 75 *Salmonella* Montevideo isolates obtained from broiler farms, along with related sampling information, and antimicrobial-resistance (AMR) profiles. The Dice coefficient was used to perform similarity analysis. [a]S, cloacal swabs; L, litter; F, feed. [b]R, resistance (dark pink); I, intermediate resistance (pink), S, sensitivity (light pink). Dotted lines indicate 90% similarity

In the present study, 17.9% of isolates showed resistance to AZM (Table 3), a macrolide antimicrobial used for the limited treatment of MDR *Salmonella* infections [8, 46]. In recent years, AZM has been used for typhoid treatment due to the increased prevalence of MDR *Salmonella* isolates, which has caused serious problems in antimicrobial selection. The sporadic occurrence of AZM-resistant *Salmonella* isolates has also caused problems in the selection of suitable antimicrobials for typhoid treatment [47, 48]. In Europe, the first case of failure of AZM treatment of *Salmonella* infection was reported [49]. Similarly, emergence of clinical AZM-resistant *Salmonella* isolates was recently reported in Asia [50, 51]; 15.24% of *Salmonella* isolates from chicken showed resistance to AZM (with an MIC ≥16 μg/mL) in another study [52]. However, reports of AZM resistance in *Salmonella* isolates from humans or animals in South Korea are rare. Considering the fact that, AZM has not been used on these farms to the best of our knowledge, the prevalence of AZM resistance in *Salmonella* is probably either due to prolonged use of antimicrobials in the same class as AZM such as erythromycin and tylosin for treating poultry diseases [53], or due to co-resistance to AZM and other antimicrobials [54]. Further, the MIC_{50} of AZM was 16 μg/mL against all isolates and 64 μg/mL against *S.* Virchow (Table 3). However, the exact reason for the emergence of AZM resistance in *Salmonella* and its mechanism require further studies. Additionally, because COL is a last-line drug used to treat MDR *Enterobacteriaceae* infections, the four COL-resistant isolates identified in the present study warrant more attention; due to the prevalence of COL-resistant *Salmonella, mcr* family genes can be easily and quickly transmitted [8, 55, 56]. In Europe, plasmid-mediated COL resistance in *Enterobacteriaceae* has already spread widely in avian and pig farms, and this has necessitated prompt international action to restrict or ban COL use in agriculture to avoid further spread of resistance, similar to the solution involving NDM-1 (New Delhi metallo-β-lactamase-1) several years ago [56]. COL resistance in *Salmonella* isolates from humans had not been reported until recently, when the first COL-resistant isolates from humans were reported in the Arabian Peninsula [57], followed by two COL-resistant *Salmonella* strains in Italy [58]. The observation of co-resistance to AZM and COL in two MDR isolates (AMR profiles: NAL + NEO + STR + FOX + COL + AZM and NAL + NEO + STR + TET + XNL + COL + AZM), both of which exhibited intermediate resistance to CIP and ENR (data not shown), is of high importance. This might pose a considerable challenge when selecting drugs to treat human *Salmonella* infections. The most important finding of the present study was that the AMR profile XNL + COL + AZM, conferring resistance to the critically important antimicrobials used to treat salmonellosis or MDR *Salmonella* infections in humans, has never been reported previously in *Salmonella* strains of animal origin.

Based on the results of distribution of *Salmonella* serotypes and genotypes, a considerable cross-contamination among the farms could be inferred. For example, isolates with the same PFGE type shared between flocks, production cycles, sample types, and among chickens of different ages within the farms were frequently identified (Fig. 2). This could be because the farms shared resources including breeders, trucks for transport, veterinarians, chicks, and feed [59]. Chicks in five farms (farms A, B, D, E, and J) were from the same breeder farm which was negative for *Salmonella* (data not shown). Moreover, there was contamination between flocks, production cycles, sample types, and among chicken of different ages within the farms. Antimicrobial-resistant *Salmonella* can be circularly transmitted between continuous production cycles. Although all-in/all-out replacement systems have been applied in the commercial poultry industry, *Salmonella* contamination still occurs, especially in the farm environment. Plausible explanations include insufficient disinfection, development of resistance to disinfectants in the first production cycle and subsequent survival, and inherent resistance in *Salmonella* [1, 60]. Another explanation is that other factors including air, unclean facilities, and vectors such as insects, wild birds, farmers, and rodents, might contribute to *Salmonella* transmission in poultry farms [61].

The dissemination of antimicrobial-resistant *Salmonella* in the farms might also have occurred via litter contamination (farm H; Fig. 2), because the litter samples had higher *Salmonella* isolation and MDR rates than the cloacal swab and feed samples (Fig. 1a, b), with no significant difference in isolation rates from litter samples between production cycles (Additional file 2). The spread of indoor *Salmonella* contamination in the broiler farms was apparently caused by litter from the broilers. *Salmonella*-positive litter samples detected in our study may have important public health implications. A recent study reported a positive correlation between prevalence of *Salmonella* in litter samples and *Salmonella* isolation from broiler carcasses [6]. Antimicrobial-resistant *Salmonella* could re-circulate in the farms because of litter contamination during different production cycles. Therefore, litter in broiler farms may be an important reservoir of *Salmonella*, consistent with a speculation in a report from the USA [62]. In summary, we concluded that serious *Salmonella* contamination occurs in farms during production cycles, as does cross-contamination among farms owned by the same company. Because of dissemination and cross-contamination between the farms that used antimicrobials and those that did not, close attention should be paid to farm-level hygiene management.

Conclusion

In conclusion, we detected considerable contamination with antimicrobial-resistant *Salmonella* in broiler farms. The litter in the farm was one of the important reservoirs for *Salmonella* showing high *Salmonella* prevalence and MDR rates. Other reservoirs of *Salmonella*, such as feed, air, fans, and vectors such as insects, wild birds, farmers, and rodents might also contribute to its transmission in chicken farms [61]. Additional measures for litter and feed management might be required to prevent the transmission of antimicrobial-resistant *Salmonella* in such farms. Moreover, ENR use may be an important factor causing ENR resistance among *Salmonella* in the farms. Our results provide useful information regarding the distribution of AMR phenotypes among

Salmonella isolates from broiler farms that use ENR and those that do not, highlighting the need for improved farming practices and more cautious use of antimicrobial agents. Further studies are required to develop protocols to prevent the contamination of litter and feed; this, together with instructions for strict all-in/all-out replacement and biosecurity systems, may markedly reduce the occurrence of antimicrobial-resistant *Salmonella* species in broiler farms.

Additional files

Additional file 1: Dendrograms showing pattern analysis on the basis of *Xba* I-PFGE of the 16 *Salmonella* Senftenberg isolates obtained from broiler farms and their association with antimicrobial-resistance. The Dice coefficient was used to perform similarity analysis. [a]S, cloacal swabs; L, litter; F, feed. [b]R, resistance (dark pink); I, intermediate resistance (pink), S, sensitivity (light pink). Dotted lines indicate 90% similarity.

Additional file 2: Isolation rates and multidrug resistant (MDR) rates in different types of samples obtained during production cycles 1 and 2. Different lowercase (a/b) and capital letters (A/B) in the same row were used to indicate significant ($P < 0.05$) differences in isolation rates and MDR rates between production cycles 1 and 2, respectively.

Abbreviations

AMC: Amoxicillin/clavulanic acid; AMP: Ampicillin; AMR: Antimicrobial resistance; AZM: Azithromycin; BHI: Brain heart infusion; CHL: Chloramphenicol; CIP: Ciprofloxacin; CLSI: Clinical Laboratory Standardization Institute; COL: Colistin; ENR: Enrofloxacin; FDA: Food and Drug Administration; FFN: Florfenicol; FOS: Fosfomycin; FOX: Cefoxitin; FQs: Fluoroquinolones; GEN: Gentamicin; MDR: Multidrug resistant; MICs: Minimum inhibitory concentrations; NAL: Nalidixic acid; NDM-1: New Delhi metallo-β-lactamase-1; NEO: Neomycin; PFGE: A pulsed-field gel electrophoresis; RV: Rappaport Vassiliadis; *S.* Edinburg: *Salmonella* Edinburg; *S.* Emek: *Salmonella* Emek; *S.* Hato: *Salmonella* Hato; *S.* Infantis: *Salmonella* Infantis; *S.* Macclesfield: *Salmonella* Macclesfield; *S.* Montevideo: *Salmonella* Montevideo; *S.* Senftenberg: *Salmonella* Senftenberg; *S.* spp.: *Salmonella* species; *S.* Vellore: *Salmonella* Vellore; *S.* Virchow: *Salmonella* Virchow; STR: Streptomycin; SXT: Trimethoprim/sulfamethoxazole; TET: Tetracycline; TIG: Tigecycline; USA: United States of America; XLD: Xylose-lysine-deoxycholate; XNL: Ceftiofur

Acknowledgements

The authors wish to thank the veterinary practitioners and farmers for their help with sample collection, and lab members for their help with internal review of the paper.

Funding

This work was supported by Korea Institute of Planning and Evaluation for technology in Food, Agriculture, Forestry and Fisheries (716002-7, 315035-5), Basic Science Research Program through the NRF funded by the Ministry of Education (2017R1D1A1B03030883), and Research of Animal and Plant Quarantine Agency (Z-1543073-2015-16-01), South Korea.

Authors' contributions

KS and BW performed experiments, analyzed the data and wrote manuscript; MK designed the experiments, analyzed the data, edited the manuscript and coordinated the whole project. All authors critically read and approved the final manuscript.

Competing interests

The authors declare that they have no competing interests.

References

1. Youn SY, Jeong OM, Choi BK, Jung SC, Kang MS. Comparison of the antimicrobial and sanitizer resistance of *Salmonella* isolates from chicken slaughter processes in Korea. J Food Sci. 2017;82(3):711–7.
2. Choi SW, Ha JS, Kim BY, Lee DH, Park JK, Youn HN, Hong YH, Lee SB, Lee JB, Park SY, et al. Prevalence and characterization of *Salmonella* species in entire steps of a single integrated broiler supply chain in Korea. Poult Sci. 2014;93(5):1251–7.
3. Ishihara K, Takahashi T, Morioka A, Kojima A, Kijima M, Asai T, Tamura Y. National surveillance of *Salmonella enterica* in food-producing animals in Japan. Acta Vet Scand. 2009;51(1):35.
4. Rayamajhi N, Jung BY, Cha SB, Shin MK, Kim A, Kang MS, Lee KM, Yoo HS. Antibiotic resistance patterns and detection of *bla*DHA-1 in *Salmonella* species isolates from chicken farms in South Korea. Appl Environ Microbiol. 2010;76(14):4760–4.
5. Gaffga NH, Barton Behravesh C, Ettestad PJ, Smelser CB, Rhorer AR, Cronquist AB, Comstock NA, Bidol SA, Patel NJ, Gerner-Smidt P, et al. Outbreak of salmonellosis linked to live poultry from a mail-order hatchery. N Engl J Med. 2012;366(22):2065–73.
6. Volkova VV, Bailey RH, Rybolt ML, Dazo-Galarneau K, Hubbard SA, Magee D, Byrd JA, Wills RW. Inter-relationships of *Salmonella* status of flock and grow-out environment at sequential segments in broiler production and processing. Zoonoses Public Health. 2010;57(7–8):463–75.
7. Berghaus RD, Thayer SG, Law BF, Mild RM, Hofacre CL, Singer RS. Enumeration of *Salmonella* and *Campylobacter* spp. in environmental farm samples and processing plant carcass rinses from commercial broiler chicken flocks. Appl Environ Microbiol. 2013;79(13):4106–14.
8. World Health Organization. Critically important antimicrobials for human medicine: ranking of antimicrobial agents for risk management of antimicrobial resistance due to non-human use. Geneva: World Health Organization; 2017.
9. Angulo FJ, Johnson KR, Tauxe RV, Cohen ML. Origins and consequences of antimicrobial-resistant nontyphoidal *Salmonella*: implications for the use of fluoroquinolones in food animals. Microb Drug Resist. 2000;6(1):77–83.
10. Randall LP, Eaves DJ, Cooles SW, Ricci V, Buckley A, Woodward MJ, Piddock LJV. Fluoroquinolone treatment of experimental *Salmonella enterica* serovar typhimurium DT104 infections in chickens selects for both *gyr*A mutations and changes in efflux pump gene expression. J Antimicrob Chemother. 2005;56(2):297–306.
11. Majowicz SE, Musto J, Scallan E, Angulo FJ, Kirk M, O'Brien SJ, Jones TF, Fazil A, Hoekstra RM, International Collaboration on Enteric Disease 'Burden of Illness S. The global burden of nontyphoidal *Salmonella* gastroenteritis. Clin Infect Dis. 2010;50(6):882–9.
12. Hong J, Kim JM, Jung WK, Kim SH, Bae W, Koo HC, Gil J, Kim M, Ser J, Park YH. Prevalence and antibiotic resistance of *Campylobacter* spp. isolated from chicken meat, pork, and beef in Korea, from 2001 to 2006. J Food Prot. 2007;70(4):860–6.
13. Park H, Kang J, Jang Y, Song JY, Lee KJ. Studies on pharmacokinetics/Pharmacodanymics of Enrofloxacin in Broiler chicken. FASEB J. 2016;30:539.
14. Han FF, Lestari SI, Pu SH, Ge BL. Prevalence and antimicrobial resistance among *Campylobacter* spp. in Louisiana retail chickens after the enrofloxacin ban. Foodborne Pathog Dis. 2009;6(2):163–71.
15. Asai T, Kojima A, Harada K, Ishihara K, Takahashi T, Tamura Y. Correlation between the usage volume of veterinary therapeutic antimicrobials and resistance in *Escherichia coli* isolated from the feces of food-producing animals in Japan. Jpn J Infect Dis. 2005;58(6):369–72.
16. Diarra MS, Silversides FG, Diarrassouba F, Pritchard J, Masson L, Brousseau R, Bonnet C, Delaquis P, Bach S, Skura BJ, et al. Impact of feed supplementation with antimicrobial agents on growth performance of broiler chickens, *Clostridium perfringens* and *enterococcus* counts, and antibiotic resistance phenotypes and distribution of antimicrobial resistance determinants in *Escherichia coli* isolates. Appl Environ Microbiol. 2007;73(20):6566–76.
17. Da Costa PM, Belo A, Gonçalves J, Bernardo F. Field trial evaluating changes in prevalence and patterns of antimicrobial resistance among *Escherichia coli* and *Enterococcus* spp. isolated from growing broilers medicated with enrofloxacin, apramycin and amoxicillin. Vet Microbiol. 2009;139(3–4):284–92.
18. USDA-FSIS. Isolation and identification of *Salmonella* from meat, poultry, pasteurized egg, and catfish products and carcass and environmental sponges. In: Microbiology laboratory guidebook; 2014. p. 4.

19. Ewing W. Edwards and Ewing's identification of *Enterobacteriaceae*. New York: Elsevier Science Publishing Co. Inc.; 1986.
20. Kim YJ, Song BR, Lim JS, Heo EJ, Park HJ, Wee SH, Oh SM, Moon JS. First report on isolation of *Salmonella enteritidis* from eggs at grocery stores in Korea. Korean J Food Sci An. 2013;33(2):239–43.
21. CLSI. Performance standards for antimicrobial susceptibility testing-26th. Wayne: Clinical and Laboratory Standards Institute; 2016.
22. CLSI. Performance standards for antimicrobial susceptibility testing. Wayne: Clinical and Laboratory Standards Institute; 2008.
23. NARMS. The National Antimicrobial Resistance Monitoring System: enteric Bacteria. In: NARMS Integrated Report: 2012–2013: National Antimicrobial Resistance Monitoring System; 2013.
24. Chen Y, Hu D, Zhang Q, Liao XP, Liu YH, Sun J. Efflux pump overexpression contributes to tigecycline heteroresistance in *Salmonella enterica* serovar typhimurium. Front Cell Infect Microbiol. 2017;7:37.
25. Biswas S, Brunel JM, Dubus JC, Reynaud-Gaubert M, Rolain JM. Colistin: an update on the antibiotic of the 21st century. Expert Rev Anti-Infect Ther. 2012;10(8):917–34.
26. Ojo OE, Ogunyinka OG, Agbaje M, Okuboye JO, Kehinde OO, Oyekunle MA. Antibiogram of *Enterobacteriaceae* isolated from free-range chickens in Abeokuta, Nigeria. Veterinarski Arhiv. 2012;82(6):577–89.
27. Foley S, Lynne A, Nayak R. Challenges: prevalence in swine and poultry and potential pathogenicity of such isolates. J Anim Sci. 2008;86(14 suppl):E149–62.
28. Kim J-Y, Park Y-J, Lee S-O, Song W, Jeong SH, Yoo YA, Lee K-Y. Bacteremia due to *Salmonella enterica* serotype Montevideo producing plasmid-mediated AmpC β-lactamase (DHA-1). Ann Clin Lab Sci. 2004;34(2):214–7.
29. Harada T, Sakata J, Kanki M, Seto K, Taguchi M, Kumeda Y. Molecular epidemiological investigation of a diffuse outbreak caused by *Salmonella enterica* serotype Montevideo isolates in Osaka Prefecture, Japan. Foodborne Pathog Dis. 2011;8(10):1083–8.
30. Lalsiamthara J, Lee JH. Pathogenic traits of *Salmonella* Montevideo in experimental infections *in vivo* and *in vitro*. Sci Rep. 2017;7:46786.
31. Shah DH, Paul NC, Sischo WC, Crespo R, Guard J. Population dynamics and antimicrobial resistance of the most prevalent poultry-associated *Salmonella* serotypes. Poult Sci. 2017;96(3):687–702.
32. Yoon RH, Cha SY, Wei B, Roh JH, Seo HS, Oh JY, Jang HK. Prevalence of *Salmonella* isolates and antimicrobial resistance in poultry meat from South Korea. J Food Prot. 2014;77(9):1579–82.
33. Kim MS, Lim TH, Jang JH, Lee DH, Kim BY, Kwon JH, Choi SW, Noh JY, Hong YH, Lee SB, et al. Prevalence and antimicrobial resistance of *Salmonella* species isolated from chicken meats produced by different integrated broiler operations in Korea. Poult Sci. 2012;91(9):2370–5.
34. Kim S. *Salmonella* serovars from foodborne and waterborne diseases in Korea, 1998-2007: total isolates decreasing versus rare serovars emerging. J Korean Med Sci. 2010;25(12):1693–9.
35. Hauser E, Tietze E, Helmuth R, Junker E, Prager R, Schroeter A, Rabsch W, Fruth A, Toboldt A, Malorny B. Clonal dissemination of *Salmonella enterica* serovar Infantis in Germany. Foodborne Pathog Dis. 2012;9(4):352–60.
36. Nogrady N, Toth A, Kostyak A, Paszti J, Nagy B. Emergence of multidrug-resistant clones of *Salmonella* Infantis in broiler chickens and humans in Hungary. J Antimicrob Chemother. 2007;60(3):645–8.
37. Dionisi AM, Lucarelli C, Benedetti I, Owczarek S, Luzzi I. Molecular characterisation of multidrug-resistant *Salmonella enterica* serotype Infantis from humans, animals and the environment in Italy. Int J Antimicrob Ag. 2011;38(5):384–9.
38. Noda T, Murakami K, Ishiguro Y, Asai T. Chicken meat is an infection source of *Salmonella* serovar Infantis for humans in Japan. Foodborne Pathog Dis. 2010;7(6):727–35.
39. Medalla F, Sjolund-Karlsson M, Shin S, Harvey E, Joyce K, Theobald L, Nygren BL, Pecic G, Gay K, Austin J, et al. Ciprofloxacin-resistant *Salmonella enterica* serotype Typhi, United States, 1999-2008. Emerg Infect Dis. 2011;17(6):1095–8.
40. Mattiello SP, Drescher G, Barth VC Jr, Ferreira CA, Oliveira SD. Characterization of antimicrobial resistance in *Salmonella enterica* strains isolated from Brazilian poultry production. Anton Leeuw Int J G. 2015;108(5): 1227–38.
41. Tamang MD, Nam HM, Kim TS, Jang GC, Jung SC, Lim SK. Emergence of extended-spectrum β-lactamase (CTX-M-15 and CTX-M-14)-producing nontyphoid *Salmonella* with reduced susceptibility to ciprofloxacin among food animals and humans in Korea. J Clin Microbiol. 2011;49(7):2671–5.
42. Kim JE, Lee YJ. Molecular characterization of antimicrobial resistant non-typhoidal *Salmonella* from poultry industries in Korea. Ir Vet J. 2017;70(1):20.
43. Jarquin C, Alvarez D, Morales O, Morales AJ, Lopez B, Donado P, Valencia MF, Arevalo A, Munoz F, Walls I, et al. *Salmonella* on raw poultry in retail markets in Guatemala: levels, antibiotic susceptibility, and serovar distribution. J Food Prot. 2015;78(9):1642–50.
44. Andersson DI, Hughes D. Antibiotic resistance and its cost: is it possible to reverse resistance? Nat Rev Microbiol. 2010;8(4):260–71.
45. Enne VI, Bennett PM, Livermore DM, Hall LM. Enhancement of host fitness by the *sul*2-coding plasmid p9123 in the absence of selective pressure. J Antimicrob Chemother. 2004;53(6):958–63.
46. Chande CA, Chopdekar KA, Pradnya V, Unnati R, Jyoti B, Ritesh S, Shazia C, Joshi SG, Abhay C. Current trend of antibiotic sensitivity of *Salmonella typhi* and other *Salmonellae* in Mumbai: a 5 years study. Indian J Med Microbiol. 2016;34(1):115–6.
47. Rai S, Jain S, Prasad KN, Ghoshal U, Dhole TN. Rationale of azithromycin prescribing practices for enteric fever in India. Indian J Med Microbiol. 2012; 30(1):30–3.
48. Hassing RJ, Goessens WH, van Pelt W, Mevius DJ, Stricker BH, Molhoek N, Verbon A, van Genderen PJ. *Salmonella* subtypes with increased MICs for azithromycin in travelers returned to the Netherlands. Emerg Infect Dis. 2014;20(4):705–8.
49. Molloy A, Nair S, Cooke FJ, Wain J, Farrington M, Lehner PJ, Torok ME. First report of *Salmonella enterica* serotype paratyphi a azithromycin resistance leading to treatment failure. J Clin Microbiol. 2010;48(12):4655–7.
50. Wong MH, Yan M, Chan EW, Biao K, Chen S. Emergence of clinical *Salmonella enterica* serovar typhimurium isolates with concurrent resistance to ciprofloxacin, ceftriaxone, and azithromycin. Antimicrob Agents Chemother. 2014;58(7):3752–6.
51. Wang J, Li Y, Xu X, Liang B, Wu F, Yang X, Ma Q, Yang C, Hu X, Liu H, et al. Antimicrobial resistance of *Salmonella enterica* serovar Typhimurium in Shanghai, China. Front Microbiol. 2017;8:510.
52. Kuang X, Hao H, Dai M, Wang Y, Ahmad I, Liu Z, Zonghui Y. Serotypes and antimicrobial susceptibility of *Salmonella* spp. isolated from farm animals in China. Front Microbiol. 2015;6:602.
53. Brennan J, Moore G, Poe SE, Zimmermann A, Vessie G, Barnum DA, Wilson J. Efficacy of in-feed tylosin phosphate for the treatment of necrotic enteritis in broiler chickens. Poult Sci. 2001;80(10):1451–4.
54. Hoge CW, Gambel JM, Srijan A, Pitarangsi C, Echeverria P. Trends in antibiotic resistance among diarrheal pathogens isolated in Thailand over 15 years. Clin Infect Dis. 1998;26(2):341–5.
55. Hu Y, Liu F, Lin IYC, Gao GF, Zhu B. Dissemination of the *mcr*-1 colistin resistance gene. Lancet Infect Dis. 2016;16(2):146–7.
56. Kempf I, Fleury MA, Drider D, Bruneau M, Sanders P, Chauvin C, Madec J-Y, Jouy E. What do we know about resistance to colistin in *Enterobacteriaceae* in avian and pig production in Europe? Int J Antimicrob Ag. 2013;42(5):379–83.
57. Olaitan AO, Dia NM, Gautret P, Benkouiten S, Belhouchat K, Drali T, Parola P, Brouqui P, Memish Z, Raoult D, et al. Acquisition of extended-spectrum cephalosporin- and colistin-resistant *Salmonella enterica* subsp *enterica* serotype Newport by pilgrims during Hajj. Int J Antimicrob Ag. 2015;45(6): 600–4.
58. Frasson I, Bettanello S, De Canale E, Richter SN, Palu G. Serotype epidemiology and multidrug resistance patterns of *Salmonella enterica* infecting humans in Italy. Gut Pathog. 2016;8(1):26.
59. Davies RH, Nicholas RA, McLaren IM, Corkish JD, Lanning DG, Wray C. Bacteriological and serological investigation of persistent *Salmonella enteritidis* infection in an integrated poultry organisation. Vet Microbiol. 1997;58(2–4):277–93.
60. Donado-Godoy P, Gardner I, Byrne BA, Leon M, Perez-Gutierrez E, Ovalle MV, Tafur MA, Miller W. Prevalence, risk factors, and antimicrobial resistance profiles of *Salmonella* from commercial broiler farms in two important poultry-producing regions of Colombia. J Food Prot. 2012;75(5):874–83.

6

Epidemiological investigation of *Mycoplasma Synoviae* in native chicken breeds in China

Shi-Kai Sun[1,2], Xin Lin[2], Feng Chen[3], Ding-Ai Wang[2], Jun-Peng Lu[2], Jian-Ping Qin[2*] and Ting-Rong Luo[1*]

Abstract

Backgroud: *Mycoplasma synoviae* (*M. synoviae*) is widely distributed around the world, and leads to serious economic losses in the world every year. Nevertheless, the incidence and epidemiology of *M. synoviae* infection in China have remained unclear.

Results: In this study we demonstrate that over 9773 broiler chicken flocks in 16 Chinese provinces were affected by *M. synoviae* between 2010 and 2015. Our epidemiological study revealed that *M. synoviae* was widely prevalent in multi-aged Chinese native breeder chickens, and the prevalence of *M. synoviae* in embryos of breeders reached up to 16.29%. In addition, our data showed that chickens aged 14 days or younger carried simultaneously high levels of maternal antibody against *M. synoviae* and high *M. synoviae* infection (10%), and low *M. synoviae* antibody levels in breeders and high proportion of *M. synoviae* infection in embryos could increase the chances of incidence in the offspring. Finally, our results also indicated that 3- to 7-week-old chickens might be most the susceptible to *M. synoviae* and, therefore, might play a key role in the horizontal transmission of *M. synoviae*.

Conclusion: Our findings suggest that *M. synoviae* is widely circulating in Chinese native chickens, accordingly, effective control measures are urgently needed to control the spread.

Keywords: *Mycoplasma synoviae*, Outbreak, Chinese native chicken breeds, Epidemiology

Backgroud

Mycoplasma synoviae (*M. synoviae*) is an important disease, causing enormous economic losses to the poultry industry worldwide. *M. synoviae* infection in chickens or turkeys results in acute or chronic infectious synovitis or air sacculitis [1].

M. synoviae was reported in Australia, South America, Asia, Europe and Africa [2], nevertheless, *M. synoviae* distribution appeared regional in most cases, and large-scale outbreaks rarely occurred. In the past ten years, it seems that *M. synoviae* had taken over the role of *Mycoplasma gallisepticum* in commercial poultry [3]. Apart from air sacculitis and synovitis, eggshell apex abnormalities (EAA) and egg drop syndrome resulting from *M. synoviae* infection have been encountered worldwide [3–6]. Hence, there has been a growing emphasis on understanding the prevalence of *M. synoviae*, especially its subclinical infection [3, 7]. Many reports [8–11] world with a varying degree of incidence in multi-age white-feather chickens or turkeys. To date, there has been no data about *M. synoviae* infection in Chinese native chicken breeds, and the incidence and epidemiology of *M. synoviae* in China are poorly understood.

Between 2010 and 2015, a disease, characteristic of infectious synovitis occurred in native-type chickens in China. The disease resulted in the loss of millions of chickens in Chinese poultry farms. In this study, we conducted systematic epidemiological analysis of this disease.

* Correspondence: jpqin@163.com; Tingrongluo@gxu.edu.cn
[2]Guangdong Enterprise Key Laboratory for Animal Health and Environmental Control, WENS Group Academy, Yunfu, Guangdong 527439, China
[1]State Key Laboratory for Conservation and Utilization of Subtropical Agro-Bioresources, Guangxi University, Nanning, Guangxi 530004, China
Full list of author information is available at the end of the article

Methods

Sample preparation

Identification of pathogen

To determine the causative agent of disease, a total of 18,063 clinical samples (joint fluid swabs and three swabs were collected from three sick chickens at each farm) were collected from infected chickens in the farms (Table 1) of 16 Chinese provinces between 2010 and 2015. In addition, from 2010 to 2013, 1696 serum samples were collected from 53 infected broiler flocks (Table 2). Thirty-two blood samples were collected from each farm, including 16 samples from sick chickens with typical clinical signs and 16 samples from chickens without any clinical symptoms.

Prevalence in multi-aged breeder farms

According to the protocol for an epidemic outbreak, a total of 5760 serum samples were collected from 180 breeder flocks in 6 large-scale multi-age native chicken farms in Guangdong, Guangxi, Hunan and Zhejiang, where the disease occurred in 2013, and detected by ELISA. Chickens ranged from 1 to 40 weeks old, 3–4 flocks were examined for each age (week), totaling 180 flocks with 32 blood samples from each flock. In addition, a total of 1332 7- to 9-day-old embryos (Table 3) were collected from farms in Guangxi, Guangdong, Zhejiang and Hunan in 2013. Allantoic fluid was collected from each embryo for detection of *M. synoviae* infection.

Natural infection in the offspring of native breeders

To determine the natural prevalence of *M. synoviae* in broilers, three batches of 1-day-old Three-Yellow chickens (according to the high incidence of *M. synoviae* infection in Three-Yellow broiler chickens) were purchased from a farm in 2014, with 100 chickens in each batch and a 30-day time interval between each batch. Throat swab samples and serum samples of all chickens were collected at 1, 5, 10, 20, 30, 40, 50, 60, 70, 80, 90 and 100 days old, respectively.

Prevalence in different native breeders

To further understand *M. synoviae* infection in different native chicken breeds from a farm in Zhejiang, serum and associated embryos collected from three different native chicken breeds (Three-Yellow, Silky, and Qingyuan Ma chickens) were examined. A total of 128 serum samples were simultaneously collected from 3 different chicken breeds (a total of 8 breeder

Table 1 Information of samples for PCR detection and strains isolation

Province	Sampling city	Breed	Age (day)	Joint fluid swabs[a]	Chicken legs[a]	No. of farm	Sampling time
Guangdong	Yunfu, YangJiang, Heyuan, Zhaoqing, Foshan, Qingyuan, Jiangmen	Three-Yellow, Chaohua, Silky, Qingyuan Ma, Wenchang, Tianlushan, Short footed	29–70	5934	61	1978	2010.10–2014.12
Guangxi	Nanning, Guilin, Hezhou, Yulin	Three-Yellow, Chaohua, Silky, Qingyuan Ma, Short footed	31–75	2004	24	668	2010.09–2014.12
Zhejiang	Quzhou, Huzhou, Jiangshan	Ditto	30–65	1233	24	411	2012.03–2014.12
Jiangsu	Zhengjiang, Suzhou, Nanjing, Yancheng, Suqian, Huaian, Nantong, Lianyungang,	Ditto	28–68	1809	14	603	2012.01–2014.12
Hunan	Hengyang, Changsha	Ditto	30–65	1407	16	469	2013.01–2014.12
Hubei	Wuhan, Jingzhou	Three-Yellow, Chaohua, Silky, Qingyuan Ma, Wenchang, Tianlushan, Short footed	42–58	468	6	156	2013.02–2014.12
Anhui	Hefei, Wuhu, Haozhou, Chuzhou	Ditto	33–71	900	4	300	2013.01–2014.12
Jiangxi	Nanchang, Jian	Ditto	30–60	1110	8	370	2013.04–2014.12
Sichuang	Meishan, Deyang, Chengdu	Ditto	32–66	921	12	307	2011.06–2014.12
Yunnan	Kunming, Dali	Ditto	26–71	504	4	168	2012.04–2014.12
Guizhou	Qingzhen, Guiyang	Ditto	31–56	156	5	52	2011.08–2014.12
Chongqing	Chongqing	Ditto	30–64	297	4	99	2012.03–2014.12
Fujian	Putian, Zhangzhou	Three-Yellow, Chaohua, Silky, Qingyuan Ma, Short footed	30–75	723	21	241	2013.02–2014.12
Shangdong	Taian	Three-Yellow, Tianlushan,	30–52	81	3	27	2013.11–2014.12
Hebei	Cangzhou	Three-Yellow, Silky, Qingyuan Ma	33–61	384	4	128	2011.06–2014.12
Hainan	Haikou, Qionghai	Three-Yellow, Wenchang	45–65	132	4	44	2014.03–2014.12
Total				18,063	215	6021	2010–2014

[a]Joint fluid swabs were collected for PCR detection from 2010; [b] Chicken legs were collected for strain isolation between 2013 and 2014

Table 2 *M. synoviae* antibody of native broilers infected by infectious synovitis in China

Group	Breed	No. of farm[a]	Age (day)	Clinical signs	Type of sample	No. of positive *M. synoviae* antibody	Positive proportion	Sampling time
A1	Tianlu	8	35–60	Arthrocele	Serum	79/103	76.70%	2012.01–2014.12
A2	Tianlu		35–60	NO	Serum	88/131	67.18%	2012.01–2014.12
B1	San Huang	17	30–75	Arthrocele	Serum	208/299	69.57%	2010.10–2014.12
B2	San Huang		30–75	NO	Serum	78/252	30.95%	2010.10–2014.12
C1	Silky	5	31–63	Arthrocele	Serum	59/62	95.16%	2012.06–2014.12
C2	Silky		31–63	NO	Serum	30/80	37.5%	2012.06–2014.12
D1	Hua	9	40–71	Arthrocele	Serum	140/155	90.32%	2012.02–2014.12
D2	Hua		40–71	NO	Serum	81/128	63.28%	2012.02–2014.12
E1	Qingyuan Ma	14	29–62	Arthrocele	Serum	197/240	82.08%	2011.03–2014.12
E2	Qingyuan Ma		29–62	NO	Serum	112/238	47.06%	2011.03–2014.12
Total		53				1072/1696	63.20%	2010.10–2014.12

[a]Thirty-two blood samples were collected from each farm, including 16 samples from sick chickens with typical clinical signs and 16 samples from chickens without any clinical symptoms

flocks) on December 11, 2013. A total of 223 7- to 9-day-old embryos corresponding to these breeder flocks were collected on December 20, 2013. Allantoic fluid was collected from each embryo. Subsequently, 12 offspring broiler flocks corresponding to the breeder flocks mentioned above were tracked for the change of *M. synoviae* antibody. A total of 478 serum samples were collected from offspring broiler flocks at 4, 5, 6, 7 weeks old, respectively.

Samples detection

The presence of *M. synoviae* DNA in swab samples and allantoic fluid was determined by PCR. The presence of *M. synoviae* antibody in serum samples was determined by ELISA.

Isolation and culture of *M. synoviae*

A total of 215 joint fluid swabs were sterile collected from the leg joints of *M. synoviae*-infected chickens from farms in Zhejiang, Hunan, Jiangsu, Anhui, Sichuan, Yunnan, Guizhou, Chongqing, Hubei, Hebei, Jiangxi, Fujian, Shandong, Hainan, Guangdong and Guangxi provinces were collected in 2013–2014 (Table 1). One swab sample was added to a tube containing 2 mL Frey mycoplasma medium (BD Biosciences) with 10% serum and 1% thallium-acetate [2]. This sample was then transferred into a 37 °C incubator (5% CO_2 atmosphere conditions) to culture for 20–24 h; the sample supernatant was removed when the liquid sample color changed from red to brown. After the 4th or 5th passage, pure culture of isolates was used as template and identified directly by PCR and sequenced.The 4th or 5th passage live cell titer was determined based on 50% color change unit (CCU_{50}), as described previously [12, 13].

DNA extraction, PCR and sequencing

Total DNA was extracted directly from joint swab samples and allantoic fluid using the RNeasy kit (AxyPrep, Union City, USA). The presence of *M. synoviae* DNA in swab samples and allantoic fluid was determined by PCR.The TaKaRa PrimeScript one-step PCR Kit Ver.2 (TaKaRa, Dalian, China) was used for PCR. All PCR reactions were carried out using a PCR machine (Biometra, Goettingen, Germany). Samples were detected by PCR as described by Ramirez et al. [14]. A partial sequence of the *vlhA* gene (1284–1325 bp) of positive samples was amplified by PCR for sequencing as described previously [15]. All sequences from independent PCR products were sequenced 3–4 times in both directions by Sangon Biotech (Shanghai) Co., Ltd.

Table 3 Infection proportion of *M. synoviae* in embryos of 7 breeder farms in 2013

Province	County	No. of farm[a]	No. of positive samples	Positive prevalence	Sampling time
Guangdong	Xinxing, Gaoming	2	92/250	36.80%	March, November
Hunan	Ningxiang	1	4/180	2.22%	March
Guangxi	Yongning	1	73/360	20.27%	March
Zhejiang	Nanxun, Wuxing	3	48/542	8.85%	March, July, August
	Total		217/1332	16.29%	March–November

[a]The 7–9 day-old embryos were collected from 7 breeder farms. Allantoic fluid was collected from each embryo for detection of *M. synoviae* infection by PCR

Sequence analysis

After sequencing PCR fragments, nucleotide sequence editing was accomplished using DNAStar (Madison, WI). Multiple-sequence alignments were generated with ClustalX 2.0, and nucleotide sequence homologies were obtained using the Clustal W method of DNAStar. All sequencing data were assembled using DNAStar (Madison, WI). The sequence of each different strain was compared to reference isolates in GenBank.

Serological study

Serum samples were tested for the presence of *M. synoviae* antibodies using the *M. synoviae* Antibody test kit (IDEXX, Liebefeld-Bern, Switzerland) according to the manufacturer's instructions. Based on the optical density (*OD*) values at 405 nm, sample to positive (S/P) ratios were calculated, and the averaged S/P ratio was used to evaluate the antibody level in each group. Serum samples with S/P ratios greater than 0.2 (titers greater than 1076) were considered to contain anti-*M. synoviae* antibodies, and a ratio of 0.2 or lower (titers less than or equal to 1076) were considered negative, based on the manufacturer's recommendations.

In vivo experiments

WVU1853 (*M. synoviae* standard strain, bought from China Veterinary Drug Supervision), 14 randomly chosen *M. synoviae* isolates and 160 20-day-old SPF chickens (specific pathogen free chickens, Guangdong Dahuanong Co., LTD) were used for the pathogenicity experiments. 160 chickens were divided into 16 groups, including 15 challenged groups (10 birds/group) and one mock/control group (10 birds). All chickens were kept in SPF chicken isolators (glove box) through the whole experiment. After adaption for three days in the isolator, the chickens were inoculated with fresh cultures of 14 selected isolates and WVU1853 by intramuscular injection with a dose of 10^7 CCU_{50}. Ten chickens served as mock controls and were mock-injected with 0.5 mL culture medium in the same manner. The chickens were monitored daily while clinical symptoms and morbidity were recorded. All chickens were necropsied on day post-inoculation.

Statistical analysis

The data obtained were analyzed by one-way analysis of variance (ANOVA), and the differences between means were compared by Duncan's multiple range test (DMRT) using SPSS 17 (SPSS Inc., Chicago, IL, USA). *M. synoviae* antibody titer and *M. synoviae* antigen-carrying rate were analyzed by Student's t-test. $P < 0.05$ was considered to be statistically significant.

Results

Identification of infectious synovitis caused by *M. synoviae* infection

In 2010, a disease with characteristic infectious synovitis appeared in chickens in the Guangdong and Guangxi provinces of China (Fig. 1a1). Anatomical examination revealed that visible yellowish purulent exudate or a yellow cheese-like substance accumulated in the joint and wing cavities of sick chickens (Figs. 1a, 2, 3, 4). This disease had a mass rate of 10–100% and a group morbidity rate reaching 1–20%, with an average of 5–8%. This disease was observed in several Chinese native chicken

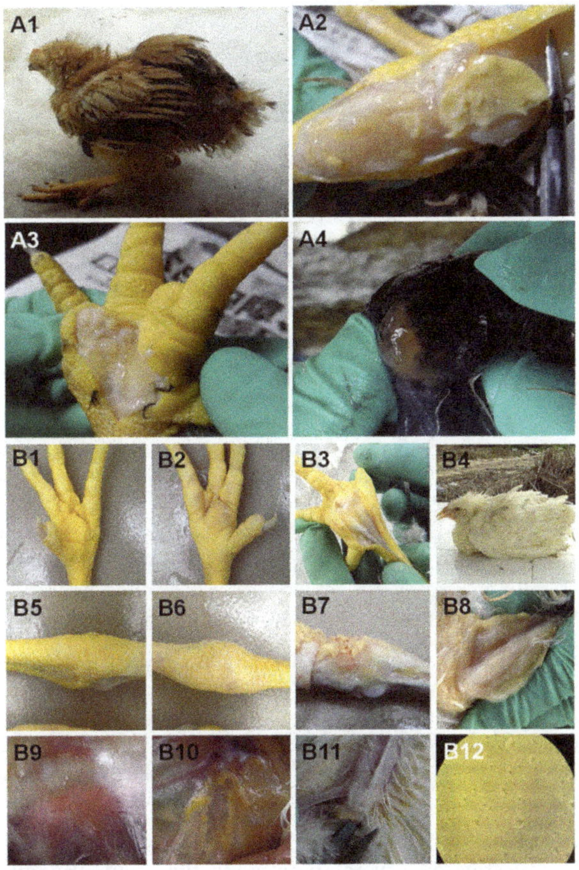

Fig. 1 Pathological studies of infected chickens and inoculated chickens.The infected chickens, which showed clinical symptoms including movement disorders, erect feathers and stunted growth (**a1**), were killed, and anatomical investigations revealed the presence of thick, clear to milky exudate in enlarged joints (**a2**, **a4**) and foot pads (**a3**). Similar to the clinically infected chickens, the inoculated chickens exhibited symptoms such as movement disorders, erect feathers and stunted growth (**b4**). The anatomical investigations showed that thick, clear to milky exudate was present in enlarged foot pads (**b2**, **b3**), joints (**b6**, **b7**), keels (**b8**) and wings (**b11**). In addition, air sacculitis (**b10**) was observed in infected chickens too. Mock-injected chickens' foot pads (**b1**), joints (**b5**) and air sacs (**b9**) were compared with those of the infection chickens. Colonies of *M. synoviae* were grown in solid medium (**b12**)

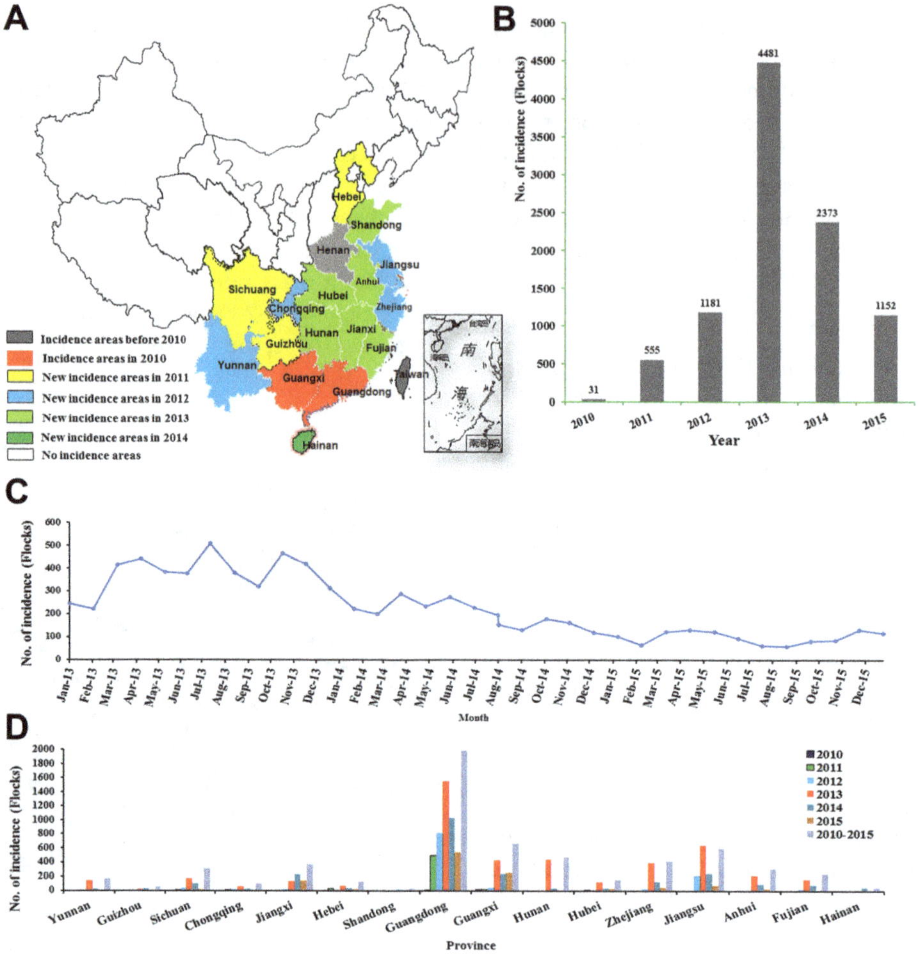

Fig. 2 The prevalence of *M. synoviae* infection in China between 2010 and 2015.**a** Incidence in various areas of China from 2010 to 2015; **b**Total cases of infectious synovitis that appeared in China from 2010 to 2015; **c** Monthly case occurrence during 2013 and 2015 in China. **d** Overall cases that appeared in different provinces with reported incidents. (We acknowledge Tang HB for providing the original map of China. The original map was published in the paper by Tang HB et al.: Re-emergence of rabies in the Guangxi province of Southern China. PLoS Negl Trop Dis 2014.8(10):e3114)

breeds, including Three-Yellow, Silky, Qingyuan Ma, Wenchang, Tianlushan, Chaohua and Short-footed chickens.

To determine the causative agent of this disease, we first investigated several pathogens that had previously been reported to associate with infectious synovitis in broiler chickens by pathogenic detection, including *Salmonella* [16], *Reovirus* [17, 18], *Streptococcus* [19, 20], *Staphylococcus* [21], and *M. synoviae* [2].The results indicated that no of these pathogens could be detected (data not shown). By contrast, we found that only *M. synoviae* DNA was detected in most of the infected chickens (17,396/18063, 96.31%). Serum antibody analysis also showed the presence of *M. synoviae* antibody in 69.57–95.16% of the infected chickens, compared to 30.95–67.18% in clinically healthy chickens of the same flocks (Table 2). The highest prevalence of *M. synoviae* antibody was observed in silky chickens and the lowest in Three-Yellow chickens (Table 2). 110 *M. synoviae* strains were successfully isolated from *M. synoviae*-infected chickens (Fig. 1b12) and sequenced (Genbank accession number: KU572280-KU572389). Subsequently, typical clinical symptoms were reproduced using part of isolates (Fig. 1b4). Similar to the clinical cases, anatomical examination of experimentally infected chickens revealed that a pale yellow or yellow purulent exudate, cheese-like substance was deposited in the feet, leg joints, wings, and chest (Figs. 1b2, 3, 6, 7, 8, 11), and a proportion of the chickens also exhibited air sacculitis (Fig. 1b10). The experimental *M. synoviae* strains were successfully recovered from the experimentally infected chickens, and the diagnosis of *M. synoviae* infection was confirmed by *M. synoviae*-specific PCR and gene sequencing in subsequent experiments (data not shown).

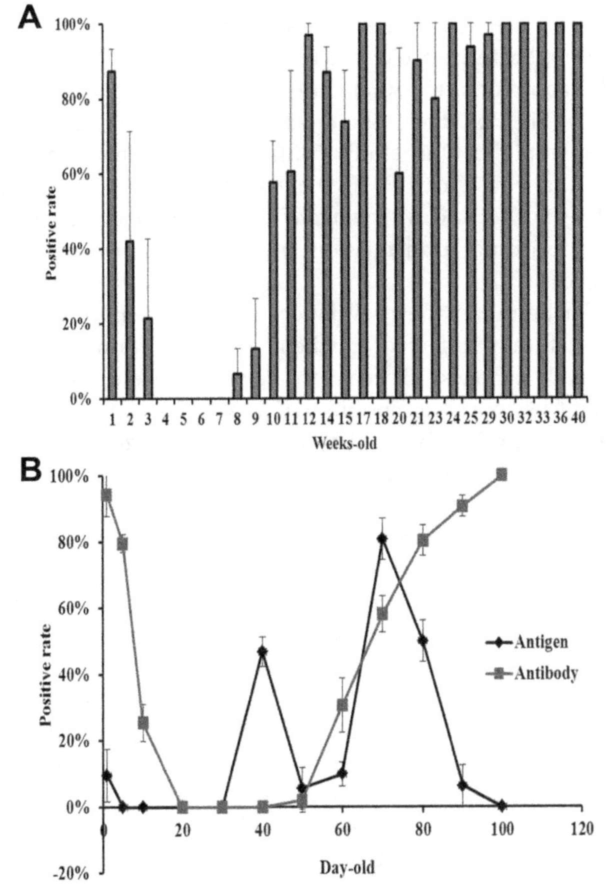

Fig. 3 Prevalence of *M. synoviae* infection in multi-aged Chinese native chicken breeds and breeder offspring. **a** *M. synoviae* antibody levels of multi-aged ages breeder farms in Hunan, Guangdong, Guangxi and Zhejiang provinces in contemporaneity, 2013; **b** Changes of *M. synoviae* infection proportion and *M. synoviae* antibody levels in monitored flocks (Three-Yellow fowl) from 1 to 100 days-old in Yunfu, Guangdong, 2013. $P < 0.05$

The results revealed that the outbreak of infectious synovitis that emerged in China from 2010 to 2015 was caused by *M. synoviae* infection.

Prevalence of *M. synoviae* infection between 2010 and 2015

In 2010, the disease initially appeared in the Guangdong and Guangxi provinces of China. The disease spread to larger areas over time, as evidenced by the observation of cases in new provinces since 2011, including Fujian, Zhejiang, Hunan, Jiangsu, Anhui, Sichuan, Yunnan, Guizhou, Chongqing, Hubei, Hebei, Jiangxi, Fujian, Shandong, and Hainan provinces (Fig. 2a). Our epidemiological study showed a dramatic increase in the number of cases from 2010 to 2013 in China, and a slight decrease in 2014 and 2015 (Fig. 2b). The increase may have been associated with a wide range of chicken breeds and a large-scale of cultivation. The disease had no obvious seasonal or breed specificity but did show a slight increase in prevalence at the turn of seasons, i.e. March, April, June, July, September and October (Fig. 2c). The most severe outbreak of the disease occurred in 2013, with the monthly average number of infected flocks reaching approximately 400 (Fig. 2d). It is worth mentioning that Guangdong province, where the incidence was the most serious, had 4417 infected flocks from 2010 to 2015; in 2013 alone, a total of 1547 flocks were infected (Fig. 2d).

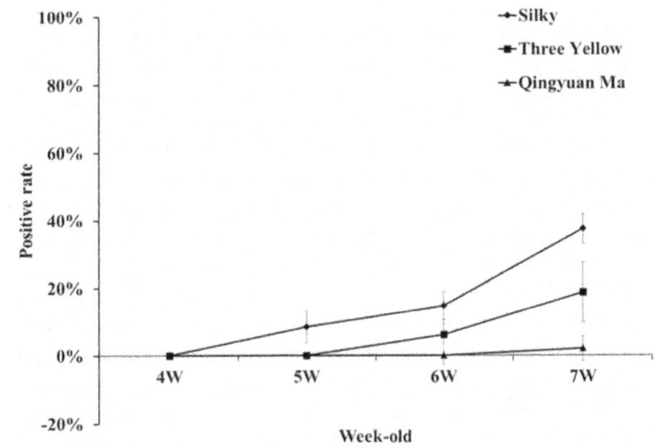

Fig. 4 Changes in *M. synoviae* antibody levels in several clinical native breed chickens.The 3 breeds of broiler (a total of 12 flocks), including Three-Yellow-, Silky-, and Qingyuan Ma-chicken, were tracking for change of *M. synoviae* antibody at 4, 5, 6, 7 weeks old in the production process, respectively. Positive *M. synoviae* antibody was first observed in the Silky chicken flocks (at 5 weeks-old), but last in the Qingyuan Ma-chicken flocks (at 7 weeks-old). $P < 0.05$

Though strong comprehensive measures for prevention and control (including the use of antibiotics and strict implementation of sanitation and disinfection) were carried out after the outbreak of *M. synoviae*, there was still relatively high morbidity in China every year (Fig. 2). Hence, further outbreaks of infectious synovitis have emerged in several native-type chickens in China.

M. synoviae infection prevalence in multi-aged breeder farms

The results indicated that chickens within a week old normally carried high levels of maternal *M. synoviae* antibody, which declined gradually over the next two weeks. *M. synoviae* antibody levels in chickens remained negative between 4 to 8 weeks old. The average positive antibody rates of 8-, 9-, 10-, 11-, 12-, 13-, 14-, 15-, 20-, 21-, 23-, 25-, and 29-week-old chickens were 7%, 13%, 58%, 60%, 97%, 87%, 74%, 60%, 90%, 80%, 94%, and 97%, respectively. The average positive antibody rates of 17-, 18-, 24-, 30-, 32-, 33-, 36-, 40-week-old chickens all reached 100%. *M. synoviae* antibody levels in 13- to 25-week-old chickens showed large antibody levels, especially between 20 and 25 weeks old. Multiple variant

M. synoviae antibody levels were observed among flocks of different ages in this period of time (Fig. 3a).

Subsequently, 1332 embryos from these 6 farms were collected for *M. synoviae* DNA detection. The results showed that the proportion of embryos from Guangdong, Guangxi, Hunan, Zhejiang provinces exhibiting *M. synoviae* infection was 36.8%, 20.27%, 2.22% and 8.85%, respectively (Table 3). In addition, the rate of occurrence of eggs with rough head in breeder flocks with lower antibody levels was higher, and the prevalence of *M. synoviae* infection in rough head eggs was significantly higher than that of normal eggs.

These results indicated that Chinese native-breed chickens could carry high levels of *M. synoviae* maternal antibody before the age of 14 days, and there was a widespread *M. synoviae* infection and varying degrees of *M. synoviae* vertical transmission in multi-age breeder farms.

Natural *M. synoviae* infection in the offspring of native breeders

The results showed that the change in *M. synoviae* antibody level in 1- to 20-day-old offspring was consistent with what was observed in the breeder flock described above. The average positive *M. synoviae* antibody rate, which was over 95% for 1-day-old chickens, gradually turned negative before the offspring were 20 days old. The chickens remained negative for the *M. synoviae* antibody between 20 and 50 days old. Then, corresponding to the data from breeders, the average positive antibody rate gradually increased from 0% to 100% in chickens 50 to 100 days old. These results revealed that *M. synoviae* antibody levels exhibited a gradual, age-dependent increase in chicken offspring (Fig. 3b).

At the same time, the results of *M. synoviae* DNA detection showed that the prevalence of infection in chickens also increased quickly as the offspring aged. There were three time points of peak discharge for excreting bacteria: 1 day old (approximately 10%), 40 days old (approximately 50%), and 70 days old (approximately 90%) (Fig. 3b). Then, the chicken stopped excreting bacteria rapidly. Our study confirmed that chickens can simultaneously carry *M. synoviae* maternal antibody and *M. synoviae*, and *M. synoviae* showed a strong capacity for horizontal transmission in Chinese native chicken breeds.

Furthermore, the prevalence of *M. synoviae* infection in naturally infected chickens showed periodic, age-related changes. Three- to seven-week-old chickens might be most susceptible to *M. synoviae* infection and, therefore, might play a key role in the horizontal transmission of *M. synoviae*.

M. synoviae infection of different native chickens on a farm

The results showed that the majority of flocks exhibited high *M. synoviae* antibody levels, and the average positive rate in seized chickens was 73.33–100%. In 30-week-old chickens, the average positive antibody rate could reach 100%, and the average titer could reach 9265 or higher. Flocks of 1005 (Silky chicken) and 1002 (Three-Yellow chicken) had the lowest antibody levels out of all 8 flocks, and the mean titer in these flocks was 2770 and 5648, respectively. Qingyuan Ma chickens had the highest average antibody levels among the three varieties of breeders, while Silky chickens had the lowest (Table 4).

The proportion of *M. synoviae* infection in embryos of flocks 1001, 1002, 1003, 1004, 1005, 1006, 1007, and 1008 was 4.2%, 0.00%, 6.67%, 7.41%, 10.34%, 8%, 3.45%, and 15.62%, respectively (Table 4). The prevalence of *M. synoviae* infection in Three-Yellow chickens, Qingyuan Ma chickens and Silky chickens was 3.78%, 9.83%, and

Table 4 Survey of *M. synoviae* infection in a farm of Zhejiang in December 2013

Flock	Week-old	Breed	Antibody of serum samples from breeder flock[b]				Infection proportion of *M. synoviae* in hatching embryos[c]		
			Positive rate	AMn[a]	CV	No. of sample	No. of positive	Infection proportion	Infection proportion of breed
1001	24 W	San Huang	100%	5648	40	16	0/27	0.00%	4.63%
1002	29 W	San Huang	100%	8634	47.6	16	2/27	7.41%	
1003	33 W	San Huang	100%	9265	48.2	16	1/24	4.20%	
1004	39 W	San Huang	100%	10,944	40.8	16	2/30	6.67%	
1005	30 W	Silky	73.33%	2770	83.3	16	3/29	10.34%	9.25%
1006	54 W	Silky	100%	16,693	39.7	16	2/25	8.00%	
1007	35 W	Qingyuan Ma	100%	11,545	34.4	16	5/32	15.62%	9.83%
1008	47 W	Qingyuan Ma	100%	11,320	34.9	16	1/29	3.45%	

[a] Arithmetic mean; [b]Serum samples were collected from the breeder flocks (1001–1008) on December 11, 2013 and detected by ELISA; [c]7–9 days-old embryos corresponding to the breeder flocks (1001–1008) were collected on December 20, 2013. Allantoic fluid was collected from embryos for *M. synoviae* antigen detection by PCR on December 20, 2013

9.25%, respectively (Table 4). The prevalence of *M. synoviae* infection in embryos across all 8 breeder flocks was 7.34%, which was substantially similar to the incidence in broiler chickens (5–8%), which were the offspring of the 8 breeder flocks in Zhejiang Province.

The progeny broilers originating from these three breeds were tracked for analysis of *M. synoviae* antibody within four weeks (in 4- to 7-week-old chickens). It was found that Silky broiler chickens were positive for the *M. synoviae* antibody earlier than the other two breeds. In Silky broilers were positive for the *M. synoviae* antibody at five weeks old, while the other two breeds were positive at six weeks old. The prevalence of the antibody in Three-Yellow chickens, Qingyuan Ma chickens and Silky chickens was 5%, 20% and 40% at seven weeks old, respectively (Fig. 4).

These results indicated that low *M. synoviae* antibody levels in breeders and high *M. synoviae* antigen positive rate in embryos could increase the chance of incidence in the offspring. Thus, *M. synoviae* maternal antibody level and *M. synoviae* infection in embryos might affect the incidence of infection in broiler chickens under threat of *M. synoviae* infection.

Discussion

In the 1950s, *M. synoviae* infection was first reported in America [22, 23], after that, researchers came to realize its importance in the poultry industry [1, 2, 24, 25]. Prior to this study, there were some single, sporadic reports about white feather chickens affected by *M. synoviae* infection in China [26], but there was a complete lack of data about *M. synoviae* infection in Chinese native chicken breeds. This study is the first to confirm the presence of large epidemics of *M. synoviae* infection in multi-aged farms of native Chinese chicken breeds.

We revealed that chickens could carry the *M. synoviae* maternal antibody and *M. synoviae* simultaneously before 3 weeks of age, and antibody levels of breeders affected the timing of antibody production and clinical incidence in broiler chickens. The incubation period of *M. synoviae* infection by horizontal transmission was usually 11–21 days but was shorter by vertical transmission (up to only 6 days) [2, 27]. Based on these results, *M. synoviae* antibody and the typical clinical signs of *M. synoviae* infection should be observed in chickens due to vertical transmission of *M. synoviae* at approximately 20 days of age. In fact, although there was vertical transmission, the presence of *M. synoviae* antibodies in chickens generally could not be detected until 50 days or even later, and the earliest onset of infection in chickens was also not earlier than 35 days in the clinic. Otherwise, although the proportion of embryos with *M. synoviae* infection could exceed 36%, there was no correspondingly high clinical incidence in broiler chickens of Chinese native breeds. It might suggest that, in addition to the potential of chickens to produce anti-*M. synoviae* antibodies [28] and virulence differences between *M. synoviae* isolates [29–31], maternal antibodies against *M. synoviae* could inhibit *M. synoviae* activity in vivo at early ages and the developmental window when chickens were negative for the *M. synoviae* antibody - might play a key role in *M. synoviae* reproduction in hosts. Therefore, improving maternal antibody levels in developing chickens and effectively controlling *M. synoviae* activity during the developmental window when chickens are negative for the *M. synoviae* antibody might an efficient strategy to prevent *M. synoviae* infection.

Many immunosuppressive pathogens can interact with *M. synoviae* and lead to serious disease in hosts [1, 32]. Co-infection of chickens with *M. synoviae* subclinical infection with avian influenza virus (AIV) H9 [11], infectious bronchitis virus (IBV) [4, 33, 34], Newcastle disease virus (NDV) [34, 35] or infectious bursal disease virus (IBDV) [32] caused more severe respiratory or systemic disease than infection with any single above-mentioned pathogen. Normally, chickens in China usually suffer from the threat of the above-mentioned situation, but the prevalence of subclinical infection in commercial broilers has not been reported. Consequently, the actual economic loss associated with *M. synoviae* infection might be more serious and larger than it is reported in the press. In this case, we need to attach more importance to *M. synoviae* prevention and control strategies in China.

Conclusions

In this study, we demonstrated that multi-aged breeder farms are widely infected with *M. synoviae* and that *M. synoviae* is circulating in Chinese native-type chickens. Chinese native-type breeders of 14 days old or younger simultaneously carried high levels of maternal antibodies against *M. synoviae*. In addition, vertical transmission of *M. synoviae* was observed, and prevalence of *M. synoviae* infection in embryos of breeders reached as high as 16.29%. Finally, our study suggested that 3- to 7-week-old chickens might be the most susceptible to *M. synoviae* infection and may therefore play a key role in the horizontal transmission of *M. synoviae*.

Abbreviations

ELISA: Enzyme-linked immunosorbent assay; PCR: Polymerase chain reaction; SPF: Specific pathogen free; CCU_{50}: 50% color change unit; SPSS: Statistic package for social science; DNA: Deoxyribose nucleic acid

Acknowledgments

The authors are thankful to Dr. Yuliang Liu for helpful comments.

Funding

This work was supported by the Scientific Research Major Project ([WENS] WSSC01-1-C-A-20120525-1419) from Guangdong WENS Foodstuff Group Co., Ltd.

Authors' contribution

SKS and XL participated in the study design, the experimental work, the analysis interpretation of the data and drafted the manuscript. XL participated in the experimental work, the analysis interpretation of the data and helped draft the manuscript. TRL, JPQ, FC participated in the study design and modify the manuscript. JPL, DAW participated in collecting samples. All authors read and approved the final manuscript.

Competing interests

The authors declare that they have no competing interests.

Author details

[1]State Key Laboratory for Conservation and Utilization of Subtropical Agro-Bioresources, Guangxi University, Nanning, Guangxi 530004, China. [2]Guangdong Enterprise Key Laboratory for Animal Health and Environmental Control, WENS Group Academy, Yunfu, Guangdong 527439, China. [3]College of Animal Science, South China Agricultural University, Guangzhou, Guangdong 510642, China.

Reference

1. Stipkovits L, Kempf I. Mycoplasmoses in poultry. Revue scientifique et technique (International Office of Epizootics). 1996;15(4):1495–525.
2. Ferguson-Noel SHKaN. Mycoplasma synoviae infection. In: Disease of Poultry (Chinese Version). Edited by Saif.Y.M, 12th edn. Beijing: Blackwell Publishing 2008;999–1014.
3. Landman WJ. Is Mycoplasma synoviae outrunning Mycoplasma gallisepticum? A viewpoint from the Netherlands. Avian Pathol. 2014;43(1):2–8.
4. Feberwee A, de Wit JJ, Landman WJ. Induction of eggshell apex abnormalities by Mycoplasma synoviae: field and experimental studies. Avian Pathol. 2009;38(1):77–85.
5. Catania S, Bilato D, Gobbo F, Granato A, Terregino C, Iob L, Nicholas RA. Treatment of eggshell abnormalities and reduced egg production caused by Mycoplasma synoviae infection. Avian Dis. 2010;54(2):961–4.
6. Gole VC, Chousalkar KK, Roberts JR. Prevalence of antibodies to Mycoplasma synoviae in laying hens and possible effects on egg shell quality. Prev Vet Med. 2012;106(1):75–8.
7. Kleven SH. Control of avian mycoplasma infections in commercial poultry. Avian Dis. 2008;52(3):367–74.
8. Hagan JC, Ashton NJ, Bradbury JM, Morgan KL. Evaluation of an egg yolk enzyme-linked immunosorbent assay antibody test and its use to assess the prevalence of Mycoplasma synoviae in UK laying hens. Avian Pathol. 2004; 33(1):93–7.
9. Dufour-Gesbert F, Dheilly A, Marois C, Kempf I. Epidemiological study on Mycoplasma synoviae infection in layers. Vet Microbiol. 2006;114(1–2):148–54.
10. Feberwee A, de Vries TS, Landman WJ. Seroprevalence of Mycoplasma synoviae in Dutch commercial poultry farms. Avian Pathol. 2008;37(6):629–33.
11. Roussan DA, Khawaldeh G, Shaheen IA. A survey of Mycoplasma gallisepticum and Mycoplasma synovaie with avian influenza H9 subtype in meat-type chicken in Jordan between 2011–2015. Poult Sci. 2015;94(7): 1499–503.
12. Branton SL, Leigh SA, Roush WB, Purswell JL, Olanrewaju HA, Collier SD. Effect of selected water temperatures used in Mycoplasma gallisepticum vaccine reconstitution on titer at selected time intervals. Avian Dis. 2008; 52(2):291–6.
13. Stemke GW, Robertson JA. Comparison of two methods for enumeration of mycoplasmas. J Clin Microbiol. 1982;16(5):959–61.
14. Ramirez AS, Naylor CJ, Hammond PP, Bradbury JM. Development and evaluation of a diagnostic PCR for Mycoplasma synoviae using primers located in the intergenic spacer region and the 23S rRNA gene. Vet Microbiol. 2006;118(1–2):76–82.
15. El-Gazzar MM, Wetzel AN, Raviv Z. The genotyping potential of the Mycoplasma synoviae vlhA gene. Avian Dis. 2012;56(4):711–9.
16. Eyigor A, Goncagul G, Gunaydin E, Carli KT. Salmonella profile in chickens determined by real-time polymerase chain reaction and bacteriology from years 2000 to 2003 in Turkey. Avian Pathol. 2005;34(2):101–5.
17. Liu HJ, Chen JH, Liao MH, Lin MY, Chang GN. Identification of the sigma C-encoded gene of avian reovirus by nested PCR and restriction endonuclease analysis. J Virol Methods. 1999;81(1–2):83–90.
18. Moradian A, Thorsen J, Julian RJ. Single and combined infections of specific-pathogen-free chickens with infectious bursal disease virus and an intestinal isolate of reovirus. Avian Dis. 1990;34(1):63–72.
19. De Herdt P, Desmidt M, Haesebrouck F, Ducatelle R, Devriese LA. Experimental Streptococcus Bovis infections in pigeons. Avian Dis. 1992; 36(4):916–25.
20. Domermuth CH, Gross WB. A medium for isolation and tentative identification of fecal streptococci, and their role as avian pathogens. Avian Dis. 1969;13(2):394–9.
21. Claire BA. Staphylococcosis. In: Disease of Poultry (Chinese Version). Edited by Saif.Y.M, 12th edn. Beijing: Blackwell Publishing. 2008:1059–70.
22. Olson NO, Bletner JK. Enlarged joint condition in poultry caused by an infectious agent. Poult Sci. 1954;33:1075.
23. Anderson GC, Bletner JK, Munro DA, Olson NO, Shelton DC. Studies of infectious synovitis in chickens. Am J Vet Res. 1956;17(65):747–54.
24. Mohammed HO, Carpenter TE, Yamamoto R. Economic impact of Mycoplasma gallisepticum and M. Synoviae in commercial layer flocks. Avian Dis. 1987;31(3):477–82.
25. Olson NO, Ken KM, Campbell A. Control of infectious synovitis. 13. The antigen study of three strains. Avian Dis. 1964;8:209–14.
26. Zhongzan Cao XL. Shengwang Liu, Jianbin he: advance in avian Mycoplasma Synoviae. Prog Vet Med. 2012;11(33):113–7.
27. Senties-Cue G, Shivaprasad HL, Chin RP. Systemic Mycoplasma synoviae infection in broiler chickens. Avian Pathol. 2005;34(2):137–42.
28. Washburn LR, Ramsay JR, Roberts LK. Characterization of the metabolism inhibition antigen of Mycoplasma arthritidis. Infect Immun. 1985;49(2):357–64.
29. Pascucci S, Maestrini N, Govoni S, Prati A. Mycoplasma synoviae in the guinea-fowl. Avian Pathol. 1976;5(4):291–7.
30. Lockaby SB, Hoerr FJ, Lauerman LH, Kleven SH. Pathogenicity of Mycoplasma synoviae in broiler chickens. Vet Pathol. 1998;35(3):178–90.
31. Narat M, Bencina D, Kleven SH, Habe F. The hemagglutination-positive phenotype of Mycoplasma synoviae induces experimental infectious synovitis in chickens more frequently than does the hemagglutination-negative phenotype. Infect Immun. 1998;66(12):6004–9.
32. Giambrone JJ, Eidson CS, Kleven SH. Effect of infectious bursal disease on the response of chickens to Mycoplasma synoviae, Newcastle disease virus, and infectious bronchitis virus. Am J Vet Res. 1977;38(2):251–3.
33. Springer WT, Luskus C, Pourciau SS. Infectious bronchitis and mixed infections of Mycoplasma synoviae and *Escherichia coli* in gnotobiotic chickens. I. Synergistic role in the airsacculitis syndrome. Infect Immun. 1974;10(3):578–89.
34. Kleven SH, Eidson CS, Anderson DP, Fletcher OJ. Decrease of antibody response to Mycoplasma synoviae in chickens infected with Marek's disease herpesvirus. Am J Vet Res. 1972;33(10):2037–42.
35. Villegas P, Kleven SH, Anderson DP. Effect of route of Newcastle disease vaccination on the incidence of airsacculitis in chickens infected with Mycoplasma synoviae. Avian Dis. 1976;20(2):395–400.

7

Methionine to cystine ratio in the total sulfur amino acid requirements and sulfur amino acid metabolism using labelled amino acid approach for broilers

Letícia G. Pacheco[1*], Nilva K. Sakomura[1], Rafael M. Suzuki[1], Juliano C. P. Dorigam[2], Gabriel S. Viana[1], Jaap Van Milgen[3] and Juliana C. Denadai[4]

Abstract

Background: Assuming that part of Methionine (Met) is converted into Cystine (Cys), but ignoring the rates with which such phenomenon occurs may lead to an excessive supply of Met in poultry diets. Such inconvenient could be easily avoided with the knowledge of the ideal Met:Cys/Total sulfur amino acids (TSAA) ratio and the rates of Met conversion into Cys.

Results: Met sources did not affect performance. Met:Cys/TSAA ideal ratio was determined using curvilinear-plateau regression model. Both optimum body weight gain and feed conversion ratio were estimated in 1007 g/day and 1.49, respectively, at 52% Met/TSAA ratio. Feed intake was not affected by Met:Cys/TSAA ratios. In the labelled amino acid assay, the rates with which Met was converted into Cys ranged from 27 to 43% in response to changes in Met:Cys/TSAA ratios, being higher at 56:44.

Conclusion: Based on performance outcomes, the minimum concentration of Met relative to Cys in diets for broilers from 14 to 28 d of age based on a TSAA basis, is 52% (52:48 Met:Cys/TSAA). The outcomes from labelled amino acid assay indicate that highest the Met supply in diets, the highest is its conversion into Cys.

Keywords: Cystine, Methionine conversion, Methionine cycle, Stable isotopes, Sulfur amino acids

Background

The composition of feedstuffs used in feed production and the amount required for different metabolic functions makes methionine (Met) more limiting in poultry diets as compared to other essential amino acids [1]. Except for tryptophan in corn protein, Met is the amino acid, whose concentration is smallest one among essential amino acids in the protein of corn and soybean meal [2]. Contrary to lysine (Lys), which is used almost entirely for body protein synthesis and does not take part in other metabolic processes [3], or threonine (Thr) and valine (Val), whose intake is deviated for few secondary functions like intestinal mucin protein synthesis [4], Met is less efficiently retained into body due to its role as a donor of methyl groups in the synthesis of compounds such as creatine, choline, polyamines and carnitine [3].

Poultry are not capable of synthetizing the carbon skeleton of Met in a sufficient amount to meet the relative needs of the organism for maintenance and growth; and therefore are completely dependent on its dietary supply to ensure the ideal rate for protein synthesis [5, 6]. An extremely important function of Met in animal metabolism, is as the precursor of cysteine [7]. As well as Met, cysteine is a sulfur amino acid formed via a transsulfuration pathway in the hepatocytes, whose synthesis requires a thiol group from Met and a nitrogen atom from serine. Due to its instability, cysteine forms disulphide bonds with another cysteine molecule to originate cystine (Cys), which confers strength and stiffness to keratin [8, 9]. Proportionally, feather proteins contain more Cys than most other tissue proteins [10, 11].

* Correspondence: legrazy@gmail.com
[1]Department of Animal Science, Unesp, Jaboticabal, SP, Brazil
Full list of author information is available at the end of the article

Emmans and Fisher [12] describe Met content in the protein of broiler carcass as being approximately two-fold higher than Cys content (2.5 vs. 1.1%), whilst in feathers such proportion markedly changes and Cys assumes a higher participation in protein composition (approximately twelve-fold higher) compared with Met (7 vs. 0.6%). Similarly, Talpaz et al. [13] reported that Cys content in the protein of broiler feathers was thirteen-fold higher than Met (6.5 vs. 0.5), but contrary to the aforementioned authors, found a different proportion between TSAA in carcass, whose value of Cys content was higher than Met (2.4 vs. 1.9%).

Based on such background about the role of Cys on feather composition and on the fact that broiler feathering rates exhibit different patterns of growth throughout growing phase, reaching the maximum rate at approximately 28 d post-hatch, it is expected that bird requirement for Cys follows the same behavior of feather growth. Commercial diets are traditionally formulated to meet broiler requirements for methionine + cystine (Met + Cys), based on the assumption that amounts of dietary Met are converted into Cys. The consequence is that Met may be provided in excess, reducing the efficiency with which the amino acid is used by the organism. This inconvenient could be easily avoided with the knowledge of the ideal Met:Cys/TSAA ratio and the amounts of Met, which are converted into Cys. Since Cys synthesis is dependent on Met, providing broiler diets supplemented with L-cystine could may spare the utilization of Met, which could increase its availability on organism for body protein accretion, and hence improve broiler performance and mitigate nitrogen excretion. Several research efforts has been made to establish broiler requirements for Met, which in turn, allowed the TSAApublication of several feed tables. Brazilian Tables for Poultry and Swine [2], recommend TSAAthe optimum daily intake of digestible Met for male broilers in 130, 401, 786, and 938 mg/bird, in the pre-starter, starter, grower, and finisher phases, respectively, whilst daily digestible Cys requirements for the these phases is described as 105, 323, 633, and 755 mg/bird.

Little is known about the ideal Met:Cys/TSAA ratio for broilers, feed tables recommend Met:Cys/TSAA ratio of 55:45 [2, 14]. With regard to Met sources, poultry diets are conventionally supplemented with DL-Methionine, which contains 50% D-Met and 50% L-Met. Nonetheless, only L-isomers are used in protein synthesis process, which forces the organism to convert all the isomers to L-form. In Met case, such conversion takes place in bird liver and kidney [2, 15, 16]. These arguments have been reinforcing the appeal for researches with the objective of evaluating the potential utilization of L-Met in replacement to DL-Met in poultry diets. The fact of whether Met sources influence or not the determination of the TSAA ratio in broiler/TSAA for broiler is unknown, which suggests that studies could focus on investigating such issue and answer this question. Additionally, besides the importance of determining the ideal Met:Cys/TSAA ratio for broilers, obtaining reliable information regarding the efficiency with which met is converted into Cys could help to achieve a clearer understanding of the dynamic of TSAA metabolism.

Studies involving labelled amino acid date from the last century [17] and since then, so many innovation have been observed in this area of science, given the improvements and developments of techniques, which allowed labelling the most varied molecules and track their final fates. Based on all these arguments, the current research was conducted in order to determine the optimum proportion between Met and CysTSAA in the total of sulfur amino acids (Met:Cys/TSAA ratio) in diets supplemented with different sources of Met for broilers in the grower feeding phase and to estimate the efficiency with Met is converted into Cys in broilers using labelled methionine and cystine.

Results

Performance assay

Irrespective of the performance response evaluated, no interactions were noticed between dietary Met:Cys/TSAA ratios and Met sources ($P > 0.05$). Methionine sources did not affect broiler performance, and thereby means are presented in Table 1 as the average of both sources. Broiler performance responses to Met:Cys/TSAA ratios and Met sources are detailed in Tables 1 and 2. Feed intake was unaffected by Met:Cys/TSAA ratios. Broilers fed 44:56 Met:Cys/TSAA ratio had the lowest BWG compared with the other dietary treatments ($P < 0.05$). Broilers fed 44:56 Met:Cys/TSAA ratio had the poorest FCR compared with broilersbroilers fed 50:50, 53:57, and 56:44 Met:Cys/TSAA ratios ($P < 0.05$),

Table 1 Body weight gain (BWG), feed intake (FI) and feed conversion ratio (FCR) from 14 to 28d of age after combination of sources

Met:Cys/TSAA ratio	BWG (g)	FI (g)	FCR (g/g)
44:56	912 ± 28.46 b	1469 ± 88.79	1.56 ± 0.10 b
47:53	970 ± 21.24 a	1482 ± 48.99	1.53 ± 0.05 ab
50:50	993 ± 23.71 a	1499 ± 62.88	1.51 ± 0.06 a
53:47	1001 ± 25.76 a	1482 ± 33.52	1.48 ± 0.06 a
56:44	1006 ± 24.24 a	1519 ± 63.01	1.51 ± 0.06 a
P-value			
Proportions	< 0.05	0.35	< 0.05
Sources	0.73	0.21	0.36
Proportions x Sources	0.96	0.84	0.97

Means in columns followed by distinct letters are different by the Tukey test ($P < 0.05$)

Table 2 Body weight gain (BWG), feed intake (FI) and feed conversion ratio (FCR) and their respective standard deviations for broilers fed different Met sources (L and DL) and Met:Cys/TSAA ratios

	14 to 28 days		
Met:Cys/TSAA			
L-Cys	BWG (g)	FI (g)	FCR (g/g)
44:56	912 ± 28.46	1466.50 ± 59.95	1.56 ± 0.07
L-Met/L-Cys			
47:53	970 ± 23.46	1489 ± 40.02	1.53 ± 0.03
50:50	991 ± 23.37	1487 ± 29.70	1.50 ± 0.04
53:47	1001 ± 30.37	1486 ± 39.90	1.49 ± 0.07
56:44	1013 ± 21.13	1546 ± 62.03	1.53 ± 0.07
DL-Met/L-Cys			
47:53	970 ± 21.03	1458 ± 44.00	1.50 ± 0.04
50:50	994 ± 26.26	1485 ± 60.61	1.49 ± 0.05
53:47	1001 ± 23.18	1478 ± 28.99	1.48 ± 0.05
56:44	1007 ± 24.20	1491 ± 55.89	1.49 ± 0.05
P-value			
Proportions	< 0.05	0.35	< 0.05
Sources	0.73	0.21	0.36
Proportions x Sources	0.96	0.84	0.97

but no difference was noticed from broilers fed TSAA47:53 Met:Cys/TSAA ratio. Curvilinear-plateau regression model was fitted to data with the purpose of estimating the TSAA ratio in broiler/TSAA ratio in broiler diets.

Labelled stable isotope assay

The data used to calculate the rates with which Met is converted into intermediates and Cys are detailed in Tables 3, 4 and 5. The recoveries of metabolic tracers (^{15}N L-Met and ^{15}N L-Cys) in the body, breast, liver, feather, and excreta samples are expressed in percentage and therefore, the effect of body weight variation among the animals was disregarded (Eqs. 2, 3 and 4). The recovery of stable nitrogen isotope recovery L-(^{15}N) Met and L-($^{15}N_2$) Cys were close to 100% in L-(^{15}N)-Met supplemented-diets, whilst in L-(^{15}N)-Cys these values were approximately 90% (Table 3). The analysis of aminogram in the different tissues and excreta of broilers fed L-(^{15}N) Met supplemented-diets indicated very similar Met recovery values irrespective of the ratio assessed, being in average approximately 86% (Table 4). Conversely, Cys recoveries in broilers fed L-(^{15}N) Cys supplemented-diets ranged from 93 to 111% as dietary Cys content decreased (Table 4). The values of methionine-specific ^{15}N recovery in broilers given bird given L-(^{15}N) Met supplemented-diets are detailed in Table 5. Broilers fed diets with the highest Cys content (44:56 and 50:50 Met:Cys/TSAA ratio) exhibited a narrow difference in Met recovery (66–68%), whilst this value in broilers fed 56:44 Met:Cys/TSAA ratio was 59%. When comparing Met and Cys deposition in the analyzed samples, both stable nitrogen isotope and aminogram analysis indicated that higher amounts of Met were incorporated into breast muscle protein compared with Cys, whilst the major deposition of Cys occurred in feathers. The rates with which Met was converted into intermediates and Cys is detailed in the Table 6. When Met:Cys/TSAA ratio increased from 44:56 to 56:44, the amount of Met converted into Cys increased from 27 to 43%, and the Met conversion into intermediates of Met cycle decreased from 24 to 12%.

Discussion

The current research was conducted with the purpose of establishing the optimum proportion between Met and Cys (Met:Cys/TSAA ratio) for broilers from to 14 to 28 d of age. Since the metabolism of Cys is particularly dependent on Met supply, it was expected that some biological events would occur in TSAA metabolism in face of the changes in the dietary supply of Met and TSAACys. Therefore, inTSAA parallel, a labelled isotope study was conducted to provide information to support the performance outcomes obtained and to understand the metabolic pathways in the organism in which Met is

Table 3 ^{15}N recoveries from stable nitrogen isotope analysis

δ^{15}N	^{15}N intake (mg)	Body	Breast	Liver	Feather	Excreta	Total	Rec (%)
		^{15}N retention ($^{15}N_R$, mg)						
L-(^{15}N)-Met								
44:56	10.73 ± 0.06	6.21 ± 0.35	1.55 ± 0.05	0.37 ± 0.02	0.49 ± 0.02	2.12 ± 0.11	10.74 ± 0.48	100 ± 4
50:50	10.27 ± 0.28	5.83 ± 0.12	1.80 ± 0.13	0.36 ± 0.01	0.40 ± 0.05	1.85 ± 0.07	10.23 ± 0.20	100 ± 1
56:44	10.25 ± 0.49	5.53 ± 0.42	1.67 ± 0.10	0.29 ± 0.03	0.47 ± 0.04	1.85 ± 0.15	9.80 ± 0.49	96 ± 9
L-($^{15}N_2$)-Cys								
44:56	12.06 ± 0.10	4.42 ± 0.26	1.04 ± 0.11	0.28 ± 0.04	2.18 ± 0.93	2.47 ± 0.23	10.39 ± 0.72	86 ± 6
50:50	11.58 ± 0.43	4.21 ± 0.70	1.07 ± 0.11	0.31 ± 0.05	2.71 ± 0.22	2.31 ± 0.14	10.60 ± 0.69	91 ± 4
56:44	11.79 ± 0.27	4.07 ± 0.29	1.07 ± 0.08	0.26 ± 0.01	2.66 ± 0.49	2.74 ± 0.50	10.79 ± 0.74	92 ± 8

Table 4 Amino acid recoveries from aminogram analysis

Aminogram	AA intake (mg)	Body	Breast	Liver	Feather	Excreta	Total	Rec (%)
		AA retention (AA_R, mg)						
Met retention								
44:56	5063 ± 176	2287 ± 106	675 ± 20	89 ± 6	676 ± 60	605 ± 45	4332 ± 70	86 ± 4
50:50	4917 ± 288	2110 ± 145	786 ± 49	88 ± 3	575 ± 10	573 ± 30	4132 ± 166	84 ± 1
56:44	4701 ± 451	2084 ± 163	724 ± 71	66 ± 7	694 ± 30	543 ± 59	4114 ± 196	88 ± 5
Cys retention								
44:56	4590 ± 131	1781 ± 92	412 ± 12	61 ± 4	869 ± 128	1146 ± 85	4268 ± 58	93 ± 3
50:50	4081 ± 399	1652 ± 126	425 ± 30	64 ± 2	1040 ± 22	996 ± 57	4176 ± 139	103 ± 2
56:44	4211 ± 234	1790 ± 142	483 ± 42	52 ± 5	1120 ± 64	1212 ± 113	4659 ± 165	111 ± 8

involved after consumed. In dose-response assays, the success in obtaining reliable estimates of nutrient requirements, among several factors, depends on the range of the levels of the nutrient studied, i.e. the optimal dose of a specific nutrient can only be estimated if the levels studied allow observing different phases of growth response. Additionally, as pointed out by Garlich et al. [18], there is the inherent decrease in sensitivity of the assay as the diets approach nutritional adequacy. Based on the latter assumption, in the current research, the amount of TSAA provided by experimental diets was fixed in 70% of the TSAA recommended by [2] and such diets differed only with respect of the Met:Cys/TSAA ratio assessed and the Met source supplemented.

In general, methionine is supplemented in poultry diets as DL-methionine, which consists of the mixture of 50% D-Met and 50% L-Met. It is well acknowledged that the organism incorporates only L-isomers into bodily proteins, which makes the conversion of the D-isomer into the L-isomer by the organism a mandatory step for the synthesis of protein [19]. Such conversion of isomers takes place in the liver and kidneys in metabolic pathways, whose effectiveness do not achieve the rate of 100% [20–25]. Considering such losses of efficiency, several research efforts have been put forward finding and validating alternatives sources to DL-Met with a higher bioavailability. Based on recent evidences published in literature that L-Met has a higher relative bioavailability than DL-Met, and that broilers fed L-Met supplemented-diets had greater performance than those fed DL-Met, this study was conducted based on the hypothesis that L-Met could indeed improve performance responses [26, 27]. However, contrary to what was previously hypothesized the Met sources tested herein did not affect broiler overall performance ($P > 0.05$). These outcomes support previous research that broilers and turkeys at different ages, fed either L or DL-Met supplemented-diets, did not differ from each with respect to performance responses [18, 28, 29].

Optimum Met:Cys/TSAA ratio

The outcomes obtained from the current research TSAA clearly indicate that the range among the ratios were indeed sufficient to induce a response, which in turn allowed fitting statistical models to collected performance data which allowed estimating ideal Met:Cys/TSAA ratio for broilers. Graber and Baker [30] highlight that the complexity of determining nutrient requirements relies on the difference in the estimates produced from statistical models. With a considerable range in the dose of a nutrient, broiler responses exhibit different phases: an increase in response, a plateau phase, and a decrease in performance, which in some cases are associated with incidence of mortality if the dose administrated exceeds the toxicity tolerance of the organism [31]. The model should therefore describe as close as possible the entire pathway at which an animal move towards its maximum response to the range of a given nutrient. Initially, collected performance data obtained in this research were fitted to polynomial quadratic, linear-plateau and

Table 5 15 N-Met recoveries from compound-specific stable isotope (CSIA) analysis

δ^{15}N-Met[a]	^{15}N-Met intake (mg)	Body	Breast	Liver	Feather	Excreta	Total	Rec (%)
		^{15}N-Met retention (^{15}N-Met_R, mg)						
L-(^{15}N)-Met								
44:56	107.41 ± 0.58	38.54 ± 1.71	7.04 ± 0.07	1.78 ± 0.20	4.94 ± 0.40	21.19 ± 1.65	73.50 ± 1.42	68.4 ± 1
50:50	102.80 ± 2.77	35.16 ± 1.71	7.30 ± 0.13	1.70 ± 0.06	3.87 ± 0.20	20.01 ± 0.87	68.04 ± 1.73	66.2 ± 0
56:44	102.59 ± 4.90	31.31 ± 2.34	6.43 ± 0.27	0.75 ± 0.50	4.09 ± 0.51	18.44 ± 1.04	61.02 ± 3.19	59.6 ± 4

[a] Results from compound-specific stable isotope analysis using labelled methionine (CSIA)

Table 6 Methionine conversion to cystine and intermediates

Met:Cys/ TSAA ratio	$^{15}N_{Rec}$ [a] (%)	$^{15}N\text{-}Met_{Rec}$ [b]	$Met_{Cys + interm.}$ [c]	$AA_{Rec\ (Cys)}$ [d]	$^{15}N_{Rec(Cys)}$ [e]	Cys_{Met} [f]	$Met_{interm.}$ [g]
44:56	100 ± 3.89	49 ± 4.62	51 ± 2.83	93 ± 10.92	66 ± 2.94	27 ± 7.83	24 ± 8.60
50:50	100 ± 2.80	47 ± 4.62	53 ± 2.41	103 ± 10.52	72 ± 3.89	31 ± 4.31	22 ± 3.61
56:44	96 ± 2.76	42 ± 3.77	54 ± 1.71	111 ± 9.50	68 ± 7.73	43 ± 4.11	12 ± 6.31

[a] $^{15}N_{Rec}$ (Table 3) from Met
[b] $^{15}N\text{-}Met_{Rec}$ (Table 5) excluding excreta
[c] $Met_{Cys + interm} = {}^{15}N_{Rec\ (Met)} - ({}^{15}N\text{-}Met_{Rec})$
[d] AA_{Rec} (Table 4) of Cys
[e] $^{15}N_{Rec}$ (Table 3) from Cys, excluding excreta
[f] $Cys_{Met} = AA_{Rec\ (Cys)} - {}^{15}N_{Rec\ (Cys)}$
[g] $Met_{interm.} = Met_{Cys + interm.} - Cys_{Met}$

curvilinear-plateau regression models. After fitting data, AIC value indicated that data fitted better to polynomial quadratic model. However, such model was ignored since when estimating requirements it assumes a bilateral symmetry in animal responses as a function of increments in the nutrient intake, which could not be biologically condoned. Estimates from curvilinear-plateau regression modelTSAA were accepted as those which better represented the responses obtained in this research. Curvilinear-plateau estimated the breakpoint for both BWG and FCR at 52:48 Met:Cys/TSAA ratio. These outcomes are consistent and support the 55:45 Met:Cys/TSAA ratio for broilers in the starter phase recommended by the [2]. Research efforts with the purpose of investigating the effects of TSAA in broiler performance usually range the levels of Met + Cys through supplementing a crystalline Met source. When both assays described herein were designed and conducted until the present date, no investigations in which the total amount of TSAA was fixed so that Met and Cys could be supplied in order to range the proportion between both amino acids were conducted. Such lack of references makes the current research innovative at some point, but concomitantly it makes harder any kind of comparisons with other results.

It was hypothesized that when supplemented in diets, Cys could spare dietary Met, which in turn could be used for body and feather protein accretion instead of being converted into Cys. However, contrary to such expected positive effects, as L-Cys was added to diets, i.e. as Met:Cys/TSAA ratio decreased, broiler performance was impaired, which indicates that broilers need to convert Met into Cys, irrespective whether diets are supplemented or not with L-Cys. Cysteine may be originated from diet, muscle proteolysis and/or in the liver from dietary Met [32]. Due to its unstable nature, it may be readily autoxidizes to cystine, which in turn is the most abundant amino acid in feather keratin. For modern broiler industry, poor feathering has been correlated to low carcass quality and carcass condemnations at slaughtering [33, 34]. Considering that broilers are generally slaughtered around the 42 day of age, the period in which feather growth rate reaches maximum values comprises 14 to 28 d post hatch. Until 28 d of age, it is expected a complete replacement of the primary feathers by the secondary ones. Since protein accretion rates in feathers are increased, higher requirements for amino acids, particularly Cys, is also expected in this period [35]. Why L-Cys supplementation or Met deficiency decreased broiler performance instead of improving it, it is still unknown. However, two hypothesis, which explain such phenomenon may be risen. The first one considers the premise that the presence of Cys in plasma may presumably act as a trigger, which could signalize to the organism that Met does not need to absorbed and then converted into Cys, since Cys status is in accordance to organism needs, i.e. Cys may downregulate the intestinal uptake of Met. Although, this hypothesis makes some sense, data collected herein cannot support or corroborate its validation. Another hypothesis to explain the performance impairments implies in considering the dynamic and importance of the Met cycle for metabolism. Although simple, the Met cycle is crucial to maintain structural and metabolic functions of the organism. The first step of this cycle consists of the formation of the derivate S-adenosilmethionine (SAM), a methyl donor, which enables the occurrence of a plethora of transmethylation reactions, which include the DNA methylation, and the synthesis of phosphatidylcholine, creatine and polyamines. In a next step, homocysteine is formed from SAM and from this step on it may follow distinct pathways: 1) being remethylated to methionine through the transfer of a methyl group from betaine by the enzyme betaine-homocysteine methyl transferase to produce dimethylglycine or via 5-methyl tetrahydrofolate [36] in a process dependent on vitamin B12; and/or 2) being catabolized to form cysteine in a vitamin-B6-dependent pathway of transsulfuration where sulfur group of homocysteine is transferred to serine [37, 38]. Since the first step of Cys synthesis involves the occurrence of methylation reactions in the Met cycle, and that these reactions are a salvage mechanism to ensure the maintenance of vital processes, if provided in low concentrations, insufficient amounts of Met would be available for optimum protein accretion rates, which could justify our results [39–42]. In other terms, a minimum amount of Met must participate on Met cycle so that the aforementioned reactions may occur in rates which warrant the adequate occurrence of secondary physiological process than growth.

Labelling amino acids with stable isotopes has proved to be a useful tool understand amino acid metabolism and, therefore, explain the results obtained in the performance assay. In so doing, the first focus of this research was to recover the amount of ^{15}N supplied in diets in the body and excreta. These outcomes indicated that the recovery rates for ^{15}NMet from Met in the in broilers fed 44:56, 50:50, and 56:44 Met:Cys/TSAA ratios was equal to 100, 100, and 96%, respectively. These higher recovery rates corroborate the effectiveness of 15 N as tracer in studies involving labelled molecules. Thereafter, it was determined specifically the labelled Met (^{15}N-Met) incorporated into body, feather, and liver protein, which after subtracted by the recovery rates above cited resulted in conversion rate of Met into Cys/ intermediates (MetCys+interm) of 51, 53, and 54% for the diets containing 44:56, 50:50, and 56:44 Met:Cys/ TSAA ratios, respectively. Whilst data collected in the current study could not support the first hypothesis raised above, the outcomes from the labelled isotopes assay provided some evidences, which prove to be true the second hypothesis raised about why L-Cys supplementation impaired broiler performance,. The fact that the conversion rates of Met into Cys/intermediate (Table 4) were very similar among dietary treatments clearly demonstrates that Met was equally converted into Cys and intermediates of Met cycle, regardless of the amount of dietary L-Cys supplied. From this answer, however, another question raises: were the Met:Cys/TSAA ratios studied herein capable of influencing the proportion of Met converted into intermediates of Met cycle and Cys?TSAAbroilersTSAA.

A detailed in Table 6, the lowest the Cys content of diets, the highest was the amount Met converted into Cys and the lowest was Met conversion into intermediates of Met cycle. The amount of Cys originated from Met (Cys_{Met}) decreased from 43 to 27% when dietary Met:Cys/TSAA ratio ranged from 56:44 to 44:56, i.e. when L-Cys supplementation increased. Once Cys was already supplied by diets, dietary Met used in transmethylation pathways increased, which can be confirmed by the increase in the amount of intermediates originated from Met ($Met_{interm.}$) when Met:Cys/TSAA ratios increased form 44:56 to 56:44. These outcomes prove that broilers utilized L-Cys from diets in which Cys content was lower. Additionally, it seems that Met will inevitably be used in Met cycle so that methyl transfer reactions ensure the adequate occurrence of the important physiological processes. It explains why broilers fed diets in which Met content was lower had lower performance responses compared with broilers fed higher Met supply: lower was the amount of Met available for body protein accretion.

Conclusion

When Cys assumes a high proportion of total TSAA in broiler diets, dietary Met is not spared to be used in body protein accretion, but rather to be used in transmethylation pathways. Dietary Met:Cys/TSAA ratios for broiler from 14 to 28 d of age must be higher than 52:48.

Methods

Broiler performance assay

Broilers, experimental design, and housing

One-d-old male Cobb 500 broilers were obtained from a local commercial hatchery (Pluma Agroavicola, BR) and housed in an environmentally controlled room with a cellulose-pad cooling system associated with exhausters. From one to 14 d post-hatch broilers were fed a standard diet formulated to meet or exceed [2] recommendations. On d 14 post-hatch, a total of 1080 broilers were weighed and initial body was similar among treatment groups (average initial body weight of 0.390 ± 0.004 kg). Six replicate pens of 20 broilers were randomly assigned to each of nine treatments. A factorial arrangement (4 × 2) consisting of four Met:Cys/TSAA ratios (56:44, 53:47, 50:50, and 47:53) supplemented with two different Met sources (DL-Met or L-Met). An additional diet TSAAin which no sources of synthetic Met was added was formulated to contain a 44:56 Met:Cys/TSAA ratio. Each pen (1.4 m × 3.0 m) was equipped with six nipple drinkers and a tubular feeder, which provided water and experimental diets (mash form) on ad libitum basis. Photoperiod was daily set at 18 L:6D (light:dark) throughout the entire experimental period. Throughout the 14-d feeding assay, broilers were raised under thermoneutral conditions and environmental temperature and humidity were controlled as recommended by the strain guideline [43]. Environmental temperature to which broilers were exposed prior to the beginning of the trial were 31 °C ± 2 °C at the first week, 27 °C ± 1.5 °C at the 2nd week, and 24 °C ± 1.5 °C during the trial. Humidity were kept constant (65% ± 4%), from the 1st d of age until the end of the trial. The broilers used in the current research were used in a subsequent performance assay, which as well as this research had the protocols previously approved by the institutional Animal Care and Use Committee of the Faculdade de Ciências Agrárias e Veterinárias (UNESP, Jaboticabal, Brazil).

Experimental diets

Experimental diets were formulated to meet or exceed broiler nutrient recommendations [10] (Table 7), except for dietary total sulfur amino acid content, which was supplied at 0.65%, regardless of the experimental treatment, which corresponded to 70% of the recommendation of Rostagno et al. [2]. Such deficiency was provided in order to ensure that both amino acids would be limiting in diets. The changes in the amount of Met relative to Cys was obtained by changing the supplemented level of L-Cystine (100%) in relation to DL/L-Met (99%) and vice versa. Prior to diet formulation, all ingredients were analyzed for total amino acids content. Standardized ileal digestible amino acids were estimated based on coefficient of digestibility described by Rostagno et al. [2]. Crystalline amino acid were assumed to be 100% standardized ileal digestible (SID). – Once, TSAA metabolism is dependent on a series of methyl donors and acceptors, all diets had equal premix, and the same batches of ingredients were used.

Data collection

At 28 d of age all broilers and feeders were weighed to determine average daily feed intake (FI, g/bird) and body weight (BW, kg). Body weight gain (BWG, g/bird) and feed conversion ratio (FCR) were calculated from such data. Mortality was daily recorded to adjust feed intake and feed conversion ratio. Met and Cys intakes (mg/bird/day) were calculated from FI and the analyzed content of Met and Cys in the diet.

Table 7 Ingredients and nutrient composition of experimental diets as-fed basis

Met:Cys/TSAA TSAAratio	Standart diet (1-14d)	44:56	47:53		50:50		53:47		56:44	
Ingredients (g.kg^{-1})		No Met	L-Met	DL-Met	L-Met	DL-Met	L-Met	DL-Met	L-Met	DL-Met
Corn (7.86%)	620.64	545.60	545.04	545.59	544.47	545.58	543.91	545.56	543.34	545.55
Soybean meal (45%)	216.68	350.03	350.10	350.03	350.17	350.03	350.24	350.03	350.31	350.04
Corn gluten meal (60%)	105.11									
Peanult meal (48.2%)		20.93	21.24	20.99	21.55	21.05	21.86	21.10	22.18	21.16
Soybean oil	9.27	45.81	46.00	45.77	46.19	45.73	46.37	45.68	46.56	45.64
Dicalcium phosphate	19.39	15.27	15.26	15.27	15.26	15.27	15.26	15.27	15.26	15.26
Limestone	9.31	9.04	9.04	9.04	9.04	9.04	9.04	9.04	9.04	9.04
Sodium bicarbonate	2.00	4.14	4.14	4.14	4.13	4.14	4.13	4.14	4.13	4.14
Salt	3.61	2.08	2.08	2.08	2.08	2.08	2.08	2.08	2.08	2.08
DL-Methionine (98%)	2.88	0.00	0.00	0.19	0.00	0.37	0.00	0.55	0.00	0.74
L-Methionine (99%)		0.00	0.18	0.00	0.37	0.00	0.55	0.00	0.73	0.00
L-Cystine (100%)		0.73	0.55	0.55	0.37	0.37	0.18	0.18	0.00	0.00
L-Lysine HCl (78%)	7.03	2.37	2.37	2.37	2.37	2.37	2.37	2.37	2.37	2.37
L-Threonine (98%)	1.58	0.93	0.93	0.93	0.93	0.93	0.93	0.93	0.93	0.93
L-Valine (96.5%)		0.47	0.47	0.47	0.47	0.47	0.48	0.47	0.48	0.47
Premix[a]	1.5	1.60	1.60	1.60	1.60	1.60	1.60	1.60	1.60	1.60
Choline Chloride (60%)	1.00	1.00	1.00	1.00	1.00	1.00	1.00	1.00	1.00	1.00
Nutrients content										
Met. energy (Mcal/Kg)	3.100	3.10 (2.9)	3.10 (2.9)	3.10 (2.9)	3.10 (2.9)	3.10 (2.9)	3.10 (2.9)	3.10 (2.9)	3.10 (2.9)	3.10 (2.9)
Crude protein (%)	22.0	21.5 (22)	21.5 (22)	21.5 (22)	21.5 (22)	21.5 (22)	21.5 (22)	21.5 (22)	21.5 (22)	21.5 (22)
Dig. Met+Cys (%)	0.953	0.65 (0.60)	0.65 (0.60)	0.65 (0.62)	0.65 (0.60)	0.65 (0.60)	0.65 (0.61)	0.65 (0.62)	0.65 (0.61)	0.65 (0.60)
Dig. Met (%)	0.639	0.29 (0.26)	0.31 (0.28)	0.31 (0.29)	0.33 (0.30)	0.33 (0.30)	0.34 (0.32)	0.34 (0.33)	0.36 (0.34)	0.36 (0.34)
Dig. Cys (%)	0.314	0.36 (0.34)	0.34 (0.32)	0.34 (0.33)	0.33 (0.30)	0.33 (0.30)	0.31 (0.29)	0.31 (0.29)	0.29 (0.27)	0.29 (0.26)
Dig. Lys (%)	1.324	1.19 (1.16)	1.19 (1.16)	1.19 (1.16)	1.19 (1.16)	1.19 (1.16)	1.19 (1.16)	1.19 (1.16)	1.19 (1.16)	1.19 (1.16)
Dig. Thr (%)	0.861	0.79 (0.75)	0.79 (0.75)	0.79 (0.75)	0.79 (0.75)	0.79 (0.75)	0.79 (0.75)	0.79 (0.75)	0.79 (0.75)	0.79 (0.75)
Dig. Val (%)	1.019	0.92 (0.89)	0.92 (0.89)	0.92 (0.89)	0.92 (0.89)	0.92 (0.89)	0.92 (0.89)	0.92 (0.89)	0.92 (0.89)	0.92 (0.89)

[a] Content (per kg of diet) – vit. A = 10,575 UI; vit. D3 = 2554 UI; vit. K = 1.8 mg; vit. E = 14.87 mg; vit. B1 = 2.00 mg; vit. B2 = 4.5 mg; vit. B6 = 2.50 mg; vit. B12 = 2.00 mg; niacin = 30.00 mg; folic acid = 0.75 mg; calcium pantothenate = 11.74 mg; biotin = 0.01; iron = 43.44 mg; zinc = 43.35 mg; copper = 8.56 mg; manganese = 56.00 mg; iodine = 0,56 mg; selenium = 0,34 mg; antioxidant 4.20 mg; Salinomycin sodium 12%; Butil hidroxy toluene BHT. Values in parentheses indicate analyzed dietary concentration of the amino acids

Statistical analysis
All collected performance data were analyzed as two-way ANOVA using the GLM procedure of SAS 9.0 (SAS Institute Inc., Cary, NC). The statistical model included the fixed main effects of dietary Met source (DL-Met or L-Met) and dietary Met:Cys ratio (65:35, 57:43, 50:50 and 43:57), as well as the interaction between both factors. Because interaction effects between Met source and Met:Cys ratio were not found for any of the evaluated parameters, data from the main effects were pooled and reanalyzed. Quadratic broken-line regression was computed by the PROC NLIN procedure of SAS [44] to estimate optimal dietary Met:Cys ratio based on performance traits and statistical differences were considered when $P < 0.05$.

Labelled stable isotope assay

Husbandry, experimental design and experimental diets
On d 14 posthatch, twenty-one male broilers were weighed (0.390 ± 0.004 kg) and transferred to individual metabolic cages. Initially, the broilers were divided into two groups of nine broilers each (group I and II) and one group of three broilers. Group I and II received the feed with labelled amino acids L-(^{15}N) Methionine and L-($^{15}N_2$) Cystine, respectively, whilst group III consisted of a reference group of 3 broilers euthanized with carbon dioxide (compressed cylinder, 100% CO_2) asphyxiation on day zero (14 d posthatch), and all efforts were made to minimize suffering. The gas was brought into the cage with a flow rate of 250 L/min for 15 min. Birds were not anaesthetized prior to euthanasia. Regarding group I, three individual replicate cages were randomly assigned to three different Met:Cys (44:56, 50:50, and 56:44) in diets daily enriched with a fixed-dose of labelled L-Met (60 mmol.kg^{-1}). The same procedure was replicated with broilers of group II, however the labelled amino acid was L-Cys (35 nmol.kg^{-1}). All diets above described consisted of the same treatments used in the dose-response assay. Throughout the 14-d feeding assay broilers were housed into stainless steel metabolic cages (0.5 × 0.5 m) in a controlled environmental room and had ad libitum access to water and feed (mash form). Throughout the assay broilers were exposed to the environmental temperature of 25 °C ± 2 °C and the humidity of 63% ± 2%. Lighting program was set at 18 L:6D (light:dark).

Enrichment with labelled amino acids
Both groups I and II were fed diets containing the Met:Cys ratios of 44:56, 50:50, 56:44, which were the same tested in the performance trial. The difference between the referred groups was the labelled amino acid added to diets. Therefore, broilers from group I, regardless of the Met:Cys tested, received diets enriched with L-(^{15}N) Methionine, whilst diets in group II was enriched L-($^{15}N_2$) Cystine and the enrichment procedure was performed daily by feed with a fixed-dose according to the weight of the bird (60 and 35 mmol.kg^{-1}, respectively). The L-(^{15}N) Met (150.21 g.mol^{-1}, 96–98 atm% abundance, Cambridge Isotope Laboratories®) was diluted in deionized water (100 mL) in a graduated reagent bottle and then, stored under refrigeration at 5 °C. The L-($^{15}N_2$) Cys (242.29 g.mol^{-1}, 98 atm% abundance, Cambridge Isotope Laboratories®), due to its insolubility in water, was weighed on a precision analytical balance and supplemented in the crystalline form. The dose (in μL.kg^{-1}) for each bird was calculated according to the the following Eq. 1:

$$\mathbf{Qtd_{aa}=(D_{aa}{}^{*}BW^{*}MW^{*}Wast)/(1000^{*}C)} \quad (1)$$

where D_{aa} is the labelled amino acid dose relative to the addition of ^{15}N amino acid (μmol); BW is the weight of the animal (kg); MW is the molecular weight of the labelled amino acid; Wast is an increase of 3% assuming a waste of feed by broilers; the value 1000 is used as the conversion factor from micrograms to milligrams; C is the concentration of labelled amino acid (mg.mL^{-1}). It is possible to transform mL into μL after this calculation by multiplying the Qtd_{aa} value by 1000 and, thus determining the dose (μL.kg^{-1}). Initially, the broilers were weighed to obtain the BW value and calculate the amount of labelled amino acid based on Eq. 1. This amount was supplied daily in the feed at various points using micropipettes of variable volume (10 to 100 μL; 100 to 1000 μL and 1000 μL to 10,000 μL) and mixed with a single plastic spoon per bird for a better homogenization. The amounts of labelled Met and labelled Cys were supplied based on [16, 20], respectively. Labelled amino acids were not considered as dietary Met or Cys, both amino acids were provided in a very low concentration.

Measurements
At 14d of age, three broilers were euthanized with carbon dioxide(compressed cylinder, 100% CO_2) asphyxiation, and all efforts were made to minimize suffering. The gas was brought into the cage with a flow rate of 250 L/min for 15 min. Birds were not anaesthetized prior to euthanasia. Enrichment with the labelled amino acid was performed daily until the end of the experimental period (28d of age). At 28d of age, broilers were euthanized with carbon dioxide (compressed cylinder, 100% CO_2) asphyxiation, and all efforts were made to minimize suffering. The gas was brought into the cage with a flow rate of 250 L/min for 15 min. Birds were not anaesthetized prior to euthanasia. The birds were used to determine ^{15}N deposition in body, breast muscle, liver

and feathers. As well as at 14d of age, birds were not anaesthetized prior to euthanasia. Excreta were daily collected and at the end of the assay, a pool for each cage was homogenized to estimate ^{15}N losses. Body samples corresponded to the remaining tissues than those above mentioned (digesta content was manually removed). All tissues were weighed and analyzed separately.

Sample preparation and chemical analysis

The samples of body, breast muscle, liver, feathers and excreta were dried in a freeze-drier (Edwards 501, Thermo) for 72 h, under 800 mbar pressure. All samples were frozen by nitrogen liquid and grounded in cryogenic mill (SPEX SamplePrep 2010 Geno/Grinder 2010) at -196 °C during 3 min to obtain homogeneity [17]. The samples, weighed and packed in tin capsules, were burned in a combustion furnace of the elemental analyzer coupled to the mass spectrometer (Thermo Fisher Scientific, Waltham, MA, USA) to determine the isotope concentration and percentage of nitrogen. The isotope concentrations were expressed in atm%^{15}N, with permissible analytical error of 0.1‰. The determination of labelled Met was performed by the derivatization and normalization procedures in the compound-specific stable isotope analysis of nitrogen in amino acids as described by [45].

Isotope/amino acid recoveries and methionine conversion

The ^{15}N retention ($^{15}N_R$) was calculated based on ^{15}N intake ^{15}N content of breast tissue, liver, feathers, excreta, and the remaining part of carcass as follows:

$$^{15}N_R = DM^{*}\%N^{*}\left[(Ab_p - Ab_{nat})/(Ab_{aa} - Ab_{nat})\right] \quad (2)$$

Where $^{15}N_R$ was the ^{15}N retention (mg) from labelled amino acid; DM was the dry mass of the sample (mg); %N was the nitrogen content (%); Ab_p was the isotope abundance of enriched tissue; Ab_{nat} was the isotope abundance of non-enriched tissue (determined at 14d of age); Ab_{aa} was the isotope abundance of the L-(^{15}N) amino acid.

The same equations were applied for the CSIA results, only replacing $\delta^{15}N$ by methionine-specific $\delta^{15}N$, as presents in Eq. 3:

$$^{15}N\text{-}Met_R = DM^{*}\%Met^{*}\left[(Ab_{pCSIA} - Ab_{natCSIA})/(Ab_{aa} - Ab_{natCSIA})\right] \quad (3)$$

Where ^{15}N-Met_R was the ^{15}N-Met retention (mg) from L-(^{15}N) Met; DM was the dry mass of the sample (mg); %Met was the Met content (%); Ab_{pCSIA} was the ^{15}N-Met abundance of enriched tissue; $Ab_{natCSIA}$ was the ^{15}N-Met abundance of non-enriched tissue (determined at 14d); Ab_{aa} was the isotope abundance of the L-(^{15}N) Met.

The amino acid retention was calculated using the aminogram results (CBO lab, Brazil), according to the Eq. 4:

$$AA_R = (DM^{*}\%AA) - AA_Ri \quad (4)$$

Where AA_R was the amino acid retention; DM was the dry mass of the sample (mg); %AA was the amino acid content (%) and AA_Ri was the amino acid retention before the experimental period.

Once calculated, the retention rates were used to calculate recovery rates of Met and Cys. The isotope/amino acid recoveries values were determined by dividing the total isotope/amino acid retention by the total isotope/amino acid intake.

$$Rec = (X_R/X_I) * 100 \quad (5)$$

Where Rec is the recovery rate; X_R is the retention of Met or Cys calculated based on the Eqs. 2, 3, or 4 and X_I is the intake of ^{15}N, methionine-specific ^{15}N, or Met and Cys.

The conversion of Met into intermediates and Cys ($Met_{Cys+interm}$) was calculated based on the difference between the recovery rates of ^{15}N ($^{15}N_{Rec}$ (Met)) and methionine-specific ^{15}N (^{15}N-Met_{Rec}) as follows:

$$Met_{Cys+interm} = {}^{15}N_{Rec}\,(Met) - ({}^{15}N - Met_{Rec}) \quad (6)$$

The amount of Met converted into Cys (Cys_{Met}) were calculated based on the difference between the recovery of Cys using the aminogram analysis (AA_{Rec} (Cys)) and the recovery of ^{15}N from Cys ($^{15}N_{Rec}$ (Cys)) as follows:

$$Cys_{Met} = AA_{Rec}\,(Cys) - {}^{15}N_{Rec}\,(Cys) \quad (7)$$

The amount of Met converted into intermediates of Met cycle ($Met_{interm.}$) was calculated based on the difference between the conversion of Met into intermediates and Cys ($Met_{Cys+interm}$) detailed in Eq. 6 and the amount Met converted into Cys (Cys_{Met}) as detailed in Eq. 7:

$$Met_{interm.} = Met_{Cys+interm.} - Cys_{Met} \quad (8)$$

Abbreviations

BWG: Body weight gain; Cys: Cystine; FCR: Feed conversion ratio; FI: Feed intake; Met: Methionine; TSAA: Total sulfur amino acids

Acknowledgments

Not applicable.

Funding

This study was financially supported by the São Paulo Research Foundation – FAPESP (BRAZIL) by financial support grant number n° 2015/25717-0 (financial support to LGP during the research) and 2013/25761-4, wich was responsible for funding the acquisition of the birds and ingredients used in this research and the ^{15}N analysis at Stable Isotopes Facitiy, Davis, California. Ajinomoto Biolatina and Evonik Industries for donating the amino acids used in the current research and for the analysis of amino acid content in diets.

Authors' contributions

Conceptualization: LGP; NKS; JCPD; RMS; Data acquisition: LGP; RMS; GSV; Data analysis: LGP; GSV; JCPD; JVM; RMS; Design of Methodology: LGP; NKS; JCPD; JCD; RMS; Writing and editing: LGP; RMS; GSV; JCPD; JVM. All authors been involved in drafting the manuscript and gave final approval of the version to be published evaluated the manuscript.

Ethics approval

All animal care procedures were approved by the Animal Care and Use Committee of the Faculdade de Ciências Agrárias e Veterinárias (UNESP, Jaboticabal, Brazil) under protocol n.º 9999/14, prior to the beginning of the trials, including euthanasia by carbon dioxide asphyxiation, where all efforts were made to minimize suffering. Birds were not anaesthetized prior to euthanasia. The birds used in current research were obtained from a local commercial hatchery (Pluma Agroavicola Hatchery, Descalvado, São Paulo State, Brazil). The hatchery is facility, whose unique purpose is obtaining profit from the sale of broiler chicks, which may be used either for research or commercial purposes. The São Paulo State Research Foundation was responsible for funding the acquisition of the birds used in this research (Process number - 2013/25761–4).

Competing interests

The authors declare that they have no competing interests.

Author details

[1]Department of Animal Science, Unesp, Jaboticabal, SP, Brazil. [2]Evonik Nutrition & Care GmbH, Hanau, Germany. [3]INRA, Agrocampus Ouest, Rennes, France. [4]Department of Animal Science, Unesp, Botucatu, SP, Brazil.

References

1. Corzo A, Kidd MT, Dozier WA, Shack LA, Burgess SC. Protein expression of pectoralis major muscle in chickens in response to dietary methionine status. Br J Nutr. 2006;95(4):703–8.
2. Rostagno HS, Albino LFT, Hannas MI, Donzele JL, Sakomura NK, Perazzo FG, Saraiva A, Abreu MLT De, Rodrigues PB, Oliveira RF De, Barreto SLT, Brito CO. Brazilian tables for poultry and swine: composition of feedstuffs and nutritional requirements, 4 ed. Viçosa: Departamento de Zootecnia, UFV; 2017. p. 488.
3. Heger J, Phung TV, Křížová L. Efficiency of amino acid utilization in the growing pig at suboptimal levels of intake: lysine, threonine, sulfur amino acids and tryptophan. J Anim Physiol Anim Nutr. 2002;86(5–6):153–65.
4. Houdijk JGM, Athanasiadou S. Direct and indirect effects of host nutrition on ruminant gastrointestinal nematodes. In VI International Symposium on the Nutrition of Herbivores (Vol. 213): Universidad Autónoma de Yucatán, Mérida; 2003.
5. Labier M, Leclercq B. Nutrition and feeding of poultry. Nottingham: Nottingham University Press; 1992. p. 305.
6. Lehninger AL, Nelson DL, Cox MM. Principles of biochemistry. 3rd ed. São Paulo: Sarvier; 2011. p. 975.
7. Mcdevitt RM, Mack S, Wallis IR. Can betaine partially replace or enhance the effect of methionine by improving broiler growth and carcase characteristics? Br Poult Sci. 2000;41(4):473–80.
8. Bannai S, Tateishi N. Role of membrane transport in metabolism and function of glutathione in mammals. J Membr Biol. 1986;89(1):1–8.
9. Barone JR, Schmidt WF, Liebner CFE. Compounding and molding of polyethylene composites reinforced with keratin feather fiber. Compos Sci Technol. 2005;65(3,4):683–92.
10. Block RJ, Bolling D. A note on the amino acids yielded by yeast; sunflower seed meal, and sesame seed after hydrolysis of the fat free tissue. Arch Biochem. 1945;6:277–9.
11. Kuppuswamy S, Srinivasan M, Subrahmanyan V. Proteins in foods. Mysore City: KA Korula At The Wesley Press; 1958.
12. Emmans GC, Fisher C. Problems in nutritional theory. Nutrient requirements of poultry and nutritional research. 1986;1:9–39.
13. Talpaz H, De La Torre JR, Sharpe PJH, Hurwitz S. Dynamic optimization model for feeding of broilers. Agric Syst. 1986;20(2):121–32.
14. Ortiz LT, Rebolé A, Alzueta C, Rodríguez ML, Treviño J. Metabolisable energy value and diges-tibility of fat and fatty acids in linseed determined with growing broiler chickens. Br Poult Sci. 1994;42(1):57–63.
15. Baker DH. Comparative species utilization and toxicity of sulfur amino acids. J Nutr. 2006;136(6):1670S–5S.
16. Thwaites DT, Anderson CMH. H+–coupled nutrient, micronutrient and drug transporters in the mammalian small intestine. Exp Physiol. 2007;92(4):603–19.
17. Schoenheimer R, Ratner S, Rittenberg D. Studies in protein metabolism VII. The metabolism of tyrosine. J Biol Chem 1939. 2015;127(1):333–44.
18. Garlich JD. Response of broilers to DL-methionine hydroxy analog free acid, DL-methionine, and L-methionine. Poult Sci. 1985;64(8):1541–8.
19. Pacheco LG, Sakomura NK, Suzuki RM, Ribeiro HV, Gonçalves CA, Leme BB, Troni AR. 15N Cystine deposition in broilers. Reunião Anual da Sociedade Brasileira de Zootecnia, 54. 2017.
20. Shen YB, Ferket P, Park I, Malheiros RD, Kim SW. Effects of feed grade L-methionine on intestinal redox status, intestinal development, and growth performance of young chickens compared with conventional DL-methionine. J Anim Sci. 2015;93(6):2977–86.
21. Licatti F. Isótopos estáveis do carbono ($^{13}C/^{12}C$) em plantas do ciclo bioquímico C_3 e C_4. [Monografia]. Botucatu (SP): Universidade Estadual Paulista; 1997.
22. Yarnes CT, Herszage J. The relative influence of derivatization and normalization procedures on the compound-specific stable isotope analysis of nitrogen in amino acids. Rapid Commun Mass Spectrom. 2017;31(8):693–704.
23. Bauriedel WR. The effect of feeding D-methionine on the D-amino acid oxidase activity of chick tissues. Poult Sci. 1963;42(1):214–7.
24. Marrett LE, Sunde ML. The effect of other D amino acids on the utilization of the isomers of methionine and its hydroxy analogue. Poult Sci. 1965; 44(4):957–64.
25. Bhargava KK, Hanson RP, Sunde ML. Effects of methionine and valine on antibody production in broilers infected with new-castle disease virus. J Nutr. 1970;100:241–8.
26. Bhargava KK, Hanson RP, Sunde ML. Effects of threonine on growth and antibody production in broilers infected with Newcastle disease virus. Poult Sci. 1971;50(3):710–3.
27. Sunde ML. Zinc requirement for normal feathering of commercial leghorn-type pullets. Poult Sci. 1972;51(4):1316–22.
28. Katz RS, Baker DH. Methionine toxicity in the chick: nutritional and metabolic implications. J Nutr. 1975;105(9):1168–75.
29. Park I, Pasquetti T, Malheiros RD, Ferket PR, Kim SW. Effects of supplemental L-methionine on growth performance and redox status of Turkey poults compared with the use of DL-methionine. Poult Sci. 2017;97(1):102–9.
30. Kong C, Park CS, Ahn JY, Kim BG. Relative bioavailability of DL-methionine compared with L-methionine fed to nursery pigs. Anim Feed Sci Technol. 2016;215:181–5.
31. Fisher C. Responses of laying hens to amino acids. Amino acids in farm animal nutrition. 1994;1:245–80.
32. Esteve-Garcia E, Khan DR. Relative bioavailability of DL and L-methionine in broilers. Open J Anim Sci. 2018;8(02):151.
33. Graber G, Baker DH. Sulfur amino acid nutrition of the growing chick: quantitative aspects concerning the efficacy of dietary methionine, cysteine and cystine. J Anim Sci. 1971;33(5):1005–11.
34. Lu SC. Glutathione synthesis. Biochimica et Biophysica Acta (BBA)-General Subjects. 2013;1830(5):3143–53.
35. Lopez-Coello C. Potential causes of broiler feathering problems. St. Louis, MO: Feathering Manual. Novus International; 2003. p. 1–46.
36. Alirezaei M, Jelodar G, Niknam P, Ghayemi Z, Nazifi S. Betaine prevents ethanol-induced oxidative stress and reduces total homocysteine in the rat cerebellum. J Physiol Biochem. 2011;67(4):605–12.
37. Zeng Z, Zhang S, Wang H, Piao X. Essential oil and aromatic plants as feed additives in non-ruminant nutrition: a review. J Anim Sci Biotechnol. 2015;6(1):7.
38. Wylie LM, Robertson GW, Hocking PM. Effects of dietary protein concentration and specific amino acids on body weight, body composition and feather growth in young turkeys. Br Poult Sci. 2003;44(1):75–87.
39. Lerner J, Taylor MW. A common step in the intestinal absorption mechanisms of D-and L-methionine. Biochimica et Biophysica Acta (BBA)-Biomembranes. 1967;135(5):991–9.
40. Hirakawa DA, Baker DH. Sulfur amino acid nutrition of the growing puppy:

determination of dietary requirements for methionine and cystine. Nutr Res. 1985;5(6):631–42.

41. Gietzen DW, Hao S, Anthony TG. Mechanisms of food intake repression in indispensable amino acid deficiency. Annu Rev Nutr. 2007;27:63–78.
42. Sell DR, Featherston WR, Rogler JC. Methionine-cystine interrelationships in broilers and rats fed diets containing suboptimal levels of methionine. Poult Sci. 1980;59(8):1878–84.
43. Cobb-Vantress I. Cobb 500 breeder management guide. 2015.
44. Robbins KR, Saxton AM, Southern LL. Estimation of nutrient requirements using broken-line regression analysis. J Anim Sci. 2006;84(13):E155–65.
45. Muramatsu T, Hiramoto K, Tasaki I, Okumura J. Effect of protein starvation on protein turnover in liver, oviduct and whole body of laying hens. Comp Biochem Physiol Jpn. 1987;87(2):227–32.

8

Evaluation of hepatic lipogenesis and antioxidant status of broiler chickens fed mountain celery

Behnam Ahmadipour[1], Hossein Hassanpour[2] and Fariborz Khajali[1*]

Abstract

Background: Fatness is an unwanted side effect of genetic selection in broiler chickens. In this study, we introduce mountain celery powder as a feed supplement to suppress lipogenesis and improve antioxidant status in broiler chickens. Male broiler chicks (Ross 308) were fed a control diet or a diet that includes mountain celery (MC) at 7.5 gkg^{-1}over 42 days.

Results: Body weight gain and feed conversion ratio significantly ($P < 0.05$) improved in chicks fed MC. A highly significant down-regulation of genes involved in hepatic lipogenesis including acetyl CoA carboxylase (ACC), fatty acid synthase (FAS), malic enzyme (ME), and lipoprotein lipase (LPL) was observed in the liver of chickens fed MC. These birds, however, had greater compensatory upregulation in antioxidative genes SOD1 and catalase in the liver compared to the birds that received the control diet. Birds received MC had significantly lower level of lipid peroxidation (1.59 μmol/L serum malondialdehyde) compared to birds from the control group (3.57 μmol/L; $P = 0.0024$). Birds fed MC had significantly ($P < 0.05$) lower circulatory concentrations of triacylglycerols, cholesterol, and LDL but higher concentrations of HDL. Relative liver weight and abdominal fat deposition were significantly reduced by feeding MC.

Conclusions: It can be concluded that feeding birds MC significantly suppresses hepatic lipogenesis by down-regulating key hepatic lipogenic enzyme genes and boosts antioxidant capacity by up-regulating hepatic antioxidantive genes.

Keywords: Antioxidant, Chicken, Celery, Hepatic lipogenesis

Background

Over the past several decades, intensive genetic selection in broiler chickens for rapid growth has led to their extraordinarily high growth rate. From 1957 to 2005, the body weight of broilers experienced a four-fold increase with a compounded annual growth rate of 3.30%. Concurrently, the 42-day feed conversion ratio (FCR) of chickens decreased by 2.55% each year over the same period [1]. The main side effect of such enormous growth rate is excessive fat accumulation in the body [2]. Modern broilers tend to mature more quickly and begin depositing fat at an earlier age. The chicken meat industry now faces the challenge of controlling excessive fat. Body fat in broiler chickens are generally not physiologically essential for body functions but impose a heavy burden on the antioxidant system [3] as majority of the fat is unsaturated, which is susceptible to oxidation [4]. Lipid peroxidation is a major cause of poor food quality and generation of off-odours and off-flavours. The incorporation of antioxidants is the most effective way to control the rate and extent of lipid peroxidation [5]. Therefore, dietary antioxidants in chickens play a pivotal role in protecting unsaturated fats as well as the bird's body during digestion and metabolism, thereby the producion of high-quality meat.

Mountain celery (MC; *Kelussia odoratissima*; Umbelliferae family) is a rich source of polyphenols and flavonoids, and has antihyperlipidemic and antioxidant properties [6, 7]. Mountain celery is a self-growing plant, growing up to 200 cm in height and is found exclusively

* Correspondence: khajali@agr.sku.ac.ir
[1]Department of Animal Science, Shahrekord University, Shahrekord 88186-34141, Iran
Full list of author information is available at the end of the article

in Iran. Flavonoids and polyphenols are strong natural antioxidants with lipid-lowering capacity [5, 8]. The essential oil of MC is also a rich source of phthalides (37.3% of total bioactive components) [9], which are reported to be hypolipidemic, hypoglycemic, hypotensive, and antioxidant. The antioxidant potency of MC is has previously been described as comparable to α-tocopherol and butylated hydroxytoluene (BHT) [10]. Feeding MC to broiler chickens at 5 and 7.5 gkg^{-1} resulted in a significant improvement in growth performance and was associated with lower rate of mortality from right ventricular hypertrophy [11]. Muscle relaxing and sedative effects of MC have been reported in mice [12]. The present study aimed to investigate antihyperlipidemic and antioxidant effects of MC in broiler chickens.

Methods

Birds and experimental facility

The experiment was conducted at the Poultry Research Center of Shahrekord University, Shahrekord, Iran (altitude 2100 m) in accordance to the recommendations of the Guide for the Care and Use Committee of Shahrekord University. All procedures were conducted in accordance with guidelines set out by Directive 2010/63/EU in Europe, and were approved by the Ethical Committee of Shahrekord University Research Council.

A total of 250 day-old male broilers (Ross 308; average body weight: 44.5 g ± 1.25) were obtained from Behjoojeh Co., Shahrekord, Iran. A total of 200 chicks within a range of 43 to 47 g (i.e within ±2 SD) were randomly selected and assigned to 10 pens, each with 20 chicks. All pens had similar initial weights. Five of these pens were randomly assigned to each treatment. No statistical power calculation was conducted prior to the study. The sample size was based on our previous experience with similar design. All pens were equipped with a bell dinker and trough feeder. All chicks were allowed free access to feed and water and were provided with 23 L:1D lighting program throughout the trial. The house temperature was conformed to the strain recommendations.

Treatments

A control diet based on corn and soybean meal was formulated according to the NRC recommendations [13] for the starting (1 to 3 wk. of age) and growing (3 to 6 wk. of age) stages (Table 1). The experimental diet was prepared by substituting wheat bran with 7.5 gkg^{-1} MC, a concentration of MC that has previously been described as the optimal level for broiler chickens [11]. Diets were provided in mash form. Mountain celery (the leaves and shoots) were cut, air-dried, and ground for use in the experimental diets. Chemical composition of the samples as gkg^{-1} DM include: protein 90, dietary fiber 75, calcium 18, phosphorous 6, sodium 0.4, chloride 2.8, potassium 20, polyphenols 102, and flavonoids 12.2. The investigators were blinded to group allocation during the farm trial and laboratory analyses.

Measurements

Following the procedures established by the Association of Official Analytical Chemists (AOAC) [14], the experimental diets were analyzed for crude protein (Method 976.06), crude fiber (Method 978.10), calcium (Method 910.01), phosphorous (Method 966.01), sodium (Method 929.03), chloride (Method 915.01), and potassium (Method 956.01). The total phenolic content was determined in MC using the Folin–Ciocalteu method. In brief, 0.5 ml of each sample was mixed with 2.5 ml of Folin-Ciocalteu's phenol reagent and kept for 5 min at 37 °C. Then, 2 ml of saturated Na_2CO_3 (7.5%) was added and the mixture was brought up to 10 ml using deionized-distilled water. The mixture was maintained at room temperature in the dark for 120 min before measuring its absorbance at 765 nm against a reagent blank. Gallic acid equivalent (GAE) was used as the reference standard and total phenolic content was expressed as mg of GAE per gram of each essential oil on the dry basis. Total flavonoid content was measured using the aluminum chloride colorimetric assay. An aliquot (1 ml) of sample extracts or quercetin standard solutions (20, 40, 60, 80 and 100 μg/ml) was added to a 10 ml volumetric flask containing 4 ml of distilled water. Then, 0.30 ml of 5% $NaNO_2$ was added, followed by the addition of 0.3 ml of 10% $AlCl_3$ after 5 min. Subsequently, 2 ml of 1 M NaOH was added and the volume was brought up to 10 ml with distilled water. The absorbance was measured against a blank at 510 nm. The total flavonoid content was expressed as mg quercetin equivalents.

Body weight gain and feed intake were recorded and feed conversion ration (FCR) was calculated for the 1- to 21-day, 21- to 42-day, and 1- to 42-day periods. At 42 days of age, 10 birds per treatment were selected for blood collection and processing. The selected birds had body weights within ~5% of the average pen body weight. Blood samples (3 mL) were collected from the brachial vein and centrifuged at 2500×g for 10 min to obtain serum for the determination of cholesterol, triglycerides (TAG), low-density lipoproteins (LDL), high-density lipoproteins (HDL) and malondialdehyde (MDA). Total cholesterol, TAG, LDL, and HDL were measured using commercial laboratory kits following the manufacturer's manuals (Pars Azmoon Kits; Pars Azmoon, Tehran, Iran). Serum MDA concentration, a biomarker of lipid peroxidation, was assayed following the method of Nair and Turner [15]. An aliquot of blood was also obtained on glass slides to prepare a blood smear to determine differential leukocyte count. The blood was stained using the May-Grunwald and Giemsa protocol, and 100 leukocytes,

Table 1 Composition of the basal diet to which mountain celery (MC) added for broiler chickens during starter (1–21 days) and grower (22-42 days) stages

Item (% unless noted)	Starter (Control)	Starter (MC diet)	Grower (Control)	Grower (MC diet)
Corn	46.85	46.85	55.75	55.75
Soybean meal (44% CP)	37.50	37.50	32.90	32.90
Fish meal (60% CP)	3.65	3.65	1.00	1.00
Soy oil	7.50	7.50	6.00	6.00
Dicalcium phosphate	1.30	1.30	1.20	1.20
Oyster shell	1.45	1.45	1.50	1.50
Salt	0.35	0.35	0.30	0.30
DL-Methionine	0.15	0.15	0.10	0.10
L-Lysine	–	–	–	–
Mineral supplement[a]	0.25	0.25	0.25	0.25
Vitamin supplement[b]	0.25	0.25	0.25	0.25
Wheat bran[c]	0.75	–	0.75	–
Mountain celery	–	0.75	–	0.75
Calculated composition				
AMEn (kcal/kg)	3200	3200	3200	3200
CP	23	23	20.00	20.00
Met	0.50	0.50	0.40	0.40
Met + Cys	0.85	0.85	0.75	0.75
Lys	1.30	1.30	1.08	1.08
Thr	1.05	1.05	0.93	0.93
Arg	1.50	1.50	1.31	1.31
Ca	1.03	1.03	0.90	0.90
Available P	0.46	0.46	0.36	0.36
Na	0.18	0.18	0.16	0.16
Cl	0.27	0.27	0.29	0.29
K	0.91	0.91	0.92	0.92
Na + K-Cl (mEq/kg)	233	233	236	236

[a]Provided the following per kilogram of diet: vitamin A (*trans* retinyl acetate), 3600 IU; vitamin D3 (cholecalciferol), 800 IU; vitamin E (dl-α-tocopheryl acetate), 7.2 mg; vitamin K3, 1.6 mg; thiamine, 0.72 mg; riboflavin, 3.3 mg; niacin, 0.4 mg; pyridoxin, 1.2 mg; cobalamine, 0.6 mg; folicacid, 0.5 mg; choline chloride, 200 mg
[b]Provided the following per kilogram of diet: Mn (from MnSO4 H2O), 40 mg; Zn (from ZnO), 40 mg; Fe (from FeSO4 7H2O), 20 mg; Cu (fromCuSO4 5H2O), 4 mg; I [from Ca(IO3)2 H2O], 0.64 mg; Se (from sodium selenite), 0.08 mg
[c]*mountain celery* powder was replaced for wheat bran to provide the level of 7.5 g/kg

including granular (heterophils) and nongranular (lymphocytes) were counted to calculate the heterophil to lymphocyte ratio (H:L). All chemical reagents were obtained from Sigma–Aldrich Co. (Sigma–Aldrich Co., St. Louis, MO, USA).

After collecting blood, the birds were euthanized by CO_2 inhalation and we collected live body weight, liver weight, and abdominal fat deposition. The livers were harvested and immediately frozen in liquid nitrogen and stored at − 70 °C for RNA analyses.

Quantitative real-time PCR analysis

Total RNA was extracted from the liver using RNX-Plus reagent (Sinaclon Bioscience, Tehran, Iran) and homogenized in digestion buffer. The liver was homogenized in digestion buffer, then mixed with chloroform and centrifuged. Total RNA in the upper aqueous phase was precipitated with isopropanol, rinsed with 75% ethanol, then resuspended in DEPC-treated water. To remove residual DNA, the RNA was treated with DNase (Sinaclon Bioscience, Tehran, Iran). The RNA was measured using a spectrophotometer. Only RNA with an absorbance ratio (A260/A280) > 1.9 was used for the synthesis of cDNA. Total RNA was reverse transcribed into cDNA using PrimeScript ™ RT Reagent Kit (Takara Bio Inc., Japan). The reverse transcription mix was heated to 85 °C for 5 s to inactivate reverse transcriptase and to denature the RNA and then stored at − 20 °C.

The levels of acetyl CoA carboxylase (ACC, EC 6.4.1.2), fatty acid synthase (FAS, EC 2.3.1.85), malic enzyme (ME, EC 1.1.1.40), lipoprotein lipase (LPL, EC 3.1.1.34), superoxide dismutase 1 (SOD1, EC 1.15.1.1), catalase (CAT, EC 1.11.1.6) and β-actin transcripts were determined by real-time reverse transcriptase (RT)-PCR using SYBR® Premix Ex Taq™ II (TliRnase H Plus; Takara Bio Inc., Japan). We used β-actin as an endogenous standard to normalize the input load of cDNA among samples. Specific primers of the transcripts were designed with Primer-Blast (www.ncbi.nlm.nih.gov/tools/primer-blast/index.cgi?LINK_LOC = BlastHome) are detailed in Table 2. PCRs were carried out in a real-time PCR cycler (Rotor Gene Q 6000, Qiagen, USA) in three replicates for each liver sample. One microliter cDNA was added to 10 μl of SYBR® Premix Ex Taq II Mix with 1 μM of each specific primer (total volume of 20 μl). The thermal profile was 95 °C for 30 s, 40 cycles of 94 °C for 40 s, 64 °C for 35 s and 72 °C for 30 s. Fluorescence was measured at the end of each phase. Gene expression data were normalized to β-actin. Data were analyzed using LinRegPCR software version 2012.0 (Amsterdam, Netherland) to obtain the threshold cycle number and reaction efficiency [16]. Relative transcript levels and fold changes in transcript abundance were calculated using efficiency-adjusted Paffl methodology [17].

Statistical analysis

Results were analyzed by ANOVA using SAS (2007) software in a completely randomized design. Treatment means were separated by t-test. In the cases of sampling within pens, data were subjected to a nested design.

Results

Mountain celery analysis shows remarkable contents of polyphenols (102 mg/g), flavonoids (12.2 mg/g) and phtalides, mainly ligustilides. Ligustilides accounted for 50% of total active products in the mountain celery.

Body weight gain and FCR responses were significantly ($P < 0.05$) improved in broilers when MC was included in diets during all feeding stages. However, feed intake only showed a significant decrease in the starter stage (Table 3).

The expression of genes responsible for hepatic lipogenesis, including ACC, FAS, ME, and LPL, significantly down-regulated in birds fed MC (Table 4). In contrast, antioxidant genes including SOD1 and CAT have been highly over-expressed in broilers fed MC (Table 4).

Broilers that received MC had significantly ($P < 0.05$) lower concentrations of total TAG, total cholesterol, LDL, malondialdehyde, and heterophil to lymphocyte ratio. Feeding MC caused a significant ($P < 0.05$) increase in HDL compared to the control group (Table 5).

Discussion

In general, live performance of the broilers in this study was lower compared to the expected performance outlined in the strain catalog (ROSS 308). This is due to the fact that the birds reared at high altitude region (2100 m), where atmospheric oxygen concentration was only 15–16%. There is a direct relationship between broiler growth performance and atmoshpheric oxygen concentration [3]. Significant improvements in body weight gain and FCR associated with feeding birds MC are in line with results reported by Ahmadipour et al. [11]. Improved performance can be attributed to the naturally-occurring polyphenols in MC, which include flavonoids and non-flavonoids that exhibit a broad spectrum of beneficial biological properties such as growth-promoting, antioxidative, and immunocompetence [8, 18]. Dietary inclusion of MC was

Table 2 Details of the primers used for quantitative real time PCR analysis of chicken mRNAs

Target	Target Primers	PCR product (bp)	Accession No.
β-Actin	F:5′-AGCGAACGCCCCCAAAGTTCT-3′	139	NM_205518.1
	R:5′-AGCTGGGCTGTTGCCTTCACA-3′		
FAS	F:5′-GCTGGCTACAGTGGTGGACT-3′	129	NM_205155.2
	R:5′-CCACCTCGAACCACCAAAGC-3′		
LPL	F:5′-CTCCGATCCCGAAGCTGAGATG-3′	106	NM_205282.1
	R:5′-CTGTCCAGGAACCAGGTAGC-3′		
ACC	F:5′-CAACGAGTCGGGCTACTACC-3′	146	J03541.1
	R:5′-GGTCCTTGGTCACGTATGGG-3′		
MAE	F:5′-GGACCTCGAAGCCTTCATCC-3′	77	AF408407.1
	R:5′-GCCAGGTGTATGAGTCAGCC-3′		
CAT	F:5′-TGGCGGTAGGAGTCTGGTCT-3′	112	NM_001031215.1
	R:5′-GTCCCGTCCGTCAGCCATTT-3′		
SOD1	F:5-CACTGCATCATTGGCCGTACCA-3′	223	NM_205064.1
	R:5-GCTTGCACACGGAAGAGCAAGT-3′		

Abbreviations: *FAS* fatty acid synthase, *LPL* lipoprotein lipase, *ACC* acetyl-coA carboxylase, *MAE* malic enzyme, *CAT* catalase, *SOD1* superoxide dismutase 1

Table 3 Effects of dietary levels of mountain celery on broiler growth performance

Variable	Control	Mountain celery (7.5 gkg^{-1})	*P*-value
Feed intake (g/bird)			
1–21 days of age	1052 ± 8.4^a	1007 ± 10.9^b	0.0127
21–42 days of age	2873 ± 44.3	2839 ± 32.4	0.5499
1–42 days of age	3925 ± 36.1	3846 ± 28.5	0.1262
Weight gain (g/bird)			
1–21 days of age	664 ± 8.6^b	704 ± 7.3 ^{a}	0.0081
21–42 days of age	1334 ± 18.3^b	1488 ± 7.1 ^{a}	0.0001
1–42 days of age	1998 ± 10.3^b	2172 ± 20.3 ^{a}	0.0001
Feed conversion ratio			
1–21 days of age	1.58 ± 0.009^a	1.44 ± 0.022^b	0.0002
21–42 days of age	2.15 ± 0.008 ^{a}	1.91 ± 0.028^b	0.0001
1–42 days of age	1.96 ± 0.011 ^{a}	1.77 ± 0.026^b	0.0001

Means in the same row with different letters are significantly different
Values are mean ± SE (*n* = 5)

associated with down-regulation of ACC, FAS, ME, and LPL in the liver. In avian species, the liver is the principal site of lipogenesis with a capacity 20 times greater than adipose tissue of equal weight [19]. The pathways involved in fatty acid synthesis in birds involve ACC catalyzing the conversion of acetyl-CoA to malonyl-CoA, and FAS catalyzing the formation of palmitate from one molecule of acetyl-CoA and 7 molecules of malonyl-CoA [20]. ME catalyzes the oxidative decarboxylation of malate to pyruvate and carbon dioxide, simultaneously generating nearly all of 14 molecules of NADPH required for palmitate biosynthesis [21]. Upon the biosynthesis of palmitate, fatty acids are transported to adipose tissue as TAG in the VLDLs, where extracellular LPL catalyses the hydrolysis of TAG so that fatty acids can be incorporated into the adipose tissue [22]. Down-regulation of these four key lipogenic enzymes by MC strongly suggests antihyperlipidemic effects as reflected in reduced levels of cholesterol, TAG, and LDL. The results of our study strongly suggest a strong nutrigenomic effect of MC on lipogenesis in birds, which may be associated with flavonoids and phthalides. A report on transcription-factor pathways mediating nutrient–gene interactions indicated ER (estrogen receptor), NFkB (nuclear factor kB), and AP1 (activating protein 1) as transcription factors associated with flavonoids [23]. However, transcription factor associated with phthalide has yet to be determined. Mountain celery is a rich source of polyphenols (102 mg/g), flavonoids (12.2 mg/g) and phtalides, manily ligustilide.

Aqueous extract of celery has been reported to reduce total cholesterol and LDL in rats by lowering hepatic LPL activity [24] – an antihyperlipidemic effect linked with phthalides. The hypolipidemic effects of flavonoids have also been previously documented [25]. Bok et al. [26] indicated that polyphenols reduce the activity of HMG-CoA reductase, a key enzyme in the synthesis of cholesterol in mice. It has also been reported that polyphenols alter cholesterol homeostasis and increase LDL-receptor activity in human, which can increase LDL levels [27].

Stimulation of expression of SOD1 and CAT by MC suggests that the herb improves antioxidant defense system in birds, which could reduce levels of malondialdehyde (MDA). Previous research has shown that phtalides

Table 4 Effect of dietary mountain celery on lipogenic and antioxidant gene expression in broiler chickens measured at 42 days of age

Variable	Control	Mountain celery (7.5 gkg^{-1})	*P*-value
Acetyl CoA carboxylase	0.025 ± 0.007^a	0.003 ± 0.001^b	0.0088
Fatty acid synthase	0.071 ± 0.018^a	0.011 ± 0.002^b	0.0049
Lipoprotein lipase	0.012 ± 0.004^a	0.002 ± 0.001^b	0.0189
Malic enzyme	0.029 ± 0.009^a	0.002 ± 0.001^b	0.0111
Superoxide dismutase 1	0.0006 ± 0.0001^b	0.014 ± 0.002^a	0.0001
Catalase	0.0005 ± 0.0001^b	0.005 ± 0.001^a	0.0007

^{a,b,c}Means in the same row with different letters are significantly different
Values are relative quantitative PCR (Target/β-actin)
Values are mean ± SE (*n* = 10)

Table 5 Effect of dietary mountain celery on serum, blood, and carcass variables in broiler chickens measured at 42 days of age

Variable	Control	Mountain celery (7.5 gkg^{-1})	*P*-value
Total cholesterol (mg/dL)	126 ± 5.7^a	109 ± 5.1^b	0.0006
Total triacylglycerols (mg/dL)	92.5 ± 5.30^a	78.6 ± 3.80^b	0.0003
HDL (mg/dL)	30.6 ± 2.75	38.6 ± 5.0	0.0031
LDL (mg/dL)	34.4 ± 2.93^a	28.3 ± 1.17^b	0.0001
Malonedialdehyde (μmol/L)	3.57 ± 1.31^a	1.59 ± 0.77^b	0.0024
Heterophils to Lymphocyte (%)	0.84 ± 0.16^a	0.59 ± 0.12^a	0.0021
Abdominal fat deposition (%)	1.80 ± 0.27^a	1.20 ± 0.15^b	0.0001
Relative weight of liver (%)	2.25 ± 0.21^a	1.87 ± 0.19^b	0.0001

[a,b]Means in the same row with different letters are significantly different
Values are mean ± SE ($n = 10$)

decrease MDA levels and increase the activities of the antioxidant enzymes such as glutathione peroxidise (GSH-Px) and SOD in mice [28]. Ligustilides (the major components of phtalides in MC) have been reported to increase the activities of GSH-Px and SOD, which can prevent oxidative stress [29]. MDA is a biomarker of lipid oxidation is an index of oxidative stress in fat tissues. Decreased oxidative stress in birds fed MC can be associated with a significant reduction in the H:L ratio, an index of stress in the chicken [30]. Taken together, feeding MC can suppress ROS production and alleviate oxidative stress in birds, and can be reflected by lower levels of MDA and H:L ratio.

Antihyperlipidemic effects of MC include reduced deposition of fat in the abdominal cavity. Abdominal fat is a crucial component of lipogenesis in birds because it grows more rapidly compared to other adipose tissues and is highly correlated to total body fat content in avian species. As it is easy to dissect and weigh, abdominal fat weight has been used extensively as an estimator of total body fat content in the chicken [31]. Reduction in abdominal fat deposition in birds fed MC indicates the strong antihyperlipidemic potency of the herb. Reduction in the proportion of liver weight to live body weight in chickens fed MC was in line with decreased lipogenesis associated with reduced abdominal fat. As the liver is the principal site of lipogenesis in chickens [19], a decline in its relative weight can account for lower lipogenesis associated with dietary inclusion of MC.

Conclusion

Feeding MC (*Kelussia odoratissima*) to broiler chickens led to a significant decrease in the expression of lipogenic enzymes and a significant increase in the expression of antioxidant enzymes, suggesting that this medicinal herb has potential to produce high quality low-fat chickens. These nutrigenomic effects are attributed to the phthalides, flavonoids, and polyphenols naturally in MC.

Abbreviations

ACC: Acetyl CoA carboxylase; cDNA: Complementary DNA; DEPC-reated water: Diethylpyrocarbonate-treated water; FAS: Fatty acid synthase; H:L: Heterophil to lymphocyte; HDL: High-density lipoproteins; LDL: Low-density lipoproteins; LPL: Lipoprotein lipase; MC: Mountain celery; MDA: Malondialdehyde; ME: Malic enzyme; PCR: Polymerase chain reaction; TAG: Triacylglycerols

Acknowledgements

This study was financially supported by Shahrekord University, Shahrekord, Iran. The authors would like to sincerely appreciate Sherry Du from McMaster University, Hamilton, Canada for editing the manuscript.

Funding

This study was financially supported by Shahrekord University, Shahrekord, Iran.

Authors' contributions

Study design: FK; Sample collection and chemical analysis: BA; Field and laboratory works: BA, FK; Data analysis: FK, HH; Manuscript preparation, BA, FK. All authors contributed to the revision of the manuscript, and approve of the final version of the article.

Competing interests

The authors declare that they have no competing interests.

Author details

[1]Department of Animal Science, Shahrekord University, Shahrekord 88186-34141, Iran. [2]Department of Basic Science, Shahrekord University, Shahrekord 88186-34141, Iran.

References

1. Zuidhof MJ, Schneider BL, Carney VL, Korver DR, Robinson FE. Growth, efficiency, and yield of commercial broilers from 1957, 1978, and 2005. Poult Sci. 2014(93):2970–82.
2. Hocking PM. Unexpected consequences of genetic selection in broilers and turkeys: problems and solutions. Brit Poult Sci. 2014;55:1–12.
3. Khajali F, Wideman RF. Nutritional approaches to ameliorate pulmonary hypertension in broiler chickens. J Anim Physiol Anim Nutr. 2016;100:3–14.
4. Khajali F, Fahimi S. Influence of dietary fat source and supplementary α-tocopheryl acetate on pulmonary hypertension and lipid peroxidation in broilers. J Anim Physiol Anim Nutr. 2010;94:767–72.
5. Shahidi F, Ambigaipalan P. Phenolics and polyphenolics in foods, beveragesand spices: antioxidant activity and health effects –a review. J Func Food. 2015;18:820–97.

References

1. Zuidhof MJ, Schneider BL, Carney VL, Korver DR, Robinson FE. Growth, efficiency, and yield of commercial broilers from 1957, 1978, and 2005. Poult Sci. 2014(93):2970–82.
2. Hocking PM. Unexpected consequences of genetic selection in broilers and turkeys: problems and solutions. Brit Poult Sci. 2014;55:1–12.
3. Khajali F, Wideman RF. Nutritional approaches to ameliorate pulmonary hypertension in broiler chickens. J Anim Physiol Anim Nutr. 2016;100:3–14.
4. Khajali F, Fahimi S. Influence of dietary fat source and supplementary α-tocopheryl acetate on pulmonary hypertension and lipid peroxidation in broilers. J Anim Physiol Anim Nutr. 2010;94:767–72.
5. Shahidi F, Ambigaipalan P. Phenolics and polyphenolics in foods, beveragesand spices: antioxidant activity and health effects –a review. J Func Food. 2015;18:820–97.
6. Pirbalouti A, Setayesh M, Siahpoosh A, Mashayekhi H. Antioxidant activity, total phenolic and flavonoids contents of three herbs used as condiments and additives in pickles products. Herb Polon. 2013;59:51–61.
7. Sajjadi SE, Shokoohinia Y, Moayedi N. Isolation and identification of ferulic acid from aerial parts of *Kelussia odoratissima* Mozzaf. Jundishapur J Nat Pharm Prod. 2012;7:159–62.
8. Surai PF. Polyphenol compounds in the chicken/animal diet: from the past to the future. J Anim Physiol Anim Nutr. 2014;98:19–31.
9. Shojaei Z, Ebrahimi A, Salimi M. Chemical composition of three ecotypes of Kelussia odoratissima Mozzaf *(Kelussia odoratissima).* J Herb Spi Med Plant. 2011;17:62–8.
10. Ahmadi F, Kadivar M, Shahedi M. Antioxidant activity of *Kelussia odoratissima Mozzaf /.* In model and food systems. Food Chem. 2007; 105:57–64.
11. Ahmadipour B, Hassanpour H, Asadi E, Khajali F, Rafiei F, Khajali F. *Kelussia odoratissima* Mozzaf- a promising medicinal herb to prevent pulmonary hypertension in broiler chickens reared at high altitude. J Ethnopharmacol. 2015;159:49–54.
12. Rabbani M, Sajjadi S, Sadeghi M. Chemicalcompositionoftheessentialoil from *Kelussia odoratissima* Mozzaf. And the evaluation of its sedative and anxiolytic effects in mice. Clinics. 2011;66:843–8.
13. National Research Council, editor. Nutrient requirements for poultry. 9th revised ed. NY: National Research Council; 1994.
14. Association of Official Analytical Chemists. (AOAC). Official methods of analysis of the AOAC. 15th ed. Arlington; 1990.
15. Nair V, Turner GA. The thiobarbituric acid test for lipid peroxidation: structure of the adduct with malondialdehyde. Lipids. 1984;19:804–5.
16. Ruijter J, Ramakers C, Hoogaars W, Karlen Y, Bakker O, Van den Hoff M, Moorman A. Amplification efficiency: linking baseline and bias in the analysis of quantitative PCR data. Nucl Acid Res. 2009;37:1–12.
17. Dorak M. Real Time PCR Taylor & Francis. Oxford: UK; 2006.
18. Kosmider B, Osiecka O. Flavonoid compounds: a review of anticancer properties and interactions with cis-Diamminedichloroplatinum(II). Drug Dev Res. 2004;63:200–11.
19. Steven L. Avian biochemistry and molecular biology. First ed. Cambridge: University Press; 1996. p. 46–56.
20. Annison EF. Lipid metabolism. In: Freeman BM, editor. Physiology and biochemistry of the domestic fowl. 1st ed. London: Academic Press; 1983.
21. Griffin H, Grant G, Perry M. Hydrolysis of plasma triacylglycerol rich lipoproteins from immature and laying hens (Gallus domesticus) by lipoprotein lipase in vitro. Biochem J. 1982;206:647–54.
22. Muller M, Kersten S. Nutrigenomics: goals and strategies. Nat Rev Genet. 2003;4:315–22.
23. Tsi D, Das N, Tan B. Effects of aqueous celery (Apiumgraveolens) extract on lipid parameters of rats fed a high fat diet. Plant Med. 1995;61:8–21.
24. Anila L, Vijayalakshmi NR. Flavonoids from *Emblica officinalis* and *Mangifera indica*—effectiveness for dyslipidemia. J Ethnopharmacol. 2002;79:81–7.
25. Chen J, Li X. Hypolipidemic effect of flavonoids from mulberry leaves in triton WR-1339 induced hyperlipidemic mice. Asia Pac J Clin Nutr. 2007; 16:290–4.
26. Bok SH, Lee SH, Park YB, Bae KH, Son KH, Jeong TS, Choi MS. Plasma and hepatic cholesterol and hepatic activities of 3-hydroxy-3-methyl-glutaryl-CoA reductase and acyl CoA: cholesterol transferase are lower in rats fed citrus peel extract or a mixture of citrus bioflavonoid. J Nutr. 1999;129:1182–5.
27. Davalos A, Fernandez-Hernando C, Cerrato F, Martinez-Botas J, Gomez-Coronado D, Gomez-Cordoves C, Lasuncion MA. Red grape juice polyphenols alter cholesterol homeostasis and increase LDL-receptor activity in human cells in vitro. J Nutr. 2006;136:1766–73.
28. Kuang X, Du JR, Liu YX, Zhang GY, Peng HY. Postischemic administration of z-Ligustilide ameliorates cognitive dysfunction and brain damage induced by permanent forebrain ischemia in rats. Pharmacol Biochem Behav. 2008; 88:213–21.
29. Peng HY, Du JR, Zhang GY, Kuang X, Liu XY, Qian ZM, Wang CY. Neuroprotective effect of Z-ligustilide against permanent focal ischemic damage in rats. Biol Pharm Bull. 2007;30:309–12.
30. Khajali F, Karimi S, Qujeq D. Probiotics in the drinking water alleviate stress of induced molting in feed-deprived laying hens. Asia Aust J Anim Sci. 2008;21:1196–200.
31. Fouad AM, El-Senousey HK. Nutritional factors affecting abdominal fat deposition in poultry- a review. Asia Aust J Anim Sci. 2014;27:1057–68.

Transcripts of antibacterial peptides in chicken erythrocytes infected with Marek's disease virus

Sheng Niu[1], Ali Raza Jahejo[1], Fa-jie Jia[1], Xin Li[1], Guan-bao Ning[1], Ding Zhang[1], Hai-li Ma[1], Wei-fang Hao[2], Wen-wei Gao[1], Yu-jun Zhao[1], Shi-min Gao[1], Gui-lan Li[1], Jian-hui Li[1], Fang Yan[1], Rong-kun Gao[1], Yu-hai Bi[1,3], Ling-xia Han[4*], George F. Gao[1,3*] and Wen-xia Tian[1*]

Abstract

Background: Chicken erythrocytes are involved in immunity through binding of toll-like receptors (TLRs) with their ligands to activate downstream signaling and lead to cytokine production in erythrocytes. Some avian β-defensins (*AvBDs*) are constitutively expressed in tissues and some others can be induced by various bacteria and viruses. However, the expression of *AvBDs* in erythrocytes has not yet been studied extensively.

Results: The transcripts of eight *AvBDs* (*AvBD*1 to *AvBD*7, and *AvBD*9) and liver-expressed antimicrobial peptide-2 (*LEAP*-2) were found in normal chicken erythrocytes. The expression levels of *AvBD*2, 4 and 7 were significantly increased ($P < 0.01$), whereas the levels of *AvBD*1, 6 and 9 were significantly decreased ($P < 0.01$) after Marek's disease virus (MDV) infection. The mRNA expression level of *LEAP*-2 was not significantly changed after MDV infection. Highest viral nucleic acid (VNA) of MDV in the feather tips among the tested time points was found at 14 days post-infection (d.p.i.). In addition, 35 MD5-related gene segments were detected in the erythrocytes at 14 d.p.i. by transcriptome sequencing.

Conclusions: These results suggest that the *AvBDs* in chicken erythrocytes may participate in MDV-induced host immune responses.

Keywords: Chicken, Erythrocytes, *AvBDs*, *TLRs*, MDV

Background

Erythrocytes are the most abundant cells in circulation and the most important cells in the transportation and exchange of gases. In addition, several studies have reported that chicken erythrocytes play an important role in immunology [1]. For example, erythrocytes of vertebrates are nucleated and contain organelles in their cytoplasm, except in mammals [2]. It has been reported that the expression of cytokine transcripts in salmon erythrocytes were induced by infection with infectious salmon anaemia virus (ISAV). Similarly, cytokines, produced by trout erythrocytes when exposed to *Candida albicans*, enhanced the phagocytic capabilities [3, 4]. Moreover, the transcripts for many toll-like receptors (*TLRs*) were constitutively expressed in erythrocytes [1]. Chicken erythrocytes can induce the expression of transcripts for some cytokines and secrete thermo-labile molecules, which can enhance the antiviral capabilities of macrophages in response to *TLR* ligand treatments, such as type I interferons (*IFNs*) and interleukin (*IL*)-8 [1, 5].

Marek's disease (MD) is a contagious lymphoproliferative disease caused by Marek's disease virus (MDV), and can lead to tumors, lymphocyte proliferation, and ultimately death in chickens [6]. MDV has been evolving worldwide over the past few decades. The emerging MDV strains have changed from mild MDV (mMDV) to very virulent MDV (vvMDV) and very virulent plus MDV (vv + MDV) [7]. Host defense peptides (HDPs),

* Correspondence: 11703120@qq.com; gaof@im.ac.cn; wenxiatian@126.com
[4]Department of Laboratory Animal and Comparative Medicine, State Key Laboratory of Veterinary Biotechnology, Harbin Veterinary Research Institute, Chinese Academy of Agricultural Sciences, Harbin 150001, China
[1]College of Animal Science and Veterinary Medicine, Shanxi Agricultural University, Taigu 030801, China
Full list of author information is available at the end of the article

also known as antimicrobial peptides, are a group of cationic amphipathic peptides that are produced by the host itself; they play a vital role in innate immunity and could kill various pathogenic microorganisms such as bacteria, fungi and even enveloped viruses [8–11]. Avian β-defensins (*AvBDs*) is the only type of defensin found in chickens [12–14]. Researchers are more interested in HDPs or their derivatives, because they believe HDPs could fight and protect against infectious diseases [15].

Recent studies have determined that *TLRs*, *AvBDs* and cytokines are involved in the response of the immune system in various tissues [16]. But the changes of expression level of *AvBDs* in chicken erythrocytes after MDV infection has not yet been investigated. The aim of the present study was to explore the transcriptional levels of *AvBDs* and liver-expressed antimicrobial peptide-2 (*LEAP*-2) in erythrocytes of chickens infected by MDV.

Results

AvBDs and *LEAP*-2 genes expression detected in chicken erythrocytes

The results of the reverse transcription-polymerase chain reaction (RT-PCR) showed that the genes expression of *AvBD*1 to *AvBD*7, *AvBD*9, and *LEAP*-2 was detected in the normal commercial chicken erythrocytes, but not for *AvBD*8, and *AvBD*10 to *AvBD*14 (Fig. 1), and no bands were found in the negative controls. *AvBD*1 to *AvBD*7, *AvBD*9 and *LEAP*-2 genes expression was also detected in specific pathogen-free (SPF) chicken erythrocytes (data not shown). These detected genes expression was further confirmed by Sanger sequencing and Real-time PCR.

Virus load in the feather tips of MDV-infected chickens

No severe clinical symptom except late phase depression was observed in 14 d experimental period. The copy numbers of MDV were tested at 4, 7, 9 and 14 days post-infection (d.p.i.) by duplex real-time PCR (Fig. 2). The results showed that the MDV was detected after 7 d.p.i., and the MDV copy number at 14 d.p.i. was significantly higher than that of all other time points. MDV nucleic acid was not detected in the control group.

Detection of MDV-related genes in erythrocytes

To determine whether MDV invaded chicken erythrocytes which could activate *AvBDs* expression against the infection, we performed a search for virus genomes in the transcriptome of erythrocytes on 14 d.p.i. (SRA accession No: SRP136403) when the MDV load was significantly higher than that of all other time points. The results showed that the transcripts of 35 MD5-related genes were detected (Table 1). These genes were mainly phosphoprotein pp24, major capsid protein, assembly protein, major DNA-binding protein, deneddylase, DNA polymerase processivity factor, tegument protein VP16, and deoxyuridine triphosphatase, in which the copy number of phosphoprotein pp24 that act as an integral component of membrane was the most abundant. These results indicated that erythrocytes had been invaded by MDV.

Quantitative PCR analysis for *AvBDs* and *LEAP-2* expression in the chicken erythrocytes in response to MDV stimulation

To further investigate the functions of these genes during MDV infections, their expression levels in the erythrocytes were analyzed. We utilized real-time RT-PCR to quantify the repertoire of *AvBDs* expressed in chicken erythrocytes in response to MDV on 14 d.p.i.(Fig. 3). The expression levels of *AvBD*2, 4 and 7 were significantly upregulated compared to the control group ($P < 0.01$). However, the gene transcript levels of *AvBD*1, 6 and 9 in the experimental group were significantly lower than those in the control group ($P < 0.01$). There was no significant difference for *AvBD*3, *AvBD*5 and *LEAP*-2 between the two groups.

Discussion

Recent studies have reported that erythrocytes play an important role in immune responses. It has been found

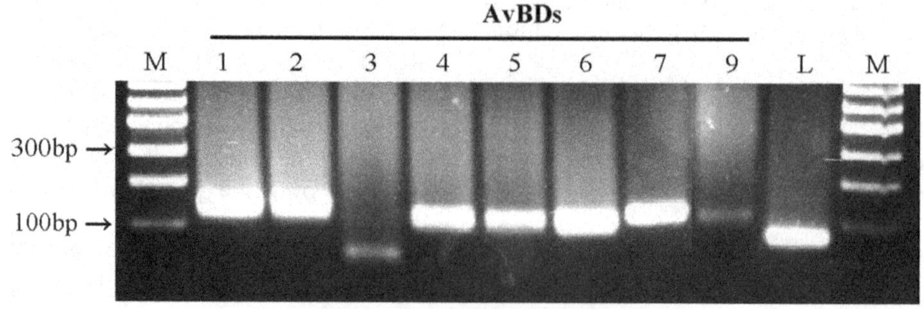

Fig. 1 Expression pattern of *AvBDs* and *LEAP*-2 genes in the erythrocytes of 3-week-old roosters. All of the *AvBDs* and *LEAP*-2 were detected by RT-PCR. M = size marker, L = *LEAP*-2

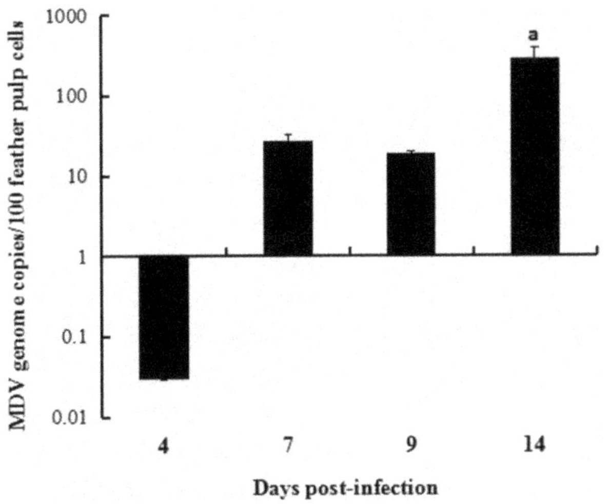

Fig. 2 MDV load in feather tips on 4, 7, 9 and 14 d.p.i. ($n = 3$). Chickens were infected with MDV. Feather tips were obtained on 4, 7, 9 and 14 d.p.i., and the virus loads were measured by duplex real-time PCR. The error bars represent standard error of the mean value. The data of virus load was subjected to one-way ANOVA and then analyzed by the Tukey's pairwise comparison to identify treatment differences, a = significant when compared to MDV-infected chickens sampled on 4, 7 and 9 d.p.i. ($P < 0.05$)

that trout and chicken erythrocytes can detect and specifically respond to different pathogen-associated molecular patterns (PAMPs) [5]. In the present study, it was observed that *AvBDs* and *LEAP*-2 transcripts are expressed not only in tissues but also in erythrocytes. This result provides a new perspective for poultry health, as the expression of these transcripts may be linked to the protection of chickens from pathogens. Primary literatures showed that *AvBDs* are not only largely associated to reproductive organs but also to bone marrow [17–19], and more generally to mucosal tissues. The transcripts for 13 types of *AvBDs* were expressed in chicken reproductive organs [20]. It has been reported that *AvBD*1, 2 and *AvBD*4 to *AvBD*7 were highly expressed in bone marrow of chicken [18, 19, 21]. Strong expression of *AvBDs* transcripts was found throughout the digestive tract such as in the esophagus, craw, proventriculus, and intestinal tissues, except for *AvBD*11 [16, 22, 23]. This is similar to our results that *AvBD*1 to *AvBD*7 and *AvBD*9 transcripts were detected in chicken erythrocytes. Moreover, many *AvBDs* also were found and regulated by both pathogenic and attenuated infectious bronchitis virus (IBV) strains in many highly vascularized tissues, such as liver, spleen and kidney [24, 25]. The expression of *AvBDs* was also observed in other tissues such as the respiratory tract and skin [12, 17, 21]. As the most abundant circulating cells, erythrocytes are distributed in all of these tissues. All of the above support the hypothesis that erythrocytes can participate in the immune response in various tissues, especially highly vascularized tissues, by releasing *AvBDs* to fight pathogens.

The results from current study indicated that MDV genome load in the feather tips was significantly increased at 14 d.p.i.. Similar results were reported by Abdul-Careem et al [26]. This could be because some latent MDV was reactivated and started the second cytolytic phase by 14 d.p.i. [6]. The transcripts of 35 MD5-related genes in erythrocytes indicated that MDV invaded the membrane or intracellular of erythrocytes. The explanation for the results could be 1) the viral PAMPs were recognized by the pattern recognition receptors (PRRs) on the surface of erythrocytes; 2) MDV invaded and replicated in the erythrocytes; 3) there was an interaction between MDV and erythrocytes. However, the molecular mechanism needs to be further elucidated.

TLRs are associated with immune cells and other tissue cells [17]. It was recently found that chicken erythrocytes constitutively express transcripts for many *TLRs* as well as for some cytokines and other immune-related genes such as *IFN-α*, *IFN-β*, and *IL-8* [1]. *AvBDs* can be induced by the interaction of *TLRs* with their ligands and proinflammatory cytokines [27, 28]. The results of the present study show that it is possible that erythrocytes can participate in some of these aspects. Furthermore, many MDV-related genes and differentially expressed *AvBD*2 were detected in the transcriptome of erythrocytes, suggesting the expression of *AvBDs* in erythrocytes may participate in the antiviral response against MDV infection. The significant upregulation of *AvBD*2, 4 and 7 was found after MDV infection, which may result from the activation of broader innate immune signaling pathways after PRRs-viral PAMPs interaction, and this still needs further researches to confirm. We found that the upregulated expression of *AvBD*2 is the most significant result obtained after MDV infection. Similarly, some studies found that *AvBD*2 was also upregulated by the infections of pathogenic IBV and pigeon paramyxovirus type 1 (PPMV-1) in most tissues [24]. Furthermore, it has been reported that *AvBD*2 is the major defensin in heterophiles [27, 29]. We speculate that *AvBD*2 may be expressed as functional proteins. In duck tissues, significant antiviral activities of three *AvBDs* (Apl_*AvBD*4, 7 and 12) were shown against duck hepatitis virus (DHV) [30]. In another study, the recombinant *AvBD*2, 6 and 12 displayed the obvious anti-viral activity against IBV in vitro [24]. These suggested that *AvBD*2, 4, 6 and 7 may have strong antiviral activities in chicken erythrocytes. Moreover, we also found that the transcripts of *AvBD*1, 6 and 9 were significantly downregulated after MDV infection; this may be associated to the viral immunomodulators that fight against these *AvBDs* after MDV infection. The *LEAP*-2 expression in

Table 1 The identities of 35 MD5-related gene transcripts detected in chicken erythrocytes after MDV infection

ID	Length(nt)	Count (Reads numbers)	Swissprot_annotation (strain Chicken/Md5/ATCC VR-987)	nr_annotation [Gallid herpesvirus 2]
MDV008	2608	17	Phosphoprotein pp24	protein pp24
MDV018	2169	1	Portal protein UL6	capsid portal protein
MDV021	2526	1	Replication origin-binding protein	DNA replication origin-binding helicase
MDV023	3786	5	Tegument protein UL11	myristylated tegument protein
MDV030	960	1	Triplex capsid protein VP23	capsid triplex subunit 2
MDV031	4182	8	Major capsid protein	HSV-1 UL19-like protein
MDV034	2442	1	Envelope glycoprotein H	envelope glycoprotein H
MDV038	1992	6	Assembly protein	capsid maturation protease
MDV040	2598	1	Envelope glycoprotein B	envelope glycoprotein B
MDV042	3576	6	Major DNA-binding protein	single-stranded DNA-binding protein
MDV043	3663	3	DNA polymerase catalytic subunit	DNA polymerase catalytic subunit
MDV044	903	1	Virion egress protein UL31	nuclear egress lamina protein
MDV046	1926	1	Packaging protein UL32	DNA packaging protein UL32
MDV047	834	1	Virion egress protein UL34	nuclear egress membrane protein
MDV049	10,029	7	Deneddylase	large tegument protein
MDV050	3141	1	Capsid assembly protein UL37	tegument protein UL37
MDV051	1413	1	Triplex capsid protein	capsid triplex subunit 1
MDV052	2469	2	Ribonucleoside-diphosphate reductase large subunit	ribonucleotide reductase subunit 1
MDV053	1032	1	Ribonucleoside-diphosphate reductase small chain	ribonucleotide reductase subunit 2
MDV055	1506	6	DNA polymerase processivity factor	DNA polymerase processivity subunit
MDV059	1707	4	Tegument protein UL46	tegument protein VP11/12
MDV060	2427	3	Tegument protein UL47	tegument protein VP13/14
MDV061	1284	6	Tegument protein VP16	transactivating tegument protein VP16
MDV062	750	1	Tegument protein VP22	tegument protein VP22
MDV063	1311	9	Deoxyuridine 5' –triphosphate nucleotidohydrolase	deoxyuridine triphosphatase
MDV065	750	1	Tegument protein UL51	tegument protein UL51
MDV068	1422	3	mRNA export factor ICP27	multifunctional expression regulator
MDV069	810	1	Uncharacterized gene 69 protein	protein LORF4
MDV070	501	1	Tegument protein UL55	nuclear protein UL55
MDV071	585	1	Uncharacterized gene 71 protein	myristylated tegument protein CIRC
MDV072	2712	5	Uncharacterized gene 72 protein	protein LORF5
MDV072.5	873	3	–	membrane protein UL56
MDV088	540	3	Transcriptional regulator ICP22	regulatory protein ICP22
MDV089	642	2	Virion protein US10	virion protein US10
MDV094	4716	3	Envelope glycoprotein D	Envelope glycoprotein D

erythrocytes during MDV infection has also been found; thus, we speculate that *LEAP*-2 might also be an effector of innate immune signaling as a response to some pathogens [31].

The modulation of *AvBDs* in response to the bacterial components or ligands for *TLRs* in chickens have already been studied [1, 20, 23], but the reaction of *AvBDs* in response to MDV in chicken erythrocytes has not yet been investigated. Based on the expressed repertoire of *AvBDs*, it is possible that erythrocytes can respond both virus and bacteria, which is supported by the fact that *AvBD*2 which is highly overexpressed in response to MDV possess a strong antibacterial activity [19]. This finding is also supported by another fact that the transcript of type I *IFNs* are also expressed in chicken erythrocytes and are upregulated after the treatment with poly I:C or CpG ODN [1], such that a second wave of signaling pathways would be activated to synthesize IFN-regulated proteins contributing to the antiviral response and inhibiting virus replication [32]. Another

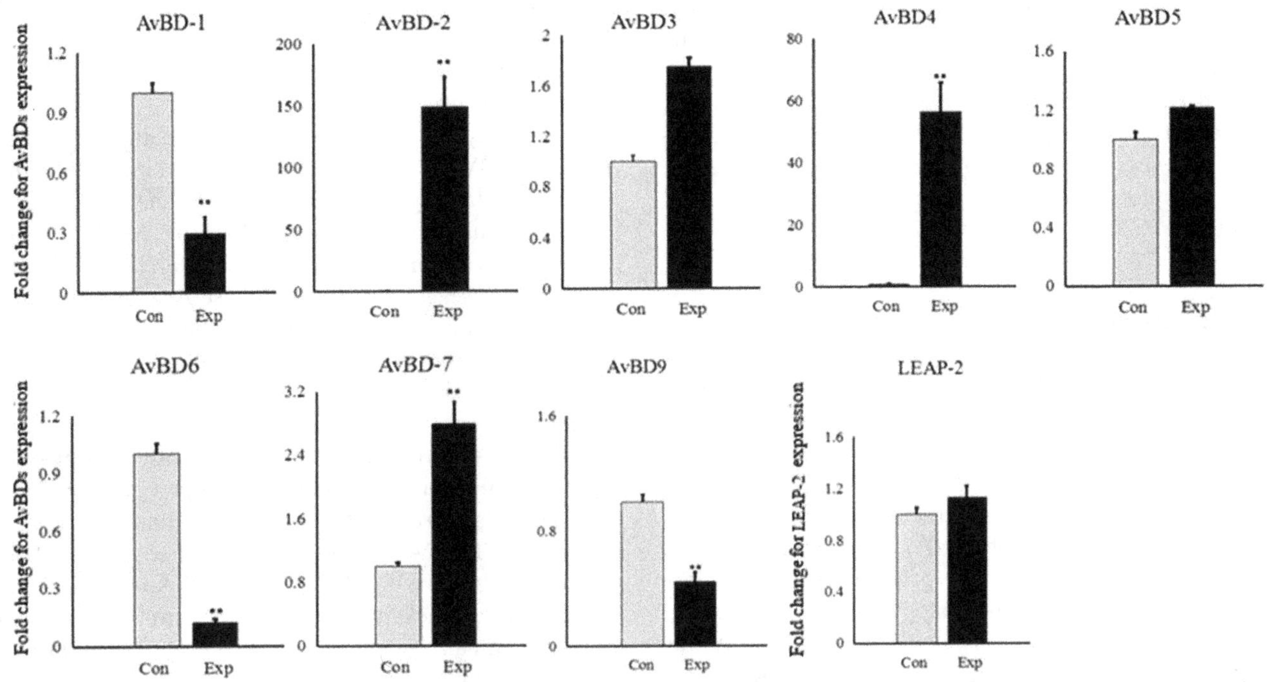

Fig. 3 The expression pattern of mRNAs of *AvBDs* in chicken erythrocytes on 14 d.p.i. The experimental groups of this study were as follows: the control group (Con, PBS-injection, $n = 3$) and the infection group (Exp, MDV-injection, n = 3). Expression levels of *AvBDs* were calculated relative to that of the housekeeping gene *18S rRNA* by using quantitative real-time PCR. The data represent mean fold changes from PBS-treated controls ± standard error. Statistical significance between the infection group and the control group was analyzed using t-test. $^*P < 0.05$, $^{**}P < 0.01$

study showed the expression of several *AvBDs* was closely correlated to the expression of *TLR*3, *TLR*7 and *TLR*15 in tissues [33]. All the above clues supported the possible defensive mechanism of chicken erythrocytes that the PAMPs could be recognized by various PRRs, such as *TLRs* and scavenger receptor (*SR*). Cytokines, *AvBDs* and other molecules will be activated and modulate host immune response when pathogens invade into the blood.

Conclusions

In conclusion, the transcripts of eight *AvBDs* and *LEAP*-2 were identified in chicken erythrocytes. The transcripts of 35 MD5-related genes were found in erythrocytes after MDV infection. MDV induced an immune response wherein *AvBD*2, 4 and 7 were significantly upregulated and *AvBD*1, 6 and 9 were significantly downregulated. Further studies are needed on the mechanism of *AvBDs* in the response of chicken erythrocytes to MDV infections, which could help to prevent and control MDV in the future.

Methods

Experiment 1. Detection of *AvBDs* and *LEAP*-2 genes expression in chicken erythrocytes

Experimental animals

Three 3-week-old roosters (Hy-Line Brown) were purchased from a commercial supplier in Taigu County. The roosters were kept in cages, under an artificial light regimen of 14 h day-time/10 h night-time. A diet appropriate for their age was provided ad libitum. Before exsanguination and necropsy, injection of pentobarbital was used along with the standard protocols of euthanasia to minimize animals suffering.

Erythrocyte collection

Venous blood was drawn from these chickens (2 mL per bird) and mixed with the same volume of Alsever's solution (Solarbio, Beijing, China). The diluted blood was added carefully into a 4 mL Histopaque-1119 solution (Sigma–Aldrich, Oakville, ON) and centrifuged at 500 *g* for 20 min; subsequently, the supernatant containing platelets and leukocytes-was removed. Subsequent procedures were performed as described earlier [34]. Isolated erythrocytes were determined to be > 99.9% pure by Wright–Giemsa staining.

Extraction of total RNA from chicken erythrocytes, and cDNA synthesis

Total RNA was purified from erythrocytes isolated from blood samples using RNAiso Plus (Takara Bio Inc., Dalian, China) according to the manufacturer instructions, and dissolved in 20 μL RNase-free water. Purified RNAs were examined by 1.5% agarose gel electrophoresis and a NanoDrop Bioanalyzer ND1000 (Labtech, Uckfield, UK).

The criteria for assessing the quality of RNA were that *28S* and *18S rRNA* could be seen in the gel and the A260/A280 ratio was between 1.9–2.1. cDNA was synthesized from 500 ng total RNAs using the PrimeScript RT reagent Kit with gDNA Eraser (Takara Bio Inc., Dalian, China) according to manufacturer instructions. The cDNA samples were diluted at 1:10 in sterile water and stored at − 20 °C.

Primer design and PCR

Primers were designed by Primer Express 3.0 (Applied Biosystems, Foster City, CA, USA) and synthesized by Shanghai Generay Biotech Co., Ltd. (Shanghai, China). GenBank accession numbers of the reference sequences used for primer design are presented in Table 2. The PCR was conducted in a volume of 20 μL of reaction mixture containing 2 μL of cDNA, 0.4 μL at 10 pmol/μL of each of forward and reverse primers, 10 μL 2 × MasterMix (Real-Times Biotechnology Co., Ltd., Beijing, China) and 7.2 μL double-distilled H_2O. The cDNA was replaced by water as negative control for each gene. Amplification for *AvBDs* or *LEAP*-2 was performed in a Gradient Thermocycler (Eppendorf AG, Germany). The optimum thermal cycling procedure was as follows: 95 ° C for 3 min; 35 cycles of 95 °C for 30 s, 55 to 57 °C (the optimal temperature of *AvBDs*, listed in Table 2) for 30 s, and 72 °C for 1 min, followed by 72 °C for 5 min. The PCR results were detected by electrophoresis in 1.5% TAE agarose gels, and imaged by GelDoc-It Imaging Systems (Ultra-Violet Products, USA).

Experiment 2. *AvBDs* and *LEAP*-2 gene expression and virus load in erythrocytes and feather tips of MDV-infected chickens

Virus and animals

Twenty 1-day-old SPF chickens (B15 haplotype), very virulent MDV strain MD5, and recombinant plasmid pMD-O containing the *ovotransferrin* gene and recombinant plasmid pMD-M containing the MDV Eco Q (*meq*) gene were provided by Dr. Lingxia Han of Harbin Veterinary Research Institute, Chinese Academy of Agriculture Sciences (Harbin, China). The strain of MDV (MD5) was preserved in liquid nitrogen until use.

Animal infection

The chickens were equally divided into two groups: an experimental group and a control group. The animal experiment was carried out in an Animal Biosafety Level 2 Laboratory. The MDV was diluted properly with sterile phosphate buffered saline (PBS) (Solarbio, Beijing, China). Each chicken in the experimental group was intra-abdominally infected with MDV at a dose of 500 plaque-forming units (PFU)/500 μL. The control group was injected with the same volume of PBS. On 4, 7, 9 and 14 d.p.i., 4–5 feather tips were collected from each bird in the experimental and control groups and preserved in liquid nitrogen immediately. Anticoagulated blood samples were collected at 14 d.p.i. for both groups and used for total RNA extraction as described in Experiment 1. Before exsanguination and necropsy, injection of pentobarbital was used along with the standard protocols of euthanasia to minimize animals suffering.

Table 2 Primer sequences and reference genes used for PCR

Target Genes	Primer sequence(5′→3′)	Annealing temperature	Sizes(bp)	GenBank Accession number of the reference gene
AvBD1	F: GGCTCTAGGAAGGAAGTCA R: GCAGCAGAGGTAAAATCTTG	57 °C	112	NM_204993.1
AvBD2	F: GCACTCCAGGTTTCTCCA R: GGCGTCCGACTTTGATTA	57 °C	110	DQ677633.1
AvBD3	F: AGAGGAGGATTCTGTCGTGTT R: CAACCTCATATGCTCTTCCA	55 °C	100	NM_204650.2
AvBD4	F: TCATGGAGCTGTGGGCTTTT R: AGCATTCCCATAAGGGCATT	55 °C	100	NM_001001610.2
AvBD5	F: ATTACCCCAGGACTGTGA R: GCAGCAGAAGTCTTCCTT	55 °C	100	NM_001001608.2
AvBD6	F: CTGCTGCTGTCTGTCCTCTT R: TGCAGACACCCCTTTGATAT	55 °C	100	NM_001001193.1
AvBD7	F: CTATTGATACTTGTTGGCTTCG R: AACTCCTCCATCCCCTTG	57 °C	122	AY621322.1
AvBD9	F: CCAGCTTACAGCCAAGAAGA R: TCCCAATGTCAACTGAAGGA	55 °C	100	NM_001001611.2
LEAP-2	F: ACTCTGGAATTCTGCCTGATGACA R: CATCTGCATCCGTGCCTGA	57 °C	66	NM_001001606.1
18S rRNA	F: TTCCGATAACGAACGACAC R: GACATCTAAGGGCATCACAG	55 °C	139	FM165414

All these tests were performed by Lingxia Han at Harbin Veterinary Research Institute, Chinese Academy of Agriculture Sciences (Harbin, China).

Library preparation and sequencing

RNA was extracted from the erythrocytes. A total of 1 μg purified RNA per sample was used for next-generation sequencing (NGS). Sequencing libraries were generated using NEBNext®Ultra™ RNA Library Prep Kit for Illumina® (NEB, USA) following the manufacturer's instruction, and index codes were added to attribute sequences to each sample as described previously [35]. The libraries were sequenced by Illumina HiSeq 2500.

Transcriptome analysis

To identify the viral sequence, the clean reads were mapped to the reference genomic sequence by Tophat2 tools software [36], and each unigene was then transformed into FPKM (Fragments Per Kilo base per Million mapped reads) by using RSEM software [37]. Annotations of the unigenes were searched using BLAST against the Nr (non-redundant) and the Swiss-Prot databases (E-value $\leq 10^{-5}$).

Detection of AvBDs and LEAP-2 mRNA expression in erythrocytes

RNA was extracted from the erythrocytes on 14 d.p.i.. The quantitative expression analysis of *AvBDs* and *LEAP*-2 genes were performed with real-time RT-PCR using a TaKaRa SYBR Premix Ex Taq™II (Takara Bio Inc.). The total volume of real-time PCR mixture contained 6 μL 1× SYBR Premix Ex TaqII, 0.1 μL ROX dye II, 1.5 pmol each of forward and reverse primers, 1 μL of diluted cDNA, and 2.6 μL of double distilled H_2O. The reaction mixture was placed into MicroAmp® Optical 8-Cap Strip (Applied Biosystems), and real-time PCR was performed following user instructions of the QuantStudio™ 6 (Applied Biosystems). Briefly, each reaction involved a pre-incubation at 95 °C for 3 min, followed by 42 cycles of 95 °C for 30 s, 55 °C - 57 °C (T_A as per primer) for 30 s, and extension at 72 °C for 10 s. The expression level of *AvBDs* and *LEAP*-2 were calculated relative to that of the housekeeping gene *18S rRNA* using the QuantStudio™ 6 Flex Real-Time PCR System Software (Applied Biosystems, USA).

DNA extraction from feather tips

DNA was extracted from feather tips collected at 4, 7, 9 and 14 d.p.i. using Trizol reagent (Invitrogen, Carlsbad, California, USA) as described previously [38]. The purity and concentration of purified DNAs were detected by a NanoDrop Bioanalyzer ND1000 (Labtech). The feather tips DNA was then diluted to 100 ng/μL for future use.

Primers and probes for meq and ovotransferrin genes

Primers and probes of the duplex real-time PCR (Table 3) for MDV *meq* gene and *ovotransferrin* gene followed that by Islam et al. [39] and Baigent et al. [40], respectively. Chicken *ovotransferrin* genomes gene was used as DNA extraction reference.

Quantification of virus loads by real-time PCR

Serial 10-fold dilutions (10^4–10^{10} copies/μL) of the recombinant plasmids of pMD-O and pMD-M were used to generate a standard curve for *ovotransferrin* and *meq* gene. The duplex real-time PCR was performed in a 25-μL volume. The reaction mixture included 2.5 μL 10 × Ex Taq buffer; 1 μL dNTPs; 0.2 μL HS Ex Taq; 8 pM *meq*-P and *ovo*-P probes; 10 pM *meq*-F, *meq*-R, *ovo*-F and *ovo*-R primers; 1 μL pMD-O and pMD-M. Sterilized H_2O was added to bring the final volume to 25 μL. The protocol for the reaction was as follows: 95 ° C for 5 min; 40 cycles of 94 °C for 15 s, 59.5 °C for 50 s. The fluorescence intensity of the reporter dye (FAM and ROX) was measured during the 59.5 °C annealing/extension step. The data analysis was performed using an iCycler Thermal Cycler (BioRad, Hercules, CA, USA) generating the standard curve. For the detection of the copy numbers of MDV in feather tips on 4, 7, 9 and 14 d.p.i., the DNAs in feather tips were used as templates for the duplex real-time PCR assay to detect the *meq* and *ovotransferrin*. The copy numbers of MDV were calculated using the standard curve.

Table 3 Sequences of the primers, probes and reference genes used for the duplex real-time PCR assay

Primer/probe name	Primer/probe sequence(5′→3′)	Target gene	GenBank Accession number
meq-F	GGAGCCGGAGAGGCTTTATG		
meq-R	ATCTGGCCCGAATACAAGGAA	MDV Eco Q	M89471
meq-P	(FAM)CGTCTTACCGAGGATCCCGAACAGG(TAMRA)		
ovo-F	CACTGCCACTGGGCTCTGT		
ovo-R	GCAATGGCAATAAACCTCCAA	ovotransfenrin	Y00407
ovo-P	(ROX)AGTCTGGAGAAGTCTGTGCAGCCTCCA(BHQ2)		

FAM: Carboxy fluorescein, TAMRA: Carboxy tetramethyl rhodamine, ROX: Carboxy-X-rhodamine, BHQ2: Black Hole Quencher 2

Statistical analysis

The $2^{-\Delta\Delta Ct}$ method was used to calculate the real-time PCR data. Data represent mean fold change from PBS-treated controls ± standard error. Statistical analyses were performed using SPSS statistical software 17.0 (IBM Company, New York, NY, USA). The data of virus load was subjected to one-way ANOVA and then analyzed by the Tukey's pairwise comparison to identify treatment differences which were considered significant when $P < 0.05$. Statistical significance of each gene expression between the experimental group and the control group was analyzed using t-test for independent samples. Differences were considered significant when the P-value was $^{*}P < 0.05$ or $^{**}P < 0.01$.

Abbreviations

AvBDs: Avian β-defensins; HDPs: Host defense peptides; IBV: Infectious bronchitis virus; *IFNs*: Type I interferons; *IL*: Interleukin; ISAV: Infectious salmon anaemia virus; *LEAP*-2: Liver-expressed antimicrobial peptide-2; MD: Marek's disease; MDV: Marek's disease virus; *meq*: MDV Eco Q; mMDV: Mild MDV; NGS: Next-generation sequencing; PAMPs: Pathogen associated molecular patterns; PBS: Phosphate buffered saline; PFU: Plaque-forming units; PPMV-1: Pigeon paramyxovirus type 1; PRRs: Pattern recognition receptors; RT-PCR: Reverse transcription polymerase chain reaction; SPF: Specific pathogen-free; *SR*: Scavenger receptor; *TLRs*: Toll-like receptors; VNA: Viral nucleic acid; vvMDV: Very virulent MDV

Acknowledgements

The authors are very thankful to Gary Wong (Institute of Microbiology, Chinese Academy of Sciences, Beijing, China) and Tuoying Ao (Alltech, USA) for help with English language quality of the manuscript.

Funding

This work was supported by the National Key R & D Program (2016YFD0500800), Research Project Supported by Shanxi Scholarship Council of China (2017-073), Key scientific and technological innovation platform of Jinzhong (P171002-3), the National Natural Science Foundation of China (31072179), Key Technologies R and D Program of Shanxi Province (20130311027-3) and "131" Leading Talent Project for Colleges and Universities of Shanxi Province.

Authors' contributions

SN drafted the basic manuscript and analyzed the data. WXT, LXH, GFG and YHB contributed to conception and design of the research and reviewed the manuscript. ARJ, FJJ, XL, GBN, DZ, HLM and WFH participated in sample collection and laboratory testing. WWG, YJZ, SMG, GLL, JHL, FY and RKG participated in the data analysis and revising the manuscript. All authors read and approved the final manuscript.

Competing interests

The authors declare that they have no competing interests.

Author details

[1]College of Animal Science and Veterinary Medicine, Shanxi Agricultural University, Taigu 030801, China. [2]Taiyuan Center for Disease Control and Prevention, Taiyuan 030024, China. [3]CAS Key Laboratory of Pathogenic Microbiology and Immunology, Collaborative Innovation Center for Diagnosis and Treatment of Infectious Disease, Institute of Microbiology, Center for Influenza Research and Early-warning (CASCIRE), Chinese Academy of Sciences, Beijing 100101, China. [4]Department of Laboratory Animal and Comparative Medicine, State Key Laboratory of Veterinary Biotechnology, Harbin Veterinary Research Institute, Chinese Academy of Agricultural Sciences, Harbin 150001, China.

References

1. Paolucci S, Barjesteh N, Wood RD, Sharif S. Chicken erythrocytes respond to toll-like receptor ligands by up-regulating cytokine transcripts. Res Vet Sci. 2013;95(1):87–91.
2. Claver JA, Quaglia AI. Comparative morphology, development, and function of blood cells in nonmammalian vertebrates. J Exot Pet Med. 2009;18(2):87–97.
3. Passantino L, Massaro MA, Jirillo F, Di Modugno D, Ribaud MR, Di Modugno G, Passantino GF, Jirillo E. Antigenically activated avian erythrocytes release cytokine-like factors: a conserved phylogenetic function discovered in fish. Immunopharmacol Immunotoxicol. 2007;29(1):141–52.
4. Workenhe ST, Kibenge MJ, Wright GM, Wadowska DW, Groman DB, Kibenge FS. Infectious salmon anaemia virus replication and induction of alpha interferon in Atlantic salmon erythrocytes. Virol J. 2008;5(1):36.
5. Morera D, Roher N, Ribas L, Balasch JC, Doñate C, Callol A, Boltaña S, Roberts S, Goetz G, Goetz FW. RNA-Seq reveals an integrated immune response in nucleated erythrocytes. PLoS One. 2011;6(10):e26998.
6. Calnek B. Pathogenesis of Marek's disease virus infection. In: Marek's Disease. Berlin, Heidelberg: Springer; 2001. p. 25–55.
7. Witter RL. Characteristics of Marek's disease viruses isolated from vaccinated commercial chicken flocks: association of viral pathotype with lymphoma frequency. Avian Dis. 1983;27(1):113–32.
8. Eade CR, Wood MP, Cole AM. Mechanisms and modifications of naturally occurring host defense peptides for anti-HIV microbicide development. Curr HIV Res. 2012;10(1):61–72.
9. Hancock RE, Diamond G. The role of cationic antimicrobial peptides in innate host defences. Trends Microbiol. 2000;8(9):402–10.
10. Brogden KA. Antimicrobial peptides: pore formers or metabolic inhibitors in bacteria? Nat Rev Microbiol. 2005;3(3):238–50.
11. Yacoub HA, Elazzazy AM, Abuzinadah OA, Al-Hejin AM, Mahmoud MM, Harakeh SM. Antimicrobial activities of chicken β-defensin (4 and 10) peptides against pathogenic bacteria and fungi. Front Cell Infect Microbiol. 2015;5:36.
12. Xiao Y, Hughes AL, Ando J, Matsuda Y, Cheng J-F, Skinner-Noble D, Zhang G. A genome-wide screen identifies a single β-defensin gene cluster in the chicken: implications for the origin and evolution of mammalian defensins. BMC Genomics. 2004;5(1):56.
13. Higgs R, Lynn DJ, Gaines S, McMahon J, Tierney J, James T, Lloyd AT, Mulcahy G, O'Farrelly C. The synthetic form of a novel chicken β-defensin identified in silico is predominantly active against intestinal pathogens. Immunogenetics. 2005;57(1–2):90–8.
14. Lynn DJ, Higgs R, Lloyd AT, O'Farrelly C, Hervé-Grépinet V, Nys Y, Brinkman FS, Yu PL, Soulier A, Kaiser P, Zhang G. Avian beta-defensin nomenclature: a community proposed update. Immunol Lett. 2007;110(1):86–9.
15. Cuperus T, Coorens M, van Dijk A, Haagsman HP. Avian host defense peptides. Dev Comp Immunol. 2013;41(3):352–69.
16. van Dijk A, Veldhuizen EJ, Haagsman HP. Avian defensins. Vet Immunol Immunopathol. 2008;124:1):1–18.
17. Yoshimura Y. Avian β-defensins expression for the innate immune system in hen reproductive organs. Poult Sci. 2015;94(4):804–9.
18. Lynn DJ, Higgs R, Gaines S, Tierney J, James T, Lloyd AT, Fares MA, Mulcahy G, O'Farrelly C. Bioinformatic discovery and initial characterisation of nine novel antimicrobial peptide genes in the chicken. Immunogenetics. 2004; 56(3):170–7.
19. Derache C, Labas V, Aucagne V, Meudal H, Landon C, Delmas AF, Magallon T, Lalmanach AC. Primary structure and antibacterial activity of chicken bone marrow-derived β-defensins. Antimicrob Agents Chemother. 2009; 53(11):4647–55.
20. Subedi K, Isobe N, Nishibori M, Yoshimura Y. Changes in the expression of gallinacins, antimicrobial peptides, in ovarian follicles during follicular growth and in response to lipopolysaccharide in laying hens (Gallus domesticus). Reproduction. 2007;133(1):127–33.
21. Zhao C, Nguyen T, Liu L, Sacco RE, Brogden KA, Lehrer RI. Gallinacin-3, an inducible epithelial β-defensin in the chicken. Infect Immun. 2001;69(4):2684–91.
22. Derache C, Esnault E, Bonsergent C, Le Vern Y, Quéré P, Lalmanach A-C. Differential modulation of β-defensin gene expression by Salmonella

Enteritidis in intestinal epithelial cells from resistant and susceptible chicken inbred lines. Dev Comp Immunol. 2009;33(9):959–66.
23. Lu L, Li S, Zhang L, Liu X, Li D, Zhao X, Liu Y. Expression of β-defensins in intestines of chickens injected with vitamin D3 and lipopolysaccharide. Genet Mol Res. 2015;14(2):3330–7.
24. Xu Y, Zhang T, Xu Q, Han Z, Liang S, Shao Y, Ma D, Liu S. Differential modulation of avian β-defensin and toll-like receptor expression in chickens infected with infectious bronchitis virus. Appl Microbiol Biotechnol. 2015; 99(21):9011–24.
25. Li S, Ouyang L, Zhou D. Effects of vitamin D3 on expression of defensins, toll-like receptors, and vitamin D receptor in liver, kidney, and spleen of silky fowl. Czech J Anim Sci. 2013;58:1):1–7.
26. Abdul-Careem MF, Hunter BD, Lee LF, Fairbrother JH, Haghighi HR, Read L, Parvizi P, Heidari M, Sharif S. Host responses in the bursa of Fabricius of chickens infected with virulent Marek's disease virus. Virology. 2008;379(2): 256–65.
27. Kannan L, Liyanage R, Lay JO, Rath NC. Evaluation of beta defensin 2 production by chicken heterophils using direct MALDI mass spectrometry. Mol Immunol. 2009;46(15):3151–6.
28. Abdel-Mageed AM, Isobe N, Yoshimura Y. Effects of different TLR ligands on the expression of proinflammatory cytokines and avian β-defensins in the uterine and vaginal tissues of laying hens. Vet Immunol Immunopathol. 2014;162(3):132–41.
29. Kannan L, Rath N, Liyanage R, Lay J. Direct screening identifies mature β-defensin 2 in avian heterophils. Poult Sci. 2009;88(2):372–9.
30. Ma D, Lin L, Zhang K, Han Z, Shao Y, Liu X, Liu S. Three novel Anas platyrhynchos avian β-defensins, upregulated by duck hepatitis virus, with antibacterial and antiviral activities. Mol Immunol. 2011;49(1):84–96.
31. Michailidis G. Expression of chicken *LEAP*-2 in the reproductive organs and embryos and in response to Salmonella enterica infection. Vet Res Commun. 2010;34(5):459–71.
32. Samuel CE. Antiviral actions of interferons. Clin Microbiol Rev. 2001;14(4): 778–809.
33. Li Y, Xu Q, Zhang T, Gao M, Wang Q, Han Z, Shao Y, Ma D, Liu S. Host avian Beta-Defensin and toll-like receptor responses of pigeons following infection with pigeon paramyxovirus type 1. Appl Environ Microbiol. 2015; 81(18):6415–24.
34. Kabanova S, Kleinbongard P, Volkmer J, Andrée B, Kelm M, Jax TW. Gene expression analysis of human red blood cells. Int J Med Sci. 2009;6(4):156.
35. Shao ES, Lin GF, Liu S, Ma XL, Chen MF, Lin L, Wu SQ, Sha L, Liu ZX, Hu XH, Guan X. Identification of transcripts involved in digestion, detoxification and immune response from transcriptome of Empoasca vitis (Hemiptera: Cicadellidae) nymphs. Genomics. 2017;109(1):58–66.
36. Trapnell C, Pachter L, Salzberg SL. TopHat: discovering splice junctions with RNA-Seq. Bioinformatics. 2009;25(9):1105–11.
37. Li B, Dewey CN. RSEM: accurate transcript quantification from RNA-Seq data with or without a reference genome. BMC bioinformatics. 2011;12(1):323.
38. Abdul-Careem MF, Hunter BD, Sarson AJ, Mayameei A, Zhou H, Sharif S. Marek's disease virus–induced transient paralysis is associated with cytokine gene expression in the nervous system. Viral Immunol. 2006;19(2):167–76.
39. Islam A, Harrison B, Cheetham BF, Mahony TJ, Young PL, Walkden-Brown SW. Differential amplification and quantitation of Marek's disease viruses using real-time polymerase chain reaction. J Virol Methods. 2004;119(2):103–13.
40. Baigent SJ, Petherbridge LJ, Howes K, Smith LP, Currie RJ, Nair VK. Absolute quantitation of Marek's disease virus genome copy number in chicken feather and lymphocyte samples using real-time PCR. J Virol Methods. 2005; 123(1):53–64.

10

Influence of silver nanoparticles on growth and health of broiler chickens after infection with *Campylobacter jejuni*

Krishna Prasad Vadalasetty[1], Charlotte Lauridsen[2], Ricarda Margarete Engberg[2], Radhika Vadalasetty[1], Marta Kutwin[3], André Chwalibog[1,3*] and Ewa Sawosz[3]

Abstract

Background: Silver nanoparticles (AgNP) have gained much attention in recent years due to their biomedical applications, especially as antimicrobial agents. AgNP may be used in poultry production as an alternative to the use of antibiotic growth promoter. However, little is known about the impact of oral administration of AgNP on the gut microbiota and the immune system. The aim of the present study was to investigate the effects of AgNP on growth, hematological and immunological profile as well as intestinal microbial composition in broilers challenged with *Campylobacter jejuni* (*C. jejuni*).

Results: AgNP did not affect the intestinal microbial profile of birds. The body weight gain and the relative weights of bursa and spleen were reduced when supplemented with AgNP. There was no difference with respect to packed cell volume. However, the plasma concentrations of IgG and IgM were lower in birds receiving AgNP compared to the non-supplemented control group. The expression of *TNF-α* and *NF-kB* at mRNA level was significantly higher in birds receiving AgNP.

Conclusions: The application of AgNP via the drinking water in the concentration of 50 ppm reduced broiler growth, impaired immune functions and had no antibacterial effect on different intestinal bacterial groups, which may limit the applicability of AgNP against *C. jejuni* in broiler chickens.

Keywords: Broiler chickens, Silver nanoparticles, Microflora, Immunoglobulins, Gene expression

Background

In broiler production, different kinds of antimicrobial agents are used for preventing and controlling diseases. Antimicrobials can affect the host intestinal flora, by reducing the colonization of intestinal bacteria, inhibiting the growth of pathogenic microorganisms, and enhancing the immune system, hence preventing diseases and improving animal performance [1–3]. However, the overuse of antimicrobial agents (antibiotics) promotes the emergence of antibiotic resistance in microorganisms [4], being harmful to animal and human health [5, 6]. For example, resistance to ciprofloxacin in *Camplylobacter jejuni* (*C. jejuni*) isolated from Danish broiler meat increased significantly from 0% in 2009 to 17% in 2010 [7]. The use of all antibiotics as growth promoter has been prohibited in the European Union since 2006 [2, 8]. Thus, there is a need to find alternatives to antibiotics in poultry production. When the conventional therapies, including antibiotics, anti-inflammatory agents, growth hormones, surgical interventions, and cytotoxic chemotherapies are ineffective in curing poultry infections, it is necessary to explore novel drug compounds. Nanoparticles have been emerging as one of the new treatment options, and their capability of penetrating normally intact physiologic barriers has reached a variety of molecular targets [9, 10].

Recent studies on antibacterial materials such as various natural (oils, acids), inorganic antimicrobial agents such as metals (Ag, Au, Cu) and metal oxides (ZnO, SiO_2, Fe_2O_3, TiO_2) have received increasing attention. Among metal

* Correspondence: ach@sund.ku.dk
[1]Department of Veterinary and Animal Sciences, University of Copenhagen, 1870 Frederiksberg, Denmark
[3]Department of Animal Nutrition and Biotechnology, Warsaw University of Life Sciences, 02-786 Warsaw, Poland
Full list of author information is available at the end of the article

nanoparticles, silver is one of the most promising components in several nanotechnology products. Currently, there are several consumer products containing various silver nanoparticles (AgNP) because of their antimicrobial properties [11–14]. AgNP have been shown to have a wide range of antibacterial activities against both Gram-positive and Gram-negative bacteria, including major foodborne pathogens [15–19]. At present, there is no study available on their antibacterial effect against *C. jejuni* being a leading cause of human gastroenteritis worldwide. It is mainly transmitted from contaminated chicken meat. *Campylobacter* infections are associated with the neurological disorder Guillain-Barre syndrome [20, 21]. A common feature of *C. jejuni* causes enterocolitis and is involved in acute inflammatory response that can lead to tissue damage and may be responsible for many of the clinical symptoms [22]. Furthermore, antimicrobial resistance was observed in *C. jejuni* and *Campylobacter coli* [23].

Although the antimicrobial effect of AgNP has been studied extensively, the mechanism of antibacterial activity specific to bacteriostatic or bactericidal activity remains unclear. Studies had shown that AgNP upon contact with water can release Ag + ions from their surface [24]. Free Ag + has a potent antimicrobial effect, which destroys microorganisms immediately by blocking the cellular respiration and disrupting the function of bacterial cell membranes. This occurs when Ag + binds to tissue proteins, causing structural changes in the bacterial cell membranes which, in turn, cause cell death. Essential protein complexes of the bacterial electron transport chains are located on the cell exterior and, therefore, are manageable for inactivation by reactive silver ions. Ag + also binds and denatures the bacterial DNA and RNA, thus inhibiting cell replication [25]. Recently, evidence has been obtained suggesting that silver nanoparticles may modulate the phosphotyrosine profile of putative bacterial peptides that could affect cellular signaling and, therefore, inhibit the growth of bacteria [26].

In recent years, several studies have been focused on anti-inflammatory therapy and on molecules which could block pro-inflammatory pathways. Nanoparticles of Ag and Au are considered anti-inflammatory agents or components of anti-inflammatory molecules [27, 28]. Moreover, in vivo studies with chicken embryos and quails showed that AgNP did not affect growth, development [29] and DNA oxidative damage to chicken embryos [30]. Results from toxicological assays have shown no in vitro cytotoxicity of AgNP (0.1, 0.5 and 1.0%) [17] but concentrations (2.5–50 μg/mL) of AgNP exert a cytotoxicity effect on human mesenchymal stem cells [31]. The most common health effects associated with chronic exposure to silver are a permanent grey or blue-grey discoloration of the skin (argyria). From the immunological perspective, it is known that phagocytosis of AgNP stimulates inflammatory signaling through the generation of reactive oxygen species (ROS) in macrophage cells, followed by the activated macrophage cell-induced secretion of TNF-α. The increase of TNF-α level causes damage to the cell membrane and apoptosis [32]. An inadvertent recognition of AgNP as a foreign particle by the immune cells may result in a multilevel immune response and finally lead to toxicity in the host. However, when the AgNP are recognized as self or an absence of immune recognition, then their ability to stimulate immune response may decide the fate of AgNP in the host. In vivo studies have demonstrated that nanoparticles are capable of promoting inflammation or suppressing immune functions [33–35]. The nanoparticle-induced inflammatory response may have an impact on immune defense, and the T-helper 1 (Th1)/T-helper 2 (Th2) balance [9, 36].

There are limited data regarding the effect of orally administered AgNP on the intestinal bacterial population and immune system of animals. In this study, we hypothesized that the antimicrobial properties of colloidal solutions of AgNP may affect the microbial population and immune responses upon challenge. The use of AgNP in poultry production may potentially function as an alternative to the use of antibiotic based growth promoters. The objective of this study was to investigate the effect of AgNP on the growth, the microbial profile of digestive tract and the immune status of broilers exposed to *C. jejuni* infection. In this study, we used hydrocolloid of AgNP because they exhibit high surface/volume ratio which may effectively enhance the bactericidal activity. In addition, chicks were used as an animal model for bacterial GI infection.

Methods

Experimental solution

The hydrocolloid of AgNP obtained from Nano-Tech (Warsaw, Poland) was produced by an electric non-explosive patented method (Polish Patent 3,883,399) from high purity metals (99.9999%) and high purity demineralized water. The concentration of the hydrocolloid was 50 mg/kg (50 ppm). The shape and size of nanoparticles were inspected by transmission electron microscopy (TEM), the particles had a crystal structure with an average size of 3.5 nm. The average surface area was 2.827×10^{-13} cm^2 and the pH of the colloidal silver solution was 7.1 to 8.1 (data provided by Nano-Tech, Poland). Furthermore, more detailed information regarding the applied AgNP are given by Sawosz et al., 2011 [18].

Experimental design

Ninety day-old male broiler chickens (Ross 308), obtained from a Danish commercial hatchery (DanHatch, Vrå, Denmark) were used. Upon arrival at the laboratory, the birds were wing-labeled, weighed and randomly distributed to two experimental groups: control, no AgNP, and provided with 50 ppm of AgNP in the drinking water. Furthermore, prior to the experiment antibacterial tests were done at in vitro level.

The chickens were housed in six individual isolators with 15 birds per isolator. The room temperature was 32 °C at the beginning of the experiment and was gradually decreased according to the demands of the growing chickens. The humidity was maintained during the experiment. Over the entire experimental period, both experimental groups were fed the same diet as described in Table 1. Drinking water was provided via nipple drinkers placed in the isolators. Birds were inspected every day and for every 2 days water and fed consumption were recorded. All birds had free access to food and water.

At day 11, all chickens were weighed and cloacal swab samples of 5 chickens per isolator were taken and examined for the presence of *C. jejuni* colonies for initial observation. Subsequently, two birds were randomly selected from each treatment and killed, and then blood samples were collected for packed cell volume (PCV) determination before samples of liver tissue were collected for gene expression studies. Afterwards, the chickens were orally challenged with 0.5 mL of an overnight culture of *C. jejuni* (4×10^7/ mL per bird). The infection strain (DVI-sc181) has been isolated from infected commercial broilers and provided by the Technical University of Denmark, National Veterinary Institute.

On days 15, 22 and 30, all chickens were weighed, and 5 chickens per isolator were randomly chosen. Blood samples were taken from the jugular vein and the chickens were immediately sacrificed by cervical dislocation. Liver, heart, spleen and bursa were collected, weighed and stored at -80 °C for further analysis. Ileum and caecum were collected to enumerate intestinal bacteria.

Table 1 Composition of the diet for broiler chickens (g/kg)

Ingredients	
Wheat	49.9
Maize	10.0
Rape seed (LL), grounded	4.00
Soybean meal (de-hulled, toasted)	29.4
Soybean oil	2.45
Calcium carbonate	0.90
Monocalcium phosphate	1.56
Sodium chloride	0.20
Natrium-bicarbonate	0.27
Lysine hydrochloride (100%)	0.25
DL-Methionine (100%)	0.35
Threonine (98%)	0.10
Vitamins and minerals (Vitfoss, Slut, 0.5%)	0.60
Total	100

Enumeration of *Campylobacter* and selected groups of intestinal bacteria

C. jejuni were enumerated in intestinal contents of ileum, caecum and faeces from 5 chickens per isolator at each sampling day. The samples (approximately 3 g) were homogenized, and serially diluted in 10-fold in phosphate buffered saline and plated on modified blood free charcoal cefoperazone deoxycholate agar base (Oxoid, CM0739). The plates were incubated with *C. jejuni* specific growth supplements at 42 °C for 48 h under microaerobic conditions (5% O_2, 5% CO_2, 5% H_2, and 85% N_2). Enterobacteria (*E. coli* and lactose negative enterobacteria) were enumerated on MacConkey agar (Merck, Darmstadt, Germany, 1.05465) incubated aerobically at 37 °C for 24 h as described by Engberg et al. [37]. Lactic acid bacteria (LAB) and *Clostridium perfringens* were counted respectively on De Man Rogosa Sharpe agar (Merck, 1.10660) incubated anaerobically at 37 °C for 48 h and tryptose sulphite cycloserine plates (TSC-Agar, Merck, 1.11072) incubated anaerobically at 37 °C for 24 h. Enterococci were counted on Slanetz and Bartely plates (Merck, 1.05289) after aerobic incubation at 37 °C for 48 h. The results are presented as microbial number (CFU/g) in ileal/caecal or faecal material. The detection limit was 10^{2} bacteria/g. During the experiment, water samples were collected from the drinkers in the isolators and the antimicrobial effect of AgNP on *C. jejuni* was investigated in in vitro using the plate count method.

Hematology

Before cervical dislocation, blood was collected by puncture of the jugular vein in disposable sodium heparinized hematocrit capillary tubes (Camlamb Ltd., Cambridge, UK). The tubes were filled up to two-thirds and sealed with cristaseal (Hawksley, Sussex, UK). The percentage of packed cell volume (PCV) was measured by using a micro-hematocrit reader (Hearaeus Reader, Osterode, Germany).

Concentrations of plasma immunoglobulins

From 5 chickens, blood samples were drawn from the jugularis vein and were collected in heparinized tubes. Blood samples were subsequently placed on ice, centrifuged at 2000 g for 10 min at 4 °C, and the plasma was stored at −20 °C until analysis of immunoglobulins. The concentrations of IgA, IgM and IgG were measured. Chicken plasma specific antibodies such as IgA (Bethyl laboratories, cat. no. E33–103), IgG (Bethyl laboratories, cat. no. E33–104) and IgM (Bethyl laboratories, cat. no. E33–102) concentrations were determined in diluted samples (1:100) by enzyme-linked immunosorbent assay

(ELISA) using microtiter plates (NuncImmunoplate 96-well, cat. no. 446612) as per manufacturer's ELISA quantitation kits (Bethyl Laboratories Inc., Montgomery, TX, USA).

Measurement of the content of specific chicken antibodies was done by indirect ELISA as follows: microtiter plate wells were coated with 100 µl of diluted coating antibody (1:200) to all wells and incubated for 2 h at room temperature (RT). After incubation, coated plates were washed with washing solution pH 8.0 (0.05 M Tris, 0.15 M NaCl, 10% Tween 20) to eliminate excess capture antibodies. Wells were incubated with 200 µl blocking buffer pH 8.0 (0.05 M Tris, 0.15 M NaCl, 1% BSA) on a shaker for 30 min at RT to block nonspecific protein binding and then washed 3 times with washing solution.

For the determination of immunoglobulins (IgG, IgM, or IgA), plasma samples were diluted and 100 µl of diluted plasma was added in triplicate. Plasma dilutions were 1:3000 for IgM, 1:1000 for IgA and 1:25,000 for IgG. Concentrations of IgG, IgM, and IgA in the standards were 6.25 mg/ml, 0.4 mg/ml, and 0.38 mg/ml respectively. Standards were diluted for IgG ranging from 200 ng/ml to 3.12 ng/ml, for IgM from 250 ng/ml to 3.9 ng/ml and for IgA from 1000 ng/ml to 15,625 ng/ml. The concentrations of immunoglobulins in test plasma samples were determined using these standard curves. The plates were incubated for 1 h at RT and were washed three times in washing solution. 100 µl of (horseradish-peroxidase) HRP detection anti-chicken IgM (A30-102P) was diluted 1:75,000 and for anti-chicken IgA (A30-103P) and IgG (A30-104P) were diluted 1:50,000 in conjugate diluent pH 8 (0.05 M Tris, 0.15 M NaCl, 10% Tween 20) was added to each well and the plates were allowed to incubate for 1 h at RT. After incubation, to remove unbound peroxidase-conjugates, each well was washed with washing solution for 3 times. 100 µl of (3,3,5,5-tetramethylbenzidine) TMB substrate was added to each well to determine bound peroxidase, and after incubation for 5–15 min in the dark at RT, 100 µl of 1 M H_2SO_4 was added to each well to stop the TMB reaction. The optical density was measured at 450 nm and expressed as ng of IgA, IgM and IgG per ml of plasma with a microplate reader. Using the mean absorbance value for each sample, determines the corresponding concentration of immunoglobulins in ng/ml from the standard curve.

Gene expression of TNF-α and NF-kB

Liver tissue samples (approximately 30 mg) were homogenized in TRIzol® Reagent (Invitrogen, Carlsbad CA, USA, cat. no. 15596–018). Then, RNA clean-up was performed and total RNA was extracted according to the manufacturer's protocol using SV Total RNA isolation system (Promega Corporation, Madison, WI, USA, cat. no. Z3105). Final RNA preparations were resuspended in RNase-free water and stored at -80 °C. The content of isolated RNA was quantified by UV-spectroscopy at 260 nm/280 nm with a NanoDrop® ND-1000 Spectrophotometer (Thermo Fisher Scientific, Waltham, MA, USA). Further quality was assessed by Agilent 2100 bioanalyser (Agilent RNA 6000 Nano kit, Waldbronn, Germany) and more than 6.5 RNA integrity number values were considered for cDNA preparation.

Two mg of total RNA was reverse-transcribed using the cDNA Reverse Transcription Kit (Promega, cat. no. A3500) in a G-storm PCR System. After this, real-time PCR was performed with complementary DNA and gene-specific primer pairs (TAG, Copenhagen A/S, Copenhagen, Denmark) mixed with LightCycler®480 SYBR Green I Master mix (Roche Applied Science, Penzberg, Germany) in a LightCycler® 480 real-time PCR system (Roche Applied Science). The cycling conditions included an initial heat step at 95 °C for 5 min, denaturation at 95 °C for 10 s, annealing temperatures described in Table 2 for each of the primers and product elongation at 72 °C for 20 s for 45 cycles. Following amplification, the melting curve was conducted on each sample to ensure that a single product was obtained using three-segment cycle of 95 °C for 5 s, 65 °C for 1 min and 95 °C no hold for continuous-acquisition mode with a heating rate of 0.11 °C/s and 5 acquisitions per 1 °C). Characterization of the product size was, furthermore, confirmed by agarose gel electrophoresis and sequencing (Bechman coulter genomics Takeley, UK). Results were quantified based on the relative expression of the *TNF-α* and *NF-kB* genes versus the housekeeping genes GAPDH and ACTB using advanced relative quantification (Efficiency method) by Light Cycler 480 Software release 1.50 SP4. The gene expression experiment was repeated 2 times chemically with consistent results and three replicates were used per sample each time.

Table 2 Genes and primers used in the study

RNA target gene	Primer sequences(5′–3′)	PCR Product		Gene bank Assess. No
		Annealing Temperature	Amplicon (bp)	
GAPDH				NM_204305.1
Forward	TGCTGCCCAGAACATCAT	61 °C	199	
Reverse	ATCAGCAGCAGCCTTCAC			
ACTB				NM_205518.1
Forward	GTCCACCTTCCAGCAGATGT	60 °C	169	
Reverse	ATAAAGCCATGCCAATCTCG			
NF-kB				M86930.1
Forward	TTGCTGCTGGAGTTGATGTC	60 °C	167	
Reverse	TTGCTGCTGGAGTTGATGTC			
TNF-α				NM_204267.1
Forward	TTCAGATGAGTTGCCCTTCC	59 °C	150	
Reverse	TCAGAGCATCAACGCAAAAG			

GAPDH Glyceraldehyde-3-phosphate dehydrogenase, *ACTB* Actin beta, *NF-kB* nuclear factor kappa B, and *TNF-α* Tumor necrosis factor alpha

Statistical analysis

The data were analyzed using the GLM procedure of SAS (SAS Institute Inc., 2008) 9.2 version for windows considering the treatments (AgNP vs. non-supplemented control), the day of age (15, 22 and 30) and interactions between treatment and age. Tukey-Kramer significant different test was employed to test the separation of the means and differences were considered statistically significant when the $p \leq 0.05$.

Results

Enumeration of bacteria

There were no differences between birds receiving AgNP and the control group (Fig. 1) with respect to the numbers of *C. jejuni,* lactic acid bacteria, *Enterococci, Clostridium perfringens, Escherichia coli,* and Lactose negative *enterobacteria* in the contents of caecum, ileum and feces of the birds.

The influence of AgNP on feed and water intake

AgNP had no effect on the average daily water intake (ADWI), average daily feed intake (ADFI) and feed conversion ratio (FCR = ADFI/ ADWG) (Table 3). The average daily weight gain (ADWG) was significantly lower in birds receiving AgNP compared to the control birds, in the periods 0–11 days ($p = 0.007$), 11–15 days ($p = 0.002$) 15–22 days ($p = 0.003$) but not in 22–30 days ($p = 0.728$). The average amount of AgNP which single chicken received in the drinking water was 8.26 mg/d (Additional file 1: Table S1).

The effect AgNP on chicken body weight and relative organ weights

The body weight of birds supplemented with AgNP was significantly lower than that of control bids (Table 4). Significant interactions between age and treatment were observed for the relative weight of bursa and spleen, where a lower relative bursa at age 15, and for spleen at age 30 were noticed for AgNP group. Relative organ weights of heart and liver were not different between the control birds and birds supplemented with AgNP.

AgNP response on PCV

At days 15, 22 and 30 of age no significant difference between control birds and birds receiving AgNP was observed with respect to blood PCV levels (Fig. 2). Likewise, no effect of age was observed.

Plasma immunoglobulin concentrations

We observed significant interactions between treatment and age for the concentrations of IgA, IgG and IgM in plasma (Fig. 3). At day 30, the plasma concentration of IgA was higher in birds supplemented with AgNP than in control birds, whereas the opposite effect was observed at age 15 and 22 days ($p = 0.05$, Fig. 3a). At day 30, the concentration of IgG in plasma was significantly lower in birds receiving AgNP as compared to the control birds (Fig. 3b). At age 15, plasma concentrations of IgM were higher in the AgNP group than in the control group (Fig. 3c), whereas at day 22 and 30, the concentrations of IgM were lower ($p < 0.05$).

Gene expression of *NF-kB* and *TNF-α*

The influence of AgNP on the mRNA *NF-κB* and *TNF-α* expression in liver tissue of *C. jejuni* infected chickens is shown in Fig. 4. Interestingly, at day 30, the levels of *NF-κB* (Fig. 4a-c) and *TNF-α* expression (Fig. 4b-d) were increased ($p < 0.05$) in the group supplemented with AgNP compared to the control group.

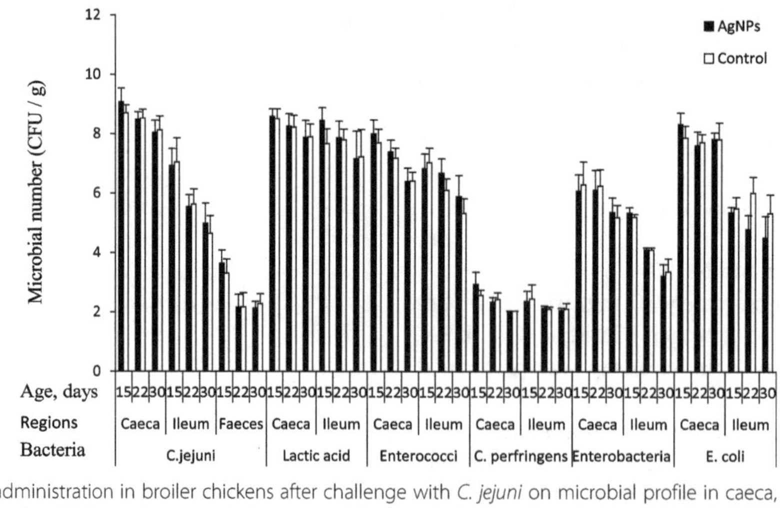

Fig. 1 Influence of AgNP administration in broiler chickens after challenge with *C. jejuni* on microbial profile in caeca, ileum and feces. At 15, 22, and 30 days of age, samples ($n = 30$ for each time point) were used for microbial count. Error bars represent the mean and standard errors of 6 isolators each with 15 birds

Table 3 Effect of silver nanoparticles (AgNP) on daily feed and water intake, weight gain and feed conversion ratio of chickens infected with *C. jejuni*

Parameters	Mean		Pooled	Treatment*Age
(g/bird)	Control	AgNP	SE	*p*-value
ADWI				
Day 0–30	161	165	0.34	0.64
ADFI				
Day 0–30	97.0	102	0.17	0.16
ADWG				
Day 0–11	21.7[a]	18.7[b]		
Day 11–15	43.7[a]	38.6[b]		
Day 15–22	62.7[a]	57.4[b]		
Day 22–30	70.5	69.4		
Day 0–30	46.6	44.9	0.03	0.34
FCR				
Day 0–30	1.97	2.22	0.04	

*Indicates interaction
Values are mean of 6 isolators each with 15 birds
SE pooled standard error, *ADWI* Average daily water intake (g per bird), *ADFI* Average daily feed intake (g per bird), *ADWG* Average daily weight gain (g per bird) and *FCR* Feed conversion ratio
[a,b] values within rows with different superscripts are significantly different at $p < 0.05$

Table 4 Effects of silver nanoparticles (AgNP) on cumulative body and relative organ weight gain (g/kg) in *C. jejuni* infected broiler chickens

Parameters (g/bird)	Mean		Pooled	Treatment*Age
	Control	AgNP	SE	p-value
AWG				
Age 15	380[a]	318[b]		
Age 22	788[a]	722[b]		
Age 30	1424[a]	1346[b]	0.305	0.013
Heart				
Age 15	0.71	0.70		
Age 22	0.62	0.66		
Age 30	0.47	0.50	9.17	0.340
Bursa				
Age 15	0.25 [a]	0.20[b]		
Age 22	0.21	0.22		
Age 30	0.20	0.20	5.08	0.003
Spleen				
Age 15	0.09	0.11		
Age 22	0.07	0.07		
Age 30	0.07[a]	0.05[b]	2.31	<0.001
Liver				
Age 15	2.71	2.82		
Age 22	2.52	2.56		
Age 30	2.01	2.00	0.003	0.657
Abnormalities	At age 30, big gall bladder was found from 2 out of 15 birds in AgNP group			

*Indicates interaction
[a, b] values within rows with different superscripts are significantly different at $p < 0.05$. SE – pooled standard error, AWG – average body weight gain. The relative organ weights (weight of organ/100 g live body weight). Values are mean of 6 isolators each with 15 birds

Discussion

In vitro antibacterial tests against *C. jejuni* showed a minimal inhibitory concentration of AgNP at the level of 40 and 50 ppm. AgNP have an effect on the target microorganism in the absence of chickens. In the present in vivo experiment, 50 ppm of AgNP was used because, prior to this experiment we carried out dose response studies to examine the antibacterial effect against *C. jejuni* of different concentrations of AgNP on chickens by chicken intestinal organ culture model (CIOC-model, Additional file 2: Table S2). The minimum inhibitory and minimum bactericidal concentration of silver nanoparticles on *C. jejuni* was determined by using the microtiter plate method (Additional file 3: Figure S1), showed bactericidal effect with minimum concentration of 50 ppm. We have evaluated samples of ileum, caecum and feces because of the uniformity of the gut microbiota in these segments. The caecum has a slow digesta passage rate, allowing to harbor a complex microbiome that has considerable effects on host nutrition and health [38]. The small intestinal region of the ileum has received attention since it is the principal site of nutrient absorption and microflora [39]. Furthermore, we focused on test samples at various time points because the caecal and ileal microflora changes in relation to age and dietary treatments [38, 39].

However, unexpectedly, AgNP did not change the microbial profile of caecum, ileum and feces of examined chickens (Fig. 1). These results are consistent with in vivo experiments with the microbial profile of young quails receiving hydrocolloids of AgNP administered with 5, 15 and 25 ppm [29]. Moreover, another study demonstrated that supplying AgNP did not effectively rescue *Salmonella*-mediated mortality in chickens but nanoscale silicate platelet (NSP) and its nanohybrid composite of AgNP/NSP effectively controlled the infection [40]. However, our present study was not similar with this study in terms of end point of examination, concentration of nanoparticles, bacterial infection time points and target bacterial species. The factors that could affect the present results might be the method of AgNP administration, although *C. jejuni* colonization in the intestinal tract depends on the number and diversity of environmental microorganisms, as well as influence of feedstuffs. The other possibilities could be the *C. jejuni* biofilm formation and adhesion to the host intestinal wall may provide protection against nanoparticles. The chicken intestinal mucus is able to attenuate *C. jejuni* virulence by inhibiting its ability to adhere and invade intestinal epithelial cells [41]. Furthermore, the antibacterial

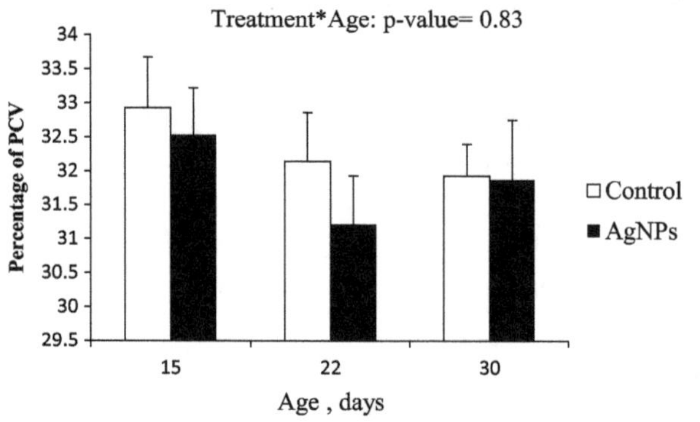

Fig. 2 Effect of silver nanoparticles (AgNP) on packed cell volume levels in chickens infected with *C. jejuni*. Error bars represent the mean values and standard errors of 6 isolators each with 15 birds

properties of AgNP could vanish inside the digestive tract environment of chickens, probably, due to AgNP agglomeration in the presence of low pH gastric acid. The obtained results may suggest that AgNP were gastro-sensitive, the stability and dispersion of AgNP in gastric acid is a critical factor for antibacterial activity. Once the particle enters a biological system, physical properties such as solubility, particle agglomeration, surface charge and particle-protein complex interactions might be different from those of the in vitro-measured properties. The loss of nanoparticles' properties may be due to the low colloidal stability and the reduction of reactive surfaces may affect the efficacy for controlling pathogens in a solution [42].

The reduction in cumulative body weight and relative organ weights of the spleen and bursa of birds supplemented with AgNP might be due to the effect of AgNP blocking the intestinal absorption of actively transported sugars and amino acids, and decreased protein digestibility through the small intestine where mainly enzymatic activity is present. Another possible reason might be the fact that the birds undergo more cellular stress and excessive cellular interactions with AgNP. It appears that *C. jejuni* colonization may lead to weight loss in grower chickens. The present results are consistent with decreased body and organ weights in chickens treated with 25 ppm of AgNP at 42 days of age [43]. Moreover, the study of Asharani et al. [44] suggested that AgNP increase ROS production and interrupt ATP synthesis, leading to DNA damage and cell cycle arrest at G2/M phase. Park et al. [32] found that AgNP induce G1 phase arrest and a complete blockage of the S phase, with the induction of apoptosis. The AgNP (10 ppm) injected into fertilized eggs on days 5, 11 and 17 of incubation did not influence the development of embryos

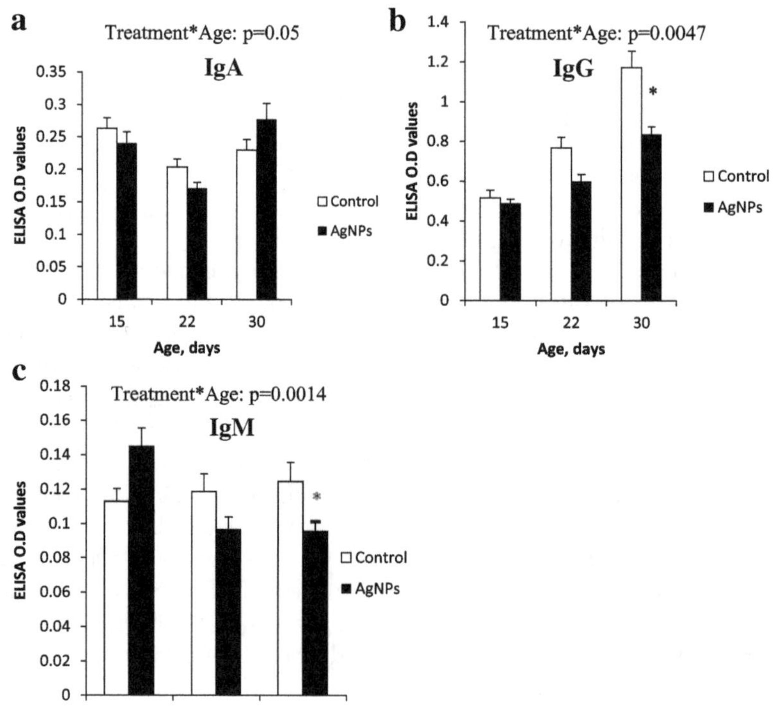

Fig. 3 Concentration of immunoglobulins (IgA, IgG and IgM) in chickens infected with *C. jejuni* Samples were analyzed at days 15, 22 and 30 of age; **a**) IgA **b**) IgG and **c**) IgM. Error bars represent the mean values and standard errors of 6 isolators, each with 15 birds. Interactions between treatment and age effect significantly ($p < 0.05$). * Indicates significant difference between control and AgNP ($p < 0.05$)

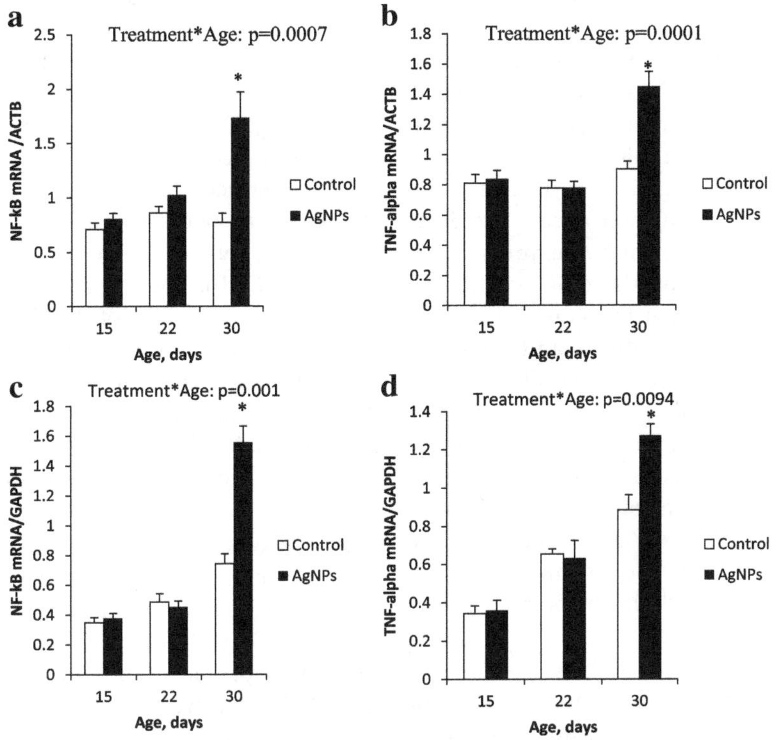

Fig. 4 The expression of mRNA *NF-κB* and *TNF-α* normalized to the housekeeping genes *ACTB* and *GAPDH* in the liver tissue of chickens infected with *C. jejuni*. Samples were analyzed at days 15, 22 and 30 of age; **a)** *NF-κB /ACTB*, **b)** *TNF-α /ACTB*, **c)** *NF-κB /GAPDH* and **d)** *TNF-α /GAPDH*. Error bars represent the mean values and standard errors of 6 isolators, each with 15 birds. Interactions between treatment and age effect significantly ($p < 0.05$). * Indicates significant difference between control and AgNP ($p < 0.05$)

but decreased the number and size of lymph follicles in the bursa of Fabricius [45].

The PCV results indicated that the provision of AgNP did not influence the percentage of red blood cells (Fig. 2). On the other hand, it was reported that the oral administration of AgNP induced some changes in the red blood compartment, such as increased red blood cell count [46] and coagulation parameters [47].

The present results demonstrated age-related differences in immunoglobulin concentrations, which could probably be ascribed to the infection with *C. jejuni* on day 11 of age. Immunoglobulins are generally associated with immune resistance to extracellular bacteria and viruses or other foreign substances. Therefore, their levels might be most appropriate to study diseases caused by extracellular microorganisms [48]. Plasma immunoglobulin levels were measured following *C. jejuni* infection of chickens on days 15, 22 and 30 of age. Noticeable, significantly lower levels of IgG and IgM were observed in AgNP supplemented chickens (Fig. 3b-c). The decrease in the levels of IgG and IgM in AgNP birds might be due to the microbial colonization and provision of AgNP via drinking water. Importantly, the impaired T−cell function might be a reason for reduced IgG levels. AgNP might impair intestinal actively transported sugars, amino acids, trace elements, and vitamins, and deficiencies of these nutrients may decrease antibody formation. Similar observations showed decreased plasma IgG levels in chickens treated with 10 ppm and 20 ppm AgNP but not infected with *C. jejuni* [49]. Reduction of bursa and spleen weights may be correlated to the decreased IgG levels. Therefore, the expression of immunological effects in this study was assumed to be the result of AgNP impairment. It could be that lymphocyte production and self- or non-self-antigen selection against *C. jejuni* infection changes with age, and could also be influenced by AgNP. Our results suggested that AgNP can diminish the activity of humoral immunity of broiler chickens by decreasing the levels of immunoglobulins in plasma.

In the present study, *NF-kB* and *TNF-α* expression was determined in chicken liver tissue to evaluate whether or not AgNP could modify inflammation at the transcriptional level (Fig. 4). NF-κB and TNF-α expressions are important in *Campylobacter* colonization as *C. jejuni* primarily colonized the lower small intestine and caecum, where remarkable histopathologic and ultrastructural changes in the epithelium were noticed. In a novel rabbit model, the pathogen induced intestinal inflammation had increased levels of IFN-γ, TNF-α, IL-1β, IL-2, IL-6, IL-8, and IL-22 gene expression. In the acute phase, the bacteria induced a significant increase in the expression of the most pro-inflammatory genes [50]. The infection of human intestinal epithelium with *Campylobacter* results in activation of *NF-kB*, which is needed for the induction of pro-inflammatory genes [51]. We have recently observed that *TNF-α* mRNA expression was consistent with 50 ppm when AgNP were injected into chicken embryos after LPS stimulation during the 19th day [52]. Results concerning the mRNA expression of *NF-kB* were not in agreement with those of the chicken embryo liver [53]. We suppose that this could be due to the time of exposure, tissue specificity and route of administration of AgNP. However, our results confirm the pro-inflammatory activity of AgNP, which has been previously observed in chickens and mice [54, 55]. Furthermore, an increased inflammatory response of the AgNP group in comparison with the control group indicates an over-immunostimulation activity of AgNP. The ability of nanoparticles to freely move into local lymphoid tissues and trigger antigen-presenting cells might be responsible for their immunostimulatory activity.

The exact mechanism and reason how and where AgNP induce pro-inflammatory effects are not known, but it has been reported that they stimulate production of reactive oxygen species (ROS), and thereby modulate intracellular calcium concentrations, activate transcription factors, and induce cytokine production [56]. The toxicity of nanoparticles is manifested by inflammation resulting from oxidative stress [57–59]. Recently, various studies with AgNP have been published, demonstrating conflicting results that silver nanoparticles are toxic [44, 59–63] or non-toxic [64–66]. Furthermore, they could be pro-inflammatory [47, 67–69] or anti-inflammatory agents [27, 28, 65]. Importantly, particle size-dependent effects of AgNP were observed with respect to cellular uptake, pro-inflammatory response and changes at the proteome level [70]. The 20 nm AgNP elucidated a higher inflammatory response than the 200 nm particles in Caco-2/TC7 cells. It might be that bigger particles have different transport rates, reduced interaction with the cellular membranes and are better retained in the gastrointestinal epithelial mucous layer than smaller particles. Thus, smaller particles may cross the mucus layer and reach the cells.

An increase in mRNA expression of inflammatory mediators and low IgG and IgM levels could be due to the nanoparticle uptake triggering cellular effects, leading to inflammatory responses. However, the conflicting results might indicate that AgNP have multiple cellular targets that vary among different cell type. These results are attributed to several confounding factors such as pH [71], continuous oral administration of AgNP, [47, 68] or high concentration, [28] or even the availability of free radicals to induce oxidative stress and damage cells [58]. We propose that nanoparticles time of exposure, route of administration, particles size, aggregate formation, and altered bio-distribution in the form of rapid clearance owing to non-specific pathogen clearance from the systemic circulation could serve as aided factors. One possible cause for the AgNP dependent initiation of inflammation could be the fact that they enhance the production of ROS. These oxygen-derived free radicals may lead to mitochondrial dysfunction, increased gene expression of inflammatory cytokines (TNF-α) and activation of specific transcription factors (NF-kB). The absorption, distribution, metabolism, excretion, and toxicity of AgNP are largely dependent on their physicochemical properties and the surrounding environmental conditions. However, we were unable to find a mechanism involved in the pro-inflammatory pathway of AgNP in the present work and could not verify the presence of AgNP in the digestive organs.

Conclusions

Orally-administered AgNP via the drinking water (50 ppm) had no effect on the intestinal colonization of *C. jejuni* following infection and did not influence the intestinal microbial profile of broiler chickens. However, AgNP reduced body weight gain, lowered concentrations of plasma immunoglobulins and upregulated mRNA expression of *TNF-α* and *NF-kB*, indicating toxicity and impaired immune response. Thus, the use of orally administered AgNP might be harmful to the chicken health.

Additional files

Additional file 1: Table S1. Average daily water intake and the concentrations of silver nanoparticles (AgNP) when chickens were infected with *C. jejuni*.

Additional file 2: Table S2. Antibacterial effect of silver nanoparticles (AgNP) against *C. jejuni* by using chicken intestinal organ culture model (CIOC-model). AgNP concentrations 10 ppm, 20 ppm were provided via drinking water, 40 ppm and 80 ppm were supplemented in intestinal organ culture. Bacterial count was done by standard plate count method.

Additional file 3: Figure S1. Effect of silver nanoparticles (AgNP) on *C. jejuni* at various concentrations, using broth microdilution method by microtiter plate. Bacterial density was measured by the plate counter. D20 to D70 ppm means diluted AgNP concentration from stock solution, C80 and C50 ppm means stock AgNP solution. Media, with bacteria were considered as negative and positive controls. The mean values of 3 repetitions. 1a) Bacterial growth (Optical density (OD)) difference = Average of bacteria - Average of media. 1b) Bacterial growth difference (%) = Bacterial growth difference / Average of bacteria.

Abbreviations

ADFI: Average daily feed intake; ADWG: Average daily weight gain; ADWI: Average daily water intake; AgNP: Silver nanoparticles; *C. jejuni*: *Camplylobacter jejuni*; CCDA: Charcoal cefoperazone deoxycholate agar; CIOC: Chicken intestinal organ culture; *E. coli*: *Escherichia coli*; ELISA: Enzyme-linked immunosorbent assay; FCR: Feed conversion ratio; HRP: Horseradish-peroxidase; Igs: Immunoglobulins; LAB: Lactic acid bacteria; MRS: de Man Rogosa Sharpe; PCV: Packed cell volume; RIN: RNA integrity number; ROS: Reactive oxygen species; RT: Room temperature; TEM: Transmission electron microscopy; TMB: 3,3,5,5-tetramethylbenzidine; TSC: Tryptose sulfite cycloserine

Funding
This research work was supported by the Danish Agency for Science Technology and Innovation # 2106-08/0025.

Authors' contributions
KPV performed the experiments, analyzed data and wrote the paper. CL and RGE designed and coordinated the study, and evaluated the performance and laboratory measurements (microbial and hematological profiles, immunological results). ES and MK contributed with interpretation of the results and helped drafting the manuscript. RKP helped in laboratory setup and gene expression studies. AC participated in the design and coordination and helped with drafting the manuscript. All authors read and approved the final manuscript.

Competing interests
All authors declare that they have no competing interests.

Author details
[1]Department of Veterinary and Animal Sciences, University of Copenhagen, 1870 Frederiksberg, Denmark. [2]Department of Animal Science, Aarhus University, 8830 Tjele, Denmark. [3]Department of Animal Nutrition and Biotechnology, Warsaw University of Life Sciences, 02-786 Warsaw, Poland.

References
1. Modi CM, Mody SK, Patel HB, Dudhatra GB, Kumar A, Sheikh TJ. Growth promoting use of antimicrobial agents in animals. J Appl Pharm Sci. 2011; 01(08):33–6.
2. Hao H, Cheng G, Iqbal Z, Ai X, Hussain HI, Huang L, Dai M, Wang Y, Liu Z, Yuan Z. Benefits and risks of antimicrobial use in food-producing animals. Front Microbiol. 2014;5:288.
3. Niewold TA. The non-antibiotic anti-inflammatory effect of antimicrobial growth promotors, the real mode of action? A hypothesis Poult Sci. 2007;86:605–9.
4. Asai T, Kojima A, Harada K, Ishihara K, Takahashi T, Tamura Y. Correlation between the usage volume of veterinary therapeutic antimicrobials and resistance in *Escherichia coli* isolated from the feces of food-producing animals in Japan. Jpn J Infect Dis. 2005;58(6):369–72.
5. Bach Knudsen KE. The nutritional significance of "dietary fibre" analysis. Anim Feed Sci Tech. 2001;90(1–2):3–20.
6. Wiesner RS, Hendrixson DR, DiRita VJ. Natural transformation of *Campylobacter jejuni* requires components of a type II secretion system. J Bacterial. 2003;185(18):5408–18.
7. DANMAP 2013 report. Use of antimicrobial agents and occurrence of antimicrobial resistance in bacteria from food animals, food and humans in Denmark. http://www.danmap.org/downloads/reports.aspx. Accessed 6 Oct 2015.
8. Van Immerseel F, De Buck J, Pasmans F, Huyghebaert G, Haesebrouck F, Ducatelle R. *Clostridium perfringens* in poultry an emerging threat for animal and public health. Avian Pathol. 2004;33(6):537–49.
9. Chang C. The immune effects of naturally occurring and synthetic nanoparticles. J Autoimmun. 2010;34(3):234–46.
10. Zhang Y, Bai Y, Yan B. Functionalized carbon nanotubes for potential medicinal applications. Drug Discov Today. 2010;15(11–12):428–35.
11. Panacek A, Kvitek L, Prucek R, Kolar M, Vecerova R. Silver colloid nanoparticles: synthesis, characterization and their antibacterial activity. J Phys Chem B. 2006; 110(33):16248–53.
12. Lok CN, Ho CM, Chen R, He QY, Yu WY, Sun H, Tam PK, Chiu JF, Che CM. Silver nanoparticles: partial oxidation and antibacterial activities. J Biol Inorg Chem. 2007;12(4):527–34.
13. Morones JR, Elechiguerra JL, Camacho A, Holt K, Kouri JB, Ramírez JT, Yacaman MJ. The bactericidal effect of silver nanoparticles. Nanotechnol. 2005;16(10):2346–53.
14. Kim JS, Kuk E, Yu KN, Kim JH, Park SJ, Lee HJ, Kim SH, Park YK, Park YH, Hwang CY, Kim YK, Lee YS, Jeong DH, Cho MH. Antimicrobial effects of silver nanoparticles. Nanomed. 2007;3(1):95–101.
15. Yoon K, Hoon Byeon J, Park J, Hwang J. Susceptibility constants of *Escherichia coli* and *Bacillus subtilis* to silver and copper nanoparticles. Sci Total Environ. 2007;373(2–3):572–5.
16. Ayala-Núñez NV, Lara Villegas HH, Ixtepan Turrent LC, Padilla CR. Silver nanoparticles toxicity and bactericidal effect against methicillin-resistant *Staphylococcus aureus*: nanoscale does matter. Nanobiotechnol. 2009;5:2–9.
17. Alt V, Bechert T, Steinrucke P, Wagener M, Seidel P, Dingeldein E, Domann E, Schnettler R. An *in vitro* assessment of the antibacterial properties and cytotoxicity of nanoparticulate silver bone cement. Biomaterials. 2004;25(18):4383–91.
18. Sawosz E, Chwalibog A, Mitura K, Mitura S, Szeliga J, Niemiec T, Rupiewicz M, Grodzik M, Sokolowska A. Visualization of morphological interaction of diamond and silver nanoparticles with *Salmonella Enteritidis* and *Listeria Monocytogenes*. J Nanosci Nanotechnol. 2011;11:1–7.
19. Chen D, Xi T, Bai J. Biological effects induced by nanosilver particles: *in vivo* study. Biomed Mater. 2007;2(3):126–8.
20. Rees JH, Gregson NA, Griffiths PL, Hughes RAC. *Campylobacter jejuni* and Guillain-Barré syndrome. Q J Med. 1993;86:623–34.
21. Nachamkin I. Microbiologic approaches for studying campylobacter in patients with Guillain-Barre syndrome. J Infect Dis. 1997;176(2):106–14.
22. Ketley JM. Pathogenesis of enteric infection by campylobacter. Microbiology. 1997;143(1):5–21.
23. Gormley FJ, Strachan NJ, Reay K, MacKenzie FM, Ogden ID, Dallas JF, Forbes KJ. Antimicrobial resistance profiles of campylobacter from humans, retail chicken meat, and cattle feces. Foodborne Pathog Dis. 2010;7(9):1129–31.
24. Sotiriou GA, Pratsinis SE. Antibacterial activity of nanosilver ions and particles. Environ Sci Technol. 2010;44(14):5649–54.
25. Dunn K, Edwards-Jones V. The role of Acticoat with nanocrystalline silver in the management of burns. Burns. 2004;30(Supp 1):S1–9.
26. Shrivastava S, Bera T, Roy A, Singh G, Ramachandrarao P, Dash D. Characterization of enhanced antibacterial effects of novel silver nanoparticles. Nanotechnol. 2007;18:225103.
27. Bhol KC, Schechter PJ. Topical nanocrystalline silver cream suppresses inflammatory cytokines and induces apoptosis of inflammatory cells in a murine model of allergic contact dermatitis. Br J Dermatol. 2005;152(6):1235–42.
28. Shin SH, Ye MK, Kim HS, Kang HS. The effects of nano-silver on the proliferation and cytokine expression by peripheral blood mononuclear cells. Int Immunopharmacol. 2007;7(13):1813–8.
29. Sawosz E, Binek M, Grodzik M, Zielinska M, Sysa P, Szmidt M, Niemiec T, Chwalibog A. Influence of hydrocolloidal silver nanoparticles on gastrointestinal microflora and morphology of enterocytes of quails. Arch Anim Nutr. 2007;61(6):441–51.
30. Sawosz E, Grodzik M, Zielinska M, Niemiec T, Olszanska B, Chwalibog A. Nanoparticles of silver do not affect growth, development and DNA oxidative damage in chicken embryos. Arch Geflügelk. 2009;73(3):S208–13.
31. Greulich C, Kittler S, Epple M, Muhr G, Koller M. Studies on the biocompatibility and the interaction of silver nanoparticles with human mesenchymal stem cells (hMSCs). Langenbeck's Arch Surg. 2009;394(3):495–502.
32. Park EJ, Yi J, Kim Y, Choi K, Park K. Silver nanoparticles induce cytotoxicity by a Trojan-horse type mechanism. Toxicol in Vitro. 2010;24(3):872–8.
33. Nygaard UC, Hansen JS, Samuelsen M, Alberg T, Marioara CD, Løvik M. Single-walled and multi-walled carbon nanotubes promote allergic immune responses in mice. Toxicol Sci. 2009;109(1):113–23.
34. Mitchell LA, Lauer FT, Burchiel SW, McDonald JD. Mechanisms for how inhaled multi-walled carbon nanotubes suppress systemic immune function in mice. Nat Nanotechnol. 2009;4(7):451–6.
35. Yamaguchi A, Fujitani T, Ohyama K, Nakae D, Hirose A, Nishimura T, Ogata A. Effects of sustained stimulation with multi-wall carbon nanotubes on immune and inflammatory responses in mice. J Toxicol Sci. 2012;37(1):177–89.
36. Wang X, Podila R, Shannahan JH, Rao AM, Brown JM. Intravenously delivered graphene nanosheets and multi-walled carbon nanotubes induce site specific Th2 inflammatory responses via the IL-33/ST2 axis. Int J Nanomedicine. 2013;8(3):1733–48.
37. Engberg RM, Hedemann MS, Leser TD, Jensen BB. Effect of zinc bacitracin and salinomycin on intestinal microflora and performance of broilers. Poult Sci. 2000;79(9):1311–9.
38. Pan D, Yu Z. Intestinal microbiome of poultry and its interaction with host and diet. Gut Microbes. 2014;5(1):108–19.
39. Knarreborg A, Simon MA, Engberg RM, Jensen BB, Tannock GW. Effects of dietary fat source and subtherapeutic levels of antibiotic on the bacterial community in the ileum of broiler chickens at various ages. Appl Environ Microbiol. 2002;68(12):5918–24.

40. Chiao SH, Lin SH, Shen CI, Liao JW, Bau IJ, Wei JC, Tseng LP, Hsu SH, Lai PS, Lin SZ, Lin JJ, Su HL. Efficacy and safety of nanohybrids comprising silver nanoparticles and silicate clay for controlling salmonella infection. Int J Nanomedicine. 2012;7:2421–32.
41. Alemka A, Whelan S, Gough R, Clyne M, Gallagher ME, Carrington SD, Bourke B. Purified chicken intestinal mucin attenuates *Campylobacter jejuni* pathogenicity in vitro. J Med Microbiol. 2010;59:898–903.
42. Sondi I, Salopek-Sondi B. Silver nanoparticles as antimicrobial agent: a case study on E. Coli as a model for gram-negative bacteria. J Colloid Interface Sci. 2004;275(1):177–82.
43. Ahmadi F, Kurdestany AH. The impact of silver nano particles on growth performance, lymphoid organs and oxidative stress indicators in broiler chicks. Global Veterinaria. 2010;5(6):366–70.
44. Asharani PV, Lian Wu Y, Gong Z, Valiyaveettil S. Toxicity of silver nanoparticles in zebrafish models. Nanotechnol. 2008;19(25):255102–10.
45. Grodzik M, Sawosz E. The influence of silver nanoparticles on chick embryo development and bursa of Fabricius morphology. J Anim Feed Sci. 2006; 15(1):111–4.
46. Hillery AM, Jani PU, Florence AT. Comparative, quantitative study of lymphoid and non-lymphoid uptake of 60 nm polystyrene particles. J Drug Target. 1994;2(2):151–6.
47. Kim YS, Kim JS, Cho HS, Rha DS, Kim JM, Park JD, Choi BS, Lim R, Chang HK, Chung YH, Kwon IH, Jeong J, Han BS, Yu IJ. Twenty-eight-day oral toxicity, genotoxicity, and gender-related tissue distribution of silver nanoparticles in Sprague–Dawley rats. Inhal Toxicol. 2008;20(6):575–83.
48. Sarker N, Tsudzuki M, Nishibori M, Yasue H, Yamamoto Y. Cell-mediated and humoral immunity and phagocytic ability in chicken lines divergently elected for serum immunoglobulin M and G levels. Poult Sci. 2000;79(12):1705–9.
49. Pineda L, Chwalibog A, Sawosz E, Lauridsen C, Engberg R, Elnif J, Hotowy A, Sawosz F, Gao Y, Ali A, Moghaddam HS. Effect of silver nanoparticles on growth performance, metabolism and microbial profile of broiler chickens. Arch Anim Nutr. 2012;66(5):416–29.
50. Shang Y, Ren F, Song Z, Li Q, Zhou X, Wang X, Xu Z, Bao G, Wan T, Lei T, Wang N, Jiao XA, Huang J. Insights into campylobacter jejuni colonization and enteritis using a novel infant rabbit model. Sci Rep. 2016;6(28737):1–12. doi: 10.1038/srep28737.
51. Zheng J, Meng J, Zhao S, Singh R, Song W. *Campylobacter*-induced interleukin-8 secretion in polarized human intestinal epithelial cells requires *Campylobacter*-secreted cytolethal distending toxin- and toll-like receptor-mediated activation of NFkappaB. Infect Immun. 2008;76:4498–508.
52. Bhanja SK, Hotowy A, Mehra M, Sawosz E, Pineda L, Vadalasetty KP, Kurantowicz N, Chwalibog A. *In ovo* administration of silver nanoparticles and/or amino acids influence metabolism and immune gene expression in chicken embryos. Int J Mol Sci. 2015;16:9484–503.
53. Sawosz E, Grodzik M, Lisowski P, Zwierzchowski L, Niemiec T, Zielińska M, Szmidt M, Chwalibog A. Influence of hydrocolloids of ag, au, and ag/cu alloy nanoparticles on the inflammatory state at transcriptional level. Bull Vet Inst Pulawy. 2010;54:81–5.
54. Loghman A, Iraj SH, Naghi DA, Pejman M. Histopathology and apoptotic effect of nanosilver in liver of broiler chickens. Afr J Biotechnol. 2012;11(22):6207–11.
55. Małaczewska J. The effect of silver nanoparticles on splenocyte activity and selected cytokine levels in the mouse serum at early stage of experimental endotoxemia. Pol J Vet Sci. 2011;14(4):597–604.
56. Brown DM, Donaldson K, Borm PJ, Schins RP, Dehnhardt M, Gilmour P, Jimenez LA, Stone V. Calcium and ROS-mediated activation of transcription factors and TNF-alpha cytokine gene expression in macrophages exposed to ultrafine particles. Am J Physiol Lung Cell Mol Physiol. 2004;286(2):L344–53.
57. Su HL, Chou CC, Hung DJ, Lin SH, Pao IC, Lin JH, Huang FL, Dong RX, Lin JJ. The disruption of bacterial membrane integrity through ROS generation induced by nanohybrids of silver and clay. Biomaterials. 2009;30:5979–87.
58. Park EJ, Yoon J, Choi K, Yi J, Park K. Induction of chronic inflammation in mice treated with titanium dioxide nanoparticles by intratracheal instillation. Toxicology. 2009;260(1–3):37–46.
59. Carlson C, Hussain SM, Schrand AM, Braydich-Stolle LK, Hess KL, Jones RL, Schlager JJ. Unique cellular interaction of silver nanoparticles: size-dependent generation of reactive oxygen species. J Phys Chem B. 2008;112(43):13608–19.
60. Ahamed M, Karns M, Goodson M, Rowe J, Hussain SM, Schlager JJ, Hong Y. DNA damage response to different surface chemistry of silver nanoparticles in mammalian cell. Toxicol Appl Pharmacol. 2008;233(3):404–10.
61. Park S, Lee YK, Jung M, Kim KH, Eun-Kyung Ahn NC, Lim Y, Lee KH. Cellular toxicity of various inhalable metal nanoparticles on human alveolar epithelial cells. Inhal Toxicol. 2007;19(1):59–65.
62. Yen HJ, Hsu SH, Tsai CL. Cytotoxicity and immunological response of gold and silver nanoparticles of different sizes. Small. 2009;5(13):1553–61.
63. Vandebriel RJ, Tonk EC, de la Fonteyne-Blankestijn LJ, Gremmer ER, Verharen HW, van der Ven LT, van Loveren H, de Jong WH. Immunotoxicity of silver nanoparticles in an intravenous 28-day repeated-dose toxicity study in rats. Part and Fibre Toxicol. 2014;11:21.
64. Ji JH, Jung JH, Kim SS, Yoon JU, Park JD, Choi BS, Chung YH, Kwon IH, Jeong J, Han BS, Shin JH, Sung JH, Song KS, Yu IJ. Twenty-eight-day inhalation toxicity study of silver nanoparticles in Sprague-Dawley rats. Inhal Toxicol. 2007;19(10):857–71.
65. Bhol KC, Schechter PJ. Effects of nanocrystalline silver (NPI 32101) in a rat model of ulcerative colitis. Dig Dis Sci. 2007;52(10):2732–42.
66. Sung JH, Ji JH, Park JD, Yoon JU, Kim DS, Jeon KS, Song MY, Jeong J, Han BS, Han JH, Chung YH, Chang HK, Lee JH, Cho MH, Kelman BJ, Yu IJ. Subchronic inhalation toxicity of silver nanoparticles. Toxicol Sci. 2009;108(2):452–61.
67. Trickler WJ, Lantz SM, Murdock RC, Schrand AM, Robinson BL, Newport GD, Schlager JJ, Oldenburg SJ, Paule MG, Slikker W, Hussain SM, Ali SF. Silver nanoparticles induced blood-brain barrier inflammation and increased permeability in primary rat brain microvessel endothelial cells. Toxicol Sci. 2010;118(1):160–70.
68. Park EJ, Bae E, Yi J, Kim Y, Choi K, Leed SH, Yoon J, Lee BC, Park K. Repeated-dose toxicity and inflammatory responses in mice by oral administration of silver nanoparticles. Environ Toxicol Pharmacol. 2010;30(2):162–8.
69. Chen X, Schluesener HJ. Nanosilver: a nanoproduct in medical application. Toxicol Lett. 2008;176(1):1–12.
70. Georgantzopoulou A, Serchi T, Cambier S, Leclercq CC, Renaut J, Shao J, Kruszewski M, Lentzen E, Grysan P, Eswara S, Audinot JN, Contal S, Ziebel J, Guignard C, Hoffmann L, Murk AJ, Gutleb AC. Effects of silver nanoparticles and ions on a co-culture model for the gastrointestinal epithelium. Part and Fibre Toxicol. 2016;13:9.
71. Nadworny PL, Wang JF, Tredget EE, Burrell RE. Anti-inflammatory activity of nanocrystalline silver-derived solutions in porcine contact dermatitis. J Inflamm. 2010;7(13):1–20.

11

Effects of dietary postbiotic and inulin on growth performance, IGF1 and GHR mRNA expression, faecal microbiota and volatile fatty acids in broilers

Karwan Yaseen Kareem[1,5], Teck Chwen Loh[1,2*], Hooi Ling Foo[3,4*], Henny Akit[1] and Anjas Asmara Samsudin[1]

Abstract

Background: Postbiotics (metabolic products by lactic acid bacteria) and prebiotics have been established as substitute to antibiotics in order to enhance immunity and growth performance in broiler chickens. Nonetheless, insufficient information is available on the effects of postbiotics and prebiotics combination on growth performance, faecal microbiota, pH and volatile fatty acids (VFA), as well as liver insulin like growth factor 1 (IGF1) and growth hormone receptor (GHR) mRNA expressions in broiler chickens. The aim of this experiment was to evaluate the effects of different types of postbiotics with different levels of prebiotic (inulin) on broiler for those parameters.

Results: The results showed that birds fed T3: (0.3 % RI11 + 0.8 % Inulin), T4: (0.3 % RI11 + 1.0 % Inulin), and T6: (0.3 % RG14+ 1.0 % Inulin) had higher ($p < 0.05$) final body weight (BW) and total weight gain (WG) than other treatments. Birds fed T3 had lower feed conversion ratio (FCR) which was significantly different from those fed with negative control diet but was similar to other treatments. Postbiotic and inulin increased ($p < 0.05$) faecal lactic acid bacteria (LAB) and reduced ($p < 0.05$) *Enterobacteriaceae* count. Birds fed T4 and T6 had higher faecal acetic acid and propionic acid respectively, and both had higher total VFA and lactic acid bacteria but lower pH and *Enterobacteriaceae* (ENT) counts compared to other treatments. The liver of birds fed T4 and T6 had higher IGF1 expression compared to other treatments while T6 had higher GHR mRNA expression compared to other treatments.

Conclusions: Results indicate that the addition of postbiotics and inulin combinations had beneficial effects on total BW, feed efficiency, mucosa architecture and IGF1 and GHR mRNA expression in broiler chickens.

Keywords: Broilers, Inulin, Prebiotic, Postbiotic, Intestinal microbiota, IGF1, GHR, Volatile fatty acid

Background

Intestinal microbiota play a vital role in the nutritional, physiological, immunological, and protective functions of the host [1] and their composition and activities can be influenced by diet [2]. The efficacy of feeding sub-therapeutic levels of antibiotics to modulate gut microbiota to enhance production performance of livestock has been espoused [3]. Unfortunately, the usage of antibiotics as feed additives for long periods in poultry diets can lead to antibiotic resistance [4] and high residue levels in poultry products such as meat and egg [5, 6]. Antimicrobial resistance encoding genes may represent risk to both human and animal health if it is transferred to other formerly susceptible bacteria [7]. Since the quest for safer and healthier chicken meat has remarkably increased in recent time, the use of natural feed additives can produce antibiotic-free chicken and can also prevent food-borne diseases [8].

* Correspondence: tcloh@upm.edu.my; hlfoo@upm.edu.my
[1]Department of Animal Science, Faculty of Agriculture, Universiti Putra Malaysia, 43400 UPM, Serdang, Selangor, Malaysia
[3]Department of Bioprocess Technology, Faculty of Biotechnology and Biomolecular Science, Universiti Putra Malaysia, 43400 UPM, Serdang, Selangor, Malaysia

In recent years, several feed additives such as prebiotics, probiotics, symbiotics, postbiotics and the combination of postbiotics and prebiotics have been used as growth promoters to replace antibiotics [9–12]. The mode of action of these additives differs. Probiotics colonize the host digestive system, increase the natural microbiota and prevent the colonization of pathogenic organisms [10]. Despite their beneficial effects, most probiotics especially the plasmids probiotics have antibiotic resistance genes which can be transferred between organisms [13]. As a consequence, probiotic as a live bacteria might not be used anymore in the near future. As a substitute to probiotics, metabolite products synthesized from probiotic known as postbiotics could be used. It is believed that postbiotics have the probiotic effects without living cells [14–16]. Prebiotics are non-living fibrous feed additives which when added to feed are preferred by harmful microbes. Prebiotics control the growth of pathogens (i.e. *Escherichia coli* and *Salmonella*) and stimulate the growth of *Bifidobacteria* and *Lactobacilli* and consequently promoting the health and performance of animals [17, 18]. A typical example of prebiotics is inulin. Postbiotics and inulin combination inhibited reproduction of pathogenic bacteria such as *Listeria monocytogenes, Salmonella enterica, Escherichia coli* and Vancomysin Resistant *Enterococci* [19]. Furthermore, addition of metabolite combinations to the feed of broilers [14, 15], laying hens [20] and pigs [16] improved the growth performance, faecal lactic acid bacteria and villus height. Various studies have examined the effects of postbiotic and prebiotics on growth performance, intestinal microbial ecology and histomorphology of broilers. However, there is dearth of information on the use of postbiotics and prebiotics combination and their synergistic effects on growth performance, intestinal microbial ecology, faecal VFA and histomorphology.

Chicken IGF1 has been identified as a biological candidate gene responsible for body composition, growth, fat deposition and metabolic activities in chickens [21]. It has been reported that the IGF-I level, feeding level, and growth rate are concurrent [22]. The dependence of nutritional and growth hormones on hepatic IGF1 production has been demonstrated [23, 24]. The pituitary releases the growth hormone which stimulates the hepatic production of IGF1 through the actions of GH activated GH receptors. However, the overall nutritional status of the animal modulates the ability of hepatic tissue to respond to GH. The IGF1 level can be affected by factors and situations that affect primary processes and controll the IGF1 production [25]. The GHR gene play vital role as a mediator of body size in bird [26, 27]. Since probiotics, prebiotics, antiobiotics and postbiotics influenced growth peroformance in poultry, a relationship between the feed additives and genes related to growth is anticipated. Thus, the aim of this work was to examine the effect of postbiotics and prebiotics on growth performance, IGF1 and GHR expression, intestinal microbial ecology, histomorphology and faecal VFA in broilers.

Methods

Postbiotics and inulin

The stock culture of *Lactobacillus plantarum* (*L. plantarum*) RG14 and *L. plantarum* RI11 were prepared at the Laboratory of Prebiotic and Probiotic Technology II at Institute of Bioscience, Universiti Putra Malaysia. The stock cultures were revived two times using de-Mann Rogosa Sharpe (MRS) broth and incubated at 30 °C for 48 and 24 h subsequently at static condition, followed by spread plate and incubation was performed in 48 h at 30 °C. A single colony was then picked and inoculated into 10 mL MRS broth and incubated for 24 h. It was followed by subculturing it into 10 mL MRS broth and incubated for 24 h at 30 °C. The culture was then ready to be used as an inoculum. An inoculum size of 1 % (v/v) was inoculated into the respective reconstituted media and incubated for 24 h at 30 °C at static condition. Centrifugation at 10,000 × g for 15 min was performed to separate the bacterial cell. The postbiotics were collected and kept at 4 °C [28] prior to feeding trials. The inulin (Frutafit IQ) was provided by Connell Bros. Company (Malaysia) Sdn. Bhd.

Animals and experimental design

Two hundred and eighty-eight day old chicks were purchased from a commercial hatchery. The broiler chickens were allocated into eight treatment groups. Each group had six replicates while each replicate had six birds. The treatment groups included basal diet (negative control), basal diet + neomycin and oxytetracycline (positive control), T1 = Basal diet + 0.3 % postbiotic RI11, T2 = Basal diet + 0.3 % postbiotic RG14, T3 = Basal diet + 0.3 % postbiotic RI11 + 0.8 % inulin, T4 = Basal diet + 0.3 % postbiotic RI11 + 1.0 % inulin, T5 = Basal diet + 0.3 % postbiotic RG14 + 0.8 % inulin, T6 = Basal diet + 0.3 %, postbiotic RG14 + 1.0 % inulin. Water and feed were offered ad libitum to the birds until 42 days of age. Starter and finisher diets (Tables 1 and 2) were offered from days 0 to 21 and days 22 to 42, respectively. The experimental animals received humane care as outlined and approved by Institutional Animal Care and Use Committee for the Care and Use of Animals for Scientific Purposes (Research Policy, Universiti Putra Malaysia).

Timing of sample collection

On weekly basis, BW and feed intake (FI) were recorded and WG and FCR were calculated. For sampling, 12 birds per treatment group were slaughtered at day 42.

Table 1 Composition and nutrient content of starter diets

Ingredients	Dietary treatment[a]							
	Negative control	Positive control	T1	T2	T3	T4	T5	T6
Corn	50.00	50.00	50.18	50.18	50.20	50.20	50.20	50.20
Soybean	29.380	29.375	30.94	30.94	29.995	30.00	29.995	30.00
Wheat pollard	6.895	6.895	4.645	4.645	4.820	4.490	4.820	4.490
CPO	3.400	3.400	3.680	3.680	3.590	3.665	3.590	3.665
Fish meal (55 %)	7.580	7.575	6.825	6.825	7.550	7.600	7.550	7.600
L-Lysine	0.25	0.25	0.25	0.25	0.25	0.25	0.25	0.25
DL-Methionine	0.20	0.20	0.15	0.15	0.20	0.20	0.20	0.20
Monodicalcium phosphate21	1.00	1.00	1.10	1.10	1.00	1.00	1.00	1.00
Calcium carbonate	0.68	0.68	1.0	1.0	0.68	0.68	0.68	0.68
Choline chloride	0.06	0.06	0.06	0.06	0.06	0.06	0.06	0.06
Salt	0.25	0.25	0.30	0.30	0.25	0.25	0.25	0.25
Mineral premix[b]	0.10	0.10	0.10	0.10	0.10	0.10	0.10	0.10
Vitamin premix[c]	0.06	0.06	0.06	0.06	0.06	0.06	0.06	0.06
Antioxidant[d]	0.01	0.01	0.01	0.01	0.01	0.01	0.01	0.01
Toxin binder[e]	0.135	0.135	0.400	0.400	0.135	0.135	0.135	0.135
Antibiotic[f]		0.01						
postbiotic RI11			0.30		0.30	0.30		
postbiotic RG14				0.30			0.30	0.30
Inulin					0.80	1.00	0.80	1.00
Calculated nutrient content (g/kg)[g]								
Crude protein	220.5	220.4	220.0	220.0	220.4	220.3	220.4	220.3
Metabolizable energy (MJ/Kg)	12.96	12.96	12.96	12.96	12.95	12.95	12.95	12.95
Calcium	9.9	9.9	10.01	10.01	9.9	9.9	9.9	9.9
Total phosphorus	8.5	8.5	8.3	8.3	8.5	8.3	8.3	8.3
Avail. P for poultry	5.1	5.1	5.1	5.1	5.1	5.1	5.1	5.1

[a]Negative control: (basal diet), Positive control: (basal diet + neomycin and oxytetracycline), T1: (0.3 % RI11), T2: (0.3 % RG14), T3: (0.3 % RI11 + 0.8 % Inulin), T4: (0.3 % RI11 + 1.0 % Inulin), T5: (0.3 % RG14 + 0.8 % Inulin), T6: (0.3 % RG14+ 1.0 % Inulin)
[b]Mineral mix contains Fe 100 mg, Mn 110 mg, Cu 20 mg, Zn 100 mg, I 2 mg, Se 0.2 mg, Co 0.6 mg
[c]Vitamin premix contains retinol 2 mg, cholicalciferol 0.03 mg, α-tocopherol 0.02 mg, menadione 1.33 mg, cobalamin 0.03 mg, thiamine 0.83 mg, riboflavin 2 mg, folic acid 0.33 mg, biotin 0.03 mg, panthothenic acid 3.75 mg, niacin 23.3 mg, pyridoxine 1.33 mg
[d]Antioxidant contains butylated hydroxyanisole (BHA)
[e]Toxin binder contains natural hydrated sodium calcium aluminium silicates
[f]A combination of oxytetracyclin and neomycin at the concentration of 100 ppm (w/w)
[g]The diets were formulated using FeedLIVE International software (Thailand)

FCR was calculated as follow: FCR = total feed consumed by birds/total weight gain.

Faecal LAB, ENT count and pH determination

The method of Foo et al. [29] was used to determine the faecal LAB and population of ENT. Faecal samples were kept at room temperature for 1 h once the 10-fold dilution (w/v) was done in sterile peptone water. Furthermore, 10-fold serial dilutions (v/v) were done after the 1 h soacking time. MRS-agar (Lactobacillus-Agar MRS) (Merck, KgaA, Darmstadt) was used to perform the enumerations of LAB. Incubation of plates was performed in anaerobic jars for 48 h at 30 °C. The incubation of ENT was performed aerobically for 24 h at 37 °C after spreading and counting them on EMB-Agar (Eosin-methyleneblue Lactose Sucrose Agar, Merck, KgaA, and Darmstadt). The base 10 logarithm of colony-forming unit (CFU) (logCFU) per g was applied to express the number of CFU. The whole samples were in triplicates. Almost 9 ml of deionized distilled water was used to homogenise about 1 g of the sample in a universal tube. Mettler-Toledo pH meter with a glass electrode (Mettler-Toledo LTD, England) was used to measure the pH. The

Table 2 Composition and nutrient content of finisher diets

Ingredients	Dietary treatment[a]							
	Negative control	Positive control	T1	T2	T3	T4	T5	T6
Corn	54.70	54.70	54.89	54.89	54.80	54.69	54.80	54.69
Soybean	29.10	29.10	27.04	27.04	29.30	29.31	29.30	29.31
Wheat pollard	5.36	5.35	5.90	5.90	3.41	3.16	3.41	3.16
CPO	3.460	3.460	3.400	3.400	3.74	3.815	3.74	3.815
Fish meal (55 %)	3.600	3.600	5.040	5.040	3.870	3.945	3.870	3.945
L-Lysine	0.25	0.25	0.25	0.25	0.25	0.25	0.25	0.25
DL-Methionine	0.20	0.20	0.20	0.20	0.20	0.20	0.20	0.20
Monodicalcium phosphate21	1.40	1.40	1.35	1.35	1.40	1.40	1.40	1.40
Calcium carbonate	1.30	1.30	1.00	1.00	1.30	1.30	1.30	1.30
Choline chloride	0.06	0.06	0.06	0.06	0.06	0.06	0.06	0.06
Salt	0.25	0.25	0.25	0.25	0.25	0.25	0.25	0.25
Mineral premix[b]	0.10	0.10	0.10	0.10	0.10	0.10	0.10	0.10
Vitamin premix[c]	0.06	0.06	0.06	0.06	0.06	0.06	0.06	0.06
Antioxidant[d]	0.01	0.01	0.01	0.01	0.01	0.01	0.01	0.01
Toxin binder[e]	0.15	0.15	0.15	0.15	0.15	0.15	0.15	0.15
Antibiotic[f]		0.01						
postbiotic RI11			0.30		0.30	0.30		
postbiotic RG14				0.30			0.30	0.30
Inulin					0.80	1.00	0.80	1.00
Calculated nutrient content (g/kg)[g]								
Crude protein	199.1	199.1	199.1	199.1	199.0	199.1	199.0	199.1
Metabolizable energy (MJ/Kg)	13.00	13.00	13.00	13.00	13.00	13.00	13.00	13.00
Calcium	10.3	10.3	10.0	10.0	10.5	10.5	10.5	10.5
Total phosphorus	8.1	8.1	8.2	8.2	8.0	8.0	8.0	8.0
Avail. P for poultry	4.7	4.7	4.9	4.9	4.7	4.7	4.7	4.7

[a]Negative control: (basal diet), Positive control: (basal diet + neomycin and oxytetracycline), T1: (0.3 % RI11), T2: (0.3 % RG14), T3: (0.3 % RI11 + 0.8 % Inulin), T4: (0.3 % RI11 + 1.0 % Inulin), T5: (0.3 % RG14 + 0.8 % Inulin), T6: (0.3 % RG14+ 1.0 % Inulin)
[b]Mineral mix contains Fe 100 mg, Mn 110 mg, Cu 20 mg, Zn 100 mg, I 2 mg, Se 0.2 mg, Co 0.6 mg
[c]Vitamin premix contains retinol 2 mg, cholicalciferol 0.03 mg, α-tocopherol 0.02 mg, menadione 1.33 mg, cobalamin 0.03 mg, thiamine 0.83 mg, riboflavin 2 mg, folic acid 0.33 mg, biotin 0.03 mg, panthothenic acid 3.75 mg, niacin 23.3 mg, pyridoxine 1.33 mg
[d]Antioxidant contains butylated hydroxyanisole (BHA)
[e]Toxin binder contains natural hydrated sodium calcium aluminium silicates
[f]A combination of oxytetracyclin and neomycin at the concentration of 100 ppm (w/w)
[g]The diets were formulated using FeedLIVE International software (Thailand)

meter was calibrated prior to measuring the pH of the samples by using buffer solutions (Merck, KgaA, Dramstadt) at pH 4 and 7.

Histomorphology

Specimens were taken from three different parts of jejunum, duodenum, and ileum. These specimens were taken from the following locations;

(i) the middle part of the duodenal loop,
(ii) midway between the end point of duodenal loop and Meckel's diverticulum (jejunum), and
(iii) midway between the Meckel's diverticulum and the ileo-caecal junction (ileum), and fixed in 10 % neutral buffered formalin.

A tissue processing machine (Leica, Japan) was used to excise, dehydrate these specimens. Then they were embedded in paraffin wax. Each sample was cut into sections (4 mm). These sections were stained with haematoxylin and eosin, fixed on slides, and then mounted and examined under the light microscope. The tip of the villus to the villus-crypt junction area was measured as the villus height. Furthermore, the crypt depth was defined as the depth of the invagination between two villi.

Determination of VFA

The modified method of Thanh et al. [14] was applied to determine the VFA concentration in the faeces. From each sample, 1 g faeces (stored at −20 °C) were weighed. Then, 1 mL of 24 % metaphosphoric acid was added which was diluted in 1.5 M sulphuric acid (BDH Laboratories, Poole, UK). First, under room temperature, this mixture was stored overnight and then the mixture was centrifuged at 10 000 × g for 20 min at 4 °C. The supernatant was kept in a 1.5-ml screw-capped vial (Kimble Glass Inc., USA). For the GLC analysis, the internal standard, 20 mM 4-methyl-valeric acid (Sigma Chemical Co., St. Louis, MO, USA) was consistently added to the supernatant to make up 10 mM and the mixture was stored at −20 °C. The VFA was separated on a Quadrex 007 Series (Quadrex Corp., New Haven, CT 06525, USA). The bonded phase was a fused silica capillary column (15 m, 0.32 mm ID, 0.25 mm film thickness) with a 6890 N (Hewlett-Packard, Avondale, PA) equipped with a flame ionization detector. The carrier gas was purified liquid nitrogen with a flow rate of 60 mL/min. The temperature of the injector and detector was set at 230 °C. The column temperature was set at 200 °C. For identification of peaks, the commercial standards of 20 mM acetic, and 10 mM each of propionic, butyric, isobutyric, valeric, isovaleric and 4-methyl-valeric acids from Sigma were used as external standards.

Total RNA isolation and reverse transcriptase polymerase chain reaction (RT-PCR) analysis of hepatic IGF1 and GHR

Lysis buffer (Qiagen, USA) received 30 mg (per sample) of frozen liver and it was properly homogenized. Following the manufacturer's instructions, Qiagen one-step RT-PCR kit (Qiagen, USA) was used to perform the RT-PCR. Real-time RT qPCR analyses were done using QuantiTect Primer Assay (200) IGF1 (QT00621334), GHR (QT00601321) and B-actin (QT00600614). Thereafter, NanoDrop was used to evaluate the purity of RNA at 260/280 OD ratio and RNA integrity. Only highly purified samples (OD260/280 > 1.8) were chosen for further manipulation. The master mix preparation was also conducted following the manufacturer's procedure. The total volume of the reaction was 25 µl for each gene of interest arranged as 2× QuantiFast SYBR Green RT-PCR Master Mix 12.5 µl, 1 µl of forward primer, 1 µl of the reverse primer, 0.25 µl QuantiFast RT Mix, 1 µl of RNA, and finally 9.25 µl of RNase-free water. The reaction was carried out in a Bio-Rad thermal cycler (MyCycler, Germany). The RT-PCR conditions were as follows: (1) reverse transcription, 30 min, 50 °C, (2) initial PCR activation step, 15 min, 95 °C, (3) 3-step cycling for 40 cycles, each cycle consisting of denaturation for 30 s at 94 °C followed by annealing for 30 s at 52–57 °C and extension for 1 min at 72 °C. The linearity of response was ensured and the saturation of the reaction was averted by optimizing the template concentration and the cycle number. For standardization of the expression data, the β-actin mRNA fragment was used as internal standard (housekeeping gene). The results were standardized to the levels obtained for the β-actin gene. It was done by taking the ratio of the value obtained for the gene of interest to that of β-actin and then relative to the control. $2^{-\Delta\Delta Ct}$ ($\Delta\Delta Ct = \Delta Ct$ Test sample-ΔCt Calibrator sample) calculated the relative mRNA expression

Statistical analysis

Data analysis was performed using the General Linear Model procedure of the Statistical Analysis System. Means were compared using the Duncan Multiple Range Test. The Bio-Rad CFX Manager 3.0 Software of the C1000 Touch thermal cycler-CFX96 Real time PCR (BIO-RAD, Foster city, California, USA) was used to calculate the relative gene expression of target genes in comparison to the β-actin reference gene.

Results

Growth performance

The growth performance of the birds fed diets containing different additives is presented in Table 3. Birds fed with T3, T4 and T6 had higher ($p < 0.05$) final BW and total WG than other treatments. The final BW and WG of birds fed the negative control diet, postive control diet, T1, T2 and T5 were similar ($p > 0.05$). There was no significant difference ($p > 0.05$) among the treatments for FI. Birds fed with T3 and T6 had lower ($p < 0.05$) FCR compared with birds fed the negative control diet. The FCR of birds fed T3, T4, T5, T6 and positive control were similar ($p > 0.05$). Similarly, the FCR of birds fed negative control, positive control, T1, T2, T4 and T5 diets did not differ.

Faecal LAB, ENT and pH

The faecal LAB, ENT and pH of birds fed various treatment groups are shown in Fig. 1. Significantly, the faecal pH for T3 were lower ($p < 0.05$) than the negative and positive controls. Dietary treatments affected ($p < 0.05$) LAB and ENT counts. Postbiotic and inulin increased ($p < 0.05$) faecal LAB and decreased ENT count when compared to negative control.

Histomorphology

The villus height and crypt depth of the duodenum, ileum and jejunum of birds fed different dietary treatments are shown in Table 4. Birds fed T3 and T6 had significantly higher ($p < 0.05$) villus height in the duodenum than the other treatments. Birds fed T2, T3, T4 and

Table 3 Growth performance at week 6 of treatments supplemented with different postbiotics and differen levels of inulin

Parameter	Dietary treatments^{e}								SEM
	Negative control	Positive control	T1	T2	T3	T4	T5	T6	
FI (g)	4245.17	4153.71	4289.19	4298.08	4199.29	4279.61	4187.49	4267.91	25.03
BW (g)	2239.59^{b}	2248.93^{b}	2267.28^{b}	2266.69^{b}	2334.90^{a}	2330.79^{a}	2264.72^{b}	2345.48^{a}	6.51
WG (g)	2189.10^{b}	2198.34^{b}	2217.79^{b}	2217.52^{b}	2284.31^{a}	2279.52^{a}	2215.59^{b}	2295.24^{a}	6.50
FCR	1.94^{a}	1.89^{abc}	1.93^{ab}	1.93^{ab}	1.84^{c}	1.88^{abc}	1.89^{abc}	1.86^{bc}	0.01

^{abc}means within a row for each parameter with different superscripts are significantly different ($p < 0.05$)
^{e}negative control: basal diet, positive control: basal diet + neomycin and oxytetracycline, T1: (0.3 % RI11), T2: (0.3 % RG14), T3: (0.3 % RI11 + 0.8 % inulin), T4: (0.3 % RI11 + 1.0 % inulin), T5: (0.3 % RG14 + 0.8%inulin), T6: (0.3 % RG14 + 1.0%inulin)

T6 had higher villus height in the ileum than the negative control. However, there were no significant differences ($p > 0.05$) for villus height in jejunum, and crypt depth in duodenum, jejunum and ileum among the treatments.

Volatile fatty acid

The faecal VFA of birds fed different dietary treatments is presented in Table 5. The result shows that acetic acid is the major VFA found in the broiler faeces followed by butyric and propionic acid. Brioler chickens fed T6 had the highest concentration of acetic acid and total VFA which was significantly different ($p < 0.05$) from other treatment groups. Birds fed T3 had the highest ($p < 0.05$) propionic acid as compare to birds fed with other treatments. No significant difference ($p > 0.05$) was observed for butyric acid in all treatment groups.

IGF1 and GHR mRNA expression

Gene expression profile of liver fed different dietary treatments is shown in Fig. 2. Birds fed T6 had the highest ($p < 0.05$) IGF1 mRNA expression. The IGF1 expression in T1, T3, T4, T5 and T6 were significantly higher ($p < 0.05$) than the negative and positive control. The GHR mRNA expression in the liver of broilers fed with T6 was significantly higher ($p < 0.05$) than that of other treatment groups.

Discussion

Postbiotics influenced the growth rate of broiler chickens. It has bacteriostatic and bactericidal ability which decreases the pathogenic bacterial load in gastrointestinal microbiota. Kareem et al. [19] reported that *L. plantarum* exhibited inhibitory effect against various pathogens. Chicory root powder which comprise 68 % inulin enhances food digestion and absorption through jejunum histomorphometry modification thereby improve growth performance in broiler chickens [30]. The current results are in agreement with those of Thanh et al. [14] who found that birds fed combinations of metabolites produced by *L. plantarum* had higher ($p < 0.05$) final BW and WG compared with those fed the negative control diet. In contrast, Rosyidah et al. [31] did not observe significant difference in BW and WG in broilers fed with metabolites and combination of metabolite and acidifier groups and those fed positive and negative control diets.

The similarity in the BW and WG of birds fed the negative control diet and postive control diet suggests that antibiotics are no longer effective as growth enhancer and may not be effective as antibacterial in the future. The

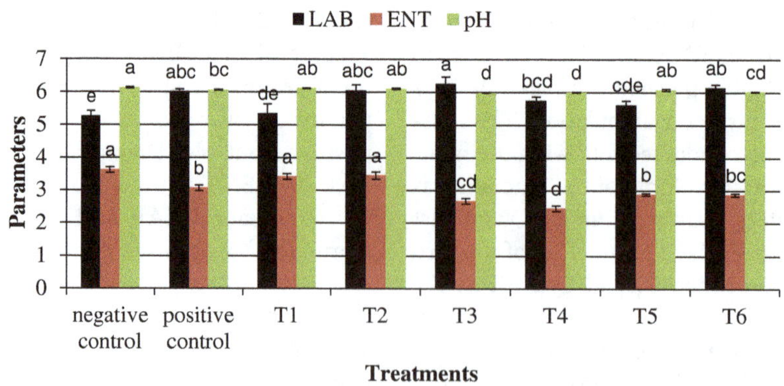

Fig. 1 Microbiota counts (log CFU/g) and pH of broiler supplemented with different treatment diets at week 6. Treatments: Negative control: basal diet, Positive control: basal diet + neomycin and oxytetracycline, T1: (0.3 % RI11), T2: (0.3 % RG14), T3: (0.3 % RI11 + 0.8 % inulin), T4: (0.3 % RI11 + 1.0 % inulin), T5: (0.3 % RG14 + 0.8 % inulin), T6: (0.3 % RG14 + 1.0 % inulin). Bars with no common letter differ significantly ($p < 0.05$)

Table 4 Villus height and crypt depth in small intestine of treatments supplemented with different postbiotics and different levels of inulin

Parameter	Dietary treatments[e]								SEM
	Negative control	Positive control	T1	T2	T3	T4	T5	T6	
Villi height week 6, μm									
Duodenum	1304.88^{b}	1304.75^{b}	1306.80^{b}	1308.55^{b}	1419.68^{a}	1311.18^{b}	1315.80^{b}	1395.38^{a}	9.08
Jejunum	837.75	875.34	887.85	894.15	942.60	911.90	845.60	943.07	8.36
Ileum	540.90^{c}	595.71^{ab}	562.90^{bc}	617.30^{a}	634.08^{a}	610.87^{a}	597.25^{ab}	622.18^{a}	7.29
Crypt depth week 6, μm									
Duodenum	168.12	165.52	177.85	166.00	174.95	143.17	170.22	173.72	3.78
Jejunum	132.75	139.20	143.07	127.85	139.12	128.85	120.72	134.85	3.34
Ileum	101.07	106.15	105.50	102.12	107.42	105.75	95.15	109.45	2.67

[abcd]means within a row for each parameter with different superscripts are significantly different ($p < 0.05$)
[e]negative control: basal diet, positive control: basal diet + neomycin and oxytetracycline, T1: (0.3 % RI11), T2: (0.3 % RG14), T3: (0.3 % RI11 + 0.8 % inulin), T4: (0.3 % RI11 + 1.0 % inulin), T5: (0.3 % RG14 + 0.8%inulin), T6: (0.3 % RG14 + 1.0%inulin)

current observation is in tandem with that of Aristides et al. [32] who observed that dietary supplementation of avilamycin did not affect WG in broiler chickens. Postbiotics added with inulin was even better than diet added with antibiotic in BW and WG.

The lower FCR in T3 and T6 birds compared with the negative control birds is consistent with the report of Liu et al. [33] who found that probiotic, prebiotic and synbiotic significantly improved feed efficiency as compare to the negative control diet. In contrast, Elrayeh and Yildiz [34] reported 0.7 % inulin had no effect on FCR of broiler chickens.

Different combinations of postbiotics and inulin decreased faecal pH and ENT count. This observation corroborates the report of earlier findings which showed that addition of prebiotics beneficially modified intestinal microbiota in animal models and human studies. The authors posited that addition of prebiotics enhanced the population of protective bacteria (i.e. *Lactobacilli* and *Bifidobacteria*) and hindered the attachment of pathogenic bacteria to the gut epithelium [35–37]. The current findings are in agreement with those of Loh et al. [10] who observed that dietary postbiotics increased the faecal LAB and reduced the faecal pH and faecal ENT in laying hens. Also, Rosyidah et al. [31] observed an increase in LAB count and a decrease in ENT count in the faeces of broilers fed with a metabolite produced by *L. plantarum*. Nabizadeh [38] found that dietary 1 % inulin reduced *E. coli* counts and pH in cecal contents in broiler chickens. Contrarily, Biggs et al. [39] found that corn-soybean meal diets containing 4 % inulin and some oligosaccharides had no effect on cecal *Lactobacillus, Bifidobacterium, Clostridium perfringens,* or *E. coli* populations in 21-d-old chicks. Abdel-Raheem et al. [40] also reported that prebiotic, probiotic and synbiotic had no effect on cecal *Lactobacilli* and *E. coli* counts in broilers chickens.

Villus height and crypt depth are reliable indicators of the gut function and health. According to Uni et al. [41] the health status of the gastrointestinal tract of an animal is a true reflection of intestinal morphology. Birds fed T3 and T6 had higher ($p < 0.05$) duodenal villus height than the control diets while birds fed T2, T3, T6 and T4 had higher ($p < 0.05$) ileal villus compared with the control diets. Our findings are consistent with the findings of earlier studies. The postbiotic and prebiotic altered the mucosal architecture in terms of longer villi and increased birds' performance [14, 40, 42]. Furthermore, probiotics and synbiotics can enhance broiler performance by improving the intestinal morphology and microbial balance which are associated with suppressing intestinal pathogens such as *E. coli Campylobacter* and

Table 5 Faecal VFA in broiler chickens fed with postbiotics and different levels of inulin

Parameter (mM)	Dietary treatments[e]								SEM
	Negative control	Positive control	T1	T2	T3	T4	T5	T6	
Acetic	27.78^{c}	29.78^{c}	30.39^{c}	34.70^{bc}	40.22^{b}	31.47^{bc}	36.85^{bc}	49.30^{a}	1.57
Propionic	0.27^{c}	0.66^{bc}	0.89^{ab}	0.71^{bc}	1.30^{a}	0.84^{ab}	0.87^{ab}	0.66^{bc}	0.07
Butyric	0.21	0.22	0.39	0.38	0.40	0.28	0.43	0.31	0.02
Total	28.25^{d}	30.65^{cd}	31.67^{cd}	35.45^{bcd}	41.92^{b}	32.59^{cd}	38.15^{bc}	50.27^{a}	1.59

[abcd]means within a row for each parameter with different superscripts are significantly different ($p < 0.05$). [e]negative control: basal diet, positive control: basal diet + neomycin and oxytetracycline, T1: (0.3 % RI11), T2: (0.3 % RG14), T3: (0.3 % RI11 + 0.8 % inulin), T4: (0.3 % RI11 + 1.0 % inulin), T5: (0.3 % RG14 + 0.8%inulin), T6: (0.3 % RG14 + 1.0%inulin)

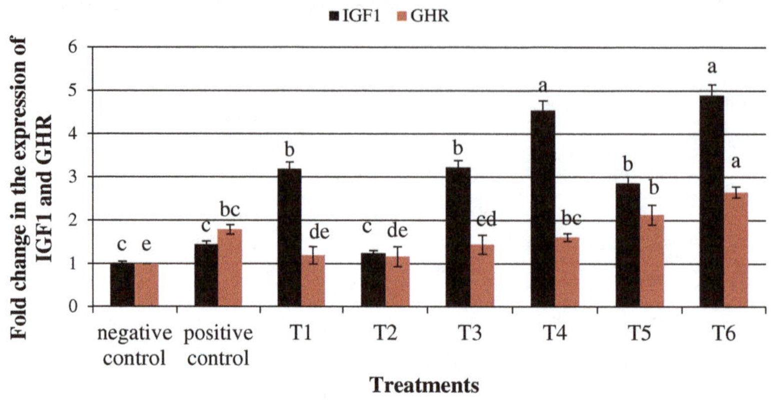

Fig. 2 IGF1 and GHR mRNA expression in the liver of broiler chicken. Treatments: Negative control: basal diet, Positive control: basal diet + neomycin and oxytetracycline, T1: (0.3 % RI11), T2: (0.3 % RG14), T3: (0.3 % RI11 + 0.8 % inulin), T4: (0.3 % RI11 + 1.0 % inulin), T5: (0.3 % RG14 + 0.8 % inulin), T6: (0.3 % RG14 + 1.0 % inulin). Bars with no common letter differ significantly ($p < 0.05$)

Salmonela and at the same time increase nutrient digestibility [43, 44]. In contrast, according to Nabizadeh [38], inulin did not affect the mophorlogy of duodenum, ileum and jejunum of broiler chickens.

The observed increase in the cocentration of acetic acid in T6, the higher propionic acid in T3 and increased in total VFA in both T3 and T6 compared with other treatments could be responsible for the increased lactic acid bacteria and decreased ENT counts and pH observed in the treatments. This finding is consistent with those of Loh et al. [10] and Thu et al. [16]. Postbiotics originating from *Lactobacillus* include valuable compounds such as organic acids and bacteriocin which enhance the growth of lactic acid bacteria [10]. Moreover, Van der Wielen et al. [45] reported that during growth of broiler chickens, VFA are responsible for the reduction in numbers of ENT in the ceca. In addition, the increased VFA could be due to the inulin as it contains polysaccharides and oligosaccharides. Gebbink et al. [46] demonstrated that fructooligosaccharide can be used to assist and maintain the healthy gastrointestinal tract environment by increasing the colonization of *Bifidobacteria* or reducting the *E. coli* in the intestinal system. Digestive enzymes do not hydrolyse the fructooligosaccharide in the small intestine of monogastrics and thus, it enters into the colon intact. Colonic microbiota metabolize it completly. Gases, lactate, and short cain fatty acid (i.e. acetate, propionate, and butyrate) are the output of carbohydrate fermentation [38]. The main fermentative chamber in broiler chicken is the caecum and this contain the largest number of bacteria compared with other gastrointestinal tract segment. Hence, the microbiota has high ability to ferment the carbohydrates [47].

Dietary postbiotic and inulin influenced the expression of mRNA IGF1 and mRNA GHR in the liver. The increase in IGF-I mRNA and GHR mRNA in birds fed T6 is a true reflection of the growth performance of the birds. It is improtant to note that the IGF1 level, feeding level, and growth rate are concurrent [22]. The dependence of nutritional and growth hormones on hepatic IGF1 production has been demonstrated [23, 25]. Moreover, Amongst the genes influencing growth, IGF1 has been demonstrated as an indicator of growth rate in chicken by several authors [48, 49]. The pituitary relseases the growth hormone which stimulates the hepatic production of IGF1 through the actions of GH activated GH receptors. However, the overall nutritional status of the animal modulated the ability of hepatic tissue to respond to GH [25]. The IGF1 level can be affected by factors and situations that affect primary processes and controll the IGF1 production. These results might provide bases for the development of IGF1 as a growth index. Results from this study corroborate the findings of Beckman et al. [50] who found significant and positive correlation between mean growth rates of juvenile Chinook salmon and mean plasma IGF1 levels. However, other studies have not demonstrated any relationship between IGF1 and growth [51, 52] which can lead to uncertainties about the consistency of IGF1 growth relationships.

Conclusion

The study demonstrated that addition of postbiotics and inulin had beneficial effects on total BW, feed efficiency, mucosa architecture and expression of IGF1 and GHR mRNA in the liver of broiler chickens. However, birds fed T4: (0.3 % RI11 + 1.0 % inulin) and T6: (0.3 % RG14 + 1.0 % inulin) had higher total BW and feed efficiency than the other treatments. The faecal pH and ENT was reduced while VFA and LAB was increased in birds fed T4 and T6 as compared to the other treatments. Thus, both

T4 and T6 could be used as a substitute for antibiotics in broiler diet to improve the growth and gut health in broiler chickens.

Abbreviations

BW, body weight; CFU, colony-forming unit; ENT, *Enterobacteriaceae*; FCR, feed conversion ratio; FI, feed intake; GHR, growth hormone receptor; IGF1, insulin like grouth factor 1; *L. plantarum*, *Lactobacillus plantarum*; LAB, lactic acid bacteria; MRS, de-Mann Rogosa Sharpe; RT-PCR, reverse transcriptase polymerase chain reaction.; VFA, volatile fatty acid; WG, weight gain.

Acknowledgements

The Long-Term Research Grant Scheme (LRGS) from Ministry of Education Malaysia has supported this project.

Funding

This project was supported by Long-Term Research Grant Scheme (LRGS) - Food Security entitled 'Enhancing the competitiveness and sustainability of the poultry industry for food security' from the Ministry of Higher Education, Malaysia.

Authors' contributions

KYK managed the chickens, TCL, HLF, HA, SAA participated in the whole design of the study and performed the statistical analysis. KYK carried out laboratory analyses and KYK, TCL, HLF, HA, SAA contributed to the preparation of the manuscript. All authors read and approved the final manuscript.

Competing interests

The authors declare that they have no competing interests.

Author details

[1]Department of Animal Science, Faculty of Agriculture, Universiti Putra Malaysia, 43400 UPM, Serdang, Selangor, Malaysia. [2]Institute of Tropical Agriculture, Universiti Putra Malaysia, 43400 UPM, Serdang, Selangor, Malaysia. [3]Department of Bioprocess Technology, Faculty of Biotechnology and Biomolecular Science, Universiti Putra Malaysia, 43400 UPM, Serdang, Selangor, Malaysia. [4]Institute of Bioscience, Universiti Putra Malaysia, 43400 UPM, Serdang, Selangor, Malaysia. [5]Department of Animal Resource, University of salah al-Din, Erbil, Iraq.

References

1. Vispo C, Karasov WH. The interaction of avian gut microbes and their host: an elusive symbiosis. In: Mackie RI, White BA, editors. Gastrointestinal microbiology. New York: Chapman & Hall; 1997. p. 116-55.
2. Rehman HU, Vahjen W, Awad WA, Zentek J. Indigenous bacteria and bacterial metabolic products in the gastrointestinal tract of broiler chickens. Arch Anim Nutr. 2007;61(5):319–35.
3. Bolu SA, Adeyemi KD, Sola-Ojo FE, Fabiyi AA, Adedeji AT, Oluyemi O, Babalola AB. Lower dietary inclusion of alphamune g may improve performance, hematology, serum biochemistry and histology of broiler chicken. Sustainable Agr Res. 2012;1(2):15.
4. Shazali N, Foo HL, Loh TC, Choe DW, Abdul RR. Prevalence of antibiotic resistance in lactic acid bacteria isolated from the faeces of broiler chicken in Malaysia. Gut Pathog. 2014;6(1):1.
5. Shareef AM, Jamel ZT, Yonis KM. Detection of antibiotic residues in stored poultry products. Iraq J Vet Sci. 2009;23(1):45–8.
6. Olatoye IO, Ehinwomo AA. Oxytetracycline residues in edible tissues of cattle slaughtered in Akure, Nigeria. Nigerian Vet J. 2010; 31(2):93-102.
7. Montagne L, Pluske JR, Hampson DJ. A review of inter-actions between dietary fiber and the intestinal mucosa, and their consequences on digestive health in young non-ruminant animals. Anim Feed Sci Tech. 2003;108:95–117.
8. Kleter GA, Marvin HJ. Indicators of emerging hazards and risks to food safety. Food Chem Toxicol. 2009;47(5):1022–39.
9. Hajati H, Rezaei M. The application of prebiotics in poultry production. Int J Poult Sci. 2010;9(3):298–304.
10. Loh TC, Choe DW, Foo HL, Sazili AQ, Bejo MH. Effects of feeding different postbiotic metabolite combinations produced by Lactobacillus plantarum strains on egg quality and production performance, faecal parameters and plasma cholesterol in laying hens. BMC Vet Res. 2014;10(1):149.
11. Kareem KY, Loh TC, Foo HL, Asmara SA, Akit H, Abdulla NR, Ooi MF. Carcass, meat and bone quality of broiler chickens fed with postbiotic and prebiotic combinations. Int J Probiotics Prebiotics. 2015;10(1):23–30.
12. Rebolé A, Ortiz L, Rodríguez ML, Alzueta C, Treviño J, Velasco S. Effects of inulin and enzyme complex, individually or in combination, on growth performance, intestinal microflora, cecal fermentation characteristics, and jejunal histomorphology in broiler chickens fed a wheat-and barley-based diet. Poultry Sci. 2010;89(2):276–86.
13. Marteau P, Shanahan F. Basic aspects and pharmacology of probiotics: an overview of pharmacokinetics, mechanisms of action and side-effects. Best Pract Res Cl Ga. 2003;17(5):725–40.
14. Thanh N, Loh TC, Foo HL, Hair-Bejo M, Azhar B. Effects of feeding metabolite combinations produced by Lactobacillus plantarum on growth performance, faecal microbial population, small intestine villus height and faecal volatile fatty acids in broilers. Brit Poultry Sci. 2009;50(3):298–306.
15. Loh TC, Thanh NT, Foo HL, Hair-Bejo M, Azhar BK. Feeding of different levels of metabolite combinations produced by *Lactobacillus plantarum* on growth performance, fecal microflora, volatile fatty acids and villi height in broilers. Anim Sci J. 2010;81(2):205–14.
16. Thu T, Loh TC, Foo H, Yaakub H, Bejo M. Effects of liquid metabolite combinations produced by *Lactobacillus plantarum* on growth performance, faeces characteristics, intestinal morphology and diarrhoea incidence in postweaning piglets. Trop Anim Health Pro. 2011;43(1):69–75.
17. Bogusławska-Tryk M, Piotrowska A, Burlikowska K. Dietary fructans and their potential beneficial influence on health and performance parametrs in broiler chickens. J Cent Europ Agri. 2012; 13(2):272-91.
18. Jung S, Houde R, Baurhoo B, Zhao X, Lee B. Effects of galacto-oligosaccharides and a *Bifidobacteria* lactis-based probiotic strain on the growth performance and fecal microflora of broiler chickens. Poultry Sci. 2008;87(9):1694–9.
19. Kareem KY, Ling FH, Chwen LT, Foong OM, Asmara SA. Inhibitory activity of postbiotic produced by strains of *Lactobacillus plantarum* using reconstituted media supplemented with inulin. Gut Pathog. 2014;6(1):23.
20. Choe D, Loh T, Foo H, Hair-Bejo M, Awis Q. Egg production, faecal pH and microbial population, small intestine morphology, and plasma and yolk cholesterol in laying hens given liquid metabolites produced by *Lactobacillus plantarum* strains. Brit Poultry Sci. 2012;53(1):106–15.
21. Kadlec J, Hosnedlová B, Řehout V, Čítek J, Večerek L, Hanusová L. Insulin-like growth factor-I gene polymorphism and its association with growth and slaughter characteristics in broiler chickens. J Agrobiology. 2011;28(2):157–63.
22. Beckman BR, Shimizu MG, Gadberry BA, Parkins PJ, Cooper KA. The effect of temperature change on the relations among plasma IGF-1, 41-kDa IGFBP, and growth rate in postsmolt coho salmon. Aquaculture. 2004;241:601–19.
23. Moriyama S. Increased plasma insulin-like growth factor-I (IGF-I) following oral and intraperitoneal administration of growth hormone to rainbow trout, Oncorhynchus mykiss. Growth Regulat. 1995;5(3):164–7.
24. Shamblott MJ, Cheng CM, Bolt D, Chen TT. Appearance of insulin-like growth factor mRNA in the liver and pyloric ceca of a teleost in response to exogenous growth hormone. P Natl Acad Sci. 1995;92(15):6943–6.
25. Beckman BR. Perspectives on concordant and discordant relations between insulin-like growth factor 1 (IGF1) and growth in fishes. Gen comp endocr. 2011;170(2):233–52.
26. Li Y, Yang X, Ni Y, Decuypere E, Buyse J, Everaert N, Grossmann R, Zhao R. Early-age feed restriction affects viability and gene expression of satellite cells isolated from the gastrocnemius muscle of broiler chicks. J Anim Sci Biotechnol. 2012;3(1):33.
27. Zhou N, Lee WR, Abasht B. Messenger RNA sequencing and pathway analysis provide novel insights into the biological basis of chickens' feed efficiency. BMC genomics. 2015;16(1):195.
28. Foo H, Loh T, Lai P, Lim Y, Kufli C, Rusul G. Effects of adding Lactobacillus plantarum I-UL4 metabolites in drinking water of rats. Pakistan J Nutr. 2003; 2(5):283–8.
29. Foo H, Loh T, Law F, Lim Y, Kufli C, Rusul G. Effects of feeding Lactobacillus plantarum I-UL4 isolated from Malaysian Tempeh on growth performance,

faecal flora and lactic acid bacteria and plasma cholesterol concentrations in postweaning rats. Food Sci Biotech. 2003;12(4):403–8.
30. Izadi H, Arshami J, Golian A, Raji MR. Effects of chicory root powder on growth performance and histomorphometry of jejunum in broiler chicks. Vet Res. 2013;4(3):169-74
31. Rosyidah MR, Loh TC, Foo H, Cheng X, Bejo MH. Effects of feeding metabolites and acidifier on growth performance, faecal characteristics and microflora in broiler chickens. J Anim and Vet Adv. 2011;10(21):2758–64.
32. Aristides LG, Paião FG, Murate LS, Oba A, Shimokomaki M. The effects of biotic additives on growth performance and meat qualities in broiler chickens. Int J Poult Sci. 2012;11:599–604.
33. Liu X, Yan H, Lv L, Xu Q, Yin C, Zhang K, Wang P, Hu J. Growth performance and meat quality of broiler chickens supplemented with *Bacillus licheniformis* in drinking water. Asian Austral J Anim. 2012;25(5):682.
34. Elrayeh AS, Yildiz G. Effects of inulin and\ beta-glucan supplementation in broiler diets on growth performance, serum cholesterol, intestinal length, and immune system. Turk J Vet Anim Sci. 2012;36(4):388–94.
35. Santos A, San Mauro M, Diaz DM. Prebiotics and their long-term influence on the microbial populations of the mouse bowel. Food Microbiol. 2006; 23(5):498–503.
36. Langlands S, Hopkins M, Coleman N, Cummings J. Prebiotic carbohydrates modify the mucosa associated microflora of the human large bowel. Gut. 2004;53(11):1610–6.
37. Langen LV, Mirjam A, Dieleman LA. Prebiotics in chronic intestinal inflammation. Inflamm Bowel Dis. 2009;15(3):454–62.
38. Nabizadeh A. The effect of inulin on broiler chicken intestinal microflora, gut morphology, and performance. J Anim Feed Sci. 2012;21:725–34.
39. Biggs P, Parsons C, Fahey G. The effects of several oligosaccharides on growth performance, nutrient digestibilities, and cecal microbial populations in young chicks. Poultry Sci. 2007;86(11):2327–36.
40. Abdel-Raheem SM, Abd-Allah SM, Hassanein KM. The effects of prebiotic, probiotic and synbiotic supplementation on intestinal microbial ecology and histomorphology of broiler chickens. Ijavms. 2012;6:277–89.
41. Uni Z, Noy Y, Sklan D. Posthatch changes in morphology and function of the small intestines in heavy-and light-strain chicks. Poultry Sci. 1995; 74(10):1622–9.
42. Yang Y, Iji P, Choct M. Effects of different dietary levels of mannanoligosaccharide on growth performance and gut development of broiler chickens. Asian Aust J Anim. 2007;20(7):1084.
43. Ashayerizadeh A, Dabiri N, Ashayerizadeh O, Mirzadeh K, Roshanfekr H, Mamooee M. Effect of dietary antibiotic, probiotic and prebiotic as growth promoters, on growth performance, carcass characteristics and hematological indices of broiler chickens. Pakistan J Biol Sci. 2009;12:52–7.
44. Li X, Piao X, Kim S, Liu P, Wang L, Shen Y, Jung S, Lee H. Effects of chito-oligosaccharide supplementation on performance, nutrient digestibility, and serum composition in broiler chickens. Poultry sci. 2007;86(6):1107–14.
45. Van der Wielen PW, Biesterveld S, Notermans S, Hofstra H, Urlings BA, van Knapen F. Role of volatile fatty acids in development of the cecal microbiota in broiler chickens during growth. Appl Environ Microb. 2000; 66(6):2536–40.
46. Gebbink G, Sutton A, Richert B, Patterson J, Nielsen J, Kelly D, Verstegen M, Williams B, Bosch M, Cobb M. Effects of addition of frutooligosaccharide (FOS) and sugar beet pulp to weanling pig diets on performance, microflora and intestinal health. Acess. 1999;12:53–9.
47. Józefiak D, Kaczmarek S, Rutkowski A. A note on the effects of selected prebiotics on the performance and ileal microbiota of broiler chickens. J Anim Feed Sci. 2008;17:392–7.
48. Jones JI, Clemmons DR. Insulin-like growth factors and their binding proteins: biological actions. Endocr rev. 1995;16(1):3–34.
49. Beccavin C, Chevalier B, Cogburn LA, Simon J, Duclos MJ. Insulin-like growth factors and body growth in chickens divergently selected for high or low growth rate. J Endocrinology. 2001;168(2):297–306.
50. Beckman BR, Larsen DA, Moriyama S, Lee-Pawlak B, Dickhoff WW. Insulin-like growth factor-I and environmental modulation of growth during smoltification of spring chinook salmon (Oncorhynchus tshawytscha). Gen comp endoc. 1998;109(3):325–35.
51. Silverstein JT, Shearer KD, Dickhoff WW, Plisetskaya EM. Effects of growth and fatness on sexual development of chinook salmon (Oncorhynchus tshawytscha) parr. Can J Fish Aquat Sci. 1998;55(11):2376–82.
52. Nankervis L, Matthews S, Appleford P. Effect of dietary non-protein energy source on growth, nutrient retention and circulating insulin-like growth factor I and triiodothyronine levels in juvenile barramundi, Lates calcarifer. Aquaculture. 2000;191(4):323–35.

12

Interaction between avian influenza subtype H9N2 and Newcastle disease virus vaccine strain (LaSota) in chickens

Hany F. Ellakany[1], Ahmed R. Gado[1], Ahmed R. Elbestawy[1*], Hatem S. Abd El-Hamid[1], Hafez M. Hafez[2], Mohamed E. Abd El-Hack[3*], Ayman A. Swelum[4,5*], Abdullah Al-Owaimer[4] and Islam M. Saadeldin[4,6]

Abstract

Background: H9N2 avian influenza virus is endemic in Egyptian poultry flocks. The role of the live viral vaccines such as LaSota in exaggeration of the clinical picture of H9N2 infection under field conditions is significantly important leading to severe economic losses due to higher mortality and lower growth performance. This experiment was designed to identify the possible interaction between experimental infection with H9N2 virus and NDV live vaccine (LaSota strain) in broiler chickens. Six groups each of 20 broiler chicks were used. Three groups (G1–3) were infected with H9N2 and vaccinated with LaSota, 3 days before, at the same day or 3 days post vaccination (dpv), while the remaining groups (G4–6) were non-vaccinated infected, vaccinated non-infected and non-vaccinated non-infected.

Results: The highest mortality rate (37.5%) was noticed in chickens of G1 (H9N2 infected 3 days prior LaSota vaccination). Also, this bird group had the most severe clinical signs, histopathological lesions and the longest viral shedding for 9 days post infection (dpi). In the 2nd and 3rd groups, the mortality rate was the similar (31.2%) with less pronounced clinical signs, histopathological lesions and H9N2 shedding was for only 6 dpi with the least shedding quantity in chickens of G3. The control non-vaccinated infected chickens (G4) had 18.7% mortality with the least degree of clinical signs, lesions and the highest viral shedding quantity but only for 6 dpi. At 35 days of age, there was a statistical significant decrease ($P < 0.05$) in chicken's body weight of all H9N2 infected groups from G1 to G4 compared to non-infected control groups, G5 and G6 respectively.

Conclusion: It was clear that laSota vaccination significantly affect H9N2 infection in broiler chickens regarding clinical signs, mortality rate, lesions, performance and viral shedding.

Keywords: Interaction, Avian influenza, H9N2, LaSota, NDV

Background

Low pathogenic avian influenza (LPAI) subtype H9N2 virus is widespread worldwide, being endemic in poultry populations in Asia and the Middle East. The first isolation of H9N2 virus in Egypt was obtained from the apparently healthy commercial bobwhite quail flock (A/quail/Egypt/113413v/2011/H9N2) in May 2011 [17]. Later on, the virus was isolated from commercial broiler, broiler breeder and layer flocks [27, 32]. H9N2 AIV infection in broilers cause reduction of growth rate and adverse the feed conversion ratio. In addition, high mortality rate appeared in case of co-infection with other pathogens such as infectious bronchitis virus (IBV), Newcastle disease virus (NDV), *Staphylococcus aureus*, Ornithobacterium rhinotracheale and *E. coli* or immune suppression, which exacerbates H9N2 influenza A virus infection in chickens [18, 23]. Lentogenic NDV strains are commonly used as live vaccines in the commercial poultry industry to protect from virulent forms of NDV. Experimental infections of specific pathogen-free (SPF) chickens with NDV vaccine strains cause little to no clinical disease. However, the application of the vaccine

* Correspondence: ahmed.elbestawy@vetmed.dmu.edu.eg; dr.mohamed.e.abdalhaq@gmail.com; m.ezzat@zu.edu.eg; aswelum@ksu.edu.sa
[1]Department of Poultry and Fish Diseases, Faculty of Veterinary Medicine, Damanhour University, Elgomhoria st, 63, Damanhour, Elbehira 22511, Egypt
[3]Department of Poultry, Faculty of Agriculture, Zagazig University, Zagazig 44511, Egypt
[4]Department of Animal Production, College of Food and Agriculture Sciences, King Saud University, P.O. Box 2460, Riyadh 11451, Saudi Arabia
Full list of author information is available at the end of the article

under field conditions can decrease productivity by inducing mild respiratory symptoms, which could be exacerbated concomitantly with other respiratory pathogens or in combination with environmental stressors [33]. Viral interference is a phenomenon explained previously [13], in which the infected cells by a virus do not permit multiplication of a second virus. Both viruses (LPAI and NDV) replicate in the epithelial cells of the respiratory and intestinal tracts (where trypsin-like enzymes is present) through competing for target cells or replicating in adjacent cells exacerbating clinical signs of disease in infected birds or produce viral interference, masking infections in one of them [9, 38].

In this experiment, we studied the interaction between LPAI subtype H9N2 infection and NDV vaccine (LaSota strain) in broiler chickens regarding clinical signs, body weights, post-mortem lesions, mortality, viral shedding, histopathological lesions and serological response for both viruses.

Results

Clinical signs (Table 1)

The observed signs in all infected groups with H9N2 virus (G1–4) were depression, decreased feed consumption and respiratory signs. The most severe clinical signs in terms of morbidity were observed in chickens of G1, while the least were in chickens of G4 as the respiratory signs started by sneezing and coughing then exaggerated as rales, nasal discharge, and head swelling till the end of observation period (21 days). Also, white greenish diarrhea was observed from 3rd dpi and subsiding at 11th -12th dpi. In addition, only in G1 exhibited thick oral mucous, which started at 7th dpi and continued until end of observation period.

Daily mortality (Table 2) and post-mortem (PM) lesions

Chickens of G1 suffered from early highest mortality 6/16 birds (37.5%) from 7th to 17th dpi compared to 31.2% (5/16 birds) in both G2 and G3 which began from 8th to 15th and 2nd to 9th dpi, respectively in both groups. While, in G4, the mortality rate was 18.7% (3/16 birds) with delayed start at day 12th dpi and continued only for 3 days.

Table 1 Degree of severity of clinical signs

Chicken groups Clinical signs	G1	G2	G3	G4	G5	G6
Sneezing	+++	++	++	+	-	-
Coughing	+++	++	++	+	-	-
Rales	+++	++	++	+	-	+
Head swelling	+++	++	+	-	-	-
Diarrhea	+++	++	++	+	-	-
Depression	+++	++	++	+	-	-
Thick oral saliva	+++	-	-	-	-	-

-: No signs +: Mild degree ++: Moderate degree +++: Severe degree

Table 2 Mortality rates in all experimental chicken groups from 15 to 35 DO after infection with LPAI subtype H9N2 alone or in combination with LaSota vaccine

Age (days)	dpi	Daily mortality/each chicken group					
		1	2	3	4	5	6
15	1	-	-	-	-	-	-
16	2	-	-	-	-	-	-
17	3	-	-	-	-		-
18	4	-	-	-	-	-	-
19	5	-	-	-	-	-	-
20	6	-	-	-	-	-	-
21	7	2	-	-	-	-	-
22	8	1	-	-	-	-	-
23	9	2	-	1	-	-	-
24	10	-	-	-	-	-	-
25	11	-	-	-	-	-	-
26	12	-	1	-	-	-	-
27	13	-	-	-	-	-	-
28	14	-	-	-	-	-	-
29	15	-	-	1	-	-	-
30	16	-	-	1	1	-	-
31	17	1	1	1	1	-	-
32	18	-	2	2	1	-	-
33	19	-	1	-	-	-	-
34	20	-	-	-	-	-	-
35	21	-	-	-	-	-	-
Mortality No.		6/16	5/16	5/16	3/16	0/16	0/16
Percent %		37.5	31.2	31.2	18.7	0	0

The most prominent PM lesions appeared were tracheitis, pneumonia, air sacculitis, splenomegaly, pancreatitis, enteritis, and nephritis. Chickens of G1 were more affected than other groups in terms of frequency and severity of lesions. No thymus, bursal lesions or caseous plug at tracheal bifurcation were observed.

No clinical signs, mortalities or PM lesions were recorded in G5 and G6.

Body weights of the experimental groups

All chicken groups had no significant difference at 14th day of age. On the other hand, a significantly decrease of body weight were observed in chickens of G1, G3 and G4 ($P < 0.05$) in compare to the birds of G2, G5 and G6 at 21th day of age. At 28th day, all vaccinated and infected chickens (G1, G2 and G3) were significantly decreased ($P < 0.05$) in body weights than G4, G5 and G6. Finally, at 35th day of age the body weights were significantly decreased ($P < 0.05$) in all H9N2 infected groups from G1 to G4 as 1540, 1509, 1474 and 1605 g, respectively than non-infected control G5 and G6 as 1769 and 1782 g, respectively (Fig. 1).

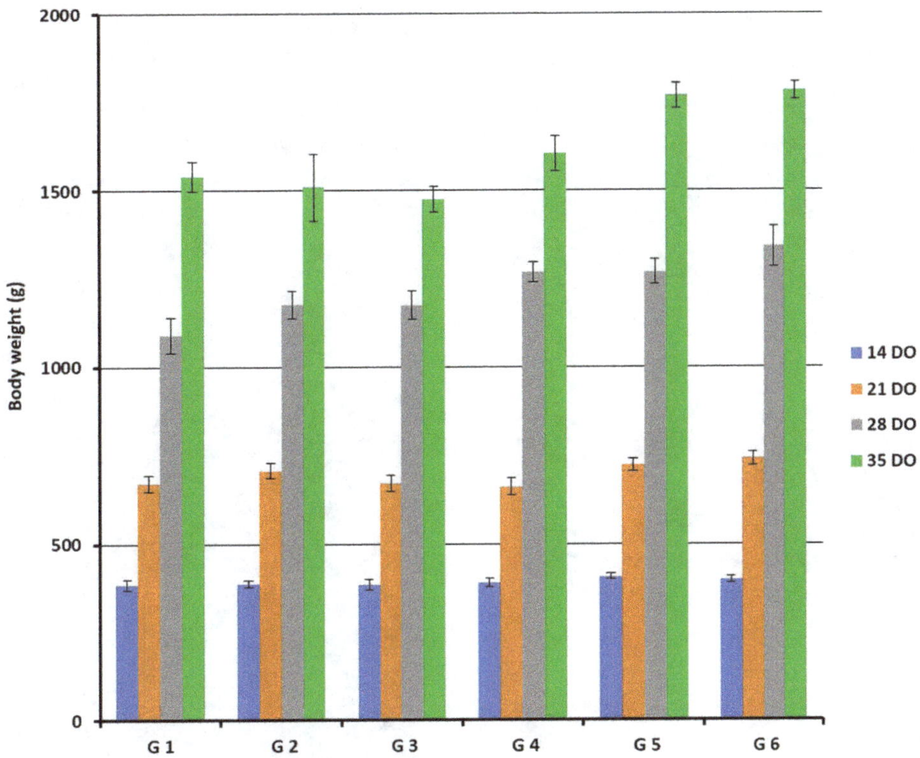

Fig. 1 Mean ± SE body weight (gram) of all experimental chicken groups at 14, 21, 28 and 35 DO

Histopathology

Trachea of bird in G1 showed severe degeneration of mucous glands with partial loss of the lining mucosa (Fig. 2a), while other H9N2 infected groups G2, G3 and G4 showed mild to moderate epithelial degeneration and hyperplasia of the mucosal lining (Fig. 2b and c).

Thymus lesions appeared as marked loss of cortical basophilic thymocytes which revealed marked lymphoid depletion was also noticed in the early infected G1 chickens (Fig. 3a). While birds of G4 showed degeneration and marked decrease of cortical thymocytes compared to control chicken G6 (Fig. 3b and c).

Bursa of Fabricius of birds of G1 (Fig. 4a) and G2 showed lesser degree of lymphoid depletion than birds of G3 and G4 (Fig. 4b) which appeared as marked lymphoid depletion with large mucin filled cysts within some follicles and thickening of the interfollicular connective tissue compared to control chicken G6 (Fig. 4c).

Spleen of chicken G1 and G4 (Fig. 5a and b) showed mild to moderate degree of lymphoid depletion, while the examined spleen of all other chicken groups was normal histology (Fig. 5c).

Renal inflammation of chickens G1 (Fig. 6a) and G4 showed lesions of glomerulonephritis which was obvious in most of examined birds. Interestingly, most of the lesions were glomerular rather than tubular and with a diffuse pattern with some tubular interstitial nephritis (Fig. 6b). Other chicken groups were histologically normal (Fig. 6c).

Mean HI titers of experimental chicken groups

The mean HI titer of the 10 sacrificed birds at day 1 for H9N2 and NDV were 6.1 and 6.7 log2, respectively.

For H9N2: At 7th dpi, the HI titers of G3 were significantly ($P < 0.05$) lower than G2, G4 or G1 for (with no significant difference between samples of chicken G1 and G4). While, the only significant lower titers ($P < 0.05$) at 14th dpi were in G1 (4.7) rather than G2, G3 and G4 (6.9, 6.2 and 6.7, respectively).

Regarding NDV: At 7th and 14th dpi, the highest significant ($P < 0.05$) HI titers were G2, G3 and G5 as 4.2, 4 and 3.9 and 3.7, 3.9 and 4.2, respectively compared to the lowest significant ($P < 0.05$) titers in G4 at the same dpi as 2.8 and 0.9, respectively (Table 3).

Quantification of H9N2 virus shedding by real-time RT-PCR

Tracheal swabs collected before challenge then at 3, 6 and 9 dpi revealed that the highest viral titers were in chicken G4 followed by G1, while the least titers were in G3 at 3 dpi. Also, the shedding of the H9N2 virus was persisted for 9 dpi in G1 only (Fig. 7).

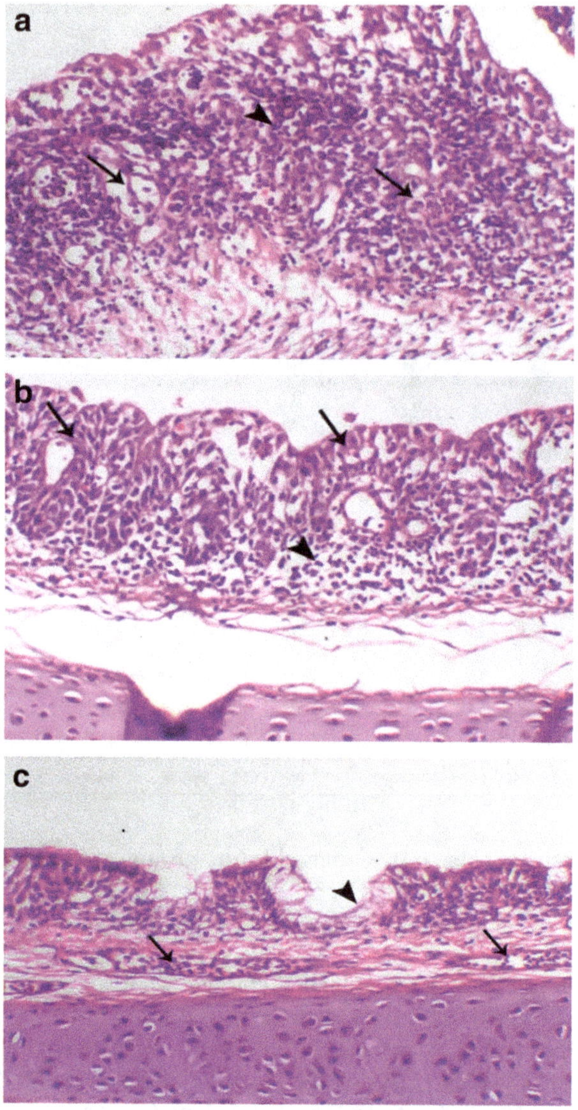

Fig. 2 **a** Trachea of birds in G1 showed severe degeneration of the tracheal mucous glands (arrow) with marked leukocytic infiltration. H&E, X 200; **b** Trachea of in G2 showed epithelial and glandular degeneration (arrow) and lower sub-epithelial leukocytic infiltration (arrowhead), H&E, X 200; **c** *trachea* of non-infected non-vaccinated chicken G6 showed normal epithelial lining (arrowhead) contained blood vessels (arrow), H&E, X 200

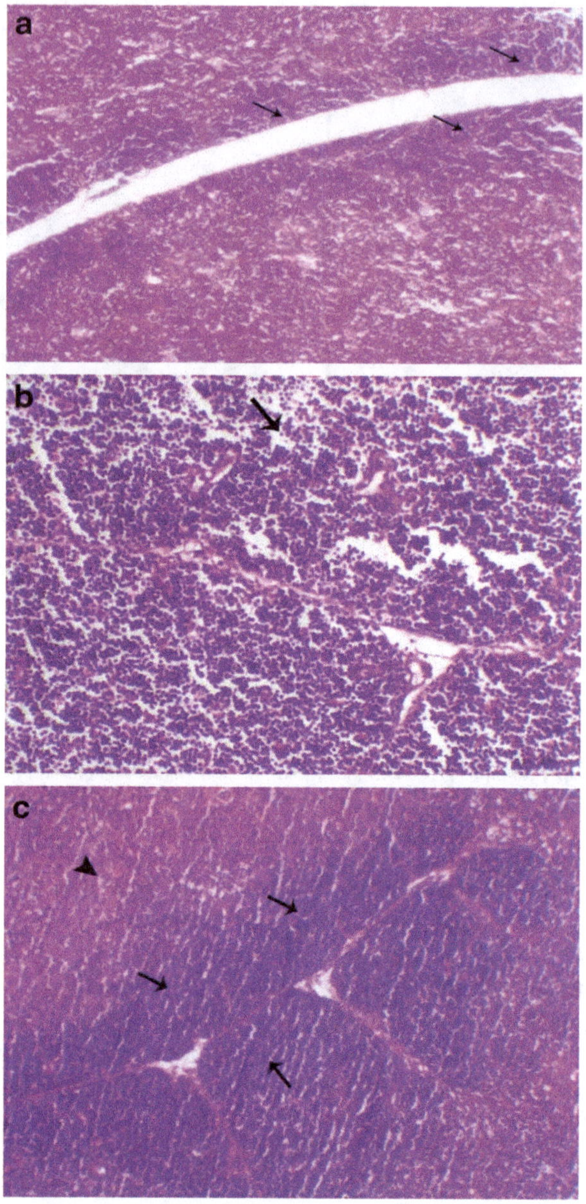

Fig. 3 **a** Thymus of bird in G1 showed severe loss of cortical basophilic thymocytes H&E, X 100; **b** Thymus of birds in G4 showed a marked decrease of cortical thymocytes (arrow), H&E, X 200; **c** Thymus of control chicken G6 revealed normal cortex and medulla with marked distinction between each. H&E, X 200

Discussion

Co-infections between LPAI subtype H9N2 and NDV usually occur, but cannot be easily diagnosed due to confusing similar clinical signs [38]. However, H9N2 infection is mostly associated with elevated mortality and severe economic losses due to many factors such as bad management, concurrent bacterial or viral diseases, live IBV or NDV vaccines, immune suppressive agents, age at the infection and the breed of chickens, which are mostly influenced the course of H9N2 infections [1, 9, 20, 29].

In the present work, we studied the interaction between both LPAI subtype H9N2 infection and vaccination with lentogenic NDV (LaSota strain) under experimental condition.

No clinical signs were observed in G5 and G6 (non-infected groups), while infected and non-vaccinated G4 exhibited mild respiratory signs (without head swelling) and diarrhea. All groups infected and vaccinated with LaSota showed more severe clinical signs started from 3 dpi and continued until the end of the observation

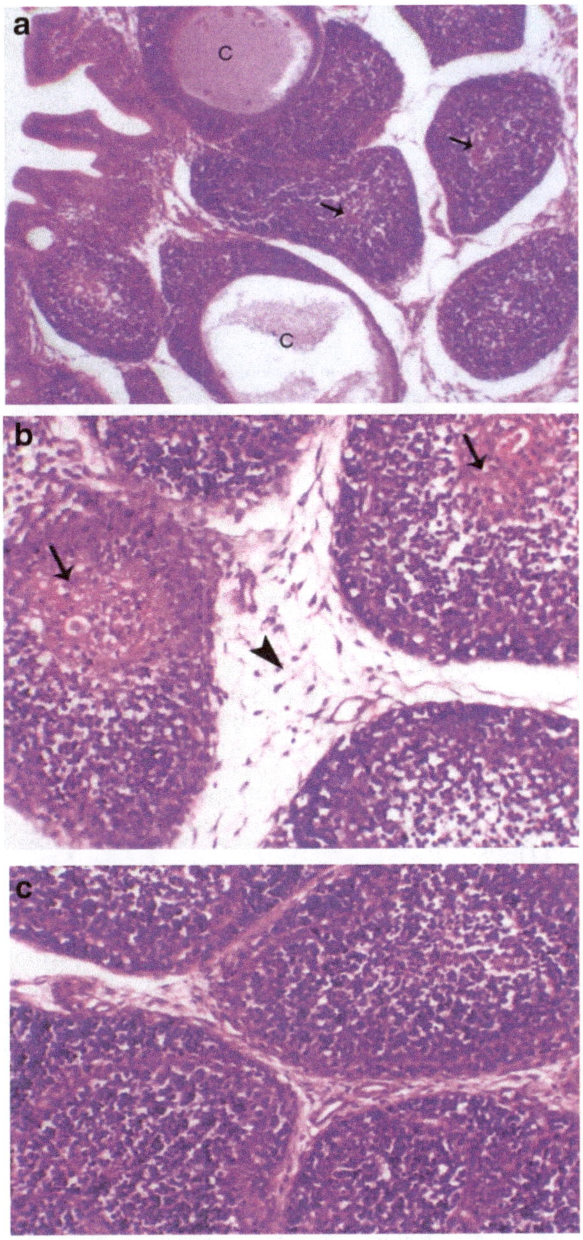

Fig. 4 **a** Bursa of birds in G1 revealed lymphoid depletion (arrow) with large mucin filled cysts within some of follicle (**c**). H&E, X 100; **b** *bursa* of birds in G4 showed lymphoid depletion with appearance of reticular fibers in the medulla of lymphoid follicle (arrow), and with marked increase in thickness of interfollicular connective tissue (arrowhead). H&E, X 200; **c** *bursa* of Fabricius of birds in G6 showed normal bursal lymphoid follicles. H&E, X200

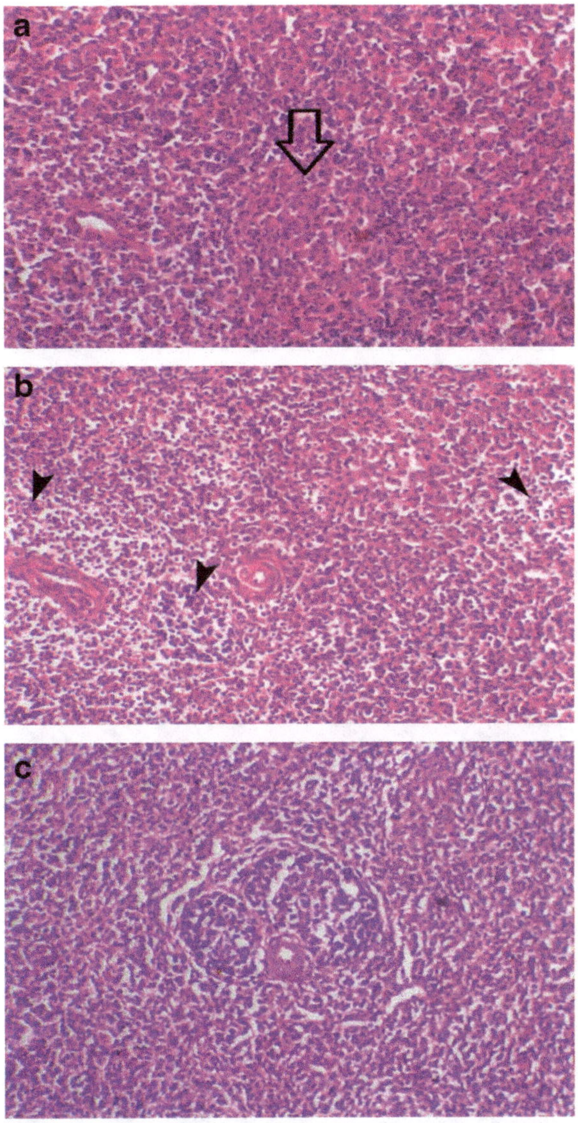

Fig. 5 **a** Spleen of G1 birds showed absence of lymphoid follicle (arrow). H&E, X 200; **b** Spleen of birds in G4 (non-vaccinated and H9N2-infected) showed decrease of lymphoid follicle (arrowhead). H&E, X 200; **c** Spleen of birds in G6 showed normal lymphoid follicle. H&E, X 200

period except diarrhea that stopped after 7 days from its appearance. Interference between H9N2 and vNDV was studied by Bonfante, et al. [8] who reported that the H9N2 challenge made birds more susceptible to the vNDV, lowering the minimum dose required to cause an infection, exacerbating the clinical outcome, while delaying the onset of the disease and time of death.

The highest severity of signs was recorded specially in chickens of G1 at 3–6 dpi concurrently with high H9N2 viral shedding in this group compared to G4 as detected by rRT-PCR, the results are in agreement with the findings detected by Kwon, et al. [24], which observed that following infection with an LPAI H9N2 virus, the clinical sign was peaked on 6 dpi. In the present study, diarrhea and dullness of the birds were observed after 2 dpi, reached its maximum at 5 dpi and started to decline at 8 dpi, which is like the findings of Aslam, et al. [4].

Regarding the mortality, the chickens of G3 were suffered from early mortality at 2nd dpi and persisted till

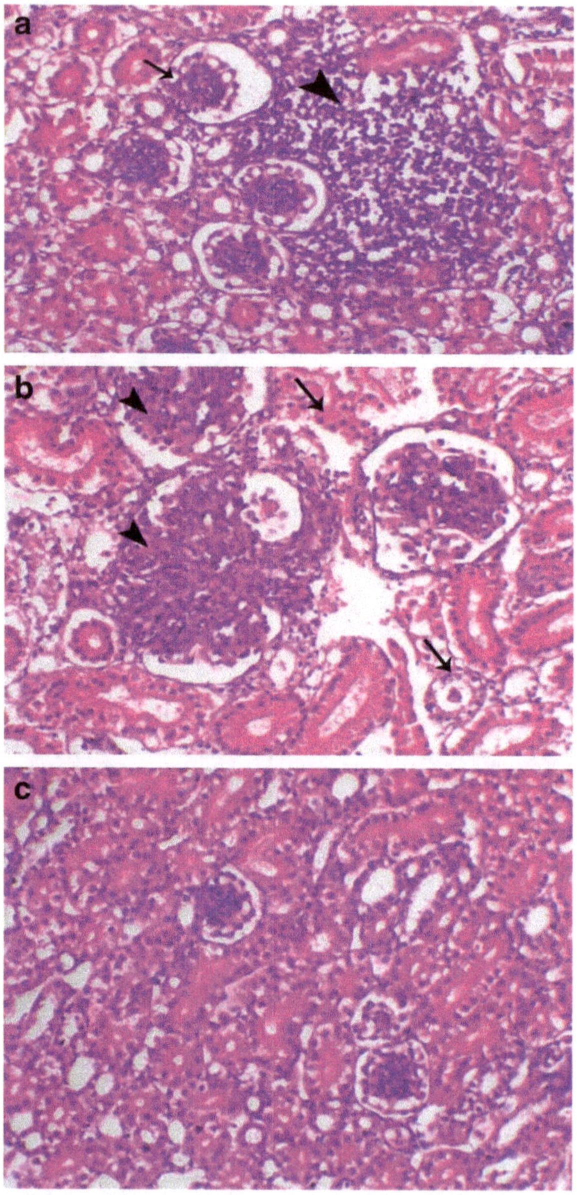

Fig. 6 a Kidney of G1 birds showed periglomerular and peritubular inflammatory cells infiltrations mainly lymphocytic (arrowhead). H&E, X 200; **b** Kidney of G4 birds showed marked glomerulonephritis features which revealed loss of capillaries and hypercellularity (arrowhead), with marked degeneration of renal tubules (arrow). H&E, X 200; **c** Kidneys of normal birds showed normal renal tubular and glomerular structures. H&E, X 200

11 dpi while the highest mortality (37.5%) was observed in G1. The least mortality (18.7%) was in chicken G4 began from 12th dpi and continued only for 3 days. Some studies reported no mortality after experimental infection of SPF-chickens with H9N2 [6, 8, 26].

This unusual high mortality rate in the present study may be due to the use of commercial broiler and not SPF chicks, which might be naturally carried some bacterial infection such as Mycoplasma and/or E coli, which are known to exaggerate the mortality of H9N2 infection. Li, et al. [25] found out that infection of SPF chicken with H9N2 virus did not induce any clinical signs or deaths after experimental infection, implying that co-infections with other pathogens might be the cause for these H9N2 viruses to be lethal in the field.

Table 3 Mean ± SE of HI titers for H9N2 and NDV (Geometric Mean) in different experimental chicken groups at 7th and 14th dpi

Group	H9N2		NDV	
	7th dpi	14th dpi	7th dpi	14th dpi
1	7.1 ± 0.2b	4.7 ± 0.2b	3.8 ± 0.3ab	2.9 ± 0.3b
2	8.2 ± 0.3a	6.9 ± 0.3a	4.2 ± 0.4a	3.7 ± 0.3a
3	5.5 ± 0.4c	6.2 ± 0.5a	4.0 ± 0.3a	3.5 ± 0.3a
4	7.3 ± 0.3b	6.7 ± 0.6a	2.8 ± 0.2bc	0.9 ± 0.2c
5	2.0 ± 0.2d	0.9 ± 0.3c	3.9 ± 0.3ab	4.2 ± 0.2a
6	2.2 ± 0.2d	0.8 ± 0.2c	2.6 ± 0.4c	0.7 ± 0.2c

$n = 10$ serum samples from each group
Mean values with different letters in the same column differ significantly at $P < 0.05$

Also, the increase in morbidity and mortality frequency after H9N2 AIV infection could be as a result of viral dissemination and spreading from the nasal cavity, which is the primary site of infection, to the lower respiratory tract, kidneys, intestines and some visceral organs [21, 28].

A significant decrease in body weights ($P < 0.05$) at the end of the experiment was reported among all infected groups in compare to G5 and G6, where the vaccinated and infected groups (specially at the same time in G3) was more affected than that only infected with H9N2 (G4). Also, the bird weights in G4 significantly ($P < 0.05$) differ by 9.2 and 9.9% than G5 and G6 respectively. In contrast, Costa-Hurtado, et al. [11] reported that no significant difference in body weight between control and groups either infected with H9N2 or vaccinated with NDV (LaSota).

Post-mortem lesions were more severe in chickens of G1 than other groups. The affected organs were trachea, thymus, bursa, spleen and kidney. These finding indicated that H9N2 not only affects respiratory and digestive systems but can also affects the immune organs (thymus gland and bursa of Fabricius) and kidney especially in cases of infected and LaSota vaccinated birds. However, the respiratory organs showed more prominent gross lesions in comparison to other organs. Similar findings were observed by Taubenberger and Morens [37]; Capua and Alexander [10] Hadipour, et al. [22] as well as Subtain, et al. [35]. However, the frequent cast formation in the tracheal bifurcation, which has been reported in field cases of H9N2 avian influenza outbreaks, was not observed in this experiment.

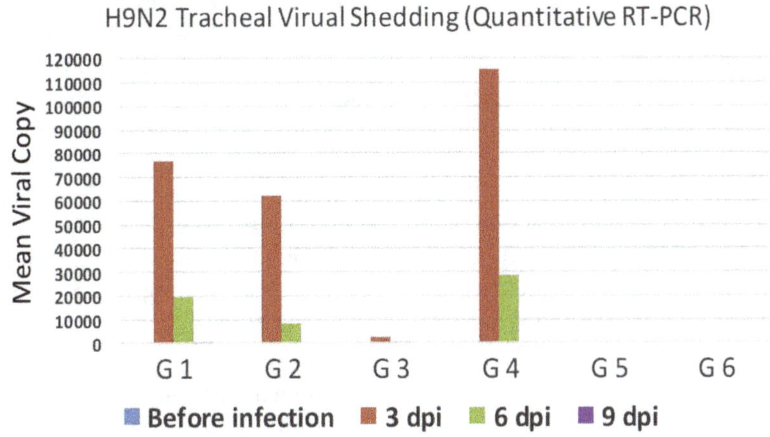

Fig. 7 Results of H9N2 shedding (just before infection, 3, 6 and 9 dpi) using quantitative RT-PCR

Alexander [2] reported that more cell injuries in the tissues of the respiratory system and gastrointestinal tract were recorded after infection with LPAIV. More severe lesions in the lungs and trachea could be attributed to the presence of a trypsin-like enzyme in the respiratory mucosa.

Histopathologically, chickens of G1 had severe of tracheal epithelial degeneration and hyperplasia of the mucosal lining compared to mild to moderate degree in G2, G3 and G4 which may be related to the higher late mortality obtained in this group along with obvious renal glomerulonephritis causing renal failure. This agreed with Aslam, et al. [4] who recorded an inflamed trachea and kidney of broilers infected with LPAI H9N2 and the histopathological lesions characterized by tracheal epithelial sloughing off with severe infiltration of inflammatory cells along with severe glomerular congestion of kidney.

Also, Doustar, et al. [15] proved in an experimental study that H9N2 virus could induce apoptosis and subsequent renal tissue damage. In addition, different mechanisms by which H9N2 can cause this effect were discussed in further investigations too [34, 40].

H9N2 infected birds (G4) also showed marked lymphoid depletion in the form of degeneration of thymocytes, follicular atrophy and cystic follicles of bursa of Fabricius and splenic lymphoid depletion. Similar results were reported by Kwon, et al. [24]. In addition, EL-bayoumi, et al. [16] reported about lymphocytic depletion and decreased immune response of birds infected H5N1, H9N2, and NDV.

Interestingly, the antibody titers against NDV (LaSota-vaccine) were significantly decreased ($P < 0.05$) by the presence of H9N2 suggesting that NDV had insufficient replication to trigger the humoral response even though they were administered the typical vaccination dose of LaSota strain. Another explanation is the immunosuppressive effect of H9N2 that proved histopathologically as degeneration of thymocytes and lymphoid depletion of bursa and spleen. Also, Allawe [3] compared between LaSota vaccinated broiler chickens with or without H9N2 infection. At 40 days of age, the birds vaccinated with live NDV vaccines showed highest titers against ND in comparison with the vaccinated infected group with H9N2 that indicated the immunosuppressive effect of H9N2 infection.

Viral shedding patterns in tracheal swabs was detected for up to 9 dpi in birds of G1, while in all other groups shedding stopped on the 6 dpi. This may indicate that in case of NDV live vaccination after accidental unnoticed field infection with LPAI subtype H9N2 leads to prolonged shedding of H9N2 viruses compared with the case of NDV live vaccination before H9N2 field infection.

The least concentration of virus shedding (2630 virus copy/swab) appeared in G3 versus the highest concentration (115,300 virus copy/swab) in the non-vaccinated and infected group (G4) at 3dpi and 614 virus copy/swab compared to 28,500 virus copy/swab at 6 dpi in both groups respectively, may turn out a possibility of some mechanism of interference between NDV live vaccines and H9N2 viruses in the respiratory tract of broiler chickens.

This may be related to the fact that LaSota vaccine and LPAI viruses replicate in cells where there are trypsin-like proteases such as in the upper respiratory and intestinal epithelium, so they might compete for the same target cells or replicate in adjacent cells [31, 36]. LaSota virus binds through the Haemagglutinin-Neuraminidase (HN) glycoprotein to sialic acid-containing receptors on the cell surface, as well as the HA glycoprotein does for LPAI-H9N2 [39]. This suggests that infection with a heterologous virus may result in temporary competition for cell receptors or susceptible cells, resulting in decreased initial replication of the second virus; but as replication of the first virus declines, the second virus increases to fill the gap [11].

Also, LaSota NDV strain is known to be a weak interferon inducer as part of their low virulent phenotype profile, local interferon production might still be able to interfere with LPAIV replication [14].

The results obtained underline the importance of co-infections, which can either exacerbate clinical disease or affect virus replication by lowering viral titers to under the levels of detection and affecting serological results and in some cases prolonged the time of virus shed which could favor for further transmission and these data was supported by Costa-Hurtado, et al. [12] and Ge, et al. [19] who concluded that, co-infection with NDV and HPAI can affect the viral replication dynamics and the disease caused by these viruses in chickens, but this effect will depend on the virulence of the viruses involved, the challenge titer of the viruses and the timing of the coinfections.

Conclusions

There was an adverse effect of combined infection with H9N2 and live LaSota vaccination in terms of clinical signs, body weights, mortality, histopathological lesions, viral shedding, and humoral immune response. Here we also recommend to diagnose the presence of H9N2 infection before ND vaccination to avoid to the exaggeration and immunosuppressive effect of H9N2 infection.

Methods

Experimental animals

A total of 130 one-day old commercial broiler chicks Cobb 500 purchased from a local poultry company (NASCO-Egypt) located in Alexandria governorate were used for the experimental infection in this study. All chicken groups were weighted at 14th, 21th, 28th and 35th day of age. Study design and allocating animals to experimental groups are explained in Table 4.

All the experimental work, tests and procedures were complied with the general guidelines of Damanhour University and were approved by the Local Ethics Commission of the Animal Health and Welfare of Damanhour University with respect to care of animals under study and all efforts were made to minimize suffering. The Ethical Approval Code was DMU/VetMed-2017/0023.

Housing and husbandry

Chicks were housed in galvanized wire cages (40 cm high × 50 cm width × 100 cm length) in (ten chicks each). Fresh feed and water were offered ad-libitum through the experimental period. All chicks were kept under the same managerial, hygienic and environmental conditions. Chicks were fed to cover their requirements according to NRC [38]. The sacrificed birds all over the experiment were previously anesthetized using sodium pentobarbital (50 mg/kg) by intraperitoneal injection to minimize suffering.

Experimental procedures

Vaccines. Birds of G1–4 were vaccinated against NDV by eye drop between days 15th to 21st (Table 4) using LaSota® (Schering plough, USA).

Challenge virus

Egyptian H9N2 strain isolated from broiler chickens in Alexandria Governorate (A/Chicken/Egypt/93/2015 and the GenBank accession No: KY872759) was used as a challenge virus in a dose 0.2 ml/bird of $10^{6.3}$ EID50/dose (0.1 ml by nasal instillation and 0.1 ml through eye drop) at 18th day of age [30].

Sample size

Ten chicks were sacrificed at the first day of age and serum samples were used to determine the level of maternally derived antibodies against H9N2 and NDV using Haemagglutination inhibition test. The remaining 120 chicks were distributed randomly into 6 equal groups of each 20 chicks. Birds were weighed and sacrificed by exsanguination after prior anesthesia.

Experimental outcomes

All the birds were examined daily for the following parameters:

a) ***Clinical signs, Mortality rates and Post-mortem (PM) lesions.*** Birds in all groups were examined daily up to 35 days of age and all disease signs,

Table 4 Experimental design

Group	Vaccination with LaSota	Day of H9N2 infection	Weighting	Pooled tracheal swabs	10 serum samples/group (7th and 14th dpi)	Histopathology of 2 sacrified birds at 4th and 9th dpi
1	18 DO	15 DO	14,21,28,35 DO	15,18,21,24 DO	22,29 DO	19,24 DO
2	18 DO	18 DO		18,21,24,27 DO	25,32 DO	22,27 DO
3	18 DO	21 DO		21,24,27,30 DO	28,35 DO	25,30 DO
4	–	18 DO		18,21,24,27 DO	25,32 DO	22,27 DO
5	18 DO	–		21,28 DO	25,32 DO	22,27 DO
6	–	–		21,28 DO	25,32 DO	22,27 DO

dpi, days post infection, *DO* Days Old

mortalities and PM lesions of dead or sacrificed birds were recorded.

a) ***Histopathological examination.*** Two birds from each group were sacrificed under anesthesia with intravenous injection of sodium pentobarbital (50 mg/kg) at 4th and 9th dpi in groups (1–4) and at 22 and 27 days of age in groups 5 and 6. Samples from trachea, thymus, bursa, spleen and kidney were collected, fixed in 10% formalin solution, stained with Hematoxylin and Eosin (HE) and examined for lesions [5].
b) ***Haemagglutination inhibition (HI) test.*** Serum samples collected at 7th and 14th dpi were tested by HI test for H9N2 and ND antibodies using standard haemagglutination antigens (2^8 titers) for both viruses. HI test was applied using 4 Haemagglutination Units (HAU) of the relevant antigens and 25 µl of 1% chicken RBCs according to Alexander [2] and the results were analyzed using SPSS
c) ***Quantification of H9N2 virus shedding by Real-Time RT-PCR (rRT-PCR).*** Pooled tracheal swabs were collected from 10 chickens in each group just before H9N2 infection then at day 3, 6 and 9 post H9N2 infections. The RNA was extracted using QIAamp viral RNA Mini-kits (QIAGEN 52904-Germany) and the test was performed using rRT-PCR (Strategen 3005P USA) with Quantitect RT-PCR Kit reaction buffer (QIAGEN 204443-Germany) and positive control RNA from A/quail/Egypt/113413v/2011(H9N2). The Sequence of H9 primers and probe as published by Ben Shabat, et al. [7] were used.
d) Thermal profile was RT at 45 °C for 10 min and 95 °C for 10 min, followed by 40 cycles of PCR at 95 °C for 15 s for denaturation and 60 °C for 45 s for annealing and extension.

Statistical methods

Data of birds' body weight were subjected to analysis of variance procedures appropriate for a completely randomized design using the GLM procedures of SPSS (2008). The differences among means were determined using the student Newman keuls test. Statements of statistical significance are based on $p < 0.05$.

Abbreviations

AIV: Avian influenza viruses; dpi: Day post infection; dpv: Day post vaccination; IBV: Infectious bronchitis virus; NDV: Newcastle disease virus; PM: Post-mortem; SPF: Specific pathogen-free

Acknowledgments

Authors thank their respective universities and institutes for their support.

Funding

The authors extend their appreciation to the Deanship of Scientific Research at King Saud University for funding this work through a research group project (# RG-1438-018).

Authors' contributions

Made substantial contributions to conception and design, or acquisition of data, or analysis and interpretation of data: HFE ARE HSA ARG. Been involved in drafting the manuscript or revising it critically for important intellectual content: ARE HFE HSA HMH MEA AAS IMS AA. Given final approval of the version to be published: HFE HSA ARE HMH ARG MEA IMS AAS AA. Agreed to be accountable for all aspects of the work in ensuring that questions related to the accuracy or integrity of any part of the work are appropriately investigated and resolved: ARE MEA HFE HSA HMH ARG IMS AAS AA.

Competing interests

The authors declare that they have no competing interests.

Author details

[1]Department of Poultry and Fish Diseases, Faculty of Veterinary Medicine, Damanhour University, Elgomhoria st, 63, Damanhour, Elbehira 22511, Egypt. [2]Institute of Poultry Diseases, Free University Berlin, Königsweg, 63, 14163 Berlin, Germany. [3]Department of Poultry, Faculty of Agriculture, Zagazig University, Zagazig 44511, Egypt. [4]Department of Animal Production, College of Food and Agriculture Sciences, King Saud University, P.O. Box 2460, Riyadh 11451, Saudi Arabia. [5]Department of Theriogenology, Faculty of Veterinary Medicine, Zagazig University, Zagazig 44511, Egypt. [6]Department of Physiology, Faculty of Veterinary Medicine, Zagazig University, Zagazig 44511, Egypt.

References

1. Aamir UB, Wernery U, Ilyushina N, Webster RG. Characterization of avian H9N2 influenza viruses from United Arab Emirates 2000 to 2003. Virology. 2007;361:45–55.
2. Alexander DJ. Avian Influenza - Diagnosis. Zoonoses Public Health. 2008;55: 16–23.
3. Allawe AB. Study of effect of immunosuppresion caused by H9N2 in broiler. Iraqi J Vet Sci. 2008;32:80–9.
4. Aslam R, Aslam A, Tipu Y, Nazir J, Ghafoor A, Fatima S. Histopathological and immunohistochemical studies for the pathogenesis of a low pathogenicity H9 avian influenza virus in experimentally infected commercial broilers. J An Plant Sci. 2015;25:45–52.
5. Bancroft JD, Layton C. The hematoxylin and eosin. In: Suvarna SK, Layton C, Bancroft JD, editors. Theory practice of histological techniques. 7th ed. Philadelphia: Churchill Livingstone of El Sevier; 2013.
6. Banks J, Speidel EC, Harris PA, Alexander DJ. Phylogenetic analysis of influenza a viruses of H9 haemagglutinin subtype. Avian Pathol. 2010; 29:353–9.
7. Ben Shabat M, Meir R, Haddas R, Lapin E, Shkoda I, Raibstein I, Perk S, Davidson I. Development of a real-time TaqMan RT-PCR assay for the detection of H9N2 avian influenza viruses. J Virol Methods. 2010;168:72–7.
8. Bonfante F, Cattoli G, Leardini S, Salomoni A, Mazzetto E, Davidson I, Haddas R, Terregino C. Synergy or interference of a H9N2 avian influenza virus with a velogenic Newcastle disease virus in chickens is dose dependent. Avian Pathol. 2017;46:488–96.
9. Capua I, Alexander DJ. Avian influenza: recent developments. Avian Pathol. 2004;33:393–404.
10. Capua I, Alexander D. Avian influenza and Newcastle disease: a field and laboratory manual. Milan: Springer-Verlag Italia; 2009.
11. Costa-Hurtado M, Afonso CL, Miller PJ, Spackman E, Kapczynski DR, Swayne DE, Shepherd E, Smith D, Zsak A, Pantin-Jackwood M. Virus interference between H7N2 low pathogenic avian influenza virus and lentogenic Newcastle disease virus in experimental co-infections in chickens and turkeys. Vet Res. 2014;45:1.
12. Costa-Hurtado M, Afonso CL, Miller PJ, Shepherd E, Cha RM, Smith D, Spackman E, Kapczynski DR, Suarez DL, Swayne DE, Pantin-Jackwood MJ. Previous infection with virulent strains of Newcastle disease virus reduces highly pathogenic avian influenza virus replication, disease, and mortality in chickens. Vet Res. 2015;46:97.

13. DaPalma T, Doonan BP, Trager NM, Kasman LM. A systematic approach to virus–virus interactions. Virus Res. 2010;149:1–9.
14. Dortmans JCFM, Koch G, Rottier PJM, Peeters BPH. Virulence of Newcastle disease virus: what is known so far? Vet Res. 2011;42:122.
15. Doustar Y, Mohajeri D, Nazeri M, Jafari B. Experimental study of lymphoid tissue apoptosis subsequent to infection by influenza virus (H9N2) in SPF chickens Asian. J Exp Biol Sci. 2012;3:87–92.
16. EL-bayoumi KM, Mahgoub KM, Mekky HM, Hassan ER, Amin Girh Z, Maatouq AM, El-Samadony HA, Rabie NS, MAA A, Kutkat MA. Molecular detection of H5N1, H9N2 and Newcastle disease viruses isolated from chicken in mixed infection in Egypt. World Appl Sci J. 2013;27:44–50.
17. El-Zoghby EF, Arafa A-S, Hassan MK, Aly MM, Selim A, Kilany WH, Selim U, Nasef S, Aggor MG, Abdelwhab EM, Hafez HM. Isolation of H9N2 avian influenza virus from bobwhite quail (Colinus virginianus) in Egypt. Arch Virol. 2012;157:1167–72.
18. Fusaro A, Monne I, Salviato A, Valastro V, Schivo A, Amarin NM, Gonzalez C, Ismail MM, Al-Ankari AR, Al-Blowi MH, Khan OA, Maken Ali AS, Hedayati A, Garcia Garcia J, Ziay GM, Shoushtari A, Al Qahtani KN, Capua I, Holmes EC, Cattoli G. Phylogeography and evolutionary history of Reassortant H9N2 viruses with potential human health implications. J Virol. 2011;85:8413–21.
19. Ge S, Zheng D, Zhao Y, Liu H, Liu W, Sun Q, Li J, Yu S, Zuo Y, Han X, Li L, Lv Y, Wang Y, Liu X, Wang Z. Evaluating viral interference between influenza virus and Newcastle disease virus using real-time reverse transcription–polymerase chain reaction in chicken eggs. Virol J. 2012;9:128.
20. Guo YJ, Krauss S, Senne DA, Mo IP, Lo KS, Xiong XP, Norwood M, Shortridge KF, Webster RG, Guan Y. Characterization of the pathogenicity of members of the newly established H9N2 influenza virus lineages in Asia. Virology. 2000;267:279–88.
21. Hablolvarid MH, Sohraby HI, Pourbakhsh SA, Gholami MR. Histopathological study of intranasally inoculated A/chicken/Iran/259/ 1998 (H9N2) influenza virus in chicken. Arch Razi Ins. 2004;58:51–62.
22. Hadipour MM, Farjadian SH, Azad F, Kamravan M, Dehghan A. Nephropathogenicity of H9N2 avian influenza virus in commercial broiler chickens following Intratracheal inoculation. J Anim Vet Adv. 2011;10:1706–10.
23. Haghighat-Jahromi M, Asasi K, Nili H, Dadras H, Shooshtari AH. Coinfection of avian influenza virus (H9N2 subtype) with infectious bronchitis live vaccine. Arch Virol. 2008;153:651–5.
24. Kwon J-S, Lee H-J, Lee D-H, Lee Y-J, Mo I-P, Nahm S-S, Kim M-J, Lee J-B, Park S-Y, Choi I-S, Song C-S. Immune responses and pathogenesis in immunocompromised chickens in response to infection with the H9N2 low pathogenic avian influenza virus. Virus Res. 2008;133:187–94.
25. Li Z, Chen H, Jiao P, Deng G, Tian G, Li Y, Hoffmann E, Webster RG, Matsuoka Y, Yu K. Molecular basis of replication of duck H5N1 influenza viruses in a mammalian mouse model. J Virol. 2005;79:12058–64.
26. Mo IP, Brugh M, Fletcher OJ, Rowland GN, Swayne DE. Comparative pathology of chickens experimentally inoculated with avian influenza viruses of low and high pathogenicity. Avian Dis. 1997;41:125–36.
27. Monne I, Hussein HA, Fusaro A, Valastro V, Hamoud MM, Khalefa RA, Dardir SN, Radwan MI, Capua I, Cattoli G. H9N2 influenza a virus circulates in H5N1 endemically infected poultry population in Egypt. Influenza Other Respir Viruses. 2013;7:240–3.
28. Mosleh N, Dadras H, Mohammadi A. Evaluation of H9N2 avian influenza virus dissemination in various organs of experimentally infected broiler chickens using RT-PCR. Iran J Vet Res. 2009;10:12–20.
29. Pantin-Jackwood MJ, Costa-Hurtado M, Miller PJ, Afonso CL, Spackman E, Kapczynski DR, Shepherd E, Smith D, Swayne DE. Experimental co-infections of domestic ducks with a virulent Newcastle disease virus and low or highly pathogenic avian influenza viruses. Vet Microbiol. 2015;177:7–17.
30. Reed LJ, Muench H. A simple method of estimating fifty per cent Endpoints12. Am J Epidemiol. 1938;27:493–7.
31. Rott R. Molecular basis of infectivity and pathogenicity of myxovirus Brief review. Arch Virol. 1979;59:285–98.
32. Shakal M, Yi Y, Sa EZ. Surveillance on avian influenza H5N1 and H9N2 subtypes in Egypt 2012-2013. Poultry Fisher Wildlife Sci. 2014;2:1.
33. Suarez D. Newcastle disease, other avian paramyxoviruses, and avian Metapneumovirus infections. In: Swayne DE, editor. Diseases of poultry. USA: WILEY Blackwell; 2013. p. 87–138.
34. Suarez DL, Schultz-Cherry S. Immunology of avian influenza virus: a review. Dev Comp Immunol. 2000;24:269–83.
35. Subtain SM, Chaudhry ZI, Anjum AA, Maqbool A, Sadique U. Study on pathogenesis of low pathogenic avian influenza virus H9 in broiler chickens. Pak J Zool. 2011;43:999–1008.
36. Swayne DE, Halvorson DA. In: Saif YM, Barnes HJ, Glisson JR, Fadly AM, McDougald LR, Swayne DE, editors. Ames: Influenza: Diseases of Poultry Iowa State Press; 2003. p. 135–60.
37. Taubenberger JK, Morens DM. The pathology of influenza virus infections. Ann Rev Pathol Mech Dis. 2008;3:499–522.
38. Umar S, Azeem T, Abid SA, Mushtaq A, Kiran A, Qayyum MR, Rehman A. Effect of lentogenic Newcastle disease virus (Lasota) on low pathogenic avian influenza virus (H9N2) infection in fayoumi chicken. J Avian Res. 2015;1:1–4.
39. Wan Z, Ye J, Xu L, Shao H, Jin W, Qian K, Wan H, Qin A. Antigenic mapping of the hemagglutinin of an H9N2 avian influenza virus reveals novel critical amino acid positions in antigenic sites. J Virol. 2014;88:3898–901.
40. Wurzer WJ, Ehrhardt C, Pleschka S, Berberich-Siebelt F, Wolff T, Walczak H, Planz O, Ludwig S. NF- B-dependent induction of tumor necrosis factor-related apoptosis-inducing ligand (TRAIL) and Fas/FasL is crucial for efficient influenza virus propagation. J Biol Chem. 2004;279:30931–7.

13

Effects of dietary supplementation with oridonin on the growth performance, relative organ weight, lymphocyte proliferation, and cytokine concentration in broiler chickens

Qiu Jue Wu[1,2,3], Xiao Chuan Zheng[1], Tian Wang[1*] and Tie Ying Zhang[2]

Abstract

Background: The present study was conducted to investigate the effects of oridonin (ORI) on growth performance, relative organ weight, lymphocyte proliferation, phagocytic function of neutrophils, and cytokine concentration in broiler chickens. A total of 240 one-day-old Arbor Acres male broilers were randomly assigned to four treatments with six replicate pens of 10 broiler chickens per pen. Broiler chickens were fed diets based on four levels of dietary ORI (0, 50, 80 and 100 mg/kg) for a 42-d feeding trial. The experimental diets were fed in three phases: 1 to 14 d, 15 to 28 d and 29 to 42 d.

Results: The results indicated that ORI has no influence on the growth performance ($P > 0.05$). However, ORI increased the relative weights of spleen and bursa, the number of proliferation peripheral blood T and B lymphocytes, the phagocytic rate of neutrophils, as well as the Interleukin-2 (IL-2), Interleukin-4 (IL-4) and tumor necrosis factor-a (TNF-a) serum concentrations in serum in broilers at days 14, 28 and 42 ($P < 0.05$).

Conclusions: In conclusion, ORI can enhance immune function and resistance to disease in broiler chickens by stimulating T and B lymphocyte formation, division, and proliferation, as well as the modulation of Th1/Th2 cytokine secretion profiles.

Keywords: Oridonin, Lymphocyte proliferation, Phagocytisis of neutrophils, Cytokine, Broiler

Background

With the rapid development of the poultry industry, high stocking densities and high yield requirements have induced broiler chickens to have higher immunological stress reactions and increased the potential risk of disease outbreaks in commercial poultry flocks. Immunological stress may cause the loss of immune homeostasis and trigger immunological stress responses, followed by enteritis, mortality, welfare problems and economic losses due to production losses and costs for treatment or prevention. Thus, exploring alternative substances and strategies to modulate the broiler immune system is of great significance.

Oridonin (ORI) as a multifunctional compound initially isolated from *Isodon rubescens* (a. k. a. "*Rabdosia rubescens*", traditional Chinese medicine name "*Donglingcao*"), is now widely used as a drug for humans. Correspondingly, it has been proven to possess considerable anti-cancer, antioxidation, and anti-inflammatory properties, it has been used to restrain bacterial growth and viral reproduction, detoxify, reduce pain, invigorate the stomach and protect other organs in Asian countries for hundreds of years [1–3].

ORI is considered to be an immune-enhancing dietary supplement, which not only improves phagocytes and inflammatory cytokine production but also exerts

* Correspondence: hdzwqj@163.com
[1]College of Animal Science and Technology, Nanjing Agricultural University, No. 6, Tongwei Road, Xuanwu District, Nanjing 210095, Jiangsu, People's Republic of China
Full list of author information is available at the end of the article

immuno-modulating functions for the body. Several studies have provided evidence that ORI is involved in inflammation and antitumor activities [4]. Studies in broilers showed that dietary ORI supplementation decreases expression of the proinflammatory cytokine in the LPS-induced inflammatory response [5]. It has been well documented that the inclusion of oridonin in rat diets can promote the development of immune organs, stimulate organ immunoglobulin secretion, and prohibit the colonization of *Escherichia coli* and *Salmonella*. Additionally, ORI component supplementation has been reported to enhance T lymphocyte responses in vivo [6].

It is well known that immune-modulating nutrients can be harmful if not utilized in moderation. However, whether oridonin could be used as an anti-inflammatory treatment, and the mechanisms underlying its anti-inflammatory activity, remains unknown. Therefore, the following study was conducted to further investigate the effect of dietary supplementation of ORI on growth performance, relative organ weights, lymphocyte proliferation, and cytokine concentration in broiler chickens and to gain a better understanding of the application of dietary ORI supplementation in the poultry industry for improving the health and preventing infectious diseases among broilers.

Methods

Animals, study design and diets

A total of 240 one-day-old male broiler chickens (Arbor Acres) were obtained from a commercial hatchery (Broiler Breeder of South Khorasan Complex Productive Co., Iran). All birds were weighed individually and were randomly assigned to four dietary treatment groups in a completely randomized design, each of which included 6 replicates with 10 birds per replicate. The 4 treatment groups were as follows: the control group, in which birds were received the basal diet. The ORI treatment groups, the basal diet was supplemented with oridonin (ORI) at 50 mg/kg, 80 mg/kg or 100 mg/kg (O1, O2 and O3 treatments), respectively. The trial lasted 42 days. Chicks were housed in three-story step cages (2 m × 1.4 m × 0.38 m) in an environmentally controlled room. The rearing room temperature and lighting cycle were provided according to procedure of broiler rearing and management during the period of experiment. The birds were fed the experimental diets in three phases 1 to 14 d, 14 to 28 d and 28 to 42 d. The basal diets were of the maize-soybean-type. The experimental diets were formulated based on the National Research Council (1994) [7] to meet or exceed the nutrient requirements for broiler chickens (Table 1). Fresh diets were prepared once a week and were stored in sealed bags at 4 °C. Feed and water were provided ad libitum throughout the experiment.

Table 1 Ingredients and chemical composition of diets used during starter (1–14 d of age), grower (15–28 d of age), and finisher periods (29–42 d of age)

Ingredients (g/kg)	1–14 d	15–28	29–42 d
Corn	428	459	476
Soybean meal (43%, crude protein)	365	250	203
Wheat	130	220	250
Soybean oil	17	8.5	10
Corn gluten meal	20	0	0
Canola meal	0	25.5	25
Na chloride	2.3	3	2.8
Dicalcium phosphate	15	12.5	11
Na bicarbonate	2.4	0.9	1
Ca carbonate	10.8	10.4	11
DL-Methionine	2.7	1.5	1.5
L-Lysine·HCl	2.2	0.6	1.1
Premix[a]	2	2	2
Multi-enzyme	0.3	0.3	0.3
Phytase	0.3	0.3	0.3
Bentonite	0.0	5.5	5
Prebiotics	2	0	0
Total	1, 000	1, 000	1, 000
Calculation of nutrients (g/kg)			
Apparent metabolism energy (MJ/kg)	12.5	12.7	12.9
Crude protein	222	201	193
Calcium	9.7	9.3	9.0
Available phosphorus	4.7	4.5	4.2
Lysine	13.8	11.3	11.0
Methionine	6.0	4.7	4.3
Methionine + cysteine	8.1	7.6	7.1

[a]Premix provided per kg of diet: Vitamin A (transretinyl acetate), 10,000 IU; Vitamin D_3 (cholecalciferol), 3000 IU; Vitamin E (all-*rac*-*α*-tocopherolacetate), 30 IU; menadione, 1.3 mg; thiamine 2.2 mg; riboflavin, 8 mg; nicotinamide, 40 mg; choline chloride, 600 mg; calcium pantothenate, 10 mg; pyridoxine·HCl, 4 mg; biotin, 0.04 mg; folic acid, 1 mg; vitamin B_{12} (cobalamine), 0.013 mg; Fe (from ferrous sulfate), 80 mg; Cu (from copper sulfate), 8 mg; Mn (from manganese sulfate), 110 mg; Zn (Bacitracin Zn), 65 mg; iodine (from calcium iodate), 1.1 mg; Se (from sodium selenite), 0.3 mg

Sample collection and procedures

In this experiment, the pen was the experimental unit, and data on body weight and feed intake were measured weekly, those data were used to calculate weight gain (WG), feed intake (FI) and feed conversion ratio (FCR).

At 14, 28 and 42 days of age, twelve broilers from each treatment were used for sample collecting. The birds were weighed, and blood samples were collected and separated by centrifugation at 3000×g for 15 min at 4 °C. Serum samples were frozen at − 80 °C until ELISA analysis. Then, all of the 24 broilers were sacrificed by exsanguination. After decapitation, internal organs (liver, spleen, pancreas, gizzard and bursa) were excised and weights of these organs were measured. Organ indexes were calculated as weight of organ (g)/100 g body weight.

Measurement of lymphocyte proliferation by the MTT method

Take fresh anticoagulated blood from the wing vein on d 14, 28 and 42 to a heparinized centrifuge tube.The peripheral blood lymphocytes were obtained using lymphocyte separation medium. Then cells were washed, resuspended, seeded and cultured according the MTT method, respectively. Subsequently, the number of proliferation peripheral blood T and B lymphocyte were calculated using the following equation: SI = OD570 (T/B lymphocyte proliferation group)/OD570 (control group).

Isolation and examination of the phagocytic function of neutrophils

Neutrophils of broiler blood were isolated using the Histopaque (Sigma–Aldrich) separating medium. The cells were centrifuged, withdrawn, washed and detached in plastic tubes, respectively. Then, the pellet were resuspended in the Hank's medium.

Cells were counted using the phase-contrast microscopy at 40.

Hank's cellular suspension (5×10^5 neutrophils/mL) were then put into the wells of plastic macrophage migration, and discarded the supernatant. The cells were stained with acridine orange. Neutrophils was phagocytosed and dead bacteria (orange) were determined. The number of particles ingested per 100 neutrophils was expressed as the phagocytosis index (PI). The percentage of cells that had phagocytosed was expressed as the phagocytosis percentage (PP).

ELISA for the determination of TNF-α, IL-2 and IL-4 production

The serum concentrations of TNF-α, IL-2 and IL-4 were determined using commercial chicken ELISA kits (Adlitteram Diagnostic Laboratories, San Diego, CA, USA).

Statistical analysis

Variance analyses were done using the General Linear Model procedure of the Statistical Package for Social Sciences 20.0 (SPSS Inc., Chicago, IL, USA) as a completely randomized design. The significant differences among different treatment means were separated using the Duncan's new multiple range test at $P \leq 0.05$.

Results

Growth performance

All growth performance parameters were not influenced ($P > 0.05$) by dietary ORI levels (Table 2).

Table 2 Effects of ORI on the growth performance of broilers

Items	Diet Treatments[a]				SEM[b]	*P*-Value
	CON	O1	O2	O3		
1–14 d						
FI[c] (g/d)	41.8	41.5	41.4	41.0	0.1	0.20
WG[c] (g/d)	30.1	30.3	30.4	31.1	0.2	0.37
FCR[c]	1.39	1.37	1.36	1.32	0.02	0.17
14–28 d						
FI[c] (g/d)	127	126	126	125	1.3	0.24
WG[c] (g/d)	81.4	82.4	84.0	83.9	1.2	0.10
FCR[c]	1.56	1.53	1.50	1.49	0.04	0.31
28–42 d						
FI[c] (g/d)	210	209	209	207	1.7	0.19
WG[c] (g/d)	99.5	100	100.5	101.5	1.0	0.14
FCR[c]	2.11	2.09	2.08	2.04	0.06	0.21
1–42 d						
FI[c] (g/d)	127	126	125	123	1.0	0.33
WG[c] (g/d)	69.4	69.2	69.4	69.1	0.9	0.21
FCR[c]	1.83	1.82	1.80	1.78	0.02	0.25

[a]Control = basal diet; O1, O2, and O3 groups = basal diet with 50, 80, and 100 mg/kg ORI, respectively
[b]Standard error of the mean based on pooled estimate of variation
[c]FI = feed intake; WG = weight gain; FCR = feed conversion ratio

Relative organ weight

Compared with the control, broilers fed basal diet supplemented with ORI (Table 3) significantly increased relative weights of spleen and bursa at day 14, 28 and 42 ($P < 0.05$). However, inclusion of ORI did not have a significant effect on the relative weights of thymus, liver, pancreas or gizzard ($P > 0.05$).

Proliferation peripheral blood T and B lymphocyte

As shown in Table 4, inclusion of ORI in the diet significantly increased the number of proliferation peripheral blood T and B lymphocytes at day 14, 28 and 42 when compared with the control group ($P < 0.05$).

Phagocytic rate and phagocytic index of neutrophil

Dietary ORI supplementation had no significant effect on the phagocytic index of neutrophils on either day 14 or 28 (Table 5). However, dietary ORI supplementation significantly increased the phagocytic index of neutrophils on day 42 ($P < 0.05$). Dietary ORI supplementation significantly increased the phagocytic rate of neutrophils of broilers on days 14, 28 and 42 ($P < 0.05$).

Cytokine levels in serum

As showen in Table 6, compared with the basal diet, IL-2, IL-4 and TNF-a levels in the serum of broilers were markedly increased by the ORI on day 14, 28 and 42 ($P < 0.05$).

Table 3 Effects of ORI on weight of internal organs of broiler chickens (g/100 g body weight of bird)

Items	Diet Treatments[1]				SEM[2]	P-Value
	CON	O1	O2	O3		
14 d						
Liver	2.73	2.76	2.74	2.79	0.24	0.333
Spleen	0.08b	0.13a	0.15a	0.14a	0.02	0.123
Pancreas	0.42	0.40	0.38	0.39	0.67	0.231
Gizzard	2.15	2.25	2.34	2.38	0.23	0.164
Bursa	0.12b	0.16a	0.17a	0.17a	0.01	0.109
28 d						
Liver	2.87	2.95	3.02	3.08	0.24	0.233
Spleen	0.10b	0.17a	0.17a	0.18a	0.01	0.03
Pancreas	0.37	0.37	0.36	0.35	0.04	0.671
Gizzard	1.67	1.92	1.73	1.78	0.23	0.254
Bursa	0.14b	0.18a	0.19a	0.20a	0.01	0.214
42 d						
Liver	3.50	3.46	3.35	2.84	0.21	0.27
Spleen	0.12b	0.21a	0.22a	0.23a	0.01	0.03
Pancreas	0.29	0.28	0.26	0.27	0.03	0.150
Gizzard	1.05	1.25	1.20	1.21	0.21	0.124
Bursa	0.16b	0.20b	0.20b	0.21a	0.13	0.104

[1]Control = basal diet; O1, O2, and O3 groups = basal diet with 50, 80, and 100 mg/kg ORI, respectively
[2]Standard error of the mean based on pooled estimate of variation
[a-b]Means within the same row that do not share a common superscript are significantly different ($P < 0.05$); $n = 8$

Discussion

The literature suggests that oridonin could increase the average daily gain of murine injected with LPS, which was probably ascribed to the anti-inflammatory potential of oridonin by suppressing the production of pro-inflammatory cytokines (Xu et al. 2009) [8]. However, the present study showed that ORI supplementation has no effect on WG, FI or FCR at any recorded ages in the chickens. These results were consistent with results by Zheng et al. (2016) [5], who reported that adding of ORI did not affect the growth performance of broilers. The discrepancies among our findings with other studies for WG and FI are primarily due to animal species.

Several studies have reported that antibacterial activity of oridonin should be regarded as having the promising effect in the healing of gastro-duodenal complaints [9]. Therefore, oridonin can be used as a protective additives of the gastrointestinal system of the body, but may not paly some valuable roles in the growth performance of

Table 4 Effects of ORI on the number of proliferation peripheral blood T and B lymphocyte of broiler chickens (%)

Items	Diet Treatments[1]				SEM[2]	P-Value
	CON	O1	O2	O3		
T lymphocyte						
14 d	0.13b	0.29a	0.33a	0.36a	0.09	0.031
28 d	0.13b	0.32a	0.36a	0.38a	0.04	0.027
42 d	0.51b	0.71a	0.77a	0.76a	0.11	0.031
B lymphocyte						
14 d	0.13b	0.27a	0.31a	0.37a	0.05	0.043
28 d	0.16b	0.24a	0.22a	0.26a	0.03	0.024
42 d	0.14b	0.47a	0.46a	0.50a	0.02	0.032

[1]Control = basal diet; O1, O2, and O3 groups = basal diet with 50, 80, and 100 mg/kg ORI, respectively
[2]Standard error of the mean based on pooled estimate of variation
[a-b]Means within the same row that do not share a common superscript are significantly different ($P < 0.05$); $n = 8$

Table 5 Effects of ORI on phagocytic rate and phagocytic index of neutrophil of broiler chickens

Items	Diet Treatments[1]				SEM[2]	P-Value
	CON	O1	O2	O3		
Phagocytic rate (%)						
14 d	16.56b	21.18a	21.65a	24.61a	0.92	0.042
28 d	25.12b	34.47a	36.27a	39.12a	1.21	0.038
42 d	25.47b	38.37a	40.28a	41.67a	1.82	0.026
Phagocytic index						
14 d	1.38	1.40	1.39	1.42	0.17	0.524
28 d	1.64	1.97	2.04	2.17	0.21	0.178
42 d	1.79b	2.67a	2.85a	2.95a	0.79	0.037

[1]Control = basal diet; O1, O2, and O3 groups = basal diet with 50, 80, and 100 mg/kg ORI, respectively
[2]Standard error of the mean based on pooled estimate of variation
[a-b]Means within the same row that do not share a common superscript are significantly different ($P < 0.05$); $n = 8$

Table 6 Effects of ORI on the concentration of serum cytokine of broiler chickens (ng/L)

Items	Diet Treatments[1]				SEM[2]	P-Value
	CON	O1	O2	O3		
14 d						
IL-2	11.37b	23.28a	25.49a	30.12a	0.62	0.037
IL-4	96.97b	253.47a	276.37a	337.37a	3.24	0.012
TNF-a	214.77b	456.32a	469.87a	496.78a	4.45	0.027
28 d						
IL-2	12.07b	21.58a	25.34a	29.72a	1.02	0.021
IL-4	103.42b	188.67a	205.67a	211.34a	3.45	0.019
TNF-a	425.38b	675.30a	692.34a	732.04a	6.72	0.039
42 d						
IL-2	14.24b	26.23a	28.37a	28.97a	0.95	0.019
IL-4	114.37b	190.24a	219.37a	286.34a	1.38	0.031
TNF-a	109.37b	385.34a	403.37a	465.35a	7.65	0.047

[1]Control = basal diet; O1, O2, and O3 groups = basal diet with 50, 80, and 100 mg/kg ORI, respectively
[2]Standard error of the mean based on pooled estimate of variation
[a-b]Means within the same row that do not share a common superscript are significantly different ($P < 0.05$); $n = 8$

broilers when broiler chickens are kept in the clean environment and provided the highly digestible diets. Additionally, the dosage and purity of oridonin, breeding density of broilers, and many other factors can be due to inconsistent results.

In our present study, no differences were observed among the ORE treatments in the relative weights of thymus, liver, pancreas or gizzard. However, the relative weights of spleen and bursa were increased in response to feeding of basal diets containing oridonin, which was in line with results of Zheng et al. (2016) [5]. The spleen and bursa are the most important immune organs in the broiler chicken body, and they can suppress immune function by various approaches and induce the status of immunosuppression in broilers. Thus, the measurement of internal immune organ weight is a common method of estimating immune status in chickens [10]. In our present study, the increased immune ability, was likely attributed to the immunosuppressive effect of oridonin [11], may also showed that oridonin can improve immune function in broilers, thus explaining the good health status of broilers in the current study.

T and B lymphocytes play central roles in immune response to any type of stress and/or aggression. Consequently, the number of T and B cells could be a fungible index of lymphocyte immunityability. Moreover, the thymus, spleen and bursa are sites of B and T cell differentiations in broilers. The relative weights of lymphoid organ are used for reflecting the immune status of birds, and changes in weight may be related with alterations in the immune function of lymphoid organs [12]. Because ORI is an immunoenhancer, it encourages faster cell proliferation and differentiation in the immune system of experimental animals. It seems that dietary ORI stimulates the T and B lymphocyte formations, proliferation, and the division, as well as activity of blood. Liu et al. (2007) found that ORI is able to accelerate proliferation of lymphocytes, and that ORI affects body's immune function and disease resistance [11].

In the present study, we chose ORI as the irritants of T lymphocytes, which can activate many intra-lymphocyte signaling pathways [13]. For the irritants, we observed an increases in immune response in the number of peripheral blood T and B cell counts. This might be associated with ORI's ability to activate T cells, which may be partially due to T cell depletion in the thymus and the peripheral immune system. Some pharmacological studies show that oridonin is quickly removed from the circulatory system (the level of oridonin in blood was reduced by 78% within 2 h) [14]. The rapid removal of oridonin may explain the different effects of oridonin in vitro and in vivo to some extent. Additional studies will be required before fully understanding its potential applications in the modern broiler breeding.

Neutrophils, as the richest peripheral circulating immune cells, constitute the first-line defense against invading pathogens and act pivotal part in the early stages of inflammation and immune responses. Neutrophil functions, including phagocytosis, degranulation, cytokine release and neutrophil extracellular traps (NETs) formation, play critical roles in the innate immune response against infection through their ability to quickly identify and remove invading germs. Therefore, the phagocytic effect of neutrophils is an important immune parameter because it is strong connections with the clearance of pro-inflammatory necrotic tissue and cell residues.

The results showed that addition of ORI significantly increased the phagocytic index of neutrophil on day 42, and the phagocytic rate of neutrophils of broilers in the whole experiment. We presume that this increase may also be attributed to T cell depletion in the peripheral blood immune system. Numerous studies have shown that ORI has an anti-inflammatory effect. ORI decreases systemic inflammation responses and the increase of phagocytosis, and may be a potential potentially novel therapeutic agent for the treatment of inflammatory disease.

The cytokine level in serum is an important index for humoral immunity in poultry. By now, it is generally known that Th_1 cells excrete proinflammatory cytokines, such as TNF-α, IL-2 and IL-12, however Th_2 cells secrete anti-inflammatory cytokines, such as IL-4, IL-5, IL-6, and IL-10 [15]. In the present study, our findings show that serum TNF-α, IL-2 (Th_1) and IL-4 (Th_2) in the broilers receiving 50, 80, and 100 mg/kg of ORI were significantly lower than those in control broilers, indicating that ORI could regulate humoral immunity of broilers. In birds, few studies have investigated the immunostimulatory effect of ORI. About our present results, the lower cytokines levels produced in ORI-treated broilers is likely related to ORI's anti-inflammatory potential through the repression of T-cell immune responses [6]. Our results are similar to previous studies and further indicate that certain terpenoid compounds may play their anti-inflammatory effects through the regulation of Th_1/Th_2 cytokine secretion levels or repression of T-cell immune responses. We assume that ORI might regulate the differentiation of Th cells; however, future animal and laboratory tests should be conducted to verify potential mechanisms of action.

Conclusions

Based on the results of the present work, it was concluded that diet inclusion of ORI increases the relative weight of spleens and bursa, the number of proliferation peripheral blood T and B lymphocytes, and the phagocytic rate of neutrophils, it may also stimulate IL-2, IL-4 and TNF-a production in broilers. Therefore, it could be concluded that dietary use of ORI as a feed additive may improve the immune function and disease resistance ability in broiler production. Further research is needed to understand and clarify the possible mechanisms involved.

Abbreviations

FCR: Feed conversion ratio; FI: Feed intake; IL-2: Interleukin-2; IL-4: Interleukin-2; MTT: 3-(4,5-dimethylthiazol-2-yl)-2,5-diphenyltetrazolium bromide; NETs: Neutrophil extracellular traps; ORI: Oridonin; PI: Phagocytosis index; PP: Phagocytosis percentage; TNF-a: Tumor necrosis factor-a; WG: Weight gain

Acknowledgements

We would like to thank Tianyue Tang, Beibei Yuan, Fengli Zhang and Lili Yuan their skillful technical assistance for this research.

Funding

Financial support for the study and its publication was provided by the Foundation of State Key Laboratory of Animal Nutrition of China (2004DA125184f1407) and the Priority Academic Program Development of Jiangsu High Education Institutions (PAPD) (BE2010356). State Key Laboratory of Animal Nutrition of China and Priority Academic Program Development of Jiangsu High Education Institutions have given final approval of the version to be published.

Authors' contributions

WQJ and ZXC designed the study and conducted the treatment phase. WT participated in study design and discussion of results. ZTY was responsible for evaluation of results and medical writing. All authors approved the manuscript.

Competing interests

The authors declare that they have no competing interests.

Author details

[1]College of Animal Science and Technology, Nanjing Agricultural University, No. 6, Tongwei Road, Xuanwu District, Nanjing 210095, Jiangsu, People's Republic of China. [2]State Key Laboratory of Animal Nutrition, NO. 2, Yuan Ming Yuan West Road, HaiDian District, Beijing 100193, People's Republic of China. [3]College of Animal Science and Technology, Henan University of Science and Technology, Luoyang 471003, Henan, People's Republic of China.

References

1. Du YF, Liu PW, Zhu H, Shi XW, Zhao CC, Wang N, et al. A sensitive analysis method for 7 diterpenoids in rat plasma by liquid chromatography electrospray ionization mass spectrometry and its application to pharmacokinetic study of Isodon Serra extract. J Chromatogr A. 2011;1218:7771–80.
2. Liu Z, Ouyang L, Peng H, Zhang WZ. Oridonin: targeting programmed cell death pathways as an anti-tumour agent. Cell Prolif. 2012;45:499–507.
3. Ming M, Sun FY, Zhang WT, Liu JK. Therapeutic effect of oridonin on mice with prostate cancer. Asian Pac J Trop Med. 2016;9(2):184–7.
4. Osawa K, Yasuda H, Maruyama T, Morita H, Takeya K, Itokawa H. Antibacterial trichorabdal diterpenes from Rabdosia trichocarpa. Phytochemistry. 1994;36:1287–91.
5. Zheng XC, Wu QJ, Song ZH, Zhang H, Zhang JF, Zhang LL, et al. Effects of Oridonin on growth performance and oxidative stress in broilers challenged with lipopolysaccharide. Poultry Sci. 2016;95(10):2281–9.
6. Guo WZ, Zheng PG, Zhang JK, Ming L, Zhou C, Zhang SJ. Oridonin suppresses transplant rejection by depleting T cells from the periphery. Int Immunopharmacol. 2013;17:1148–54.
7. National RC. Nutrient Requirements of Poultry. 9th Rev. Ed. Washington DC: Natl. Acad. Press; 1994.
8. Xu Y, Xue Y, Wang Y, Feng DC, Lin ST, Xu LY. Multiple-modulation effects of Oridonin on the production of proinflammatory cytokines and neurotrophic factors in LPS-activated microglia. Int Immunopharmacol. 2009;9:360–5.
9. Kadota S, Basnet P, Ishii E, Tamura T, Namba T. Antibacterial activity of tri-chorabdal a from Rabdosia trichocarpa against helicobacter pylori. Zentralblatt Für Bakteriologie. 1997;286(1):63–7.
10. Heckert RA, Estevez I, Russek-Cohen E, Pettit-Riley R. Effects of density and perch availability on the immune status of broilers. Poultry Sci. 2002;81(4):451–7.
11. Liu J, Yang F, Zhang Y, Li J. Studies on the cell-immunosuppressive mechanism of Oridonin from Isodon serra. Int Immunopharmacol. 2007;7:945–54.
12. Cooper MD, Raymond DA, Peterson RD, South MA, Good RA. The functions of the thymus system and the bursa system in the chicken. J Exp Med. 1966;123:75–102.
13. Ding Y, Ding CY, Ye N, Liu ZQ, Wold EA, Chen HY, et al. Discovery and development of natural product oridonin-inspired anticancer agents. EurJ Med Chem. 2016;122:102–17.
14. Lou H, Gao L, Wei X, Zhang Z, Zheng D, Zhang D, et al. Oridonin nanosuspension enhances anti-tumor efficacy in SMMC-7721 cells and H22 tumor bearing mice. Colloid Surfaces B: Biointerfaces. 2011;87:319–25.
15. Romagnani S. Biology of human TH_1 and TH_2 cells. J Clin Immunol. 1995;15:121–9.

14

The effects of different velogenic NDV infections on the chicken bursa of Fabricius

Y. W. Kristeen-Teo[1], S. K. Yeap[1], S. W. Tan[1], A. R. Omar[1], A. Ideris[2], S. G. Tan[3] and N. B. Alitheen[1,3*]

Abstract

Background: Virulent Newcastle disease virus (NDV) was reported to cause rapid depletion of chicken bursa of Fabricius. Severe pathological condition of the organ is commonly associated with high levels of virus replication, intense inflammatory response and also the degree of apoptosis. In this study, the responses of chicken bursa of Fabricius infected with two different strains of velogenic NDV, namely AF2240 and IBS002, were investigated by observing cell population changes, oxidative stress, viral replication and cytokine expression in the organ. Subsequently, apoptosis of enriched bursal IgM+ cells was determined to help us elucidate possible host pathogen relationships between the chicken bursa of Fabricius and NDV infection.

Results: The depletion of IgM+ cells and infiltration of macrophages were observed to be higher in bursa infected with AF2240 as compared to IBS002. In line with the increment of the macrophage population, higher nitric oxide (NO) and malondialdehyde (MDA) contents which indicated higher oxidative stress were also detected in bursa infected with NDV AF2240. In addition, higher pro-inflammatory cytokines and chemokine gene expression such as chicken CXCLi2, IL-18 and IFN-γ were observed in AF2240 infected bursa. Depletion of IgM+ cells was further confirmed with increased cell death and apoptosis of the cells in AF2240 infected bursa as compared to IBS002. However, it was found that the viral load for NDV strain IBS002 was comparatively higher than AF2240 although the magnitude of the pro- inflammatory cytokines expression and cell apoptosis was lower than AF2240.

Conclusion: The results of our study demonstrated that infection of NDV strains AF2240 and IBS002 caused apoptosis in bursa IgM+ cells and its severity was associated with increased expression of pro-inflammatory cytokines/chemokine, macrophage infiltration and oxidative stress as the infection duration was prolonged. However, of the two viruses, we observed that NDV AF2240 induced a greater magnitude of apoptosis in chicken bursa IgM+ cells in comparison to IBS002. This might be due to the high level of oxidative stress and inflammatory cytokines/chemokine as well as lower IL10 expression which subsequently led to a high rate of apoptosis in the chicken bursa of Fabricius although the detected viral load of AF2240 was lower than IBS002.

Keywords: Newcastle disease virus, B lymphocytes, Apoptosis, Oxidative stress

Background

Newcastle disease virus is classified under the order Mononegavirales in the family Paramyxoviridae, which is divided into two subfamilies, Paramyxovirinae and Pneumovirinae [1]. It was reported that a few of the first outbreaks of Newcastle disease occurred in Newcastle Upon Tyne, England and also Java, Indonesia during the early 1900s [2]. The outbreak in Southeast Asia during 1926 spread to most of the other regions of the world and this outbreak was thought to be caused by three different genotypes of NDV which were types II, III and IV [3]. However, a similar pattern of outbreaks was also observed earlier in some areas of Central Europe [4]. In the late 1970s, NDV genotypes V and VI caused a panzootic outbreak in the Middle East [5]. In the early 2000s, NDV outbreaks in Asia were recorded in Taiwan and China caused by NDV genotype VII [6]. This virus was shown to survive when airborne in small particles, either in the laboratory or open air [7]. When infected, birds may show variable symptoms such as loss of appetite, weight loss, weakness,

* Correspondence: noorjahan@upm.edu.my
[1]Institute of Bioscience, Universiti Putra Malaysia, 43400 Serdang, Selangor, Malaysia
[3]Faculty of Biotechnology and Biomolecular Sciences, Universiti Putra Malaysia, 43400 Serdang, Selangor, Malaysia
Full list of author information is available at the end of the article

diarrhea and nervous signs [2, 8]. Post mortem examination showed severe hemorrhagic lesions of the intestine and proventriculus [9]. The pathological manifestations observed upon NDV infections are contributed by factors such as the strain of the virus, species of birds, concurrent diseases and preexisting immunity which can affect the severity of the disease [10].

Avian bursa of Fabricius had been identified to play an important role in the development of B lymphocytes as studies showed that the surgical removal of this organ during the embryonic stage of a bird compromised its antibody production thus leading to immunosuppresion [11, 12]. The organ serves as an essential site for generation of antibody diversity by gene conversion. It was reported to serve as a primary lymphoid organ until the age of five to 6 weeks and after that as a peripheral lymphoid organ [13]. Primarily this organ consisted of 98% B lymphocytes [14] but infiltration of other cells, such as T cells and macrophages was observed in response to viral infection [15]. The bursa of Fabricius was involved in host responses against poultry virus not only by Infectious bursal disease virus (IBDV) [16–18], in which the virus infection caused cell population and cytokine expression changes but also upon NDV infection. Several reports showed that virulent NDV strains cause rapid depletion of the bursa of Fabricius [19–21] and the severe pathology in immune organs caused by virulent NDV strains is associated with high levels of virus replication and an intense inflammatory response [22, 23]. Also, apoptosis was reported to play a role in the pathogenesis of NDV as well. It was demonstrated that the amount of apoptosis was proportional to the severity of the clinical disease elicited by the strains [20, 24]. In this study, we investigated the response of chicken bursa of Fabricius upon infection with two different strains of Malaysian isolates of velogenic NDV. Oxidative stress caused by the virus infection was also evaluated by detecting the nitric oxide content in the organ. Apoptosis studies on B lymphocytes, which are the largest cell population in the organ, was done and correlated with viral load detection in the organ to help us elucidate possible host pathogen relationships between the chicken bursa of Fabricius and Newcastle disease virus.

Methods

Experimental design

The study was approved by the Institutional Animal Care and Use Committee (IACUC), Universiti Putra Malaysia (UPM). SPF layer chicken eggs (n = 80) were obtained from the Veterinary Research Institute (Ipoh, Malaysia), hatched in the Laboratory of Vaccines and Immunotherapeutics, Institute of Biosciences, UPM and raised until the age of 35 days old. The trial was conducted in UPM Faculty of Veterinary Medicine experimental animal facility where the different groups of birds were kept separately in different rooms and cages. The birds were then randomly divided into three groups with the first two groups (n = 30) being infected with 10^6 ELD50 of NDV strain AF2240 and IBS002, respectively. The viruses were delivered to the chickens via ocular (50 μl) and nasal (50 μl) routes using 1 ml syringe. The remaining birds were treated with PBS and served as the control group. Ten birds from each infected group were then observed, sacrificed and necropsized at days 1, 3 and 4 post infection whereas the control chickens were sacrificed at day 4. Chicken bursa of Fabricius was removed and cut into two parts, which one for oxidative stress study and RNA extraction for gene expression purposes. Meanwhile the other part of the organ was used for enrichment of IgM+ cells, which were subsequently subjected to apoptosis related assays such as acridine orange/propidium iodide (AO/PI) double staining, Annexin V apoptosis and DNA cell cycle assays. Prior to cell isolation, a cell count was performed using a hemacytometer and a portion of the cells (2×10^6 cells) was subjected to flow cytometry analysis to for cell population study.

Virus isolates

The viruses used in this study are velogenic isolates Chicken/Malaysia/Perak/AF2240/1960 (AF2240) and Commercial broiler chicken/Malaysia/Johor/IBS002/2011 (IBS002). NDV AF2240 is a viscerotropic velogenic, genotype VIII strain that is used as a vaccine challenge virus in Malaysia [25]. The intracerebral pathogenicity index (ICPI) for AF2240 was 1.9 [26] while its mean death-time (MDT) was 48 h [27]. On the other hand, NDV isolate IBS002 was characterized as a virulent genotype VII strain based on the multiple basic amino acid motif of the fusion (F) cleavage site 112RRRKGF117 and length of the C-terminus extension of the hemagglutinin-neuraminidase (HN) gene with the ICPI and MDT at 1.76 and 51.2 h respectively [28]. Both viruses are belonged to and owned by Institute of Bioscience, Universiti Putra Malaysia.

Immunophenotyping of SPF chicken bursa of Fabricius

To study the cell population changes in chicken bursa when infected with NDV AF2240 and IBS002, immunophenotyping using flowcytometry was conducted. Single cells were isolated from chicken bursa of Fabricius at days 1, 3 and 4 post infection. In brief, the organs were removed (n = 5), washed with PBS and cut into pieces using a scalpel, before being meshed through 70 μm sterile wire mesh screen (SPL Life Sciences, China). The single cells were then collected and washed twice in PBS-BSA-EDTA. After that, the pellet was subjected to

Histopaque 1119 (Sigma, USA) at 400 g for 40 min. The lymphocyte layers were collected and washed twice again with PBS-BSA-EDTA. A cell count was performed using a hemacytometer and a number of 1×10^6 cells/tube were incubated with 10 μg/10 μL mouse anti-chicken CD3-FitC, clone CT-3 and mouse anti-chicken CD4-PE, clone CT-4 (tube 1), mouse anti-chicken CD3-FitC, clone CT-3 and mouse anti-chicken CD8α-PerCp, clone CT8 (tube 2), mouse anti-chicken IgM-PE, clone M-1 (tube 3) and mouse anti-chicken monocyte/macrophage-PE, clone KUL01 (tube 4) antibodies (Southern Biotech, USA) for 45 min in the dark. The cells were then washed again with PBS and lastly resuspended in 1 ml PBS. Flow cytometer analysis was carried out using a FACSCalibur machine with CellQuest Pro software (BD Bioscience, USA). A total of 20,000 events were analyzed via gating the chicken lymphocytes using forward and side scatter before the data for positive staining of different fluorescence were collected. The average cell numbers of CD4, CD8, IgM and macrophages are listed in Table 1.

Oxidative stress (NO and MDA detection)

Upon NDV AF2240 and IBS002 infection, oxidative stress in chicken bursa of Fabricius was observed and compared by measuring nitric oxide (NO) and malondialdehyde (MDA) level. Chicken bursas of Fabricius at different time point after being infected with NDV AF2240 and IBS002 were cut and meshed before the organ homogenate was subjected to NO and MDA detection (n = 5 from each group). For NO evaluation, 150 μL of the sample homogenate was added with 20 μL of Griess Reagent (Invitrogen, USA) and 130 μL of deionized water in a microplate. The mixture was incubated for 30 min at room temperature. After the incubation, the sample absorbance was measured at 548 nm using μQuant ELISA Reader (Bio-Tek Instruments, USA). Meanwhile, MDA activity was determined by measuring the thiobarbituric acid-reactive substance (TBARS). In brief, 100 μL of bursa homogenate was added with 12.5 μL butyhydroxytoulene (8.8 mg/mL), 250 μL 30% trichloroacetic acid and 400 μL of PBS before the mixture was mixed and incubated on ice for 2 h. Subsequently, the mixture was spun at 300 g for 15 min and the supernatant was collected and added with 37.5 μL of 0.1 M EDTA and 125 μL 1% thiobarbituric acid before being subjected to boiling for 15 min. The mixture would then turn pink and the absorbance was read at 532 and 600 nm using a μQuant ELISA Reader (Bio-Tek Instruments, USA). The absorbance measured was then compared with the standard made with malonaldehyde tetramethylacetal solutions at different concentrations and the MDA activity was expressed as nmol MDA/g protein.

Detection of viral load and cytokine expression by real time PCR

TaqMan probe based real time PCR was used to determine viral load in chicken samples infected with NDV AF2240 and IBS002. RNA was isolated from both crude cells and enriched IgM+ cells. Bursas of Fabricius from the control and infected chickens were removed (n = 5 from each group) and RNA from each sample was extracted using TRIZOL (Invitrogen, USA) following the instruction recommended by the manufacturer. The concentration and purity of the extracted RNA were measured by using a spectrophotometer (DU®730, Beckman Coulter, USA) at the wavelength of 260 nm over 280 nm before being subjected to cDNA synthesis using iScriptcDNA Kit (Biorad, USA) incubated in a thermal cycler for 5 min at 25 °C, 30 min at 42 °C and 5 min at 85 °C. The cDNA was then subjected to NDV viral load detection using usingiQ-Supermix (Biorad, USA) together with forward primer 5′-TCCGCAAGATCCAAGGGTCT-3′, reverse primer 5′-CGCTGTTGCAACCCCAAG-3′ and Taqman probe 5′-

Table 1 Average cell number of different lymphocytes subsets after infected with NDV AF2240 and NDV IBS002

Groups		Average cell numbers			
		Control	Day 1	Day 3	Day 4
CD4	NDV AF2240	$0.79 \times 10^5 \pm 50$	$0.34 \times 10^5 \pm 12^{*}$	$0.46 \times 10^5 \pm 21^{**}$	$1.46 \times 10^5 \pm 52^{***a}$
	NDV IBS002	$0.79 \times 10^5 \pm 50$	$0.33 \times 10^5 \pm 22^{*}$	$0.46 \times 10^5 \pm 12^{**}$	$1.08 \times 10^5 \pm 20^{***b}$
CD8	NDV AF2240	$0.27 \times 10^5 \pm 12$	$0.71 \times 10^5 \pm 22^{*a}$	$1.10 \times 10^5 \pm 115^{**a}$	$1.40 \times 10^5 \pm 21^{***a}$
	NDV IBS002	$0.27 \times 10^5 \pm 12$	$0.23 \times 10^5 \pm 10^{*b}$	$0.65 \times 10^5 \pm 13^{**b}$	$1.58 \times 10^5 \pm 11^{***b}$
IgM	NDV AF2240	$1.78 \times 10^6 \pm 36$	$1.76 \times 10^6 \pm 83$	$1.43 \times 10^6 \pm 29^{*a}$	$1.24 \times 10^6 \pm 101^{**a}$
	NDV IBS002	$1.78 \times 10^6 \pm 36$	$1.73 \times 10^6 \pm 53$	$1.64 \times 10^6 \pm 24^{*b}$	$1.37 \times 10^6 \pm 115^{**b}$
Macrophage	NDV AF2240	$0.69 \times 10^5 \pm 21$	$0.35 \times 10^5 \pm 24^{*}$	$0.95 \times 10^5 \pm 47^{**a}$	$2.86 \times 10^5 \pm 16^{***a}$
	NDV IBS002	$0.69 \times 10^5 \pm 21$	$0.40 \times 10^5 \pm 66^{*}$	$1.08 \times 10^5 \pm 63^{**b}$	$1.70 \times 10^5 \pm 15^{***b}$

Data representing average cell number of CD4+, CD8+, IgM and macrophages with standard deviation from chicken bursa of Fabricius cells of five chickens from each group.
Different asterisk signs show significant difference between infection time points ($P \leq 0.05$).
Different alphabets show significant difference between different strains of NDV ($P \leq 0.05$).

(FAM)-AAGCGTTTCTGTCTCCTTCCTCCA-(BHQ)-3′. The cycling program consists of 95 °C for 5 min and followed by 39 cycles of 95 °C for 10 s, 56 °C for 30 s and 72 °C for 20 s. Standard curves for detection of NDV AF2240 and IBS002 was provided in Additional file 1: Figure S1. As for cytokines expression, following primers and probes were used: forward primer 5′-GTGAAGAAGGTGAAAGATATCATGGA-3′, reverse primer 5′- GCTTTGCGCTGGATTCTCA-3′ and probe 5′-(FAM)-TGGCCAAGCTCCCGATGAACGA-(BHQ1)-3′ for IFN- γ expression; forward primer 5′-GCCCTCCTCCTGGTTTCAG-3′, reverse primer 5′-TGGCACCGCAGCTCATT -3′ and probe 5′-(FAM)-TCTTTACCAGCGTCCTACCTTGCGACA-(BHQ1)-3′ for CXCLi2 expression; forward primer 5′-AGGTGAAATCTGGCAGTGGAAT-3′, reverse primer 5′-ACCTGGACGCTGAATGCAA-3′ and probe 5′-(FAM)-CCGCGCCTTCAGCAGGGATG-(BHQ1)-3′ for IL18 expression; forward primer 5′-CATGCTGCTGGGCCTGAA-3′, reverse primer 5′-CGTCTCCTTGATCTGCTTGATG-3′ and probe 5′-(FAM)-CGACGATTCGGCGCTGTCACC-(BHQ1)-3′ for IL10 expression; as well as forward primer 5′-GAACGGGAAACTTGTGAT-3′, reverse primer 5′-GACTCCACAACATACTCA-3′ and probe 5′-(FAM)-CGCCATCACTATCTTCCAGG-(BHQ1)-3′ for GAPDH expression as reference gene. The real time PCR cycling program was set up as 1 cycle of 95 °C for 5 min, followed by 40 cycles of 95 °C for 20 s, 58 °C for 20 s (IFN-γ, CXCLi2, IL18 and IL10) or 60 °C for 30 s (GAPDH), depending on primer annealing temperature. Viral load was calculated by using standard curve (Additional file 1: Figure S1) with the following formula:

$$Log\ Quantity = \frac{Ct - b}{m}$$

Ct = cycle threshold value.
b = y-intercept.
m = slope.

Relative fold change for CXCLi2, IFN- γ and IL18 was calculated using $2^{-\Delta\Delta Ct}$ method. Meanwhile, IL10 was expressed as relative ΔCt gene expression normalized with GAPDH because no IL10 was detected on day 1 post infection and uninfected controls.

Isolation of enriched B lymphocyte population and cell viability assessment

At days 0, 1, 3 and 4 post infection, single cells were harvested from chicken bursa of Fabricius of SPF chickens (n = 10 from each group). The cells were then stained with 10 μg/10 μL PE-labelled IgM for 45 min and washed twice with PBS. Subsequent to that, the cells were labelled with 20 μL of anti-PE microbeads for 30 min prior to magnetic isolation of pure B (IgM^+) cell population using a LS column (Miltenyi Biotec, Germany). The enriched B cells were then washed twice with PBS before proceeding to cell viability testing by using the colorimetric 3-(4,5-dimethylthiazol-2-yl)-2,5-diphenyl tetrazolium bromide (MTT) assay as described previously [29].

Acridine orange/propidium iodide (AO/PI) double staining assay

The enriched IgM + B cell suspension was washed with PBS twice and 10 μL of the cells were put on a glass slide. Acridine orange (50 μg/ml) and propidium iodide (50 μg/ml) were mixed together at a 1:1 ratio to a total volume of 1 ml. After that, 10 μl of the mixture was added to the cells on a glass slide and resuspended carefully. The slide was then analyzed using a fluorescent microscope (Nikon, Japan) within 30 min. This assay permitted clear differentiation of viable and nonviable cells using two different fluorescent dyes. Through counting a total population of 200 cells, the percentage of viable, apoptotic and necrotic cells were determined. Besides, cells with condensed chromatin and blebbing were also considered as apoptotic cells (Fig. 1). The experiment was repeated using three biological replicates for each group.

Annexin V apoptosis assay

Enriched IgM + B lymphocytes of every group were harvested and subjected to Annexin V apoptosis assay. The cells were washed with PBS and stained with Annexin V FITC (BD Biosciences, USA) for 20 min prior to analysis using a FACSCalibur machine with CellQuest Pro Software (BD Bioscience, USA). The assay was conducted in triplicates.

DNA cell cycle analysis

A total of 1 × 10^6 enriched IgM+ B cells was collected from each sample group and fixed in 80% ethanol overnight at a temperature of −20 °C. The samples were then re-pelleted and washed twice by using PBS-Sodium-Azide-EDTA buffer. Finally, the pellet was dissolved in 1 mL of PBS buffer containing 0.1% Triton X-100, 10 mM EDTA, 50 μg/mL RNase and 2 μg/ml SYTOX green in the dark, followed by incubation for at least 30 min on ice. The samples were then subjected to analysis by flow cytometry using a FACSCalibur machine with CellQuest Pro software (BD Biosciences, USA).

Statistical analysis

The Graphpad Prism 6.0 software was used to aid the statistical analysis of the results. The results were expressed as mean ± standard deviation and subjected to one way ANOVA or two-way ANOVA test, followed by the Bonferroni procedure with infection

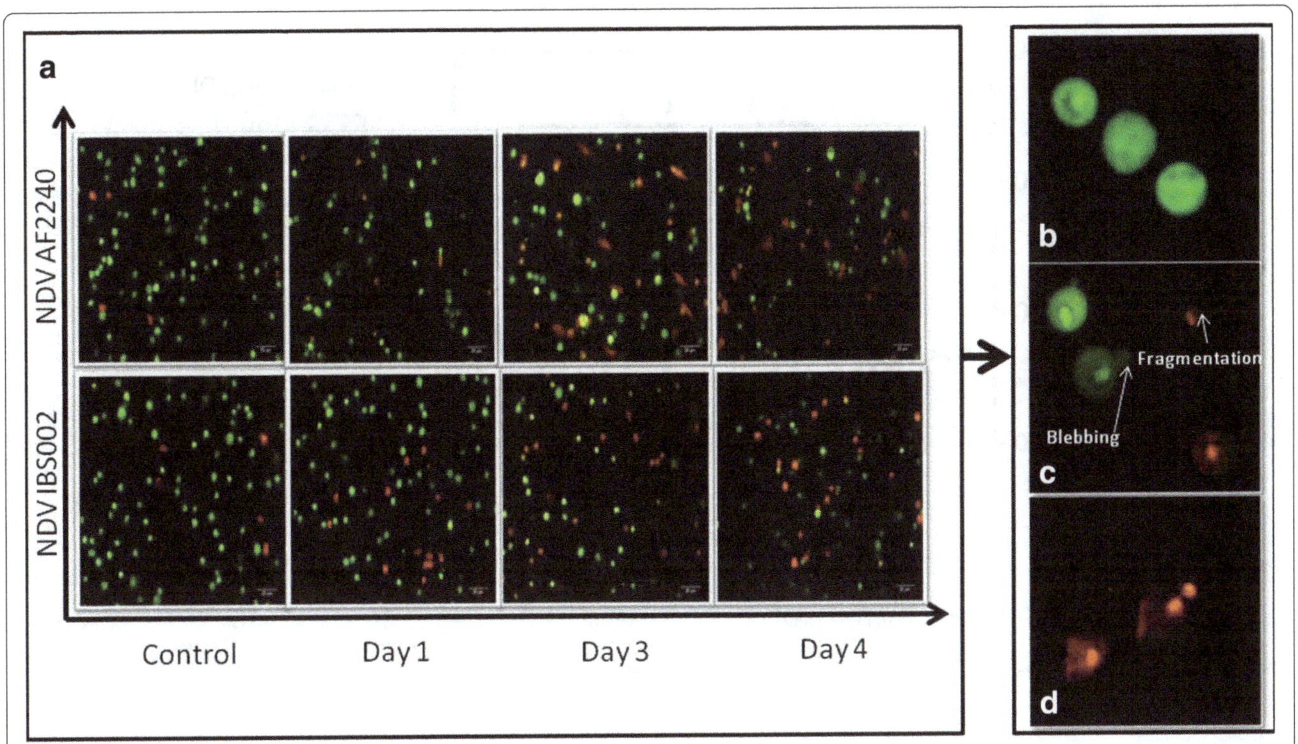

Fig. 1 AO/PI assay of chicken enriched B lymphocytes infected with (**a**) NDV AF2240 and NDV IBS002 at day 1, 3 and 4 post infections, magnification 400×; **b** Viable cells; **c** Early Apoptotic cells where *arrow* indicating membrane blebbing and chromation condensation, red cells are late apoptotic cells; **d** Late apoptotic or necrotic cells, magnification 400×; AO (excitation: 488 nm, emission: 545 nm), PI (excitation: 535 nm, emission: 617 nm)

time points and virus genotypes as main effects. Results with $P < 0.05$ were considered as being statistically significant.

Results

Immunophenotyping

In chicken bursa, cell population changes were observed when the animals were infected with different strains of velogenic NDV. The average numbers of different lymphocyte subsets were counted and the results are shown in Table 1. Similar patterns of CD4+ cell population changes were observed upon infection by both NDV AF2240 and IBS002. The infection caused depletion of CD4+ T cells on days 1 and 3 post infection, but on the fourth day, NDV AF2240 caused the CD4+ cell number to increase to 1.46×10^5 from the initial 0.79×10^5 cells while NDV IBS002 increased the number to 1.08×10^5 cells. CD8+ cells were found to gradually increase along with the time of infection by both viruses, with higher increments during day 1 and day 3 post infection caused by NDV AF2240 infection, whereas on the fourth day of infection, higher CD8+ cell number was observed in the chicken bursa infection by NDV IBS002. B lymphocytes, which were measured by the expression of IgM, were reported to decline significantly in cell number upon infection with both genotypes of NDV starting on day three after infection. However, infection with NDV AF2240 resulted in a greater decline in IgM+ cell number from an initial figure of $1.78 \times 10^6 \pm 36$ cells to $1.24 \times 10^6 \pm 101$ cells (Table 1) compared to NDV IBS002 for which the cell number of IgM+ cells reached $1.37 \times 10^6 \pm 115$ cells at day four post infection (Table 1). Also, it was observed that the macrophage population increased upon infection with both genotypes of NDV but NDV AF2240 resulted in a higher peak in macrophage cell number at $2.86 \times 10^5 \pm 16$ cells compared to NDV IBS002 at $1.70 \times 10^5 \pm 15$ cells on the fourth day after infection.

Oxidative stress evaluation (NO and MDA detection)

Upon infection with virus, NO is considered as a pro-inflammatory mediator due to its over production. The level of NO in the chicken bursa of Fabricius was measured at 1, 3 and 4 days post infection. Figure 2 shows the NO levels detected in the chickens' bursa after infection with the two viral genotypes respectively. According to Fig. 2, the level of nitric oxide content in chicken bursa of Fabricius upon NDV AF2240 infection increased from an initial of 33.06 ± 4.2 μM of nitric oxide to the highest level of 92.89 ± 2.5 μM on the fourth day of infection whereas NDV IBS002 caused increment of nitric oxide level to 81.02 ± 6.4 μM in the organ. A

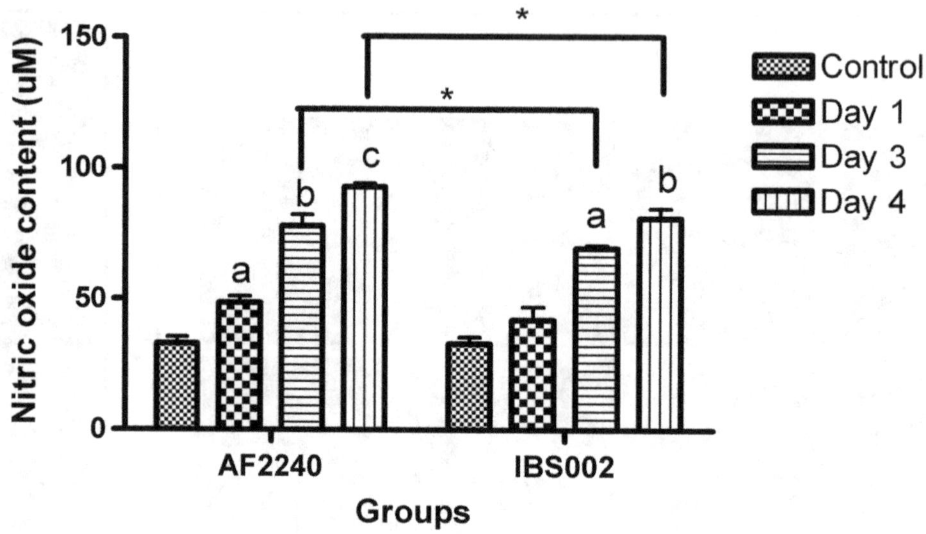

Fig. 2 The nitric oxide content in the bursa of 3-week-old chicken infected with NDV AF2240 and IBS002 at day 1, 3 and 4 post infection. Significant differences were observed using two-ways ANOVA by using strains of NDV and infection time points as effectors. *Asterisk signs* show significant difference between different strains of NDV ($P \leq 0.05$) while different *alphabets* show significant difference between different infection time points ($P \leq 0.05$)

similar trend of increment in MDA level was also observed in the chicken bursa of Fabricius after infection with the two different genotypes of NDV. Figure 3 shows that NDVAF2240 caused a significant elevation of MDA content in chicken bursa starting on day 3 post infection compared to the control group and reached its highest peak during day 4 at 28.26 ± 1.6 nM of MDA/ g of protein. Meanwhile, IBS002 caused a significant increase of the MDA level at day 1 post infection and reached 24.84 ± 0.5 nM of MDA/g of protein at day 4.

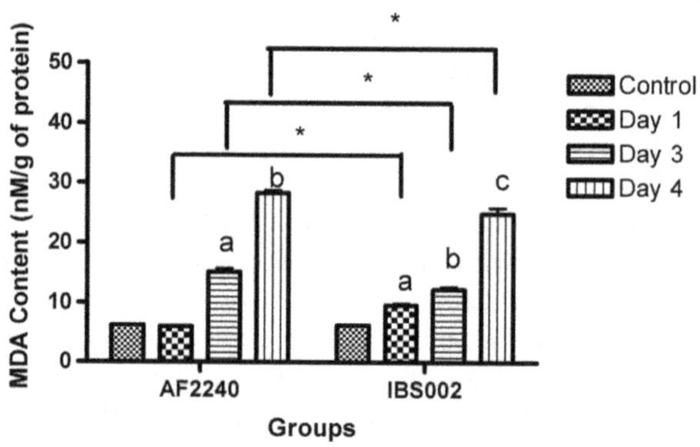

Fig. 3 The MDA content in the bursa of 3-week-old chicken infected with NDV AF2240 and IBS002 at day 1, 3 and 4 post infection. Significant differences were observed using two-ways ANOVA by using strains of NDV and infection time points as effectors. Asterisk signs show significant difference between different strains of NDV ($P \leq 0.05$) while different alphabets show significant difference between different infection time points ($P \leq 0.05$)

Detection of viral load and cytokines by real time PCR

The results showed that both viruses RNA could be detected in the samples starting on day 3 post infection and that the viral load increased on the following day. Table 2 shows that a higher viral load was detected in both chicken bursa and IgM + cells infected by NDV IBS002 compared to AF2240. On the other hand, Fig. 4 shows the expressions of IFN-ɣ, CXCLi2 and IL-18 detected in chicken bursa upon infection. As is shown in Fig. 4, the infection of NDV AF2240 significantly increased expression of IFN-ɣ starting on day 1 post infection as compared to the control group and then this was further elevated to 83 – fold change on day 3. On day 4 post infection, expression of IFN-ɣ was recorded at 22-fold compared to the control group. Both IL-18 and CXCLi2 gene expressions were also up-regulated during the infection of AF2240. The IL-18 expressions were up-regulated at 8-, 42- and 14 - folds at days 1, 3 and 4 post infection while the CXCLi2 expressions at 7-, 21- and 12- folds (Fig. 4). Significant changes of IFN-ɣ, IL-18 and CXCLi2 gene expressions were observed with NDV AF2240 infection at all different time points of infection. Meanwhile, no significant changes of IFN-ɣ and CXCLi2 were observed in chicken bursa upon NDV IBS002 infection. The expression of IL-18 was increased to 25 – fold at days 1 and 3 post infection, but decreased to 11-fold on the next day. Conversely, IL-10 was detected on day 3 and day 4 post infection of NDV AF2240 and IBS002 in the bursa (Fig. 4) but none on day 1 and uninfected control.

Table 2 TaqMan real time PCR result of detection of NDV AF2240 and IBS002 virus in the chicken bursa and IgM$^+$ cells population isolated from chicken bursa of Fabricius

	Samples	C (t) mean		Viral copy number (log 10)	
		AF2240	IBS002	AF2240	IBS002
	Control	N/D	N/D	N/D	N/D
Bursa of Fabricius	Day 1	N/D	N/D	N/D	N/D
	Day 3	29.85 ± 0.28	26.24 ± 0.88	10.34^{*a}	11.43^{*b}
	Day 4	28.33 ± 0.18	23.20 ± 0.53	10.80^{*a}	12.34^{**b}
IgM + cells	Control	N/D	N/D	N/D	N/D
	Day 1	N/D	N/D	N/D	N/D
	Day 3	35.10 ± 0.27	28.74 ± 0.03	8.76^{*a}	10.68^{*b}
	Day 4	28.48 ± 0.16	26.82 ± 1.3	10.10^{**a}	11.25^{**b}

N/D stands for not detectable. No amplification of NDV gene was detected throughout whole qPCR cycling program.
All results represent the mean ± standard deviation
Different asterisk signs show significant different between infection time points ($P \leq 0.05$).
Different alphabets show significant different between different strains of NDV ($P \leq 0.05$).
NDV AF2240, R2 = 0.984, Slope = 3.13

Cell viability and apoptosis assay

The cell viability of enriched B lymphocytes, which are the largest cell population in chicken bursa of Fabricius was examined upon infection with the NDV viruses (Table 3). The results showed that the infection of NDV virus caused increment of cell death events in IgM cells in chicken bursa of Fabricius as the duration of infection increased (Table 3). MTT assay was performed and the percentages of viable cells of virus infected groups were normalized with the control group. It was found that NDV AF2240 caused higher cell death in the enriched B lymphocytes compared to NDV IBS002 (Table 3). These results were further confirmed with the apoptosis assays using AO/PI and Annexin V binding assays. As is shown in Table 4, it was found that infection of the chicken bursa by NDV AF2240 caused a higher percentage of apoptotic IgM+ cells in the organ. A similar trend of results was observed in the Annexin V binding assay where a greater number of Annexin V positive IgM+ cells were detected in chicken bursa during infection with NDV AF2240 (Table 3). In agreement with these results, it was observed that the infection of NDV viruses caused increment of the sub G0/G1 population in the enriched IgM+ cells. DNA fragmentation, which is one

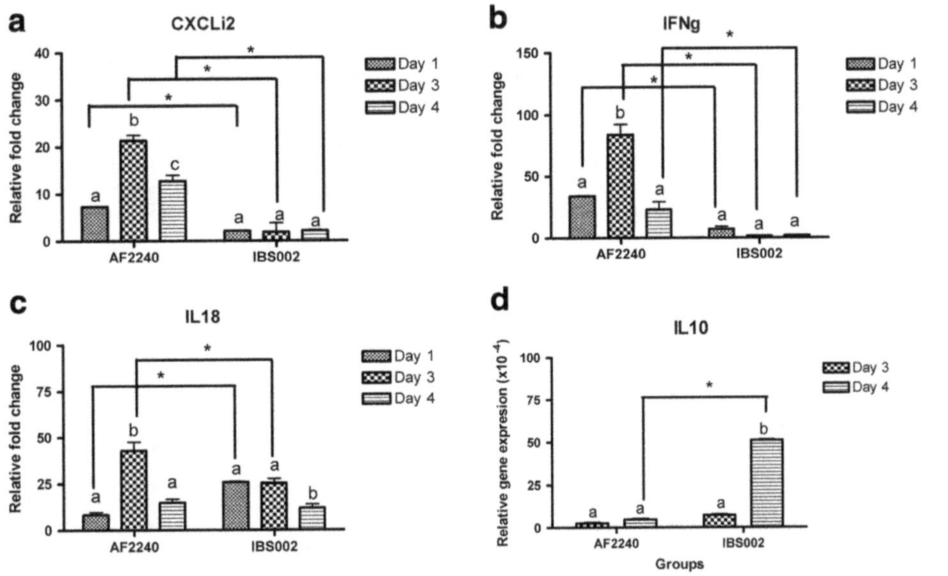

Fig. 4 The normalized fold expression of (**a**) CXCLi2, (**b**) IFN-ɣ and (**c**) IL-18 and normalized Cq value of (**d**) IL10 in the bursa of 3-week-old chickens infected with NDV AF2240 or IBS002 at 1, 3 and 4 day post infection. Significant differences were observed using two-ways ANOVA by using strains of NDV and infection time points as effectors. *Asterisk signs* show significant difference between different strains of NDV ($P \leq 0.05$) while different *alphabets* show significant difference between different infection time points ($P \leq 0.05$)

Table 3 MTT assay, AO/PI apoptotic population and Annexin V apoptotic population of chicken bursa of Fabricius enriched IgM^{+} B lymphocytes infected with NDV AF2240 and IBS002. The differences between the control group and infected group were determined by one-way ANOVA

Assays	NDV strains	Control	1 d.p.i	3 d.p.i	4 d.p.i
MTT viable IgM cells (%)	AF2240	100 ± 0.32	81.48 ± 0.36^{*}	60.86 ± 0.44^{**}	42.06 ± 0.19^{***a}
	IBS002	100 ± 0.32	85.42 ± 0.22^{*}	65.78 ± 0.12^{**}	53.15 ± 0.24^{***b}
AO/PI apoptotic IgM cell count (%)	AF2240	2.02 ± 0.90	4.37 ± 0.30^{*}	15.34 ± 0.80^{**}	17.81 ± 3.70^{**}
	IBS002	2.02 ± 0.90	3.29 ± 6.70	10.98 ± 4.40^{*}	14.91 ± 3.10^{**}
Annexin V positive IgM cell population (%)	AF2240	3.29 ± 1.44	19.72 ± 2.13^{*a}	14.35 ± 1.26^{**}	34.49 ± 2.50^{***a}
	IBS002	3.29 ± 1.44	9.22 ± 2.20^{*b}	12.15 ± 1.35^{*}	22.81 ± 1.14^{**b}

All results represent the mean ± standard deviation (n = 3)
Different asterisk signs show significant different between infection time points ($P \leq 0.05$).
Different alphabets show significant different between different strains of NDV ($P \leq 0.05$).

of the apoptosis hallmarks, can be indicated by the occurrence of the sub G0/G1 population. As is shown in Fig. 5, the population of sub G0/G1 increased over the infection time of NDV AF2240 and IBS002. It was also shown that as in the previous findings, NDV AF2240 caused higher frequency of DNA fragmentation compared to IBS002.

Discussion

Immune reaction to viral invasion is important to combat virus growth and to minimize any damage that can be imposed on the host cells [30]. In this study, the host immune response in the bursa of Fabricius, an organ mainly colonized by B lymphocytes, was assessed upon infection by the avian virus NDV genotype VII IBS002 and genotype VIII AF2240.

The cell population in the chicken bursa was evaluated by flow cytometry and the results showed that in normal chickens, this organ was predominantly occupied by B lymphocytes while the T lymphocyte and macrophage populations were at minimal cell numbers. Different from the cell population changes in chicken spleen upon NDV infection where CD4+ and CD8+ cells declined in their cell numbers when invaded by the two different NDV genotypes [31], our results showed infiltration of both CD4+ and CD8+ T lymphocytes in NDV AF2240

Table 4 Summary of results

	Infection of NDV AF2240	Infection of NDV IBS002
Cell population changes	• Increment of T lymphocytes and macrophages, depletion of B lymphocytes compared to uninfected control. • Higher increment of macrophages and depletion of B lymphocytes compared of IBS002 infection.	• Increment of T lymphocytes and macrophages, depletion of B lymphocytes compared to uninfected control. • Lower increment of macrophages and depletion of B lymphocytes compared of AF2240 infection.
Oxidative stress	• Increment of NO and MDA content compared to uninfected control. • Higher increment of NO and MDA content compared to IBS002 infection.	• Increment of NO and MDA content compared to uninfected control. • Lower increment of NO and MDA content compared to AF2240 infection.
Expression of pro-inflammatory and anti-inflammatory cykotines/chemokine (CXCLi1, IFNɣ, IL18 and IL10)	• Increment of pro-inflammatory cytokines/chemokines (CXCLi2, IFNɣ, IL18) compared to uninfected control. • Anti-inflammatory IL10 was detected in the bursa at day 3 and 4 post infection but not at day 1 and uninfected control. • Higher increment of pro-inflammatory cytokines/chemokines (CXCLi2, IFNɣ, IL18) compared to IBS002 infection.	• Increment of pro-inflammatory cytokines/chemokines (CXCLi2, IFNɣ, IL18) compared to uninfected control. • Higher anti-inflammatory IL10 was observed in day 3 and day 4 bursa compared to AF2240 infection. • Lower increment of pro-inflammatory cytokines/chemokines (CXCLi2, IFNɣ, IL18) compared to AF2240 infection.
Viral load	• Viral load was detected in in the bursa at day 3 and 4 post infection but not at day 1 and uninfected control. • Lower detected viral load compared to IBS002 infection.	• Viral load was detected in in the bursa at day 3 and 4 post infection but not at day 1 and uninfected control. • Higher detected viral load compared to AF2240 infection.
Apoptosis of IgM+ cells	• Increment of IgM+ apoptotic population over the infection time. • Higher increment of IgM+ cell apoptosis compared to IBS002 infection.	• Increment of IgM+ apoptotic population over the infection time. • Lower increment of IgM+ cell apoptosis compared to AF2240 infection.

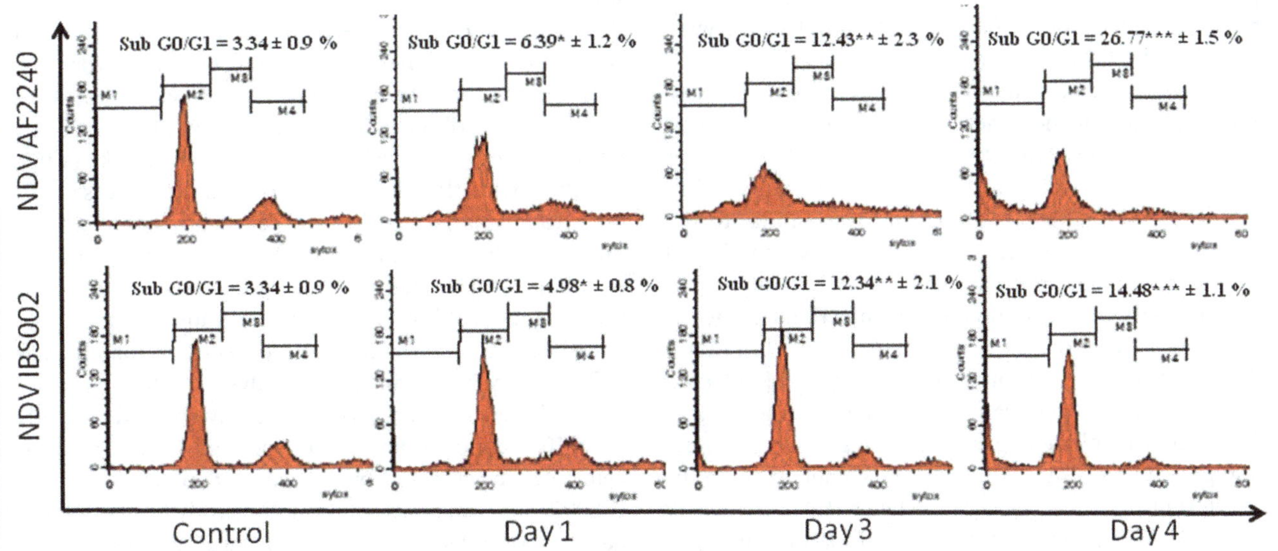

Fig. 5 Pertubations of cell cycle phases of B lymphocytes in the bursa of chicken infected with NDV AF2240 and NDV IBS002 at day 1, 3 and 4 post infections. The sub G0/G1 population differences between the control group and treated group were determined by one-way ANOVA. Groups labelled with different superscript are significantly different ($P \leq 0.05$)

and IBS022 infected chicken bursa. Such a scenario was reported earlier in cases where infections by IBDV and Marek's disease virus caused infiltration of T cells in the chicken bursa consequently indicating that T cells, especially CD8+ T cells play a role in combating virus in infected cells in the bursa [15, 17, 32]. It was reported by Rasoli et al. [31] that NDV AF2240 and IBS002 caused decrease of the T cell population in chicken spleen. Thus, we postulated that migration of T cells from chicken spleen to the bursa of Fabricius may have taken place during NDV AF2240 and IBS002 infection because in our study, we found that the T cell number was increased in the chicken bursa during infection. Similar to what was reported earlier, participation of T cells in virus pathogenicity was observed by the migration of the T cells from the chicken spleen to the bursa during infection by IBDV [33]. The higher accumulation of intrabursal T cells and macrophages detected in chicken bursa infected by NDV AF2240 than NDV IBS002 was followed by a similar trend of nitric oxide increment in the chicken bursa upon virus infections by both strains of NDV. It was reported that inflammation caused by bacterial and viral infections subsequently caused oxidative damage to both the host cells and the invading microorganisms as result of the action of host defense [34]. A high level of oxidative stress, as indicated by a rise of NO level, was caused by NDV on the organs (Fig. 2). Similarly to what was reported previously, NDV is able to induce the activation of macrophages, which subsequently triggered iNos gene expression and production of NO at the infected site [35]. The production of NO at the infected site may also be related to the cytokines being expressed. Rasoli et al. [31] reported that IBS002 NDV strain induced higher level of IL10 which inhibited the production of IFNγ upon infection. Not only that, the lower macrophage population, which resulted in a lower NO production was also observed in the case of IBS002 infection. Meanwhile, the product of lipid peroxidation known as MDA can cause damage to the cell at the enzyme, protein or DNA levels subsequently leading to cell death [36, 37]. The content of MDA was measured and the results showed significant activity of lipid peroxidation being detected in chicken bursa infected with NDV AF2240 starting on day 3 post infection while NDV IBS002 caused a significant increase of the MDA level after day 1 post infection. However, the subsequent days showed that the infection by NDV AF2240 caused greater activity of lipid peroxidation in chicken bursa compared to NDV IBS002. As lipid peroxidation can be caused by free radicals such as nitric oxide, we postulate that the increment of MDA level in the chicken bursa upon infection with NDV AF2240 and NDV IBS002 is closely related to the increment of the macrophage population which leads to excess production of nitric oxide. This phenomenon suggests that reactive nitrogen species (RNS) and oxygen species (ROS) caused by excess production of NO and lipid peroxidation, respectively play crucial roles in NDV infection as was reported previously [38].

Other than NO and MDA detections, we also evaluated the viral load and cytokine expression in chicken bursa after infection with the two different strains of NDV. The Taqman based real time PCR results showed that no virus was detected from the chicken bursa cell population on the first day of infection as well as in the control chickens. Both NDV AF2240 and IBS002 were detected in the IgM+ cell population starting from day 3 post-infection and the amount of virus increased as the duration of the infection increased. Of the two virus genotypes, NDV AF2240 was reported to have lower amount of virus residing in the cells when compared to infection by NDV IBS002. This result was similar to the recent report by Hu et al. [23] where higher viral load was able to be detected in the bursa of genotype VII NDV infected chicken compared to infections by genotypes IV and IX. However, Hu et al. [23] reported that a higher viral load in the bursa of genotype VII infected chicken was always associated with a higher fold of pro-inflammatory cytokine expression. However, in our case, genotype VIII NDV AF2240 induced higher expressions of the tested proinflammatory cytokines (CXCLi2, IFN-γ and IL-18) (Fig. 4) but with lower viral load (Table 2) in the chicken bursa than in the case of NDV IBS002 infection. Chicken CXCLi2, a proinflammatory chemokine, is expressed by and is chemotactic for monocytes [39]. On the other hand, IL-18 is a pro-inflammatory cytokine secreted by various immune and none-immune cells including monocytes, macrophages and lymphocytes that promotes expression of IFN-γ by T cells [40], which subsequently activates macrophages to produce NO [41, 42]. In our study, higher expressions of CXCLi2, IL-18 and IFN-γ were observed in infection by NDV AF2240, especially on day 3 post infection of the infected bursa correlating with the increased number of $CD8^+$ T cells. The drastic increase of IFN-γ in NDV AF2240 infected bursa further activated macrophages to produce NO on days 3 and 4 post infection. However, lower expression levels of CXCLi2, IL-18 and IFN-γ were observed on day 4 post infection particularly in the bursa of NDV AF2240 infected chicken. This may be accounted for by the immune system attempting to restore the drastic elevation of the NO content [43] in the chicken bursa by expressing anti-inflammatory cytokine IL10 [31]. IL10 was not detected in the uninfected and day 1 NDV infected bursa but was detected on day 3 and 4 post NDV infection. Similar to the expression of IL10 in the spleen [31], IL10 expression was higher in the NDV IBS002 infected bursa indicating this cytokine may help to control and lower the level of pro-inflammatory cytokines and NO in the bursa of the NDV IBS002 infected chicken compared to NDV AF2240 infection. Although IL10 expression was associated with a reduction of the proinflammatory cytokines level on day 4, this reaction was not able to relieve the inflammation in the bursa of NDV IBS002 and AF2240 infected chicken.

Infection by both genotypes of NDV was observed to cause decreased viability of IgM^+ B cells. IgM^+ B cell was then further enriched from the chicken bursa to further understand the effects of inflammatory stress towards the development of humoral immunity that plays an important role to induce protection via the synthesis of effective antibodies against the virus [44]. As B lymphocytes were observed to be depleted in the bursa upon infection with the virus, the apoptosis assay was carried out. Apoptosis is a controlled cell death mode which is implicated in the pathogenesis of viruses that causes diseases [45]. The hallmarks of apoptosis including cell membrane blebbing, DNA fragmentation and Phosphatidylserine (PS) externalization [46] were evaluated by the AO/PI assay and Annexin V analysis in this study. Our results based on the trypan blue assay showed that NDV AF2240 reduced viability and induced higher apoptotic cell rate in chicken bursa when compared to NDV IBS002. The changes were significantly observed from day 3 post infection. This result was in line with the Annexin V study which also showed that NDV AF2240 induced higher rate of apoptosis in the bursa than NDV IBS002. In the AO/Pi assay, cells showing characteristics of apoptotic cells such as membrane blebbing, chromatin condensation and up taking propidium iodide stain were evaluated. Similar to the Annexin V results, our AO/Pi results demonstrated that the time course infection of NDV AF2240 induced greater magnitude of apoptosis in chicken bursa of Fabricius compared to IBS002. From our results, the cell cycle analysis showed that NDV AF2240 resulted in a higher population of sub G0/G1 compared to IBS002, showing that apoptotic events were higher during NDV AF2240 infection as sub G0/G1 indicated DNA fragmentation, which is also one of the apoptotic cells' characteristics. The lower degree of B cell depletion in the bursa of NDV IBS002 infected chicken through apoptosis might contribute to the delayed mortality of IBS002 infected chicken compared to NDV AF2240 [31].

Considering that drastic decline in viability and higher apoptotic rate of B cells were detected in chicken bursa infected with NDV AF2240, which is genotype VIII compared to NDV strain IBS002, which belongs to genotype VII with lower virus residue detected in the organ infected by the former, we deduced that oxidative stress induced by the nitric oxide and lipid peroxidation processes which were noticed in the organ during infection might be the primary cause. Furthermore, NO had been reported to cause events of apoptosis in a variety of cells [47, 48], whereas lipid peroxidation could induce apoptosis by depleting ATP production in the cells [49, 50]. Similarly, the significant infiltration of $CD8^+$ T cells and macrophages, which led to the elevation of IFN-γ in NDV AF2240 infected chicken bursa caused higher magnitude of organ damage, failure of antibody production and higher mortality in chicken which suggested that the infection of NDV strain AF2240 was far more acute than the infection of NDV strain IBS002 [31]. A summary of the results is listed in Table 4.

Conclusion

The results of our study demonstrated that the time course infection of NDV AF2240 induced a greater magnitude of apoptosis in chicken bursa IgM$^+$ B cells that may further cause impairment of the humoral immunity when compared to IBS002. This might be due to the high level of oxidative stress which led to organ damage in the chicken bursa of Fabricius regardless of the viral load.

Abbreviations

ANOVA: Analysis of variance; AO/PI: Acridine orange/propidium iodide; BSA: Bovine serum albumin; CD: Cluster of differentiation; cDNA: Complementary DNA; Ct: Cycle threshold; CXCLi2: Chemokine (C-X-C motif) ligand 2; DNA: Deoxyribonucleic acid; EDTA: Ethylenediaminetetraacetic acid; ELD50: Embryo Lethal Dose 50; ELISA: Enzyme-linked immunosorbent assay; FITC: Fluorescein isothiocyanate; GAPDH: Glyceraldehyde 3-phosphate dehydrogenase; IBDV: Infectious bursal disease virus; ICPI: Intracerebral pathogenicity index; IFN- γ: Interferon gamma; IgM: Immunoglobulin M; IL10: Interleukin-10; IL18: Interleukin-18; MDA: Malondialdehyde; MDT: Mean death-time; MTT: 3-(4,5-dimethylthiazol-2-yl)-2,5-diphenyl tetrazolium bromide; NDV: Newcastle Disease Virus; NO: Nitric oxide; PBS: Phosphate-buffered saline; PCR: Polymerase chain reaction; PE: Phycoerythrin; PerCp: Peridinin-chlorophyll-protein complex; PS: Phosphatidylserine; RNA: Ribonucleic acid; SPF: Specific-pathogen-free; TBARS: Thiobarbituric acid-reactive substance

Funding

The study was supported by an Institute of Bioscience, Higher Institution Centre of Excellence (IBS HICoE) grant from the Ministry of Higher Education, Government of Malaysia.

Authors' contributions

KTYW, YSK, OAR, IA, TSG, ANB designed the experiment. KTYW, YSK, TSW carried out the animal trials. KTYW and TSW propagated NDV and performed the qPCR assays. KTYW, YSK performed flow cytometry, and other experiments. KTYW, YSK, OAR, ANB prepared the manuscript. KTYW, YSK, TSW, OAR, IA, TSG, ANB performed data analysis and proof-read the manuscript. All the authors read and approved the final manuscript.

Competing interests

The authors declare that they have no competing interests.

Author details

[1]Institute of Bioscience, Universiti Putra Malaysia, 43400 Serdang, Selangor, Malaysia. [2]Faculty of Veterinary Medicine, Universiti Putra Malaysia, 43400 Serdang, Selangor, Malaysia. [3]Faculty of Biotechnology and Biomolecular Sciences, Universiti Putra Malaysia, 43400 Serdang, Selangor, Malaysia.

References

1. Mayo MA. A summary of taxanomic changes recently approved by ICTV. Arch Virol. 2002;147:1655–6.
2. Alexander D. Newcastle disease and other Paramyxoviridae infections. In: Calnek B, editor. Diseases of poultry. 10th ed. Ames: Iowa State University Press; 1997. p. 541–69.
3. Alexander D. Newcastle disease and other paramyxovirus infections. In: Calnek B, Barnes H, Beard C, Reid W, editors. Diseases of poultry. 9th ed. Ames: Iowa State University Press; 1991. p. 469–519.
4. MacPherson LW. Some observation on the epizootiology of Newcastle disease. Can J Comp Med. 1956;20:155–68.
5. Masaji M, Imai K, Sanada Y, Sanada N, Yuasa N, Imada T, et al. Phylogenetic analysis of Newcastle disease virus genotypes loslated in Japan. J Clin Microbiol. 2002;40(10):3826–30.
6. Ke MG, Liu JH, Lin YMH, Chen J, Tsai SS, Chang CP. Molecular characterization of Newcastle disease viruses isolated from recent outbreaks in Taiwan. J Virol Methods. 2001;97:1–11.
7. Hugh-Jones M, Allan WH, Dark FA, Harper GJ. The evidence for the airborne spread of Newcastle disease. J Hyg (Lond). 1973;71:325–39.
8. Alexander D. Newcastle disease: methods of spread. In: Alexander D, editor. Newcastle disease. 1st ed. Boston: Springer US; 1988. p. 256–72.
9. Momayez R, Gharahkhani P, Pourbakhsh SA, Toroghi R, Shoushtari AH, Banai M. Isolation and pathogencity identification of avian paramyxovirus serotype 1 (Newcastle disease) virus from a Japanses quail flock in Iran. Arch Razi Inst. 2007;62(1):39–44.
10. Alexander D. Newcastle disease, other avian paramyxoviruses and pneumovirus infections. In: Saif Y, editor. Disease of poultry. 11th ed. Ames: Iowa State Press; 2003. p. 63–92.
11. Glick B. Historical perspective: the bursa of Fabricius and its influence on B-cell development, past and present. Vet Immunol Immunopathol. 1991;30:3–12.
12. Alitheen NB, McClure JS, Yeap SK, Kristeen-Teo YW, Tan SW, McCullagh P. Establishment of an in vitro system representing the chicken gut-associated lymphoid tissue. PLoS One. 2012;7(11):e49188.
13. Kozuka Y, Nasu T, Murakami T, Yasuda M. Comparative studies on the secondary lymphoid tissue areas in the chicken bursa of Fabricius and calf ileal Peyer's patch. Vet Immunol Immunopathol. 2010;133:190–7.
14. Imre O, Lonneke V. Structure of the avian lymphoid system. In: Schat K, Kaspers B, Kaiser P, editors. Avian Immunology. 2nd ed. London: Academic Press; 2013. p.13–50.
15. Abdul-Careem MF, Hunter BD, Sarson AJ, Parvizi P, Haghighi HR, Read L, et al. Host responses are induced in feathers of chickens infected with Marek's disease virus. Virology. 2008;370:323–32.
16. Khatri M, Palmquist JM, Cha RM, Sharma JM. Infection and activation of bursal macrophages by virulent infectious bursal disease virus. Virus res. 2005;113:44–50.
17. Kim IJ, You SK, Kim H, Yeh HY, Rautenschlein S, Sharma JM. Chracteristics of bursal T lymphocytes induced by infectious bursal disease virus. J Virol. 2000;74(19):8884–92.
18. Rautenschlein S, Yeh HY, Njenga MK, Sharma JM. Role of intrabursal T cells in infectious bursal disease virus (IBDV) infection: T cells promote viral clearance but delay follicular recovery. Arch Virol. 2002;147(2):285–304.
19. Brown C, Kind DJ, Seal BS. Pathogenesis of Newcastle disease in chickens experimentally infected with viruses of different virulence. Vet Pathol. 1999;36:125–32.
20. Kommers GD, King DJ, Seal BS, Carmicheal KP, Brown CC. Pathogenesis of six pigeon-origin isolates of Newcastle disease virus for domestic chickens. Vet Pathol. 2002;39:352–62.
21. Wakamatsu N, King DJ, Kapczynski DR, Seal BS, Brown CC. Experimental pathogenesis for chickens, turkeys, and pigeons of exotic Newcastle disease virus from an outbreak in California during 2002-2003. Vet Pathol. 2006;43:925–33.
22. Rue CA, Susta L, Cornax I, Brown C, Kapczynski DR, Suarez DL, et al. Virulent Newcastle disease virus elicits a strong innate immune response in chickens. J Gen Virol. 2011;92:931–9.
23. Hu Z, Hu J, Hu S, Song Q, Ding P, Zhu J, et al. High levels of virus replication and an intense inflammatory response contribute to the severe pathology in lymphoid tissues caused by Newcastle disease virus genotype VIId. Arch Virol. 2015;160:639–48.
24. Harrison L, Brown C, Afonso C, Zhang J, Susta L. Early occurrence of apoptosis in lymphoid tissues from chickens infected with strains of Newcastle disease virus of varying virulence. J Comp Pathol. 2011;145:327–35.
25. Murulitharan K, Yusoff K, Omar AR, Molouki A. Characterization of Malaysian velogenic NDV strain AF2240-I genomic sequence: a comparative study. Virus Genes. 2013;46:431–40.
26. Lai MC, Ibrahim A. Velogenic viscerotropic Newcastle disease virus. In: Copland JW, editor. Disease in Poultry: A New Food Pellet Vaccine 5th ed. Canberra: ACIAR. 1987; p. 33–34.
27. Lee TC, Yusoff K, Nathan S, Tan WS. Detection of virulent Newcastle disease virus using a phage-capturing dot blot assay. J Virol Methods. 2006;136:224–9.

28. Roohani K, Tan SW, Yeap SK, Ideris A, Bejo MH, Omar AR. Characterisation of genotype VII Newcastle disease virus (NDV) isolated from NDV vaccinated chickens, and the efficacy of la Sota and recombinant genotype VII vaccines against challenge with velogenic NDV. J vet Sci. 2015;16(4):447–57.
29. Mosmann T. Rapid colorimetric assay for cellular growth and survival: application to proliferation and cytotoxicity assays. J Immunol Methods. 1983;65(1–2):55–63.
30. Luca GG, Francis VC. Noncytolytic control of viral infections by the innate and adaptive immune response. Annu rev Immunol. 2001;19:65–91.
31. Rasoli M, Yeap SK, Tan SW, Moeini H, Aini I, Hair-Bejo M, et al. Alteration in lymphocytes responses, cytokine and chemokine profiles in chickens infected with genotype VII and VIII velogenic Newcastle disease virus. Comp Immunol Microbiol Infect Dis. 2014;37(1):11–21.
32. Rauf A, Khatri M, Murgia MV, Saif YM. Expression of perforin-granzyme pathway genes in the bursa of infectious bursal disease virus-infected chickens. Dev Comp Immunol. 2011;35:620–7.
33. Yu X, Rui L, Shao Q, Liu H, Lu Y, Zhang Y, et al. Changes of CD4+CD25+ cells ratio in immune organs from chickens challenged with infectious bursal disease virus strains with varying virulences. Viruses. 2015;7(3): 1357–72.
34. Akaike T. Role of free radicals in viral pathogenesis and mutation. Rev med Virol. 2001;11:87–101.
35. Victor U, Vladimir AS, Volker L, Volker S. Induction of NO synthesis in macrophages by Newcastle disease virus is associated with activation of nuclear factor- KB. Int Immunol. 1996;8(4):491–8.
36. Das U. Selective enhancement of free radicals in tumour cells as a strategy to kill tumour cells both in vitro and in vivo. In: Reddy C, Hamilton G, Madyastha K, editors. Biological oxidation systems. 1st ed. San Diego: Academic Press; 1990. p. 607–24.
37. Kumar SG, Das UN. Free radical dependent suppression of growth of mouse myelomia cells by alpha-linolenic and eicosapentaenoic acid in vitro. Cancer Lett. 1995;92:27–38.
38. Kadiam C, Ventaka S, Rajendra W, Lokanatha V. Newcastle disease virus (NDV) modulates pro/antioxidant status in different brain regions of chicken. Free Radicals Antiox. 2013;3:81–6.
39. Poh TY, Pease J, Young JR, Bumstead N, Kaiser P. Re-evaluation of chicken CXCR1 determines the true gene structure. J Biol Chem. 2008;283:16408–15.
40. Gobel TW, Schneider K, Schaerer B, Mejri I, Puehler F, Weigend S, et al. IL-18 stimulates the proliferation and IFN-γ release of $CD4^+$ T cells in the chicken: conservation of a Th1-like system in a nonmammalian species. J Immunol. 2003;171:1809–15.
41. Qureshi MA, Marsh JA, Dietert RR, Sung YJ, Nicolas-Bolnet C, Petitte JN. Profiles of chicken macrophage effector functions. Poult Sci. 1994;73(7):1027–34.
42. Weining KC, Schultz U, Munster U, Kasper B, Staeheli P. Biological properties of recombinant chicken interferon-gamma. Eur J Immunol. 1996;26:2440–7.
43. Akaike T, Maeda H. Nitric oxide and virus infection. Immunology. 2000;101(3):300–8.
44. Rauw F, Gardin Y, Palya V, van Borm S, Gonze M, Lemaire S, et al. Humoral, cell-mediated and mucosal immunity induced by oculo-nasal vaccination of one-day-old SPF and conventional layer chicks with two different love Newcastle disease vaccines. Vaccine. 2009;27:3631–42.
45. Roulston A, Marcellus RC, Branton PE. Viruses and apoptosis. Annu rev Microbiol. 1999;53:577–628.
46. Kerr JFR, Wyllie AH, Currie AR. Apoptosis: a basic biological phenomenon with wide ranging implications in tissue kinetics. Br J Cancer. 1972;26:239–57.
47. Albina JE, Cui S, Mateo R, Reichner JS. Nitric oxide-mediated apoptosis in murine peritoneal macrophages. J Immunol. 1993;150:5080–5.
48. Cui S, Reichner JS, Mateo RB, Albina JE. Activated murine macrophages induce apoptosis in tumor cells through nitric oxide-dependent or independent mechanisms. Cancer res. 1994;54:2462–7.
49. Fernandes G, Chandra SB, Luan X, Troyer DA. Modulation of antioxidant enzymes and programmed cell death by n-3 fatty acids. Lipids. 1996;31:591–6.
50. Southgate J, Pitt E, Trejdoesiewicz LK. The effects of dietary fatty acids on the proliferation of normal human arothelial cells in vitro. Br J Cancer. 1996;74:728–34.

15

Quantifying the effect of swab pool size on the detection of influenza A viruses in broiler chickens and its implications for surveillance

Amos Ssematimba[1,3], Sasidhar Malladi[1], Peter J. Bonney[1], Cristian Flores-Figueroa[2], Jeannette Muñoz-Aguayo[2], David A. Halvorson[1] and Carol J. Cardona[1*]

Abstract

Background: Timely diagnosis of influenza A virus infections is critical for outbreak control. Due to their rapidity and other logistical advantages, lateral flow immunoassays can support influenza A virus surveillance programs and here, their field performance was proactively assessed.

The performance of real-time polymerase chain reaction and two lateral flow immunoassay kits (FluDETECT and VetScan) in detecting low pathogenicity influenza A virus in oropharyngeal swab samples from experimentally inoculated broiler chickens was evaluated and at a flock-level, different testing scenarios were analyzed.

Results: For real-time polymerase chain reaction positive individual-swabs, FluDETECT respectively detected 37% and 58% for the H5 and H7 LPAIV compared to 28% and 42% for VetScan. The mean virus titer in H7 samples was higher than for H5 samples. For real-time polymerase chain reaction positive pooled swabs (containing one positive), detections by FluDETECT were significantly higher in the combined 5- and 6-swab samples compared to 11-swab samples. FluDETECT detected 58%, 55.1% and 44.9% for the H7 subtype and 28.3%, 34.0% and 24.6% for the H5 in pools of 5, 6 and 11 respectively.

In our testing scenario analysis, at low flock-level LPAIV infection prevalence, testing pools of 11 detected slightly more infections while at higher prevalence, testing pools of 5 or 6 performed better. For highly pathogenic avian influenza virus, testing pools of 11 (versus 5 or 6) detected up to 5% more infections under the assumption of similar sensitivity across pools and detected less by 3% when its sensitivity was assumed to be lower.

Conclusions: Much as pooling a bigger number of swab samples increases the chances of having a positive swab included in the sample to be tested, this study's outcomes indicate that this practice may actually reduce the chances of detecting the virus since it may result into lowering the virus titer of the pooled sample. Further analysis on whether having more than one positive swab in a pooled sample would result in increased sensitivity for low pathogenicity avian influenza virus is needed.

Keywords: Influenza A virus, Surveillance, Real-time polymerase chain reaction, Antigen detection, Sample pooling, Lateral flow immunoassays

* Correspondence: ccardona@umn.edu
[1]Secure Food Systems Team, University of Minnesota, Veterinary and Biomedical Sciences, 301C Veterinary Science Building, 1971 Commonwealth Avenue, Saint Paul, MN 55108, USA
Full list of author information is available at the end of the article

Background

Influenza type A viral (IAV) infections in poultry may either be of low or high pathogenicity. IAV evolve rapidly [1] and occasionally, once introduced into poultry, low pathogenicity avian influenza (LPAI) viruses (LPAIV) of the H5 or H7 subtypes from the wild bird reservoir can directly evolve into highly pathogenic avian influenza (HPAI) viruses (HPAIV) [2, 3]. Although infections by both viruses largely have the same host range, they cause different diseases in terms of virulence [4, 5] and hence result in different needs for control and surveillance. For example, while infections with HPAIV are associated with high morbidity and mortality, LPAI is usually asymptomatic or responsible for a mild respiratory disease that may cause a reduction in egg production or moderately increased mortality [4]. Unlike LPAI where the infection can be detected via the presence of either virus or antibodies since infected birds mostly recover, almost no HPAIV infected poultry will survive [5] and infection is mostly detected via testing sick or dead birds for virus. Also, during outbreaks, HPAIV has been shown to easily spread by proximity [3] while LPAIV often spreads through contact networks [6].

The essential components for avian influenza control are education of flock owners and workers, biosecurity, prevention, surveillance [6], and vaccination (depending on the country). There are both passive and active surveillance systems in place in most countries to ensure early detection of IAV circulation in poultry with the former well suited for HPAIV detection due to its clear clinical manifestation. Laboratory testing using sensitive testing procedures is key for an efficient surveillance system. Although polymerase chain reaction (PCR) is extensively used during IAV outbreaks [7], the shorter turnaround time and other logistical advantages render lateral flow immunoassays (LFI) desirable for field screening; even though LFI have a lower sensitivity relative to PCR or virus isolation [8, 9].

Individual test characteristics, sample collection and composition all play a vital role in determining the efficacy of IAV active surveillance systems. Precisely, given a pooled sample with at least one positive swab, the probability that the sample tests positive depends on test characteristics, the virus titer in the positive swab, the volume of the diluent and on the number of negative swabs it is pooled with. Swab-pooling has been shown to allow a single test for multiple samples. In other words, not only does swab-pooling increase testing throughput, it also reduces the number of tests needed to meet sample size requirements and consequently the testing costs. However, due to dilution, pooling may also significantly reduce test sensitivity and lead to missing low prevalence infections [10, 11].

Recommendations to improve the sensitivity of LFI in IAV surveillance have included testing large numbers of samples before confirming the absence of the virus and avoiding excessive dilution of samples by pooling [9]. Indeed, Loth et al. [12] demonstrated that as the percentage of birds in the flock shedding virus increased, the number of animals to be tested to attain a given flock sensitivity was reduced. Thus, proactive evaluation of LFI performance under different sampling and surveillance protocols is important.

Since LFI target the type A influenza virus nucleoprotein, they are suitable for any type A influenza virus [9, 13, 14]. Here, using H5N2 and H7N2 low pathogenicity influenza A viruses as models, we evaluated the impact pooling one positive swab with 4 or 5 versus 10 negative swabs on the ability of LFI kits to detect IAV in pools and also to compare individual tests' analytical sensitivities on single swabs.

We compared the effectiveness of two LFI kits, namely, FluDETECT (Synbiotics Corporation, San Diego, California, USA) and VetScan (Abaxis, Inc., Union City, California, USA), in detecting two influenza A viruses and assess the effect of sample pooling on test performance. Lastly, we also compared the effectiveness of proposed antigen capture-based testing protocols involving different swab-pooling schemes during IAV field surveillance.

Methods

The studies described were fully compliant with the University of Minnesota Institutional Animal Care and Use and the Institutional Biosafety Committee policies and were conducted under protocols 1409-31802A and 1610-34222H respectively.

Animals

Three hundred thirty (330) eleven-day old commercial broiler chickens (free of maternally derived antibodies against IAV) purchased from a commercial farm were transported to the Animal Resources Unit at the University of Minnesota. One hundred twenty-six (126) of the chicks were housed in six (6) isolator units, i.e., 21 birds/isolator under Biosafety Level 2 conditions while 204 were kept in Biosafety Level 1 floor housing. All birds were provided standard broiler grow ration and clean water which were provided ad libitum.

Challenge study

Six isolators were randomly assigned into two challenge groups. Birds in three isolators ($n = 63$ birds) were inoculated intranasally with 10^6 EID_{50}/ml of H5 LPAIV (Chicken/PA/13609/93, H5N2), while the birds in the other three isolators (n = 63 birds) were inoculated intranasally with 10^6 EID_{50}/ml of H7 LPAIV (Guinea hen/MA/148081/2002, H7N2). Re-titration from one dose of the inocula yielded titers of $10^{5.4}$ EID_{50}/ml and $10^{5.8}$ EID_{50}/ml for the H5 and H7 viruses respectively. Bird comfort and clinical signs were observed and recorded daily. Dead birds (2 total) were recorded and removed but were used for the swab samples for that day. At the end of the experiment, all remaining animals were euthanized by inhalation of carbon dioxide.

Oropharyngeal (OP) swab samples were taken from all birds on days 5–8 post-challenge using sterile polyester tipped applicators (Puritan Medical Products Company LLC, Guilford, Maine, USA). One hundred thirty-three (69 from H5-inoculated group and 64 from the H7-inoculated group) individual swab samples were used for direct comparison of RT-PCR, FluDETECT and VetScan tests. For the pooled sample analysis, 298 samples from the H5-inoculated group and 227 samples from the H7 inoculated group were used. One swab was placed into a 4-, 5- or 10- pool of swabs from uninoculated birds in brain heart infusion (BHI) or individually into a tube with 1.0 ml of BHI broth. All tubes were kept on ice until they were transported to the laboratory where they were stored at – 80 °C until they were tested.

Negative swab collection to create pool

Twice daily, uninoculated birds held in a floor unit were swabbed. The swabs were placed into pools of 4, 5, or 10. The pools of 4 and 5 swabs were placed into 3 ml of BHI and the 10-swab pool was placed into 5.5 ml of BHI. The pools were kept on ice until they were refrigerated in the laboratory.

Antigen detection testing

Two commercially available lateral flow immunoassay kits FluDETECT and VetScan were used for the detection of Influenza A virus in single swabs and only FluDETECT was used for the detection of IAV in pooled swab samples. For both tests, the manufacturer's instructions were followed.

RT-PCR

RNA was extracted from 50 μl of each sample with the MagMAX-96 Viral RNA Isolation Kit (Ambion) using MagMAX Express magnetic particle processor (Applied Biosystems) according to manufacturer's instructions. For IAV RT-PCR, each 25 μl reaction contained: 8 μl RNA, 0.83 μl of ultra-pure water, 12.5 μl of 2X buffer, 0.25 μl of 20 μM of each primer, 0.25 μl of 6 μM probe, 1.67 μl detection enhancer, and 1.0 μl 25X enzyme mix AgPath-ID ™ One-Step RT-PCR Kit (Ambion).Cycling conditions on Stratagene Mx 3005P were: 45 °C for 10 min, 95 °C for 10 min, 45 cycles of 94 °C for 15 s, 60 °C for 45 s [15, 16].

Standard curve generation

Each of the challenge viruses was titered in 9–11 day old embryonating chicken eggs (5 eggs were inoculated per dilution) following standard procedures [17]. Viral titers were calculated for each isolate using the Reed and Muench formula. The H5 titer was 1.49 * 10^8/ml while the H7 titer was 4.93 * 10^9/ml. Viral stocks were log fold diluted and tested with RT-PCR. CT values were plotted against virus titers to create a standard curve. The standard curve was used to calculate viral particle numbers based on quantitative RT-PCR CT values.

Data analysis

Data were transferred to a spreadsheet (Excel, Microsoft Corporation, Redmond, WA, USA) and the Statistical software R [18] was used to perform the analyses. For the single-swab samples, the proportions positive in both lateral flow immunoassays as well as the mean and lowest virus titer detected were determined. For the pooled-swab samples, the three swab pooling schemes were evaluated both within and between subtypes to assess the effect of swab-pooling on the performance of FluDETECT. Its ability to detect the virus in pools of 11 was compared with that in individual pools of 5 or 6 and in combined pools of 5 or 6 for separate H5 and H7 subtypes as well as for merged H5 and H7 sample groups. Only samples with CT ≤ 35 were considered RT-PCR positive [11, 19] and in the comparison of virus positive proportions, Fisher's exact test at significance level of $p \leq 0.05$ was used [20].

Influenza A virus testing scenario analyses

Using FluDETECT as a model test, we evaluated the potential use of LFI-based testing protocols as a surveillance tool for IAV infections in poultry. As application examples, we assessed the detection abilities of LFI alone for LPAIV and in combination with a 0.3% daily mortality trigger and targeted-sampling of only dead birds for HPAIV.

We performed 6000 simulations using a re-parameterized version of the mathematical model described by [21]. For IAV transmission dynamics in a broiler flock, the variables and parameters used are presented in Table 1. The outputs of interest at different days post infection (dpi) were mean infection prevalence in the flock and the mean percentage of outbreaks that are detected. The assessed testing protocols involved testing three OP swab samples comprising of either 5, 6 or 11 swabs.

For the model, disease-induced mortality [22, 23], distributions for the adequate contact rate [5, 24–28], and latent [5, 26, 29–31] and infectious [32–36] period durations were estimated from the data in the indicated literature. FluDETECT positive proportions for LPAIV were determined from this study. For HPAI, since no testing data was available in our desired format, we used the mean estimate from a study on HPAI H5N1 virus in chickens [12] with the assumption that FluDETECT sensitivity does not vary across sample pool sizes (Scenario

Table 1 Summary of the variables and parameter values or distributions used in the preliminary IAV testing scenario analyses in a broiler chicken flock

Parameter name	Parameter description	Distribution	
		LPAIV	HPAIV
Infectious period distribution	Length of the infectious period for IAV in chickens	Gamma: shape = 8.1388, scale = 0.9596	Weibull: shape = 1.965, scale = 2.90
Latent period distribution	Length of the latent period for IAV in chickens	Gamma: shape = 0.8248, scale = 0.4446	Gamma: shape = 0.89, scale = 0.7145
Adequate contact rate	The number of contacts per unit time that a broiler has with other broilers that are adequate to transmit IAV	Uniform (0.69–0.77)	PERT: minimum = 2.5, mode = 4.77, maximum = 9.0
Disease mortality	Proportion of IAV infected birds that succumbs to disease	Fixed: 0.5%	Fixed: 100%
Flock size	The number of broiler chickens per house	Log-normal: log mean = 10.0212, log SD = 0.3883 truncated at 13,000 and 50,000 birds	Same as for LPAIV
FluDETECT positive proportions	The fraction of positives detected by FluDETECT in different pooling schemes	Fixed: 58.0%, 55.1% and 44.9% for pools of 5, 6 and 11 respectively	Scenario 1: 71% [12], Scenario 2: 71%, 68.1%, 57.9% for pools 5, 6 and 11 respectively

1) or it varies by similar magnitudes as for LPAIV pools (Scenario 2).

Results

All 330 chickens were free of maternally derived antibodies against IAV at the start of the experiment and two birds died during the course. They were removed upon being recorded and were used for the swab samples for that day.

Descriptive summary of test results of the single-swab samples

A total of 133 (i.e., 69 for the H5 and 64 for the H7 virus) individual swabs were tested using all three tests. By our definition of an RT-PCR positive sample (i.e., with CT ≤ 35), we obtained 43 (62.3%) RT-PCR positive samples with a mean titer of $10^{2.43}EID_{50}$/ml for the H5 subtype and 62 (96.9%) with mean titer of $10^{3.73}EID_{50}$/ml for the H7 subtype. The highest virus titers detected were $10^{3.18}$ and $10^{4.7}$ EID_{50}/ml for the H5 and H7 viruses respectively. Table 2 presents the summary of single-swab samples testing results for both LFI for only the RT-PCR positive samples i.e., those with CT ≤ 35.

Descriptive summary of test results of the pooled-swab samples

The collected swabs were differently pooled into 525 samples (of which 298 and 227 were respectively of H5 and H7 subtype) for testing and further analysis. Figures 1 and 2 as well as Table 3 present the results on the proportions of samples that were positive in antigen detection and RT-PCR as well as their calculated virus titer.

In Table 3, we observe that, of all tested samples of a given subtype and pool size, RT-PCR positive proportions were up to 97.5% for the H7 pooled samples (observed in pools of 6) while the highest in H5 samples was 63.3% observed in pools of 11. We observe from Fig. 1 that there were fewer FluDETECT positive pooled samples of the H5 virus that the H7 across all pool sizes. Figure 2 summarizes the distribution of virus titer for all RT-PCR positive samples and indicates that, although not tested for statistical significance, the H7 samples had a relatively higher virus titer overall. We observe (Table 3) that, for RT-PCR positive samples, when compared within subtype, the highest proportion of H5 FluDETECT positive samples was 34%, which was estimated for pools of 6 and it was 9.4% more than the least estimate (obtained for the pools of 11). Similarly, for the H7 subtype, the highest proportion was 58% estimated for the pools of 5, and it was 13.1% more than the least estimate (corresponding to pools of 11). Between subtypes, for all corresponding pool sizes, the H7 subtype had higher FluDETECT positive proportions than H5, and the biggest difference was 29.7%, observed in the pools of 5. The mean titer of

Table 2 Summary results on positive proportions and virus titer present for the single-swab samples for only samples with CT ≤ 35 totaling *n* = 43 and 62 for H5 and H7 viruses respectively[a]

Test	Subtype	Number positive (%)	Mean virus titer for LFI detected samples in EID_{50}/ml	Lowest virus titer for LFI detected samples in EID_{50}/ml
FluDETECT	H5	16 (37%)	$10^{2.98}$	$10^{2.32}$
	H7	36 (58%)	$10^{3.62}$	$10^{2.65}$
VetScan	H5	12 (28%)	$10^{3.03}$	$10^{2.39}$
	H7	26 (42%)	$10^{3.73}$	$10^{2.65}$

[a]For all PCR positive single-swab samples (i.e., those with CT ≤ 35), the mean virus titer was $10^{2.43}EID_{50}$/ml and $10^{3.73}EID_{50}$/ml for the H5 and H7 viruses respectively

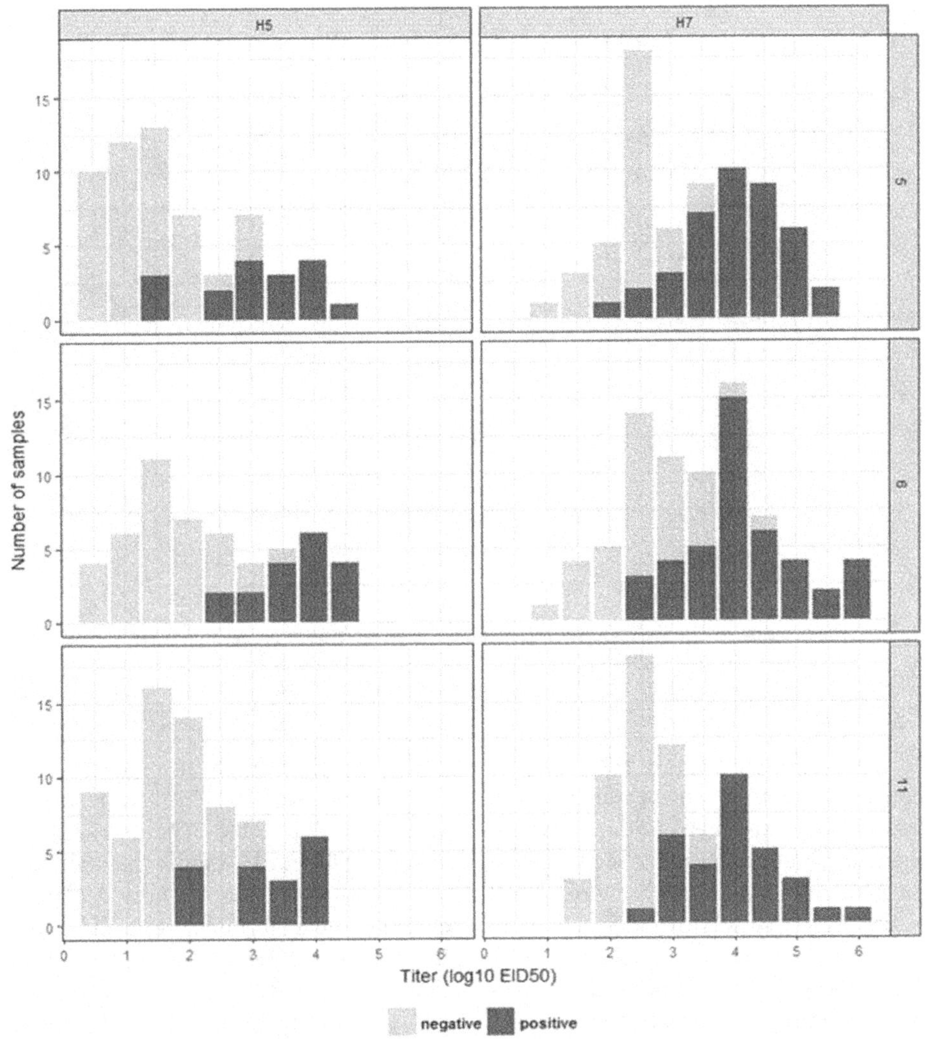

Fig. 1 Distribution of samples by virus titer (EID_{50} log_{10}/ml) present and FluDETECT test results. Left panels depict results for the H5 samples and the right panels are for H7 samples. The results for pools of 5, 6 and 11 are presented from top to bottom respectively. Note that these are results for samples that had CT ≤ 35 whose individual totals are indicated in Table 3

RT-PCR positive samples was also more than one log_{10} higher for H7 subtype samples for all corresponding pool sizes (Table 3).

Performance of FluDETECT under different sample pooling schemes

Results from the statistical comparisons of the positive proportions to assess the performance of FluDETECT in pools of 5 and/or 6 versus 11 for PCR positive samples (CT ≤ 35) are presented in Table 4. We observe that pooling 11 swabs resulted in a significantly lower proportion of samples that test positive using LFI compared to pools of 5 or 6 for all RT-PCR positive H5 and H7 samples combined ($p = 0.026$) and pools of 6 for all RT-PCR positive H7 samples ($p = 0.033$). The other noteworthy comparisons (i.e., those significant at $p \leq 0.1$) in which pools of 11 had lower positive proportions were between pools of 11 and pools of 5 for all PCR positive H5 and H7 samples combined ($p = 0.074$) and pools of 5 or 6 ($p = 0.075$) and pools of 5 ($p = 0.086$) for only H7 RT-PCR positive samples.

IAV testing scenario analysis outcomes

For LPAIV (Fig. 3), there was a negligible difference in test outcomes for pools of 5 and 6 swabs throughout the simulated period (38 days). Compared to pools of 5 and 6, testing pools of 11 resulted in a slightly higher percentage of detections between 5 and 15 dpi and after 36 dpi when infection prevalence was low and the trend is reversed on the other days when prevalence was high.

For HPAIV (Table 5), we found that all evaluated pooling protocols achieved 100% detection rate by 7 dpi. Since we could not test the differences for statistical significance, we use the phrase "substantive difference" to refer to situations where the difference between detected

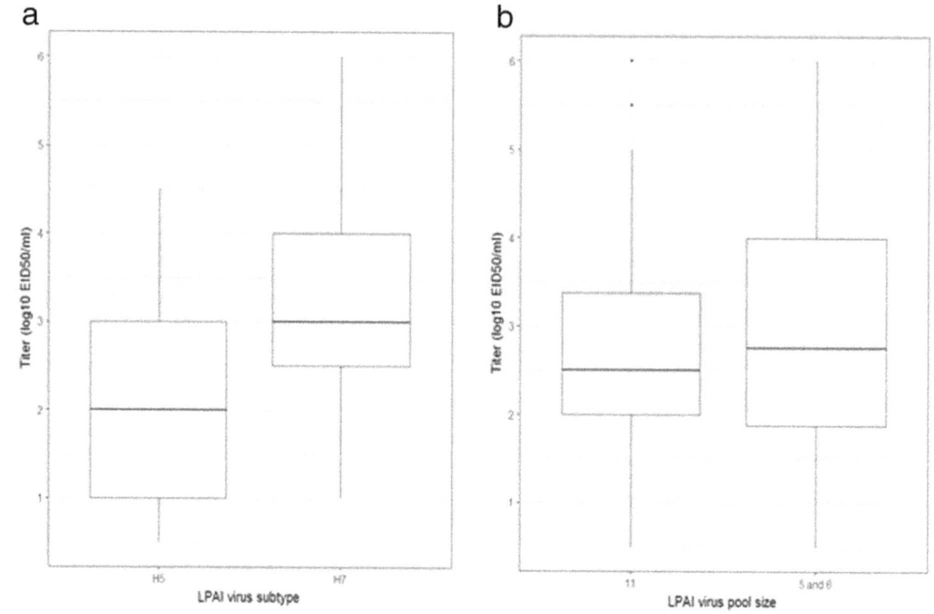

Fig. 2 Virus titer (EID_{50} log_{10}/ml) distribution of the pooled-swab samples. Panel **a** depicts a comparison of virus titers by the subtype for all pool sizes combined and **b** depicts a comparison of virus titers by pool sizes for combined subtypes. The boxplots depict the minimum, 25th percentile, median, 75th percentile and the maximum virus titer present. These are results for samples that had CT value ≤35

proportions is 2% or more. Within either of the scenarios assessed, the predicted detection percentages for pools of 5 and 6 swabs were in close agreement (either equal or within a 1% difference) in all but one (under scenario 1 at 2 dpi where testing pools of 6 detected 2% more cases). Otherwise, under scenario 1, substantive differences in proportion detected were observed at 2 and 3 dpi with testing pools of 11 detecting 4% and 5% more than pools of 5 and 2% and 4% more than for pools of 6 respectively. Under scenario 2, the only substantive difference was on 4 dpi where testing pools of 11 resulted in detecting 3% fewer cases than pools of 5 and 6 swabs.

Discussion

Using data from experimental infections of chickens with either H5 or H7 viruses, the performance of RT-PCR, FluDETECT and VetScan in detecting IAV was evaluated. For individual OP RT-PCR positive swabs (Table 2), FluDETECT detected 37% of H5 samples and 58% of H7 samples (9% and 16% more H5 and H7 virus PCR positive samples than VetScan). These detected proportions from clinically normal LPAIV-infected birds are lower than those reported for samples taken from HPAIV-infected sick or dead birds. Note however that even lower sensitivities (ranging from 0 to 5%) were reported for other LFI kits on individual swabs of H6N2 LPAIV [8].

In this study (conducted with LPAIV), the mean virus titers in the individual swabs detected by both LFI kits were comparable, and H7 samples had higher titers and more LFI positive proportions among RT-PCR positive samples than the H5 virus. The observed subtype effect may be due to differences in replication efficiency in the

Table 3 Summary results on positive proportions and virus titer present in the pooled-swab samples; unless stated otherwise, the results are for samples that are considered RT-PCR positive i.e., those with CT ≤ 35

Subtype	Pool size	Number RT-PCR positive/Total (% [a])	Mean titer for PCR positive samples in EID_{50}/ml	Highest titer in EID_{50}/ml	Number FluDETECT positive (% [b])	Mean titer for FluDETECT positive in EID_{50}/ml
H5	5	60/98 (61.2%)	$10^{1.60}$	$10^{4.26}$	17 (28.3%)	$10^{2.86}$
	6	53/91 (58.2%)	$10^{2.09}$	$10^{4.32}$	18 (34.0%)	$10^{3.46}$
	11	69/109 (63.3%)	$10^{1.78}$	$10^{4.00}$	17 (24.6%)	$10^{2.99}$
H7	5	69/74 (93.2%)	$10^{3.13}$	$10^{5.33}$	40 (58.0%)	$10^{3.87}$
	6	78/80 (97.5%)	$10^{3.19}$	$10^{5.94}$	43 (55.1%)	$10^{3.92}$
	11	69/73 (94.5%)	$10^{2.87}$	$10^{5.55}$	31 (44.9%)	$10^{3.73}$

[a]As a percentage of all samples of the specific virus subtype and pool size
[b]As a percentage of samples with CT ≤ 35

Table 4 Pairwise comparison of FluDETECT positive sample proportions for different sample pooling schemes for RT-PCR positive samples (i.e., those with $CT \leq 35$) using one-sided Fisher's exact test

Subtype	Pool size	Fraction FluDETECT positive (%): Fraction for 11-swab pools (%)[a]	Fisher's test: AC positive greater for 11-swab pools
H5 and H7	5 or 6	118/260 (45.4%): 48/138 (34.8%)	0.026
	5	57/129 (44.2%): 48/138 (34.8%)	0.074
	6	61/131 (46.6%): 48/138 (34.8%)	0.033
H5	5 or 6	35/113 (31.0%): 17/69 (24.6%)	0.228
	5	17/60 (28.3%): 17/69 (24.6%)	0.391
	6	18/53 (34.0%): 17/69 (24.6%)	0.177
H7	5 or 6	83/147 (56.5%): 31/69 (44.9%)	0.075
	5	40/69 (58.0%): 31/69 (44.9%)	0.086
	6	43/78 (55.1%): 31/69 (44.9%)	0.142

[a]In parentheses is the FluDETECT positive fraction for pools of size 11 for that subtype grouping

host as a result of differences in the viruses or inoculation dose used to challenge the birds [37, 38].

Our FluDETECT test results for the pooled swab samples (Table 3) revealed a subtype effect as H7 samples were found to have relatively higher virus titer and higher proportion of LFI positive samples (and consequently higher LFI sensitivity) than their H5 counterparts. On the potential effect of swab pooling, our findings for combined virus subtypes indicate that pooling up to 10 negative swabs with one positive swab reduced the viral antigen concentration and consequently the proportion of FluDETECT positive samples. This result is largely consistent with those of Ladman et al. [11]. However, a study by Spackman et al. [39] did not find differences in virus detection between a single swab from an inoculated bird, 5 swabs (1 from an inoculated bird, 4 from unexposed birds), or 11 swabs (1 from an inoculated bird and 10 from unexposed birds) by real time RT-PCR. This is not surprising because virus isolation and real time RT-PCR both have a high analytic sensitivity (i.e., low limit of detection) and hence the potential reduction in virus concentration due to pooling would have a relatively lesser impact.

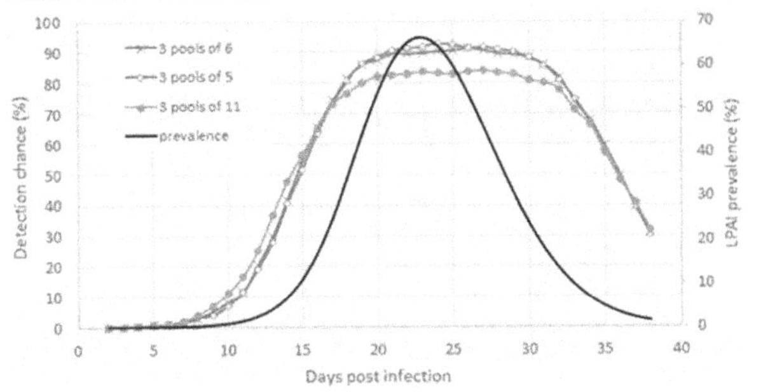

Fig. 3 LPAIV testing scenario analysis results with lateral flow immunoassay FluDETECT. The primary and secondary y-axes present outcomes on for the different pooled sample compositions and LPAI infection prevalence at different days post infection of a broiler flock. The assessed scenarios include testing 3 pools each of 5, 6 and 11 pooled samples on different days post infection. The test sensitivities used are estimated in this study for when only one swab in the pooled swab sample is positive

On test sensitivities for HPAIV, the ability of LFI kits to detect H5N1 virus subtype has been previously assessed. For FluDETECT in particular, examples of the reported detection levels include: 55.1% infections in cloacal and combined tracheal and cloacal swabs and 76.9% in tissues from infected chickens [13]; 100% and 33.3% for doses of 10^6 and 10^5 EID_{50}/ml in a serial dilution study [14] and, 71% (95% CI: 58–82%) in chicken samples from an outbreak [12]. These results, compared to those from this study, generally indicate that LFI kits perform better in detecting HPAIV infections. Due to data limitations, in one of our assessed scenarios, we assumed that LFI test sensitivity would vary with pool size by similar magnitudes for HPAIVs as those obtained

Table 5 Predicted HPAIV detection percentage at different days post infection of a broiler flock by testing three (3) dead-bird pooled swab samples of sizes 5, 6 or 11 using lateral flow immunoassay FluDETECT with varying test sensitivities (se) together with 0.3% daily mortality trigger*

Days post flock infection	1	2	3	4	5	6	7
Flock infection prevalence (%) [#]	0.0	0.1	0.4	2.7	13.0	32.5	48.8
1[a]							
3 pools of 5: se = 71%	6	27	64	88	96	99	100
3 pools of 6: se = 71%	5	29	65	88	96	99	100
3 pools of 11: se = 71%	6	31	69	89	97	99	100
2[b]							
3 pools of 6: se = 68.1%	5	27	65	88	97	99	100
3 pools of 11: se = 57.9%	5	26	63	85	95	99	100

*All results i.e. infection prevalence and detected proportions are presented as percentages
#The mean simulated flock size was estimated as 24,111 birds
[a]Scenario 1: assuming that LFI test sensitivity for HPAIV is the same across pool sizes
[b]Scenario 2: assuming that LFI test sensitivity for HPAIV is different across pool sizes

for LPAIVs in this study. This may not necessarily be true and we recommend that studies to quantify this effect for HPAIV be performed.

For our study, since swab pooling is often practiced in the field partly as a means to cut surveillance costs [10], we compared IAV field surveillance outcomes for protocols involving testing differently pooled swab samples. We found an infection prevalence dependent swab-pooling effect on FluDETECT test performance. Specifically, for LPAIV, testing pools of 5 and 6 performed better than those of 11 at a high infection prevalence and pools of 11 performed better at low infection prevalence. For HPAIV, the evaluated protocols generally detected similar proportions or were within a 1% of each other with a few notable differences on specific days (between 2 and 4) post infection. The main observation was that, at a high and similar test sensitivity of 71%, testing pools of 11 dominates (with up to 5% more detections) and it lags by up to 4% when its sensitivity was lower compared to the pools of 5 and 6. Throughout, outcomes from testing pools of 5 and 6 were almost similar.

We hypothesize that, during disease surveillance, the implication of a negative LFI result differs between lowly and highly pathogenic avian influenza viruses due to their differences in within flock transmission dynamics and pathogenicity-related factors. Note that, unlike LPAIV, HPAIV, which is associated with mortality, targeted-sampling of only dead birds as well as the use of a mortality trigger were applied and both increased the chances of detecting IAV in a sample.

It is likely that, for LPAIV at low prevalence, pools of 11 have a greater chance of containing a positive swab and this advantage (over pools of 5 and 6) diminishes at high prevalence and the influence of sample dilution takes over [10]. Note however that there may be additional benefits of pooling more samples that we have not analyzed in this study. For example, there is generally a higher chance of including more than one positive swab in pools of 11 compared to pools of 5 and 6 and this may lead to a higher flock level test sensitivity for the 11 pool samples. In the context of LPAIV, further analysis with more swab pooling schemes, e.g., those involving more than one positive swab in the pooled samples, is needed.

Conclusion

Despite their relatively low sensitivity, the logistical advantages of LFI support their use as part of influenza A virus outbreak management tools preferably during preliminary outbreak investigation or routine surveillance in an AI free country. Moreover, flock surveillance protocols would likely be based on multiple pooled samples. Hence, although the diagnostic sensitivity for a single pooled sample is not great, the flock level likelihood of detection can be adequate when several pooled samples are tested.

Since they are best suited for high virus titer samples, targeted sampling based on the clinical status of infected birds may improve their sensitivity. Consequently, based on factors such as high virus titer shed, high morbidity and fatality rate and fast within flock transmission, HPAIV infections may provide a better platform for LFI field deployment to determine the disease status of a flock. Note that the current diagnosis of HPAIV in control areas may only involve LFI as a supplement to PCR testing, although this may be specific to some countries depending on the regulations. Thus, our assessment generates insight into how these tests would perform when used (particularly) outside of a control area during high risk periods of HPAIV spread. We recommend that experimental and field studies evaluating additional pooling schemes involving more than one positive swabs per pooled sample be performed.

Abbreviations

IAV: Influenza type A Virus; LPAI: Low Pathogenicity Avian Influenza; LPAIV: Low Pathogenicity Avian Influenza Virus; HPAI: Highly Pathogenic Avian Influenza; HPAIV: Highly Pathogenic Avian Influenza Virus; LFI: Lateral Flow Immunoassay; BHI: Brain Heart Infusion; PCR: Polymerase Chain Reaction; RT-PCR: Reverse Transcription Polymerase Chain Reaction

Acknowledgements

We thank Erica Spackman for providing us with the viruses used in the experiments, Zoetis for their partial financial support and Dan Domingo for his advice and guidance. Ssematimba is grateful to Gulu University for granting him the opportunity to work with the University of Minnesota's Secure Food Systems team. Cardona is the B. S. Pomeroy Chair of Avian Health at the University of Minnesota, College of Veterinary Medicine.

Funding

Zoetis provided part of the funding for this study under award number CON000000053388 to Cardona. They played no role in the design of the study and collection, analysis, and interpretation of data and in writing the manuscript.

Authors' contributions

CJC conceived the ideas of the study; AS, SM, PJB, DAV, CJC conceived the ideas for the analysis; AS, SM, PJB performed the analyses; CJC, CF, JM performed the experiments; AS wrote the manuscript; SM, CJC were major contributors in writing the manuscript and all other authors commented on the manuscript. All authors read and approved the final manuscript.

Competing interests

The authors declare that they have no competing interests.

Author details

[1]Secure Food Systems Team, University of Minnesota, Veterinary and Biomedical Sciences, 301C Veterinary Science Building, 1971 Commonwealth Avenue, Saint Paul, MN 55108, USA. [2]Mid Central Research and Outreach Center, 1802 18th St NE, Willmar, MN 56201, USA. [3]Department of Mathematics, Faculty of Science, Gulu University, P.O. Box 166, Gulu, Uganda.

References

1. Webster RG, Bean WJ, Gorman OT, Chambers TM, Kawaoka Y. Evolution and ecology of influenza A viruses. Microbiol Mol Biol Rev. 1992;56(1):152–79.

2. Röhm C, Horimoto T, Kawaoka Y, Süss J, Webster RG. Do hemagglutinin genes of highly pathogenic avian influenza viruses constitute unique phylogenetic lineages? Virology. 1995;209(2):664–70.
3. Efsa Panel on Animal Health Welfare, More S, Bicout D, Bøtner A, Butterworth A, Calistri P, Depner K, Edwards S, Garin-Bastuji B, Good M, et al. Avian influenza. EFSA J. 2017;15(10):e04991.
4. Alexander D. Avian influenza–diagnosis. Zoonoses Public Health. 2008;55(1):16–23.
5. Van der Goot JA, De Jong MCM, Koch G, Van Boven M. Comparison of the transmission characteristics of low and high pathogenicity avian influenza A virus (H5N2). Epidemiol Infect. 2003;131(2):1003–13.
6. Halvorson DA. Control of Low Pathogenicity Avian Influenza. In: Swayne D. Ames, IA, editors. In: Avian Influenza. 1st ed. Blackwell Publishing Ltd.; 2008. p. 513-536.
7. OIE: Avian Influenza (infections with Avian Influenza viruses). 2015 available online https://www.oie.int/fileadmin/Home/eng/Health_standards/tahm/2.03.04_AI.pdf. Accessed 25 Oct 2016.
8. Woolcock PR, Cardona CJ. Commercial immunoassay kits for the detection of influenza virus type A: evaluation of their use with poultry. Avian Dis. 2005;49(4):477–81.
9. Cattoli G, Terregino C. New perspectives in avian influenza diagnosis. Zoonoses Public Health. 2008;55(1):24–8.
10. Arnold ME, Slomka MJ, Coward VJ, Mahmood S, Raleigh PJ, Brown IH. Evaluation of the pooling of swabs for real-time PCR detection of low titre shedding of low pathogenicity avian influenza in turkeys. Epidemiol Infect. 2013;141(6):1286–97.
11. Ladman BS, Spackman E, Gelb J. Comparison of pooling 11 or 5 oropharyngeal swabbings for detecting avian influenza virus by real-time reverse transcription-PCR in broiler chickens. Avian Dis. 2012;56(1):227–9.
12. Loth L, Prijono WB, Wibawa H, Usman TB. Evaluation of two avian influenza type A rapid antigen tests under Indonesian field conditions. J Vet Diagn Investig. 2008;20(5):642–4.
13. Chua TH, Ellis TM, Wong CW, Guan Y, Ge SX, Peng G, Lamichhane C, Maliadis C, Tan SW, Selleck P, et al. Performance evaluation of five detection tests for avian influenza antigen with various avian samples. Avian Dis. 2007; 51(1):96–105.
14. Marche S, van den Berg T. Evaluation of rapid antigen detection kits for the diagnosis of highly pathogenic avian influenza H5N1 infection. Avian Dis. 2010;54(1 Suppl):650–4.
15. Spackman E, Senne DA, Myers T, Bulaga LL, Garber LP, Perdue ML, Lohman K, Daum LT, Suarez DL. Development of a real-time reverse transcriptase PCR assay for type a influenza virus and the avian H5 and H7 hemagglutinin subtypes. J Clin Microbiol. 2002;40(9):3256–60.
16. Slomka MJ, Pavlidis T, Coward VJ, Voermans J, Koch G, Hanna A, Banks J, Brown IH. Validated RealTime reverse transcriptase PCR methods for the diagnosis and pathotyping of Eurasian H7 avian influenza viruses. Influenza Other Respir Viruses. 2009;3(4):151–64.
17. Reed LJ, Muench H. A simple method of estimating fifty per cent endpoints. Am J Epidemiol. 1938;27(3):493–7.
18. R Core Team. R: A language and environment for statistical computing. Vienna: R Foundation for Statistical Computing; 2015. URL https://www.r-project.org/
19. Takekawa JY, Hill NJ, Schultz AK, Iverson SA, Cardona CJ, Boyce WM, Dudley JP. Rapid diagnosis of avian influenza virus in wild birds: use of a portable rRT-PCR and freeze-dried reagents in the field. J Vis Exp. 2011;54:2829.
20. Fisher RA. On the interpretation of $\chi 2$ from contingency tables, and the calculation of P. J R Stat Soc. 1922;85(1):87–94.
21. Weaver TJ, Malladi S, Bonney PJ, Patyk KA, Bergeron JG, Middleton JL, Alexander CY, Goldsmith TJ, Halvorson DA. A simulation-based evaluation of premovement active surveillance protocol options for the managed movement of turkeys to slaughter during an outbreak of highly pathogenic avian influenza in the United States. Avian Dis. 2016;60:132–45.
22. Mutinelli F, Capua I, Terregino C, Cattoli G. Clinical, gross, and microscopic findings in different avian species naturally infected during the H7N1 low- and high-pathogenicity avian influenza epidemics in Italy during 1999 and 2000. Avian Dis. 2003;47(s3):844–8.
23. De Benedictis P, Manuel Joannis T, Hannatu Lombin L, Shittu I, Serena Beato M, Rebonato V, Cattoli G, Capua I. Field and laboratory findings of the first incursion of the Asian H5N1 highly pathogenic avian influenza virus in Africa. Avian Pathology. 2007;36(2):115–7.
24. Bos M, Nielen M, Koch G, Bouma A, De Jong M, Stegeman A. Back-calculation method shows that within-flock transmission of highly pathogenic avian influenza (H7N7) virus in the Netherlands is not influenced by housing risk factors. Preventative Vet Med. 2009;88:278–85.
25. van der Goot JA, Koch G, de Jong MCM, van Boven M. Quantification of the effect of vaccination on transmission of avian influenza (H7N7) in chickens. Proc Natl Acad Sci U S A. 2005;102(50):18141–6.
26. Poetri O, Bouma A, Claassen I, Koch G, Soejoedono R, Stegeman A, van Boven M. A single vaccination of commercial broilers does not reduce transmission of H5N1 highly pathogenic avian influenza. Vet Res. 2011;42(1):74.
27. Verdugo C, Cardona CJ, Carpenter TE. Simulation of an early warning system using sentinel birds to detect a change of a low pathogenic avian influenza virus (LPAIV) to high pathogenic avian influenza virus (HPAIV). Prev Vet Med. 2009;88(2):109–19.
28. Gonzales JL, Elbers AR, van der Goot JA, Bontje D, Koch G, de Wit JJ, Stegeman JA. Using egg production data to quantify within-flock transmission of low pathogenic avian influenza virus in commercial layer chickens. Prev Vet Med. 2012;107(3–4):253–9.
29. Spackman E, Weaver JT, Malladi S: Detection of H5 and H7 highly pathogenic avian influenza virus with lateral flow devices: Performance with healthy, sick and dead chickens. Oral presentation presented at: American Association of Veterinary Laboratory Diagnosticians, 57th annual meeting; October 16–22, 2014; Kansas City, Missouri. 2014.
30. Das A, Spackman E, Thomas C, Swayne DE, Suarez DL. Detection of H5N1 high-pathogenicity avian influenza virus in meat and tracheal samples from experimentally infected chickens. Avian Dis. 2008;52(1):40–8.
31. Bouma A, Claassen I, Natih K, Klinkenberg D, Donnelly CA, Koch G, Van Boven M. Estimation of transmission parameters of H5N1 avian influenza virus in chickens. PLoS Pathog. 2009;5:e1000281.
32. Gonzales JL, van der Goot JA, Stegeman JA, Elbers ARW, Koch G. Transmission between chickens of an H7N1 low pathogenic avian influenza virus isolated during the epidemic of 1999 in Italy. Vet Microbiol. 2011; 152(1–2):187–90.
33. Gonzales JL, Elbers ARW, Bouma A, Koch G, de Wit JJ, Stegeman JA. Transmission characteristics of low pathogenic avian influenza virus of H7N7 and H5N7 subtypes in layer chickens. Vet Microbiol. 2012;155(2–4):207–13.
34. Claes G, Welby S, Van Den Berg T, Van Der Stede Y, Dewulf J, Lambrecht B, MarchÉ S. The impact of viral tropism and housing conditions on the transmission of three H5/H7 low pathogenic avian influenza viruses in chickens. Epidemiol Infect. 2013;141(11):2428–43.
35. Kalthoff D, Bogs J, Grund C, Tauscher K, Teifke JP, Starick E, Harder T, Beer M. Avian influenza H7N9/13 and H7N7/13: a comparative virulence study in chickens, pigeons, and ferrets. J Virol. 2014; https://doi.org/10.1128/JVI.01241-14.
36. Swayne DE, Beck JR. Experimental study to determine if low-pathogenicity and high-pathogenicity avian influenza viruses can be present in chicken breast and thigh meat following intranasal virus inoculation. Avian Dis. 2005;49(1):81–5.
37. Shao H, Zhou X, Fan Z, Wan Z, Qian K, Perez D, Qin A, Ye J. Impact of a potential glycosylation site at neuraminidase amino acid 264 of influenza A/H9N2 virus. Vet Microbiol. 2016;196:9–13.
38. Swayne DE, Slemons RD. Using mean infectious dose of high- and low-pathogenicity avian influenza viruses originating from wild duck and poultry as one measure of infectivity and adaptation to poultry. Avian Dis. 2008;52(3):455–60.
39. Spackman E, Pedersen JC, McKinley ET, Gelb J. Optimal specimen collection and transport methods for the detection of avian influenza virus and Newcastle disease virus. BMC Vet Res. 2013;9(1):35.

16

The effect of dietary bacterial organic selenium on growth performance, antioxidant capacity, and Selenoproteins gene expression in broiler chickens

A. M. Dalia[1,2], T. C. Loh[1], A. Q. Sazili[1,3], M. F. Jahromi[3] and A. A. Samsudin[1,3*]

Abstract

Background: Selenium (Se) is an essential trace mineral in broilers, which has several important roles in biological processes. Organic forms of Se are more efficient than inorganic forms and can be produced biologically via Se microbial reduction. Hence, the possibility of using Se-enriched bacteria as feed supplement may provide an interesting source of organic Se, and benefit broiler antioxidant system and other biological processes. The objective of this study was to examine the impacts of inorganic Se and different bacterial organic Se sources on the performance, serum and tissues Se status, antioxidant capacity, and liver mRNA expression of selenoproteins in broilers.

Results: Results indicated that different Se sources did not significantly ($P \leq 0.05$) affect broiler growth performance. However, bacterial organic Se of T5 (basal diet +0.3 mg /kg feed ADS18 Se), T4 (basal diet +0.3 mg /kg feed ADS2 Se), and T3 (basal diet +0.3 mg /kg feed ADS1 Se) exhibited significantly ($P \leq 0.05$) highest Se concentration in serum, liver, and kidney respectively. Dietary inorganic Se and bacterial organic Se were observed to significantly affect broiler serum ALT, AST, LDH activities and serum creatinine level. ADS18 supplemented Se of (*Stenotrophomonas maltophilia*) bacterial strain showed the highest GSH-Px activity with the lowest MDA content in serum, and the highest GSH-Px and catalase activity in the kidney, while bacterial Se of ADS2 (*Klebsiella pneumoniae*) resulted in a higher level of GSH-Px1 and catalase in liver. Moreover, our study showed that in comparison with sodium selenite, only ADS18 bacterial Se showed a significantly higher mRNA level in GSH-Px1, GSH-Px4, DIO1, and TXNDR1, while both ADS18 and ADS2 showed high level of mRNA of DIO2 compared to sodium selenite.

Conclusions: The supplementation of bacterial organic Se in broiler chicken, improved tissue Se deposition, antioxidant status, and selenoproteins gene expression, and can be considered as an effective alternative source of Se in broiler chickens.

Keywords: Broiler, Selenium, Organic, Bacteria, Antioxidant, Selenoprotein

Background

In living organisms, natural antioxidant system protects cells from the action of harmful free radicals [1]. The antioxidant system components include: natural fat-soluble antioxidants such as vitamin E, water-soluble antioxidants like ascorbic acid, and antioxidant enzymes: glutathione peroxidase (GSH-Px), catalase (CAT) and superoxide dismutase (SOD) [2]. Moreover, Selenium (Se) micronutrient is considered a functional part of the antioxidant system, which acts via selenoproteins. At least 25 selenoproteins have been identified in chicken, which contain a selenocystein amino acid as a unique structural part [3]. Consequently, selenocysteine has a specific role in different selenoenzymes as active site for their catalytic activity. Selenoenzymes which have been identified in animals and humans include: glutathione peroxidases, thioredoxin reductases, and iodothyronine deiodinases [4]. Glutathione

* Correspondence: anjas@upm.edu.my
[1]Department of Animal Science, Faculty of Agriculture, Universiti Putra Malaysia, 43400 Serdang, Selangor, Malaysia
[3]Institute of Tropical Agriculture, Universiti Putra Malaysia, 43400 Serdang, Selangor, Malaysia

peroxidase is a Se-dependent enzyme involved in the antioxidant system; it is the main enzyme which helps to control free radical formation via reduction of hydrogen peroxide and lipid peroxide to water and the corresponding alcohol [5]. Addition to that, selenoprotein W plays an antioxidant function in chicken myoblasts [6]. Four endoplasmic reticulum resident selenoprotein genes (Sepn1, Selk, Sels, and Selt) were related to oxidative damages in broiler chicken muscles [7].

Selenoproteins synthesis is affected by the nutritional level of the Se. Many studies have proven that dietary Se supplementation regulates the expression of selenoproteins in most animal tissues. It has been shown that Se deficient diet decreases the expression of 25 selenoproteins in the chicken muscular stomach [3]. In broiler kidney, the mRNA levels of 14 selenoprotein genes (Dio1, Dio2, GSH-Px3, Sepp1, SelH, SelI, SelK, Sepn1, SelO, SELW1, Sep15, SelT, SelU, and SelS) were down-regulated, and 9 selenoprotein genes (GSH-Px1, GSH-Px2, GSH-Px4,SelPb, Txnrd1, Txnrd2, Txnrd3, SPS2, and SelM) were up-regulated due to low Se diet, while Dio3 and Sepx1 mRNA levels were not affected [8]. In addition, Se supplemented diet resulted in a significant elevation in mRNA level of SELW1 in broiler liver [9]. However, broiler liver gene expression of GSH-Px4 was down-regulated as a consequence of Se-enriched diet [10]. Usually, Se is supplemented to the broiler diets in the form of inorganic Se (sodium selenite), or organic Se (natural Se sources). According to Surai [1], organic Se is more bioavailable in the animal tissues than inorganic forms, and has significant biochemical and physiological benefits. Previous studies have demonstrated that different sources of Se may result in different metabolic effects in the animal tissues [11, 12]. According to Yuan et al. [11], supplementation of organic Se in the form of Se-yeast and selenomethionine to broiler chicken showed significant up-regulation in the liver GSH-Px1 and TXNDR1 mRNA levels compared to sodium selenite. Therefore, the present study sought to determine the impact of various bacterial sources of Se as an alternative organic Se compared to the inorganic Se on growth performance, antioxidant capacity, biochemical status, and mRNA expression of some selected selenoproteins in broiler chicken.

Methods

Chemicals

The antioxidant assay kits were purchased from bioassay system (USA). The inorganic sodium selenite Na_2SeO_3, ≥99%, was sourced from Sigma-Aldrich Chemical Co., St. Louis, MO, USA. Reagents used for real time PCR were purchased from Qiagen Biotechnology Malaysia Sdn. Bhd and Thermo Scientific Fermentas. All reagents for the Se assay were of analytical grade.

Bacterial strains

The bacterial strains used in this study were isolated as Se enriched bacteria from rumen fluid (ADS1 and ADS2) and hot spring water (ADS18) (Selangor, Malaysia), and identified as *Enterobacter cloacae* (ADS1), *Klebsiella pneumoniae* (ADS2), and *Stenotrophomonas maltophilia* (ADS18). All of them had high ability in accumulating organic Se in their cells when grown in Se-enriched medium according to our previous finding.

Preparation of bacterial organic se

The 30% glycerol stock culture of ADS1, ADS2, and ADS18 strains was used to prepare aliquots fresh culture (24 h) after three times reviving at the Laboratory of Microbiology, Department of Animal Science at the Faculty of Agriculture, Universiti Putra Malaysia (UPM). The commercially available media, nutrient broth supplemented with 10 µg/mL sodium selenite were used for all strains inoculation. Then, ADS1 and ADS2 strains were incubated at 39 °C and ADS18 at room temperature for 24 h. This was followed by single colony sub-culturing using spread plate technique and incubation for 24 h at the same stated temperature. A single colony was then inoculated into 10 mL inorganic Se-enriched nutrient broth and subjected to incubation for 24 h. It was then sub-cultured two times, after which it was ready for use as an inoculum. An inoculum containing 1×10^6 of isolated bacterial cells was inoculated into the same media and followed by incubation for 24 h at static temperature. The next step was centrifuging the culture at 6000 rpm for 15 min to harvest the bacterial pellets enriched with Se which were then washed two times using deionized water to remove any inorganic Se in the bacterial cells [13]. Selenium-enriched bacterial cells were collected and lyophilized at −20 °C. Furthermore, the collected bacterial biomass was subjected to ultra-sonication to disrupt the bacterial cell walls and release their organic Se-content. Sonication was performed using ice water bath for 90 cycles, with 5 s on and 5 s off. Then the sonicated Se- enriched biomass was lyophilized and kept at −20 °C prior to use it as Se source in the feeding trial.

Birds and experimental procedure

A total of 180 one-day-old female (Cobb 500) broiler chicks averaging 40 ± 0.13 g in body weight were sourced from a commercial hatchery and randomly allocated to five treatments, each of which was replicated six times with 6 birds per replicate. The treatment groups included T1 = basal diet (negative control), T2 = basal diet +0.3 mg/Kg feed inorganic Se Na_2SeO_3 (positive control), T3 = basal diet +0.3 mg /kg feed ADS1 Se, T4 = basal diet +0.3 mg /kg feed ADS2 Se, T5 = basal diet +0.3 mg /kg feed ADS18 Se. Starter and finisher basal diets (Table 1) were prepared in line with the nutritional requirements of broilers and according to NRC (National Research Council) (1994) standards, except for Se which were supplemented as 0.3 mg/kg feed according to Surai, [1]. Starter diet was offered from 0 to 3 weeks old and finisher from 4 to 6 weeks old. Water and feed were given ad libitum to all the chickens until 42 days of age. The study was conducted in compliance with the research policy guidelines of UPM on Animal Welfare and Ethics.

Growth performance

The body weight (BW) and pen feed intake (FI) of individual birds were recorded weekly, and weight gain (WG) and feed conversion ratio (FCR) were calculated.

Table 1 Ingredients and nutrient content of the basal diet

Ingredients	Starter	Finisher
	%	%
Corn	52.5	56.250
Palm oil	5.00	6.00
Soybean meal (44% cp)	32.50	30.00
Fish meal (58% cp)	5.15	3.25
L-Lysine	0.25	0.25
DL-Methionine	0.25	0.25
Dicalcium phosphate 18% [a]	1.60	1.85
Calcium carbonate	0.60	0.35
Salt	0.30	0.30
Mineral Premix[b]	0.15	0.15
Vitamin Premix[c]	0.10	0.10
Toxin Binder[d]	0.15	0.15
Choline Chloride	0.10	0.10
Wheat pollard (QL)	0.135	1.00
Calculated nutrient content (g/kg DM) [e]		
ME (MJ/Kg)	12.9	13.20
Crude protein	22.04	20.09
Crude fat	7.57	8.004
Calcium	1.189	1.0440
Phosphorus	0.786	0.768
Avail. P for Poultry	0.472	0.450
Analyzed Se (mg/kg)[f]	<0.09	<0.09

[a] di calcium phosphate provides phosphorus and calcium in a ratio of 1:1
[b]Mineral premix provided the following per kg diet: iron 120 mg, manganese 150 mg, copper 15 mg, zinc 120 mg, iodine 1.5 mg, and cobalt 0.4 mg
[c]Vitamin premix provided the following per kg diet: Vitamin A (retinyl acetate) 10.32 mg, cholecalciferol 0.250 mg, vitamin E (DL-tocopheryl acetate) 90 mg, vitamin K 6 mg, cobalamin 0.07 mg, thiamine 7 mg, riboflavin 22 mg, folic acid 3 mg, biotin 0.04 mg, pantothenic acid 35 mg, niacin 120 mg and pyridoxine 12 mg
[d]Toxin binder contains natural hydrated sodium calcium aluminium silicates
[f]The Se content measured using ICP.MS
[e]The diets were formulated using feedlive International software (Thailand)

FCR was calculated as follows: FCR = total feed consumed by birds/total weight gain.

Slaughtering, blood and tissues sampling

At day 42 of feeding trial, 12 birds per treatment were taken as representative samples and were slaughtered to collect blood and tissue samples. Blood samples were collected directly into plain serum bottles, kept at room temperature for 10 min, and then centrifuged at 4000 rpm for 10 min; resultant supernatant was collected and stored at −80 °C until further analysis. Liver and kidney tissue samples were obtained and frozen directly in liquid nitrogen and stored at −80 °C to await analysis. Around 100 mg of liver tissues was collected immediately in RNA-later Stabilization Reagent (Qiagen, Germany) and processed for storage at −80 °C, following the manufacturer's instructions.

Determination of biochemical parameters

Biochemical parameters, such as serum total protein (TP), albumin (AL), aspartate aminotransferase (AST), alanine aminotransferase (ALT), lactate dehydrogenase (LDH) blood urea nitrogen (BUN), and serum creatinine (Cr), were established with the auto-blood biochemical analyzer (Automatic Analyzer 902, Hitachi, Germany) using the appropriate kit. Serum globulin (G) and albumin/ globulin ratio (A/G) were calculated as follows: G = total protein - albumin, A/G = albumin/globulin. All samples were tested in duplicate.

Assay of se content in tissues

For Se analysis in serum and tissue samples, 0.5 mL of serum and 0.5 g tissue samples were digested by mixing with 5 mL HNO_3 and 1 mL H_2O_2 in a digestion tube using microwave digestion system. To the remaining 1 mL of the digests, deionized water was added to produce a 10 mL solution. The Se concentration was measured immediately in the solution using the inductively coupled plasma mass spectrometer (Agilent Technologies, Santa Clara) following the method reported by Wahlen et al. [14].

Determination of antioxidant enzymes activity

Total antioxidant (T-AOC), glutathione peroxidase activity (GSH-Px), superoxide dismutase activity (SOD), catalase activity, and the concentration of malondialdehyde (MDA) were measured in the serum and liver, and kidney tissues. Liver and kidney tissues were homogenized on ice using TBS buffer and centrifuging at 3000×g for 10 min at 4 °C, and the resultant supernatant was collected for enzymes measurement. Antioxidant enzymes analysis was performed using BioAssay Systems Commercial kit.

Determination of selenoprotein mRNA expression

For selenoproteins mRNA determination, 30 mg of RNAlater preserved liver samples were used for total RNA isolation with the high pure RNA Tissue Kit (RNeasy mini kit, Qiagen, USA). The samples were homogenized appropriately in the lysis buffer (Qiagene) following the manufacturer's instructions. Qualitative and quantitative assessments of the isolated RNA were carried out on ND-1000 NanoDrop (NanoDrop Technologies, USA) spectrophotometer. Only samples with more than 100 ng RNA and absorbance ratios of A260/280 and A260/230 of around (> 1.8) were selected for further manipulation. The RNA was reverse transcribed into cDNA with the transcription first strand cDNA Synthesis Kit (one-step RT-PCR kit, Qiagen, USA) following the procedure recommended by the manufacturer.

Primers for gene expression were designed (First Base, Malaysia) based on published *Gallus gallus* sequences (Table 2). The relative mRNA abundances of 6 genes were assayed (glutathione peroxidase 1, GSH-Px1; glutathione peroxidase 4, GSH-Px4; deiodinasel, DIO1; deiodinase 2, DIO2; selenoprotein W1, SELW1; thioredoxin reductase 1, TXNRD1). The reaction was done in a Bio-Rad thermal cycler (MyCycler, Germany). The RT-PCR conditions included: (1) reverse transcription, 30 min, 50 °C; (2) initial PCR activation step, 15 min, 95 °C; (3) 3-step cycling for 40 cycles, each cycle consisting of denaturation for 30 s at 94 °C followed by annealing for 30 s at 52–57 °C and extension for 1 min at 72 °C. The linearity of response was ensured and the saturation of the reaction was prevented through optimization of the template concentration and the cycle number. To standardize the expression data, the GADPH mRNA fragment was employed as internal standard (housekeeping gene). The results were standardized to the levels achieved for the β-actin gene. It was carried out by taking the ratio of the obtained value for the gene of interest to that of GADPH and then related to the control. 2-ΔΔCt (ΔΔCt = ΔCt Test sample-ΔCt Calibrator sample) to calculate the relative mRNA expression.

Table 2 Genes and primers used for relative quantification by real time PCR (qPCR) in the liver of chicken

Gene[a]	Primer sequence (5′ – 3)[b]	Fragment bp
GAPDH	Forward: 5′-AATGAGAGGTTCAGGTGCCC-3′	150
	Reverse: 5′-ACCAGACAGCACTGTGTTGG-3′	
GSH-Px1	Forward: 5′-GCGACTTCCTGCAGCTCAACGA-3′	99
	Reverse: 5′-CGTTCTCCTGGTGCCCGAAT-3′	
GSH-Px4	Forward: 5′-CGGTGAATTACACTCAGCTCGT-3′	123
	Reverse: 5′-CTTTGATCTGCGCGTCGTCC-3′	
DIO1	Forward: 5′-AAGCTGCACCTGACCTTCATT-3′	138
	Reverse: 5′-TTGTTTCTGAAGGCCCATCCA-3′	
DIO2	Forward: 5′-CAGTGTAATCCACATAGCCA-3′	137
	Reverse: 5′-CTGAGCCAAAATTAACCACC-3′	
SELW1	Forward: 5′-CTCCGCGTCACCGTGCTCT-3′	155
	Reverse: 5′-CTGCCCACCGTCACCTCGAAC-3′	
TXNDR1	Forward: 5′-ACTGGATGACTATGACCGAA-3′	103
	Reverse: 5′-TATGCATTCTCATACGTGAC-3′	

[a] Abbreviation: GAPDH, glyceraldehyde-3-phosphate dehydrogenase; GSH-Px1,Glutathione peroxidase1; GSH-Px4, Glutathione peroxidase4; DIO1, iodothyronine deiodinase1; DIO2, iodothyronine deiodinase2; SELW1, selenoproteins w; TXNDR1, thioredoxin reductase

[b]primers used for qPCR designed based on pupplished sequences [54]

Statistical analysis

An ANOVA was conducted using six replicates per means. Differences between treatments were scrutinized with one-way ANOVA (SAS Institute, 1996). Duncan test was used to determine the significant differences among the treatment groups at a significant level ($P < 0.05$). The data of serum and tissues Se concentration and antioxidant enzymes were analyzed employing GLM procedure applicable for Completely Randomized Design (SAS, 1996). Treatment differences were established by orthogonal contrasts

(1) Basal diet vs. Se supplemented diets,
(2) Sodium selenite vs. bacterial organic Se,

Values of $P < 0.05$ were considered significant.

Results

Growth performance of broiler chicken

The growth performance of birds fed diverse sources of Se is presented in Table 3. No significant differences ($P > 0.05$) among the dietary treatments were observed during the experimental period. The results demonstrated that supplementation of inorganic and bacterial organic Se did not affect body weight, weight gain, feed intake, and FCR ratio.

Serum and tissues se concentration

Table 4 shows the Se content in the serum, liver, and kidney of broiler chicken supplemented with inorganic Se and different sources of bacterial organic Se for 42 days. Selenium supplemented diets versus basal diet showed significant increase and deposition of Se in serum and tissue samples compared to negative control (T1). Moreover, bacterial organic Se in broiler feed resulted in a significant ($P < 0.05$) Se deposition in the liver and kidney tissues compared to inorganic Se (T2). Serum Se concentrations were significantly ($P < 0.0001$) higher in T2, T4, and T5 than T1, but, T5 showed the highest level among the dietary treatments. In liver and kidney tissues, the highest Se levels were observed in T4 and T3 bacterial organic Se for liver and kidney respectively. Additionally, Se de osition in the liver showed no difference between birds receiving dietary supplementation

Table 3 Growth performance (means ± SE) at week 6 of treatments supplemented with different sources of bacterial organic Se

Parameters[b]	Dietary Treatments[a]					SEM	P
	T1	T2	T3	T4	T5		
BW (g)	2007.9	2082.1	2054.8	2075.4	2093.9	21.65	NS
DWG (g)	1965.3	2039.5	2012.2	2064.2	2081.3	21.6	NS
FI (g)	3169.1	2983.6	2969.2	3011.1	3248.0	69.3	NS
FCR	1.61	1.46	1.47	1.46	1.56	0.04	NS

NS No significant differences
[a]T1; basal diet, T2; basal diet +0.3 mg/ kg feed sodium selenite, T3; basal diet +0.3 mg/ kg feed ADS1 Se, T4; basal diet +0.3 mg/ kg feed ADS2 Se, T5; basal diet +0.3 mg/ kg feed ADS18 Se
[b]BW; body weight, DWG; daily weight gain, FI; feed intake, FCR; feed conversion ratio

of inorganic Se and bacterial organic Se of T3 and T5 compared to negative control, but in kidney all Se supplemented diets showed significant ($P < 0.0001$) difference compared to birds supplemented with basal diet.

Serum biochemical profile in broiler chicken fed different sources of se

The serum biochemical parameters of broiler fed different sources of Se are shown in Table 5. Selenium supplementation as inorganic or bacterial organic forms did not affect serum total protein, albumin, globulin, albumin/globulin ratio, and urea. However, the differences were significant ($P < 0.05$) among the treatments in serum AST, ALT, LDH and Creatinine levels. The activities of AST, ALT, LDH enzymes and creatinine level in serum were decreased in birds fed Se compared to basal diet, ALT level was decreased significantly ($P < 0.05$) in T4 and T5 compared to T1 and T2, while T3 showed insignificant effect among all treatments. AST level was the highest in T1 with no significant difference compared to inorganic Se (T2) and bacterial organic Se (T4 and T5), but the level was significantly higher than T3. In addition, T3 showed the lowest ($P < 0.05$) level of LDH compared to all treatments. Moreover, dietary Se as inorganic and bacterial organic form reduce the serum creatinine level compared to negative control group T1, and the difference was significant ($P < 0.05$) compared to T2, T3, and T4 while no significant differences were observed compared to T5.

Antioxidant status of serum, liver, and kidney of broiler chicken

As shown in Table 6, the effect of bacterial organic Se supplementation on the antioxidant variables of serum, liver and kidney of broiler chicken were varied according to Se source and type of tissue. Se supplementation in contrast to basal diet induced a notable elevation ($P < 0.05$) in serum, liver, and kidney GSH-Px and catalas activity, with substantial reduction in MDA concentration, while the supplementation of bacterial organic Se in contrast to inorganic source had no obvious effect on antioxidant parameters, except in the kidney tissue where there was a significant ($P < 0.05$) increase in GSH-Px, SOD, and catalase activity with significant decrease in MDA level. Also serum SOD showed significant difference compared to inorganic Se.

In this study, the T-AOC activity showed insignificant differences among the treatment groups in serum and liver, but the difference was significant in the kidney tissue with the highest activity in T3 and T2 compared to other groups. The GSH-Px activity was highest in all Se supplemented groups compared to the negative control in serum and examined tissues. In serum and kidney, the highest activity was observed in T5 group (13.61 and 72.62 U/L) respectively, while T4 showed the highest activity in liver (154.60 U/L). Regarding SOD activity, the level was not affected by bacterial organic Se supplementation except in kidney tissue where both T5 and T4 showed a significantly ($P < 0.05$) higher activity compared to the other groups. Moreover, bacterial organic Se showed significant elevation in catalase activity in liver and kidney, while the highest activity in liver appeared in T4 with significant difference compared to T1 and T2, while T5 and T3 showed the highest activity in kidney with significant differences compared to T1 and T2.

The serum and tissues content of MDA was significantly ($P < 0.05$) decreased by Se supplementation. Bacterial organic Se of T5 showed the significantly lowest level in serum, as well as a significantly lower level in T3, T4, and T5 compared to T1 and T2 in kidney. However, in the liver, MDA content in T2 and T4 was significantly lower than T1 group.

Table 4 Selenium concentration in serum and tissues of broiler chicken fed different Se sources

Parameters	Dietary treatments[a]					SEM	P value		
	T1	T2	T3	T4	T5		Anova	B	O
Serum µg/l	38.79[c]	50.94[b]	42.07[c]	58.61[ab]	61.21[a]	2.14	<.0001	0.0002	0.3705
Liver µg/kg	124.9[b]	146.3[ab]	153.5[ab]	170.4[a]	133.5[b]	5.03	0.0246	0.0226	0.0476
Kidney µg/kg	114.9[c]	136.5[b]	178.5[a]	148.4[b]	143.2[b]	4.98	<.0001	<.0001	0.0155

B Basal diet VS Se supplemented diets, *O* Organic Se VS inorganic Se, $p < 0.05$ = significant differences
[a–c] Means with different letter within a row differed significantly
[a]T1; basal diet, T2; basal diet +0.3 mg/ kg feed sodium selenite, T3; basal diet +0.3 mg/ kg feed ADS1 Se, T4; basal diet +0.3 mg/ kg feed ADS2 Se, T5; basal diet +0.3 mg/ kg feed ADS18 Se

Table 5 Effects of dietary supplementation of inorganic and bacterial organic Se on serum biochemical profiles in broiler

Parameters	Dietary Treatments[a]						
	T1	T2	T3	T4	T5	SEM	*P*
T. Protein (g/L)	29.0 [a]	27.8 [a]	21.6 [a]	21.3 [a]	28.3 [a]	1.29	NS
Albumin (g/L)	19.5 [a]	17.6 [a]	16.7 [a]	17.0 [a]	21.4 [a]	0.66	NS
Globulin (g/l)	9.5 [a]	10.2 [a]	4.9 [a]	4.3 [a]	6.9 [a]	0.87	NS
Albumin/Globulin ratio	2.5	2.1	4.6	3.8	3.70	0.98	NS
ALT (U/L)	7.02 [a]	5.28 [ab]	5.66 [ab]	4.00 [b]	5.12 [b]	0.35	*
AST (U/L)	293.2 [a]	264.3 [ab]	203.9 [b]	234.0 [ab]	267.1 [ab]	10.8	*
LDH (U/L)	1848.3 [a]	1755.8 [a]	1456.3 [b]	1749.5 [a]	1927.6 [a]	47.0	*
Creatinine (umol/L)	28.3 [a]	21.5 [bc]	17.8 [c]	21.4 [bc]	25.0 [ab]	1.1	*
Urea (umol/L)	0.64 [a]	0.53 [a]	0.53 [a]	0.58 [a]	0.67 [a]	0.02	NS

[a]T1; basal diet, T2; basal diet +0.3 mg/ kg feed sodium selenite, T3; basal diet +0.3 mg/ kg feed ADS1 Se, T4; basal diet +0.3 mg/ kg feed ADS2 Se, T5; basal diet +0.3 mg/ kg feed ADS18 Se

[a,b,c] Means in the same row with different superscripts are significantly different

*: significant differences ($P < 0.05$). NS: No significant differences

Table 6 Effect of dietary supplementation of bacterial organic Se on serum, liver, and kidney antioxidant status of broilers

Parameters	Dietary treatments [1]					SEM	*P* value		
	T1	T2	T3	T4	T5		Anova	B	O
Serum									
T-AOC[2]	647.35	675.78	705.45	655.18	653.95	12.90	0.654	0.465	0.904
GSH-Px[3]	8.50[b]	11.78[a]	11.46[ab]	11.28[ab]	13.61[a]	0.536	0.029	0.005	0.764
SOD[4]	0.159[b]	0.166[a]	0.160[b]	0.160[b]	0.161[b]	0.001	0.027	0.112	0.004
Catalase[5]	12.064	12.642	12.588	12.325	12.363	0.081	0.149	0.039	0.271
MDA[6]	1.158[a]	0.929[b]	0.959[b]	0.869[b]	0.736[c]	0.035	<.0001	<.0001	0.1180
Liver									
T-AOC	1250.8	1306.6	1576.1	1342.7	1484.9	41.35	0.053	0.059	0.092
GSH-Px,	99.37[b]	111.23[b]	107.68[b]	154.60[a]	103.45[b]	6.117	0.009	0.039	0.376
SOD	0.1672	0.1670	0.1673	0.1673	0.1672	0.0001	0.781	0.927	0.2512
Catalase	12.279[c]	12.770[b]	12.831[ab]	13.037[a]	12.857[ab]	0.066	<.0001	<.0001	0.139
MDA	0.989[a]	0.640[b]	0.773[ab]	0.733[b]	0.800[ab]	0.039	0.044	0.007	0.141
Kidney									
T-AOC	1120.10[c]	1222.63[ab]	1263.41[a]	1122.10[c]	1182.73[cb]	16.05	0.003	0.012	0.253
GSH-Px	41.12[c]	51.26[b]	55.32[b]	47.84[bc]	72.62[a]	2.679	<.0001	0.001	0.042
SOD	0.1664[cb]	0.1663[c]	0.1665[cb]	0.1671[a]	0.1670[ab]	0.0001	0.027	0.192	0.021
Catalase	13.159[b]	13.160[b]	13.259[a]	13.217[ab]	13.284[a]	0.016	0.013	0.031	0.008
MDA	7.202[a]	7.171[a]	5.358[b]	5.124[b]	5.757[b]	0.229	<.0001	0.0003	<.0001

B = basal diet versus Se supplemented diets, O = organic Se versus inorganic Se diet,

[a,b,c] means having different superscript along the same row for each factor are significantly different ($P < 0.05$)

[1]T1; basal diet, T2; basal diet +0.3 mg/ kg feed sodium selenite, T3; basal diet +0.3 mg/ kg feed ADS1 Se, T4; basal diet +0.3 mg/ kg feed ADS2 Se, T5; basal diet +0.3 mg/ kg feed ADS18 Se

[2]T-AOC expressed as μM Trolox Equivalents

[3]Glutathione peroxidase activity is expressed as U/L (one unit is the amount of GSH-Px that produces 1 μmole of GS-SG per min)

[4]SOD: One unit corresponds to the amount of enzyme needed to scavenges dismutation of the superoxide radical

[5]Catalase activity is expressed as U/L (one unit is the amount of catalase that decomposes 1μmole of H_2O_2 per min)

[6]MDA is expressesd as μM MDA equivalents

Effects of bacterial organic se supplementation on the mRNA level of hepatic selenoproteins in broiler chicken

To examine the effect of bacterial organic Se sources and inorganic Se form on mRNA expression of some selenoproteins, the hepatic expressions of GSH-Px1, GSH-Px4, DIO1, DIO2, TXNDR1, and SELW1 genes were investigated. Our results revealed that the expression levels of GSH-Px1 and GSH-Px4 (Fig. 1) were affected by Se supplementation. Bacterial organic Se of ADS18 (T5) showed a superior level of both GSH-Px1 and GSH-Px4 mRNA expression with significant difference ($P < 0.05$) in comparison with all other treatments. However, inorganic Se and bacterial Se of ADS1 (T3) also increased GSH-Px1 mRNA level significantly ($P < 0.05$) compared to negative control. The expression levels of DIO1 and DIO2 mRNA level in liver tissue are shown in Fig. 2. The highest expression of both genes was observed in T5 with significant difference compared to other treatment groups. However, a significant increase in DIO2 mRNA level was observed in liver of chickens fed all sources of bacterial organic Se compared to chickens fed inorganic Se and basal diet. Fig. 3 shows the mRNA expression of TXNDR1 and SELW1 genes in experimented livers. The greatest increase in TXNDR1 mRNA level was observed in the liver of T5 treatment group which was significantly ($P < 0.01$) higher than the other groups. Moreover, a very significant stability of the hepatic SELW1 mRNA level was observed in all Se supplemented groups, with no significant effect compared to basal diet group. However, the treatment group of T4 which was supplemented with bacterial Se of ADS2) showed substantial difference ($P < 0.05$) compared to negative control.

Discussion

In the present study, supplementation of inorganic and bacterial organic Se to broiler chicken did not affect the birds' growth performance. This finding is in agreement with the results of Perić et al. [15], Wang et al. [16], Oliveira et al. [17], and Göçmen et al. [18] who reported that different sources and levels of Se in the diet had no influence on bird BW, WG, FI, and FCR ratio, However, other studies indicated positive effects on growth performance after Se supplementation. Jiang et al. [19] reported that selenomethionine supplementation at the 0.225 mg/ kg increased the broiler's final body weight and weight gain significantly compared to birds receiving basal diet. Zhou and Wang, [20] revealed that the supplementation of nano elemental Se in chicken for 90 days improved final BW, daily WG and FCR. Yang et al. [21] found that organic Se increased daily weight gain and feed intake in broiler chicken after 42 days of feeding. The differences in previous results and our result may be due to the fact that their basal diet was deficient in Se, whereas, in the present study, chickens received the basic requirement of Se in the basal diet. The National Research Council, (1994) maintains that the minimum dietary Se level for optimum growth and performance in broiler is 0.1 mg/kg. The Se concentration in our basal diet was around 0.09 mg/kg and this may be the reason there was no Se deficiency symptoms in our negative control.

Dietary supplementation of different sources of Se in broiler chicken affected serum, liver, and kidney Se levels. Bacterial organic Se of T5, T4, and T3 showed significantly highest Se concentration in serum, liver, and kidney respectively, compared to the basal diet and inorganic Se supplemented diet. This may be due to the bacterial Se containing proteins such as selenomethionine and selenoscystein which have the ability to be concentrated in some tissues including liver, kidney, and meat, while, inorganic Se is absorbed less efficiently and excreted in the urine at a higher level than organic Se because of their different metabolic pathways [22]. Previous studies also reported that, dietary Se resulted in an elevation of the Se concentration in the broiler liver, kidney, and breast muscle, but organic Se in the form of Se yeast showed more deposition than sodium selenite [17, 23–25]. Differences in serum and tissues Se levels between the experimented bacterial strains may attributed to the fact that all the strains in this study are able to accumulate Se- containing proteins associated with other Se- fractions such as nanoparticle elemental Se produced by *Stenotrophomonas maltophilia* and *Klebsiella pneumonia* [26, 27], and exopolysaccaride that can be produced by *Enterobacter cloacae* [28]. The abundant organic Se-molecules are Se containing proteins such as selenomethionine which is a well absorbed form and can be incorporated into body proteins in place of methionine [29]. The balance between the Se-containg proteins and other Se- fractions in the bacterial supplement may affect tissues Se deposition.

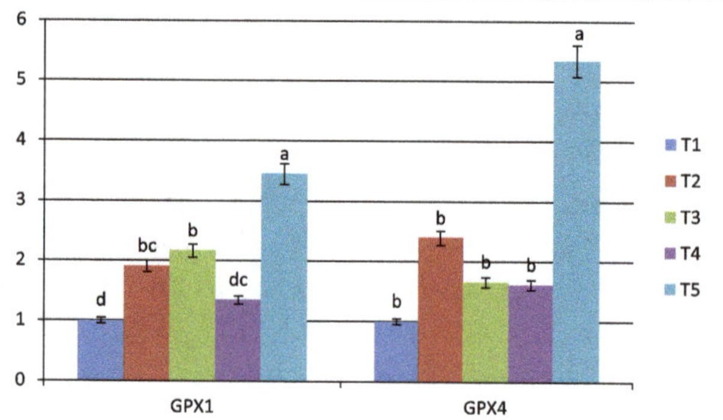

Fig. 1 GSH-Px1 and GSH-Px4 mRNA expression in the liver of broiler chicken. Treatments: T1; basal diet, T2 basal diet +0.3 mg/kg sodium selenite, T3: basal diet +0.3 mg/kg Se of ADS1, T4; basal diet +0.3 mg/kg Se of ADS2, T5: basal diet +0.3 mg/kg Se of ADS18. Bars with no common letter differ significantly ($P < 0.05$)

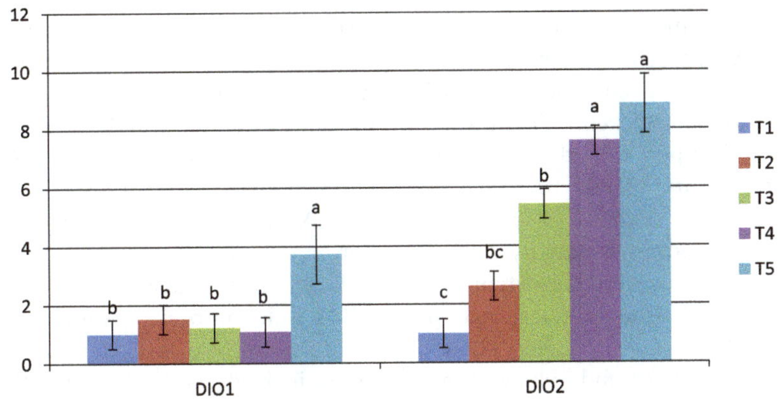

Fig. 2 DIO1 and DIO2 mRNA expression in the liver of broiler chicken. Treatments: T1; basal diet, T2 basal diet +0.3 mg/kg sodium selenite, T3: basal diet +0.3 mg/kg Se of ADS1, T4; basal diet +0.3 mg/kg Se of ADS2, T5: basal diet +0.3 mg/kg Se of ADS18. Bars with no common letter differ significantly ($P < 0.05$)

Selenium supplementation as inorganic or bacterial organic forms did not affect serum total protein, albumin, globulin, albumin/ globulin ratio, and urea. Similarly, the supplementation of Se either as Se nanoparticles or sodium selenite in the rat basal diet had no effect on serum total protein and albumin while globulin level showed improvement after Se supplementation [30]. In the same way, Yang et al. [21] reported that broiler chicks supplemented with 0.3 ppm organic Se for 42 days didn't affect serum globulin level compared to control group. Contrary to our finding Mohapatra et al. [31] stated that supplementation of 0.3 ppm nano Se in layer chicks up to 8 weeks significantly increased total protein and serum globulin levels and also significantly lowered A:G ratio compared to control. Moreover, in the present study, dietary inorganic Se and bacterial organic Se significantly affected broiler serum ALT, AST, LDH activity and serum creatinine level. This is supported by the finding of Perić et al. [15] who found substantial reduction in both ALT and AST enzymes activity in chicken fed organic Se. The same results were obtained by Biswas et al., [32], who found a decrease in ALT and AST activities in chicks supplemented with 0.5 mg and 1 mg/kg of Se in their diet. However, Okunlola et al. [33] and Gružauskas et al. [34] indicated that serum ALT and AST increased with no differences in total protein, Albumin, Urea and creatinine in poultry supplemented with 0.5 mg of sodium selenite and others supplemented with 0.15 mg of inorganic Se and 0.35 mg of organic Se. The blood enzymes ALT, AST, LDH are used as indicators of liver and kidney oxidative damage, the serum reduction of the enzymes and creatinine levels means increasing protection against oxidative damage through an improved redox status.

Animal antioxidant system is greatly influenced by animal nutrition, and dietary Se supplementation is necessary to up-regulate the body's glutathione pool and its

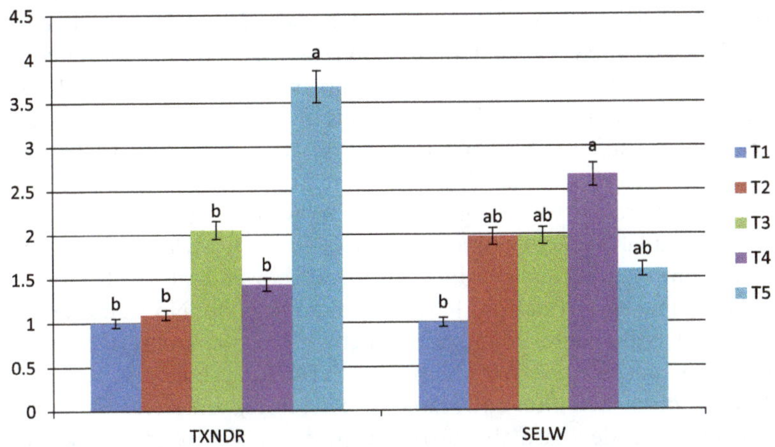

Fig. 3 TXNDR1 and SELW1 mRNA expression in the liver of broiler chicken. Treatments: T1; basal diet, T2 basal diet +0.3 mg/kg sodium selenite, T3: basal diet +0.3 mg/kg Se of ADS1, T4; basal diet +0.3 mg/kg Se of ADS2, T5: basal diet +0.3 mg/kg Se of ADS3. Bars with no common letter differ significantly ($P < 0.05$)

Se-containing antioxidant enzymes [19]. Retention of organic Se in poultry given organic Se is better than in those given sodium selenite. Accordingly, dietary organic Se can improve antioxidant system and increase GSH-Px activity in all tissues of broiler chicken [35]. However, Glutathione peroxidase and superoxide dismutase are the main enzymatic antioxidants against toxic oxygen reduction metabolites [36]. The findings of increasing serum and tissues antioxidant enzymes activity by bacterial organic Se in this experiment were consistent with earlier results of Teo et al. [37], who revealed that the dietary supplementation of Se-enriched bacteria showed higher level of GSH-Px in heart, liver, and kidney of rats. Additionally, Se-enriched yeast as organic source enhanced antioxidative status of broilers by increasing antioxidant enzyme levels compared to sodium selenite [19]. Besides that, a study by Chen et al. [38] showed that organic Se supplementation in broiler chicken increased the activity of serum GSH-Px, SOD and total antioxidant capacity (T-AOC) more significantly than dietary sodium selenite. According to Boostani et al. [39] Se supplementation raised glutathione peroxidase (GSH-Px) activity and lowered malondialdehyde (MDA) in comparison with the control group. On the contrary, Payne and Southern [40] reported that glutathion peroxidase activity was unaffected by organic, inorganic and concentrations of Se. Some serum and tissues antioxidant enzymes in this study showed no differences between bacterial organic Se and sodium selenite, which could refer to the fact that bacterial organic Se used in this study was extracted as Se-containing proteins which mainly comprise seleno-amino acids (selenomethionine and selenocystein), therefore part of selenomethionine can be merged directly with body proteins to replace methionine instead of entering selenoproteins synthesis, while, sodium selenite can be converted to Se-Cys directly before it can be incorporated into selenoproteins enzymes [41]. Moreover, the fluctuation observed between the bacterial organic sources may be attributed to the variation in the amount of Se-Cys to Se-Met accumulated by each bacterial strain. Supplementation of bacterial organic Se also caused a significant decrease in serum and kidney MDA content compared to dietary sodium selenite, although the difference was insignificant in kidney tissues. Malondialdehyde, considered a marker of oxidative stress, is one of the final products of cell polyunsaturated fatty acid peroxidation [42]. Therefore, the decreasing of MDA by bacterial organic Se is due to the presence of Se-Met and Se-Cys which are more bioavailable than sodium selenite and can raise the levels of antioxidants and decrease the production of lipid peroxidation products.

The present study investigated the expression of selenoproteins (GSH-Px1, GSH-Px4, DIO1, DIO2, TXNDR1, and SELW1) when bacterial organic Se from three bacterial strains was supplemented to the broiler chicken and compared with inorganic source. Numerous studies have shown that expression of these genes is regulated by dietary Se intake and down-regulated in Se deficiency [43]. In Se deficiency, lower selenoprotein's transcript levels in chicken erythrocytes was observed, while, GSH-Px, TXNDR1, selenoprotein P1 (SELP), and selenoprotein synthetase (SPS2) were highly expressed [44]. As well as, in chickens thyroid gland, DIO1, DIO2, and DIO3 selenoproteins were down-regulated [45], and GSH-Px mRNA level fell to 35–39% of Se-adequate levels in birds [46]. However, the effect of different Se sources on the expression of these genes was not fully investigated, and to our best knowledge, no study has examined the effect of bacterial Se as an organic source. In this study, the supplementation of inorganic Se has no effect on mRNA expression of all examined genes except GSH-Px1 gene. GSH-Px1 mRNA levels in sodium selenite supplemented group showed a significant difference compared with un-supplemented group. Previous studies reported that GSH-Px1 and SELW1 mRNA levels increased in responding to Se intake in poultry [47], sheep [48], and pig [49]. Dietary inorganic Se had no effect on GSH-Px4 in rat liver [43], However, Se supplementation in the form of sodium selenite resulted in increased GSH-Px1, SELW1, and TXNRD1 mRNA levels in the liver of lamb, while no change was observed in GSH-Px4 [50]. Therefore, dietary Se can regulate the expression of selenoproteins, which can eliminate reactive oxygen species through their antioxidant properties. GSH-Px is the most abundant selenoprotein in the liver, including GSH-Pxl, GSH-Px2, GSH-Px3, and GSH-Px4, and most of them are involved in the catabolism of peroxides. Our results indicate that GSH-Px1 mRNA could have higher sensitivity to regulation by Se status than GSH-Px4, and different response of mRNA expression to dietary Se might exist between selenoproteins GSH-Px1, and GSH-Px4. Moreover, our study showed that supplementation of bacterial organic Se up-regulated the mRNA expression of most examined genes significantly ($P \leq 0.05$) compared to negative control. However, in comparison with sodium selenite, just ADS18 bacterial Se showed significantly higher mRNA level in GSH-Px1, GSH-Px4, DIO1, and TXNDR1, while both ADS18 and ADS2 showed high level of mRNA of DIO2 compared to sodium selenite. No significant differences were observed between all the bacterial Se sources and sodium selenite in SELW1 expression.

Therefore, different Se sources change the mRNA expression of broiler liver selenoproteins, and the effects differ substantially between different selenoproteins, suggesting that some are more sensitive to changes in Se intake than others [51]. This finding is confirmed by the study of Yuan et al. [11], who studied the effect of various Se sources on the expression of liver (GSH-Px1 and TXNDR1) in broiler chicken, and reported that Se yeast and selenomethionin as a sources of organic Se increased GSH-Px1 and TXNDR1 mRNA in the liver compared with sodium selenite. Besides that, there was a considerable increase ($P < 0.01$) in TXNDR1 and SELW1 mRNA level in the group of broiler chickens fed selenomethionin compared to the group that received sodium selenite [52]. The observed difference between organic and inorganic Se may be due to the fact that Se supplied via organic forms has a higher bioavailability and thus enhances the Se level, leading to the stimulation of selenopeoteins gene expression [16]. However, the mechanisms of how different sources of Se can regulate the expression of selenoproteins are still unclear and need more investigation. We note that all Se sources in this study resulted in significant changes in gene expression of liver selenoproteins. However, the bacterial organic Se showed significant differences between the different strains. This may be due to the variation in the type of organic Se compound accumulated in bacterial strains. Organic Se compounds include selenomethionine, selenocystein, and Se-methyl-Secysteine [53], which vary according to their availability to the body.

Conclusion

To summarize, our study showed that basal diets supplemented with 0.3 mg/kg of different sources of bacterial organic Se and sodium selenite as inorganic source increased the serum and tissues anti-oxidative capacity and Se concentration, up-regulated some selenoproteins mRNA levels, and reduced serum AST, ALT and creatinine level. However, bacterial organic Se showed better effect than sodium selenite in most investigated parameters, and Se extracted from ADS18 bacterial strain had a superior action in improving antioxidant system and expression of selenoproteins compared to ADS1 and ADS2 bacterial Se.

Abbreviations

A/G: Albumin/ globulin ratio; ADS1: *Enterobacter cloacae*; ADS18: *Stenotrophomonas maltophilia*; ADS2: *Klebsiella pneumonia*; AL: Albumin; ALT: Alanine aminotransferase; AST: Aspartate aminotransferase; BUN: Blood urea nitrogen; BW: Body weight; CAT: Catalase; Cr: Creatinine; DIO1: Iodothyronine deiodinase1; DIO2: Iodothyronine deiodinase2; FI: Feed intake; G: Globulin (G); GAPDH: Glyceraldehyde-3-phosphate dehydrogenase; GSH-Px: Glutathione peroxidase; GSH-Px1: Glutathione peroxidase1; GSH-Px4: Glutathione peroxidase4; LDH: Lactate dehydrogenase; MDA: Malondialdehyde; Se: Selenium; SELW1: Selenoproteins w; SOD: Superoxide dismutase; T-AOC: Total antioxidant; TP: Total protein; TXNDR1: Thiroxine reductase; WG: Weight gain

Acknowledgements

A.M. was a recipient of scholarship from the Ministry of Higher Education and Scientific Research of Sudan and University of Khartoum, Sudan.

Funding

This study was financed by the Fundamental Research Grant Scheme (FRGS 5524272) granted by Malaysian Ministry of Higher Education.

Authors' contributions

A.M analysed and interpreted data regarding birds performance and carried out laboratory analysis. M.F was a major contributer in the part of gene expression and qPCR. A.M, T.C, A.Q and A.A participated in the whole design of the study and performed the statistical analysis and contributed to the preparation of the manuscript. All authors read and approved the final manuscript.

Competing interests

The authors declare that they have no competing interests.

Author details

[1]Department of Animal Science, Faculty of Agriculture, Universiti Putra Malaysia, 43400 Serdang, Selangor, Malaysia. [2]Department of Animal Nutrition, Faculty of Animal Production, University of Khartoum, Khartoum, Sudan. [3]Institute of Tropical Agriculture, Universiti Putra Malaysia, 43400 Serdang, Selangor, Malaysia.

References

1. Surai PF. Selenium in nutrition and health. Nottingham: Nottingham UniversityPress; 2006.
2. Surai PF, Fisinin VI, Karadas F. Antioxidant systems in chick embryo development. Part 1. Vitamin E, carotenoids and selenium. Animal Nutrition. 2016;2(1):1–11.
3. Huang X, Sun B, Zhang J, Gao Y, Li G, Chang Y. Selenium deficiency induced injury in chicken muscular stomach by Downregulating Selenoproteins. Biol Trace Elem Res. 2017:1–7.
4. Chen J, Berry MJ. Selenium and selenoproteins in the brain and brain diseases. J Neurochem. 2003;86(1):1–12.
5. Rotruck J, Pope A, Ganther H, Swanson A, Hafeman DG, Hoekstra W. Selenium: biochemical role as a component of glutathione peroxidase. Science. 1973;179(4073):588–90.
6. Yao HD, Wu Q, Zhang ZW, Li S, Wang XL, Lei XG, et al. Selenoprotein W serves as an antioxidant in chicken myoblasts. Biochimica et Biophysica Acta (BBA)-General Subjects. 2013;1830(4):3112–20.
7. Yao HD, Wu Q, Zhang ZW, Zhang JL, Li S, Huang JQ, et al. Gene expression of endoplasmic reticulum resident selenoproteins correlates with apoptosis in various muscles of se-deficient chicks. J Nutr. 2013;143(5):613–9.
8. Zhang J-l, Xu B, Huang X-d, Gao Y-h, Chen Y, Shan A-s. Selenium deficiency affects the mRNA expression of inflammatory factors and selenoprotein genes in the kidneys of broiler chicks. Biol Trace Elem Res. 2016;171(1):201–7.
9. Sun B, Wang R, Li J, Jiang Z, Xu S. Dietary selenium affects selenoprotein W gene expression in the liver of chicken. Biol Trace Elem Res. 2011;143(3):1516–23.
10. Zoidis E, Pappas A, Georgiou C, Komaitis E, Feggeros K. Selenium affects the expression of GPx4 and catalase in the liver of chicken. Comp Biochem Physiol B: Biochem Mol Biol. 2010;155(3):294–300.
11. Yuan D, Zhan X, Wang Y. Effect of selenium sources on the expression of cellular glutathione peroxidase and cytoplasmic thioredoxin reductase in the liver and kidney of broiler breeders and their offspring. Poult Sci. 2012; 91(4):936–42.
12. Boiago MM, Borba H, Leonel FR, Giampietro-Ganeco A, Ferrari FB, Stefanil LM, et al. Sources and levels of selenium on breast meat quality of broilers. Ciência Rural. 2014;44(9):1692–8.
13. Zhang B, Zhou K, Zhang J, Chen Q, Liu G, Shang N, et al. Accumulation and species distribution of selenium in se-enriched bacterial cells of the Bifidobacterium animalis 01. Food Chem. 2009;115(2):727–34.
14. Wahlen R, Evans L, Turner J, Hearn R. The use of collision/reaction cell ICP-MS for the determination of elements in blood and serum samples. Spectroscopy. 2005;20(12):1–8.
15. Perić L, Milošević N, Žikić D, Kanački Z, Džinić N, Nollet L, et al. Effect of selenium sources on performance and meat characteristics of broiler chickens. The journal of applied poultry research. 2009;18(3):403–9.
16. Wang Y, Zhan X, Zhang X, Wu R, Yuan D. Comparison of different forms of dietary selenium supplementation on growth performance, meat quality, selenium deposition, and antioxidant property in broilers. Biol Trace Elem Res. 2011;143(1):261–73.
17. Oliveira T, Rivera D, Mesquita F, Braga H, Ramos E, Bertechini A. Effect of different sources and levels of selenium on performance, meat quality, and tissue characteristics of broilers. The Journal of Applied Poultry Research. 2014;23(1):15–22.

18. Göçmen R, Yazgan O, Cufadar Y. Effect of different organic and inorganic selenium levels on performance, selenium concentration of some tissues, glutathione peroxidase enzyme activity and meat quality in broilers. JAPS: Journal of Animal & Plant Sciences. 2016;26(4):916–23.
19. Jiang Z, Lin Y, Zhou G, Luo L, Jiang S, Chen F. Effects of dietary selenomethionine supplementation on growth performance, meat quality and antioxidant property in yellow broilers. J Agric Food Chem. 2009;57(20):9769–72.
20. Zhou X, Wang Y. Influence of dietary nano elemental selenium on growth performance, tissue selenium distribution, meat quality, and glutathione peroxidase activity in Guangxi yellow chicken. Poult Sci. 2011;90(3):680–6.
21. Yang Y, Meng F, Wang P, Jiang Y, Yin Q, Chang J, et al. Effect of organic and inorganic selenium supplementation on growth performance, meat quality and antioxidant property of broilers. Afr J Biotechnol. 2012;11(12):3031–6.
22. Mahan D, Parrett N. Evaluating the efficacy of selenium-enriched yeast and sodium selenite on tissue selenium retention and serum glutathione peroxidase activity in grower and finisher swine. J Anim Sci. 1996;74(12):2967–74.
23. Wang Y-B, Xu B-H. Effect of different selenium source (sodium selenite and selenium yeast) on broiler chickens. Anim Feed Sci Technol. 2008;144(3):306–14.
24. Yuan D, Zhan X, Wang Y. Effects of selenium sources and levels on reproductive performance and selenium retention in broiler breeder, egg, developing embryo, and 1-day-old chick. Biol Trace Elem Res. 2011;144(1–3):705–14.
25. Liao X, Lu L, Li S, Liu S, Zhang L, Wang G, et al. Effects of selenium source and level on growth performance, tissue selenium concentrations, antioxidation, and immune functions of heat-stressed broilers. Biol Trace Elem Res. 2012;150(1–3):158–65.
26. Fesharaki PJ, Nazari P, Shakibaie M, Rezaie S, Banoee M, Abdollahi M, et al. Biosynthesis of selenium nanoparticles using Klebsiella Pneumoniae and their recovery by a simple sterilization process. Braz J Microbiol. 2010;41(2):461–6.
27. Lampis S, Zonaro E, Bertolini C, Cecconi D, Monti F, Micaroni M, et al. Selenite biotransformation and detoxification by Stenotrophomonas maltophilia SeITE02: novel clues on the route to bacterial biogenesis of selenium nanoparticles. J Hazard Mater. 2017;324:3–14.
28. Lu Z, Jin M, Huang M, Wang Y, Wang Y. Bioactivity of selenium-enriched exopolysaccharides produced by Enterobacter Cloacae Z0206 in broilers. Carbohydr Polym. 2013;96(1):131–6.
29. Butler J, Beilstein M, Whanger P. Influence of dietary Methionine on the metabolism of Selenomethionine in rats. J Nutr. 1989;119(7):1001–9.
30. Bunglavan S, Garg A, Dass R, Shrivastava S. Effect of supplementation of different levels of selenium as nanoparticles/sodium selenite on blood biochemical profile and humoral immunity in male Wistar rats. Vet World. 2014;7:1075–81.
31. Mohapatra P, Swain R, Mishra S, Behera T, Swain P, Mishra S, et al. Effects of dietary nano-selenium on tissue selenium deposition, antioxidant status and immune functions in layer chicks. Int J Pharmacol. 2014;10(3):160–7.
32. Biswas A, Ahmed M, Bharti V, Singh S. Effect of antioxidants on physio-biochemical and hematological parameters in broiler chicken at high altitude. Asian Australas J Anim Sci. 2011;24(2):246–9.
33. Okunlola DO, Akande TO, Nuga HA. Haematological and serum characteristics of broiler birds fed diets supplemented with varying levels of selenium powder journal of biology. Agriculture and Healthcare. 2015;5(1):107–10.
34. Gružauskas R, Barštys T, Racevičiūtė-Stupelienė A, Kliševičiūtė V, Buckiūnienė V, Bliznikas S. The effect of sodium selenite, selenium methionine and vitamin E on productivity, digestive processes and physiologic condition of broiler chickens. Veterinarija ir Zootechnika. 2013;65(87):22–9.
35. Zhang L, Wang Y, Zhou Y, Zheng L, Zhan X, Pu Q. Different sources of maternal selenium affect selenium retention, antioxidant status, and meat quality of 56-day-old offspring of broiler breeders. Poult Sci. 2014;93(9):2210–9.
36. Del Maestro R. Free radicals as mediators of tissue injury. In: Dreosti IE, editor. Trace elements, micronutrients, and free radicals. Berlin: Springer; 1991. p. 25–51.
37. Teo A, Hon S-M, Se C-Y, Ian H, Inventors; , assignee. Production of organic and inorganic selenium compounds by lactic acid bacteria. U.S.Patent application no 11/243,391. 2005.
38. Chen G, Wu J, Li C. Effect of different selenium sources on production performance and biochemical parameters of broilers. J Anim Physiol Anim Nutr. 2014;98(4):747–54.
39. Boostani A, Sadeghi A, Mousavi S, Chamani M, Kashan N. Effects of organic, inorganic, and nano-se on growth performance, antioxidant capacity, cellular and humoral immune responses in broiler chickens exposed to oxidative stress. Livest Sci. 2015;178:330–6.
40. Payne R, Southern L. Comparison of inorganic and organic selenium sources for broilers. Poult Sci. 2005;84(6):898–902.
41. Henry PR, Ammerman CB. Selenium bioavailability. In: Ammerman CB, Baker DP, Lewis AJ, editors. Bioavailability of nutrients for animals: amino acids, minerals, vitamins. New York: Academic Press; 1995. p. 303–36.
42. Gaweł S, Wardas M, Niedworok E, Wardas P. Malondialdehyde (MDA) as a lipid peroxidation marker. Wiadomosci lekarskie (Warsaw, Poland: 1960). 2004;57(9–10):453–5.
43. Hadley KB, Sunde RA. Selenium regulation of thioredoxin reductase activity and mRNA levels in rat liver. J Nutr Biochem. 2001;12(12):693–702.
44. Luan Y, Zhao J, Yao H, Zhao X, Fan R, Zhao W, et al. Selenium deficiency influences the mRNA expression of selenoproteins and cytokines in chicken erythrocytes. Biol Trace Elem Res. 2016;171(2):427–36.
45. Lin SL, Wang CW, Tan SR, Liang Y, Yao HD, Zhang ZW, et al. Selenium deficiency inhibits the conversion of thyroidal Thyroxine (T4) to Triiodothyronine (T3) in chicken thyroids. Biol Trace Elem Res. 2014;161(3):263–71.
46. Sunde RA, Li JL, Taylor RM. Insights for setting of nutrient requirements, gleaned by comparison of selenium status biomarkers in turkeys and chickens versus rats, mice, and lambs. Advances in Nutrition: An International Review Journal. 2016;7(6):1129–38.
47. Liu C, Fu J, Lin S, Wang X, Li S. Effects of dietary selenium deficiency on mRNA levels of twenty-one selenoprotein genes in the liver of layer chicken. Biol Trace Elem Res. 2014;159(1–3):192–8.
48. Humann-Ziehank E, Renko K, Mueller AS, Roehrig P, Wolfsen J, Ganter M. Comparing functional metabolic effects of marginal and sufficient selenium supply in sheep. J Trace Elem Med Biol. 2013;27(4):380–90.
49. Zhou J-C, Zhao H, Li J-G, Xia X-J, Wang K-N, Zhang Y-J, et al. Selenoprotein gene expression in thyroid and pituitary of young pigs is not affected by dietary selenium deficiency or excess. J Nutr. 2009;139(6):1061–6.
50. Juszczuk-Kubiak E, Bujko K, Cymer M, Wicińska K, Gabryszuk M, Pierzchała M. Effect of inorganic dietary selenium supplementation on selenoprotein and lipid metabolism gene expression patterns in liver and loin muscle of growing lambs. Biol Trace Elem Res. 2016;172(2):336–45.
51. Sunde RA, Raines AM. Selenium regulation of the selenoprotein and nonselenoprotein transcriptomes in rodents. Advances in Nutrition: An International Review Journal. 2011;2(2):138–50.
52. Zhan X, Wang H, Yuan D, Wang Y, Zhu F. Comparison of different forms of dietary selenium supplementation on gene expression of cytoplasmic thioredoxin reductase, selenoprotein P, and selenoprotein W in broilers. Czech Journal of Animal Science. 2014;59(12):571–8.
53. McSheehy S, Yang L, Sturgeon R, Mester Z. Determination of methionine and selenomethionine in selenium-enriched yeast by species-specific isotope dilution with liquid chromatography– mass spectrometry and inductively coupled plasma mass spectrometry detection. Anal Chem. 2005;77(1):344–9.
54. Li J-L, Sunde RA. Selenoprotein transcript level and enzyme activity as biomarkers for selenium status and selenium requirements of chickens (Gallus Gallus). PLoS One. 2016;11(4):e0152392.

17

Modulation of intestinal microbiota, morphology and mucin composition by dietary insect meal inclusion in free-range chickens

Ilaria Biasato[1], Ilario Ferrocino[2], Elena Biasibetti[1], Elena Grego[1], Sihem Dabbou[1], Alessandra Sereno[1], Francesco Gai[3], Laura Gasco[2,3], Achille Schiavone[1,4*], Luca Cocolin[2] and Maria Teresa Capucchio[1]

Abstract

Background: Gut health in poultry depends on the balance between the host, intestinal microbiota, intestinal microscopic features and diet. The effects of insect meal (a promising alternative protein source for poultry feed) on chicken gut morphology have recently been reported, but no data about intestinal microbiota and mucin composition modulation are available. The present study evaluated the effects of dietary *Tenebrio molitor* (TM) meal inclusion on gut health of free-range chickens by intestinal microbiota, morphology and mucin composition characterization.

Results: One hundred forty female medium-growing hybrids were divided into 2 dietary treatments (control feed [C] and 7.5% TM inclusion, with 5 replicate pens/treatment and 14 birds/pen) and slaughtered at 97 days of age (2 birds/pen for a total of 10 chickens/diet). The gut microbiota assessment on cecal content samples by 16S rRNA amplicon based sequencing showed higher alpha (Shannon, $P < 0.05$) and beta (Adonis and ANOSIM, $P < 0.001$) diversity in birds fed TM diet than C. In comparison with C group, TM birds displayed significant increase and decrease, respectively, of the relative abundances of *Firmicutes* and *Bacteroidetes* phyla, with higher *Firmicutes:Bacteroidetes* ratios (False Discovery Rate [FDR] < 0.05). The relative abundance of *Clostridium, Oscillospira, Ruminococcus, Coprococcus* and *Sutterella* genera was higher in TM chickens than C (FDR < 0.05). On the contrary, TM birds displayed significant decrease of the relative abundance of *Bacteroides* genus compared to the C group (FDR < 0.05). Gut morphology evaluation by morphometric analysis on small intestine revealed similar villus height, crypt depth and villus height to crypt depth ratio between C and TM birds. Characterization of gut mucin composition by periodic-acid Schiff, Alcian Blue pH 2.5 and high iron diamine staining on small and large intestine showed unaffected mucin staining intensity in TM chickens when compared to C group.

Conclusions: Dietary TM meal inclusion may positively modulate the gut microbiota of the free-range chickens without influencing the intestinal morphology and mucin composition. Since the rapid growth of chickens directly depends on morphological and functional integrity of the digestive tract, the gut health assessment by a post mortem multidisciplinary approach appears to be fundamental.

Keywords: 16S rRNA, Gut health, Microbiota, Morphometry, Mucin, Poultry, *Tenebrio molitor*

* Correspondence: achille.schiavone@unito.it
[1]Department of Veterinary Sciences, University of Turin, Largo Paolo Braccini 2, 10095 Grugliasco, TO, Italy
[4]Institute of Multidisciplinary Research on Sustainability, University of Turin, Via Accademia Albertina 13, 10100 Turin, Italy
Full list of author information is available at the end of the article

Background

Gut health has been a focus of major research efforts in production animals, since it can be considered a synonymous to animal health and is of vital importance to animal performance [1]. The latter aspect has a key role in poultry industry, which requires animals capable of growing rapidly within a short period of time.

Gut health depends on the maintenance of the delicate balance between the host, intestinal microbiota, intestinal barrier (in terms of microscopic structure) and dietary compounds [2]. In particular, gut microbiota benefits the host by providing nutrients from otherwise poorly utilized dietary substrates and modulating the development and function of the digestive and immune system [3]. Firstly, gut microbiota can affect intestinal morphology through modifications of villus height and crypt depth [4], which are considered the main indicators of gut development, health and functionality [5]. There is also evidence that gut microbiota may modulate synthesis and composition of mucins [4], which constitute a digestion- and absorption-assisting medium and represent the first line of defense for intestinal epithelium against foreign bacteria and other pathogens [6]. Several feed substances have been reported to widely affect gut health in poultry, either by directly modifying intestinal morphology [7, 8] and mucin composition [9, 10], or indirectly by modulating intestinal microbiota [3, 11].

Feed may be considered the most important entity in the poultry industry in terms of animal health and producers revenue and profit. As a first aspect to consider, feed substances are capable of exposing the birds to potentially harmful organisms and/or components through the gastrointestinal tract [12]. Feed is also the major component of the total cost of production, with protein sources representing the primary one [13]. Corn and soybean meal constitute the main ingredients of choice for poultry diets worldwide [12], but the search for more sustainable, less food-competing and alternative protein feedstuffs is progressively increasing [13].

Insect meal utilization as alternative protein source for animal feeding has already been pointed out [14, 15], because of the excellent nutritive properties, the low competitiveness with human food and the reduction of the environmental contamination [15]. Compared to the conventional protein feedstuff (in particular soybean meal), insects contain higher or similar protein content, being also richer in essential amino acids [14, 15]. Furthermore, mass production of insects is currently promising in an ecological perspective, since it can lead to a significant reduction of the environmental impact in terms of energy cost, land area utilization and footprints. Indeed, insects can easily grow on different organic side streams, whose elimination has an economic and environmental cost. As a consequence, insects rearing may promote an advantageous reutilization of by-products, also transforming waste into high-protein feed that can replace increasingly more expensive compound feed ingredients [15]. Among insect species, meal obtained from larvae of *Tenebrio molitor* (TM) has recently been tested in diets for broiler [16–20] and free-range chickens [21] as protein source partially or totally replacing the only soybean meal [16, 18], soybean meal, corn gluten meal and soybean oil together [19, 20], or corn gluten meal alone [21]. In particular, TM larvae meal has been reported to be a suitable dietary ingredient in terms of unaffected or partially improved growth performance, haematochemical parameters, carcass traits and histological features [16, 18–21]. The implications of TM larvae meal utilization have also been investigated on gut health by intestinal morphology assessment [19–21], but no data about the modulation of gut microbiota and mucin composition by their inclusion in poultry diets are currently available.

The present study aims to investigate the effects of TM meal utilization on intestinal microbiota, morphology and mucin composition of free-range chickens, also proposing a standardized multidisciplinary post mortem approach for the assessment of gut health in poultry when dietary modifications occur.

Results

Cecal microbiota characterization

A total of 530,550 raw reads (2x250bp) were obtained after sequencing. After joint and quality filtering, a total of 104,081 reads passed the filters applied through QIIME, with an average value of 10,408 reads/sample. In order to avoid biases due to different sequencing depths, all samples were rarefied at 3600 reads after raw read quality filtering.

The rarefaction analysis and the Good's coverage indicated a satisfactory coverage for all the samples (average Good's coverage of 84%) (Additional file 1). The diversity of the cecal microbiota between C and TM diets was assessed initially through measures of α-diversity and β-diversity. The Chao1, Phylogenetic Diversity (PD) Whole Tree and Shannon indexes and observed species richness were used to assess α-diversity (Additional file 1). Shannon index showed higher diversity ($P < 0.05$) in free-range chickens fed with TM compared to C diet, whereas Chao1 and PD Whole Tree indexes and observed species richness showed no significant differences ($P > 0.05$) between C and TM groups (Additional file 1). Weighted UniFrac distances were utilized as a measure of β-diversity to assess the effect of TM diet on bacterial community composition. Adonis and ANOSIM statistical tests based on Weighted UniFrac distance matrix showed significant differences among C and TM groups ($P < 0.001$). These differences were demonstrated by Principal Component Analysis (PCA), which showed a clear separation of the microbiota as a function of the diet (Fig. 1).

Relative abundances of the main phyla and genera in the free-range chickens of the present study obtained by 16S rRNA gene sequencing are summarized in Fig. 2 (phyla and genera per pools) and Additional file 2 (overall phyla and genera). *Bacteroidetes* represented the dominant phylum of the cecal community in both C and TM groups, outnumbering the *Firmicutes* and *Proteobacteria* phyla (Fig. 2a, Additional file 2). Within phylum *Bacteroidetes*, *Bacteroides*, unclassified members (U. m.) of *Bacteroidales* order, *Alistipes*, *Parabacteroides* and *Coprobacter* were identified as predominant OTUs in both birds fed with C and TM diets (Fig. 2b, Additional file 2). *Clostridium*, *Ruminococcus*, *Oscillospira*, L-*Ruminococcus*, *Faecalibacterium* and U. m. of *Lachnospiraceae* family were the dominant members of the *Firmicutes* phylum in both C and TM groups (Fig. 2b, Additional file 2). Within phylum *Proteobacteria*, U. m. of *Alphaproteobacteria* class and U. m. of *Succinivibrionaceae* family were observed as predominant OTUs in both chickens fed with C and TM diets (Fig. 2b, Additional file 2).

Compared to the C group (Fig. 3), the birds fed with TM displayed significant increase of the relative abundance of *Firmicutes* phylum (FDR < 0.05). On the contrary, the relative abundance of *Bacteroidetes* was lower in TM chickens than C (FDR < 0.05). The birds fed with TM also showed higher *Firmicutes:Bacteroidetes* ratios compared to the C group (FDR < 0.05). At genus level (Fig. 4), the relative abundance of *Bacteroides* was lower in TM chickens than C (FDR < 0.05). On the contrary, the birds fed with TM displayed significant increase of the relative abundance of *Sutterella*, *Ruminococcus*, *Oscillospira* , *Clostridium* and *Coprococcus* genera compared to the C group (FDR < 0.05).

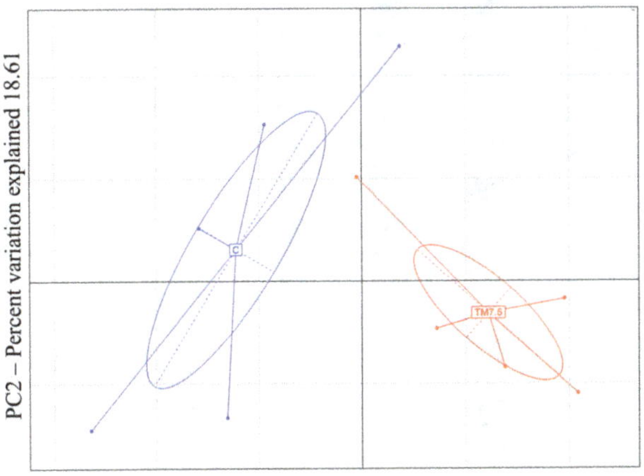

Fig. 1 Bacterial community composition (weighted UniFrac beta diversity, PCA plots) in cecal samples of free-range chickens fed with control (C) and 7.5% inclusion level of *Tenebrio molitor* meal (TM7.5) diets. PC1 and PC2 components account for 28.48 and 18.61%, respectively, of the total variation (47.09)

Intestinal morphology

Dietary TM meal inclusion did not affect the intestinal morphometric indices of the free-range chickens of the present study ($P > 0.05$, Fig. 5). Independently of TM meal utilization, the duodenum showed greater Vh (C and TM7.5 groups, $P < 0.01$) and Cd (TM7.5 group, $P < 0.05$) than the other gut segments and higher Vh/Cd ratio (C group, $P = 0.01$) than the ileum. The detailed results of the intestinal morphology of the chickens of the current trial are reported by Biasato et al. [21].

Intestinal mucin composition

Mucin staining intensity in the intestinal crypts of free-range chickens of the present study significantly depended on mucin type, gut segment and crypt fragment ($P < 0.001$). However, there was no significant influence of dietary TM meal inclusion ($P > 0.05$) on the histochemical findings (Table 1). In particular, crypts showed higher acidic sialylated mucins staining intensity ($P < 0.001$) than neutral and acidic sulfated. Furthermore, lower mucin staining intensity was found in the caecal crypts ($P < 0.001$) compared with the other gut segments and in the jejunal crypts ($P < 0.001$) compared with the duodenum and ileum, respectively. Crypt base also showed greater mucin staining intensity ($P < 0.001$) than the midsection and tip (Table 2, Fig. 6).

There was no significant effect of dietary TM meal inclusion or mucin type ($P > 0.05$) on mucin staining intensity for the intestinal villi, whereas both gut segment and villus fragment significantly influenced ($P < 0.001$ and $P < 0.01$, respectively) the histochemical findings (Table 1). In particular, villi showed higher mucin staining intensity in the ileum ($P < 0.001$) compared with the other gut segments and in the jejunum ($P < 0.001$) compared with the duodenum, respectively (Fig. 7). Lower mucin staining intensity was also observed in the villus tip ($P = 0.001$) than the base (Table 2).

Discussion

A clear definition for "gut health" that encompasses all the intestinal physiological and functional features, such as nutrient digestion and absorption, host metabolism and energy generation, a stable microbiome, mucus layer development, barrier function, and mucosal immune responses, has never been reported [1]. However, the crosstalk between the diet composition,

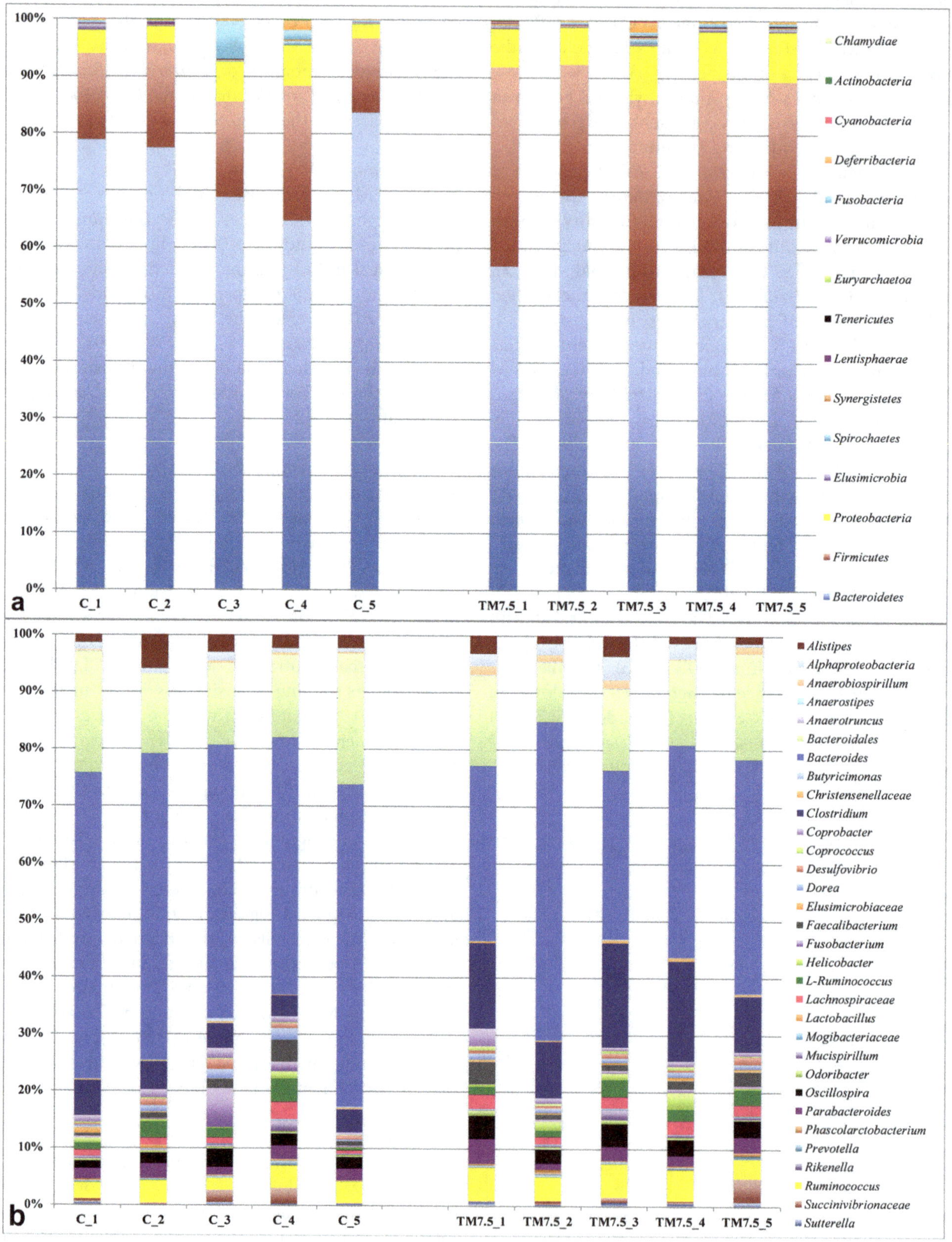

Fig. 2 (See legend on next page.)

(See figure on previous page.)
Fig. 2 Relative abundance of the main bacterial phyla (**a**) and genera (**b**) in cecal samples of free-range chickens fed with control (C) and 7.5% inclusion level of *Tenebrio molitor* meal (TM7.5) diets. Taxa summary bar plots within the pooled cecal contents from the 5 replicate pens of control (C_1, C_2, C_3, C_4 and C_5) and 7.5% inclusion level of *Tenebrio molitor* meal (TM7.5_1, TM7.5_2, TM7.5_3, TM7.5_4 and TM7.5_5) dietary treatments

intestinal barrier (formed by the mucus layer together with the epithelial cell layer) and intestinal microbiota seems to be crucial for gut health determination [2]. For such reasons, several authors have recently focused their attention on the relationship between gut microbiota and mucin composition, gut microbiota and morphology, or gut morphology and mucin composition when dietary modifications occurred in poultry. However, systematic studies about the assessment of gut health by the evaluation of all the three aspects together (gut microbiota, morphology and mucin composition) in chickens facing to new diets are still very limited. Furthermore, no information about the modulation of intestinal microbiota and mucin composition by dietary insect meal inclusion in poultry are currently available.

Cecal microbiota characterization

Characterization of the cecal microbiota of free-range poultry has already been reported in previous studies about Chinese Dagu chickens [22], Italian local breeds [23] and Bermuda free-range chickens [24]. However, the present study is the first one investigating the cecal microbiota of Label Hubbard chickens fed with different diets. The attention was herein focused on cecum as representative gut segment, since it harbors the highest microbial cell densities (up to 10^{11} cells/g) and diversity, has the longest residence time (12–20 h) of digesta in the gastrointestinal tract, and is an important site for recycling of urea, water regulation, and carbohydrate fermentations contributing to intestinal health and nutrition [25].

In the free-range chickens fed with basal diet of the present study, cecal microbiota was mainly colonized by

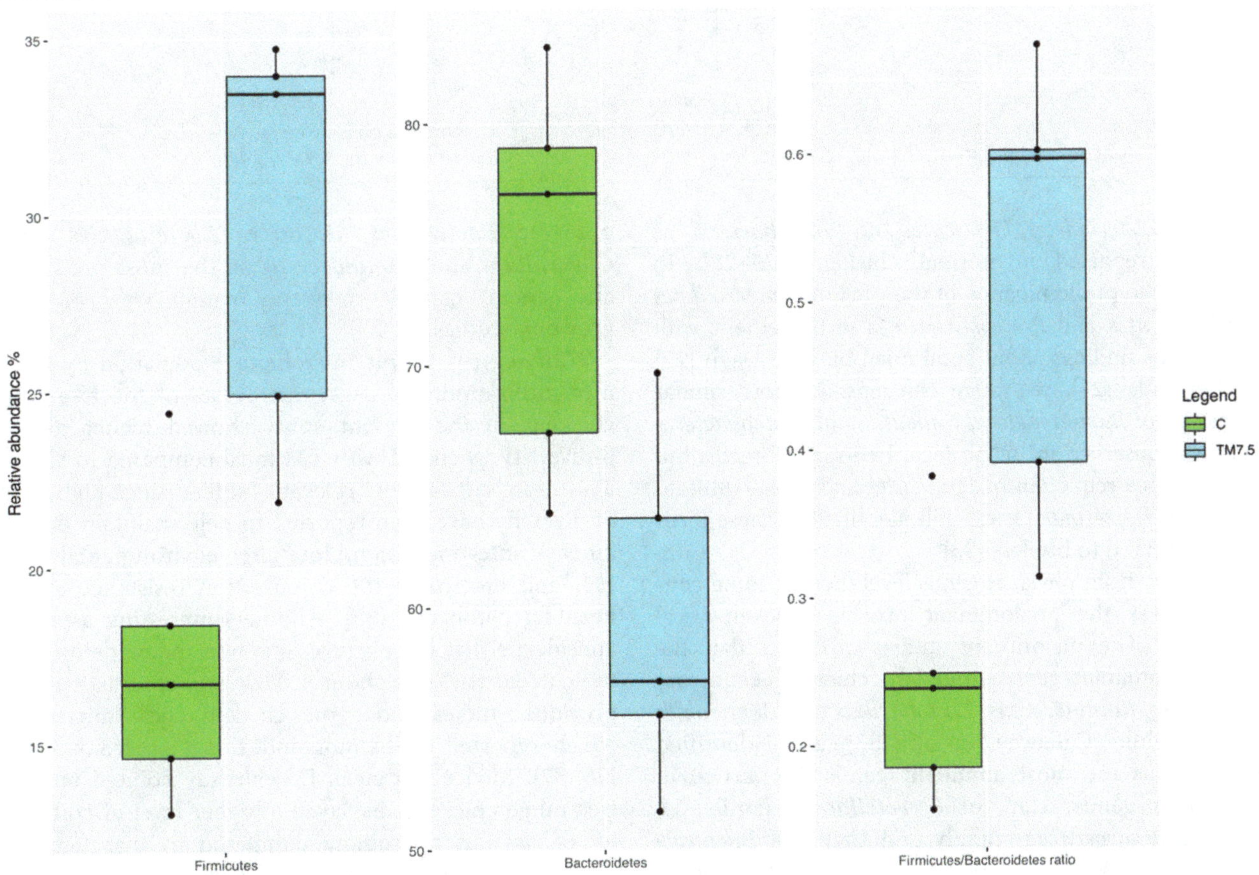

Fig. 3 Boxplots showing the relative abundance at phylum level of OTUs differentially abundant based on Pairwise Kruskal-Wallis test (FDR < 0.05) in cecal samples of free-range chickens fed with control (C) and 7.5% inclusion level of *Tenebrio molitor* meal (TM7.5) diets

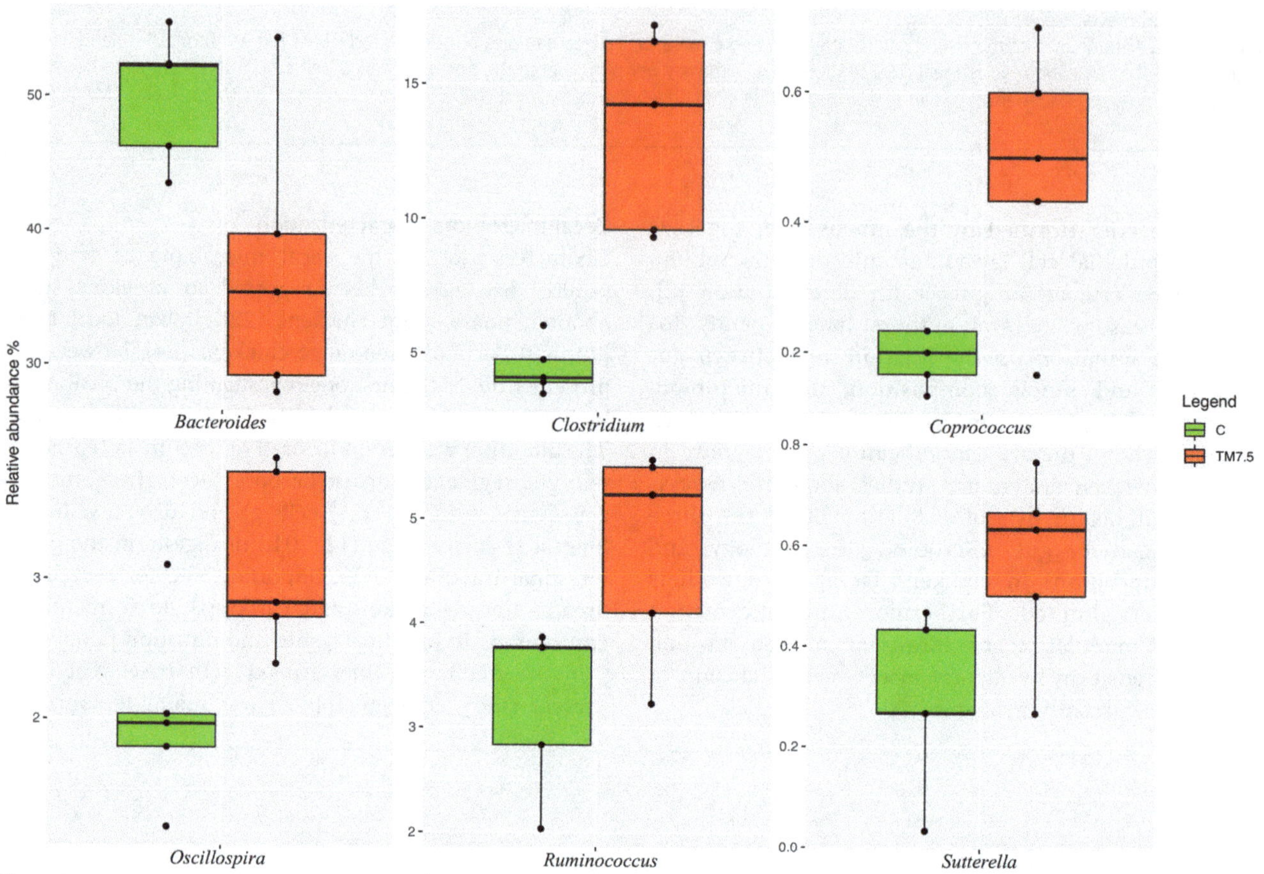

Fig. 4 Boxplots showing the relative abundance at genus level of OTUs differentially abundant based on Pairwise Kruskal-Wallis test (FDR < 0.05) in cecal samples of free-range chickens fed with control (C) and 7.5% inclusion level of *Tenebrio molitor* meal (TM7.5) diets

the phyla *Bacteroidetes*, *Firmicutes* and *Proteobacteria*, as previously reported in normal chickens [25–28]. In particular, the predominance of the phylum *Bacteroidetes* over *Firmicutes* and *Proteobacteria* is in agreement with the previous findings about cecal microbiota of Dagu [22] and Bermuda [24] free-range chickens. A more similar prevalence of *Bacteroidetes*, *Firmicutes* and *Proteobacteria* phyla was observed in Italian local free-range breeds, but a comparative reduction of *Firmicutes* and a concomitant increase of *Bacteroidetes* was still identified in these birds when compared to broilers [23].

Differently from phyla, at genus level there is some controversy over the predominant taxa in chicken cecal microbiota. The majority of studies observed that the most predominant genera found in chicken cecum are *Clostridium*, *Ruminococcus*, *Lactobacillus* and *Bacteroides* [28–32], while Callaway et al. [33] also identified *Prevotella* as the most abundant genus. In particular, *Bacteroides* genus, U.m. of *Prevotellaceae* family, U. m. of *Ruminococcaceae* family and U. m. of *Bacteroidales* order have been reported as the most predominant members of cecal microbiota of Bermuda free-range chickens [24]. The current trial, which observed *Bacteroides*, U. m. of *Bacteroidales* order, *Clostridium* and *Ruminococcus* as the most predominant genera, can fit into the overall view of these previous studies.

With regards to gut microbiota modulation by insect meal utilization, the cecal microbiota of the free-range chickens of the present study showed higher α- and β-diversity when fed with TM meal compared to C feed. This represents a quite relevant finding, since high levels of diversity have been reported to help maintain the stability of intestinal microbiota after environmental stress [34] and determine the colonization resistance against invading pathogens [35]. Another interesting aspect to consider is that the average Shannon index identified in the current trial was about 8. This is in contrast with the previous studies about the chicken cecal microbiota, which reported a Shannon index varying from 3 to 6 [36, 37]. Mancabelli et al. [23] already pointed out that free-range chickens displayed a higher level of complexity of the gut microbiota compared to that found in broilers, with a Shannon index slightly higher than 6. Interestingly, Ferrario et al. [24] also observed similar microbial diversity between free-range and feral (i.e.,

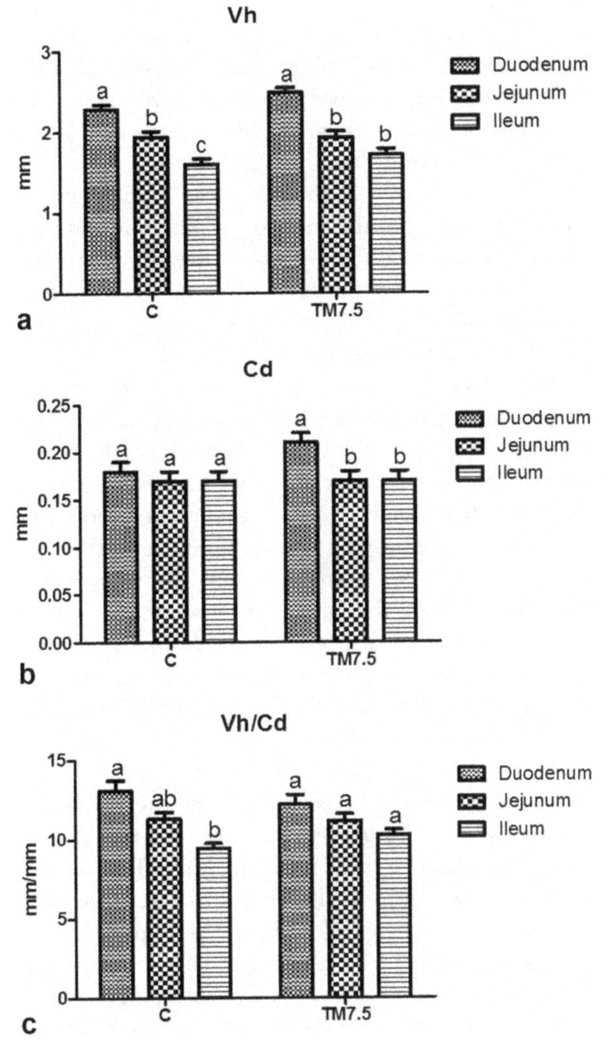

Fig. 5 Graph bars of (**a**) villus height (Vh), (**b**) crypt depth (Cd) and (**c**) villus height to crypt depth ratio (Vh/Cd) in duodenum, jejunum and ileum of free-range chickens fed with control (C) and 7.5% inclusion level of *Tenebrio molitor* meal (TM7.5) diets. Graph bars with different superscript letters (a, b, c) within each dietary treatment differ significantly ($P < 0.05$)

Table 1 Effects of diet, mucin type, gut segment and crypt-villus fragment on mucin staining intensity in free-range chickens

Factor	d.f.[f]	Chi-square	*P*-value[g]
Crypts			
Diet[a]	1	0.77	ns
Mucin type[b]	2	34.61	***
Gut segment[c]	3	145.01	***
Fragment[d]	2	71.38	***
Villi			
Diet	1	0.00	ns
Mucin type	2	3.03	ns
Gut segment[e]	2	272.02	***
Fragment	2	10.99	**

[a] Two dietary treatments: *C* control, *TM7.5* 7.5% inclusion level of *Tenebrio molitor*
[b] Three types: neutral, acidic sialylated and acidic sulfated mucins
[c] Four segments: duodenum, jejunum, ileum and cecum
[d] Three fragments: base, midsection and tip
[e] Three segments: duodenum, jejunum and ileum
[f] Degrees of freedom
[g] Statistical significance: $P < 0.05$ (*), $P < 0.01$ (**) and $P < 0.001$ (***). ns = not significant

formerly domesticated, wild-living) chickens, with the last being characterized by a Shannon index around 7. Therefore, it could be hypothesized that the semi-wild and wild rearing conditions may progressively modulate the complexity of the intestinal microbiota, thus underlying the role of environment and human influence on the bacterial communities within the chicken gastrointestinal tract.

Similarly to what observed for C birds, the cecal microbiota of the free-range chickens fed with TM diet of the present study was mainly colonized by the phyla *Bacteroidetes*, *Firmicutes* and *Proteobacteria*. Furthermore, the free-range chickens fed with TM showed *Bacteroides*, U. m. of *Bacteroidales* order, *Clostridium* and *Ruminococcus* as the most predominant genera of their cecal microbiota. These findings suggest that TM meal utilization does not alter the physiological cecal community at both phylum and genus levels.

Investigating the compositional differences in the cecal microbiota between C and TM birds, the free-range chickens fed with insect meal showed increased and decreased abundances of *Firmicutes* and *Bacteroidetes* phyla, respectively, along with higher *Firmicutes:Bacteroidetes* ratios. This represents a quite relevant finding, since bacteria within *Firmicutes* phylum have an important role in the digestion of feed and the host health [38] and greater *Firmicutes:Bacteroidetes* ratios have been associated with bacterial profile with higher capacity of energy harvesting [39]. At genus level, the free-range chickens fed with TM showed an increase in the abundance of *Clostridium, Oscillospira, Ruminococcus, Coprococcus* and *Sutterella* genera in their cecal community. As already pointed out, *Clostridium* and *Ruminococcus* are some of the most predominant genera found in chicken cecum [28–30]. On the other hand, *Clostridium* genus, along with *Oscillospira* and *Coprococcus*, also encompasses bacteria capable of producing butyrate [40, 41]. Butyrate has been demonstrated to have a positive role on growth performance, intestinal villus structure and pathogen control, as well as anti-inflammatory properties [42]. Furthermore, bacteria belonging to *Ruminococcus* genus can also produce other short chain fatty acids (i.e., acetic and succinic acid) through glucose metabolism and digest cellulose in food [43]. It is well known that short chain fatty acids are an important source of energy for enterocytes and are vital for intestinal health [44]. Another interesting finding of the cecal community of the birds fed

Table 2 Least square means of mucin staining intensity in the intestinal crypts and villi of free-range chickens in relation to diet, mucin type, gut segment and fragment

	Factor	Factor levels	Mucin staining intensity[a,b]
Crypts	Diet	C	1.11 ± 0.03
		TM7.5	1.14 ± 0.03
	Mucin type	Neutral	1.04 ± 0.03[b]
		Acidic sialylated	1.28 ± 0.04[a]
		Acidic sulfated	1.06 ± 0.03[b]
	Gut segment	Duodenum	1.38 ± 0.04[a]
		Jejunum	1.05 ± 0.04[b]
		Ileum	1.31 ± 0.04[a]
		Cecum	0.83 ± 0.03[c]
	Fragment	Base	1.36 ± 0.04[a]
		Midsection	1.03 ± 0.03[b]
		Tip	1.01 ± 0.03[b]
Villi	Diet	C	1.65 ± 0.04
		TM7.5	1.65 ± 0.04
	Mucin type	Neutral	1.63 ± 0.05
		Acidic sialylated	1.72 ± 0.05
		Acidic sulfated	1.61 ± 0.05
	Gut segment	Duodenum	1.09 ± 0.04[c]
		Jejunum	1.91 ± 0.06[b]
		Ileum	2.16 ± 0.06[a]
	Fragment	Base	1.78 ± 0.05[a]
		Midsection	1.64 ± 0.05[ab]
		Tip	1.55 ± 0.05[b]

C control, *TM7.5* 7.5% inclusion level of *Tenebrio molitor*
[a] Data are represented as mean of counts ± SEM
[b] Means with different superscript letters (a, b, c) within the same column per factor (i.e. diet, mucin type, gut segment or fragment) differ significantly ($P < 0.01$)

with TM in the current trial is the identification of increased abundance of *Sutterella* genus, which has been reported to be a potential probiotic present in pigeon "milk" that can improve the rate of growth and feed conversion ratio in chickens [45]. The increase of the above mentioned bacterial taxa suggests that TM meal utilization may positively modulate the cecal microbiota of birds. In particular, the increase of butyrate- and short chain fatty acids-producing bacteria may have important implications. However, further studies also evaluating the microbial metabolites and metabolic pathways are needed to better contextualize these OTUs changes.

Despite this overall positive modulation, the reduction of *Bacteroides* genus within the cecal community of the free-range chickens fed with TM meal of the present study may be considered a potential negative finding. Indeed, apart from being one of the most predominant genera found in chicken cecum [28, 31, 33], this taxon has been reported to be an important contributor to the intestinal health of the birds, because of its beneficial role for weight gain and growth performance [46] and the inhibition of *Clostridium perfringens* sporulation by its fermentation products [47]. As a consequence, depletion of *Bacteroides* genus could be problematic for chickens, since it may predispose the animal gut to *Clostridium perfringens* infection and gastroenteritis [47]. However, *Bacteroides* genus still outnumbered the other taxa in TM group of the present study. The birds fed with insects also showed similar growth performance to those fed with C diet [21], thus further mitigating this potential negative result. As a final aspect to underline, in spite of the overall positive cecal microbiota modulation due to TM meal utilization, the growth performance of the chickens remained unaffected. However, since a clear cause-effect relationship between diversity and composition of cecal microbiota and bird performance has not yet been confirmed, the gut microbiota findings need to be contextualized with those related to mucin composition and morphology.

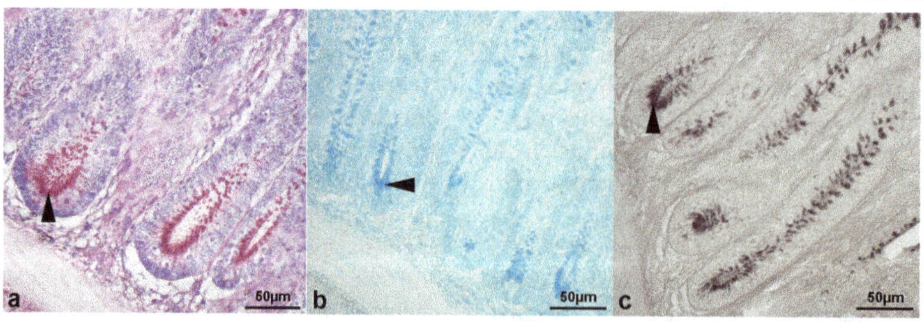

Fig. 6 Histological pictures of (**a**) duodenal crypts stained with periodic-acid Schiff (40× magnification), (**b**) jejunal crypts stained with Alcian Blue pH 2.5 (40× magnification) and (**c**) ileal crypts stained with high iron diamine (40× magnification). Crypt bases (arrowheads) show higher mucin staining intensity than midsection and tip fragments

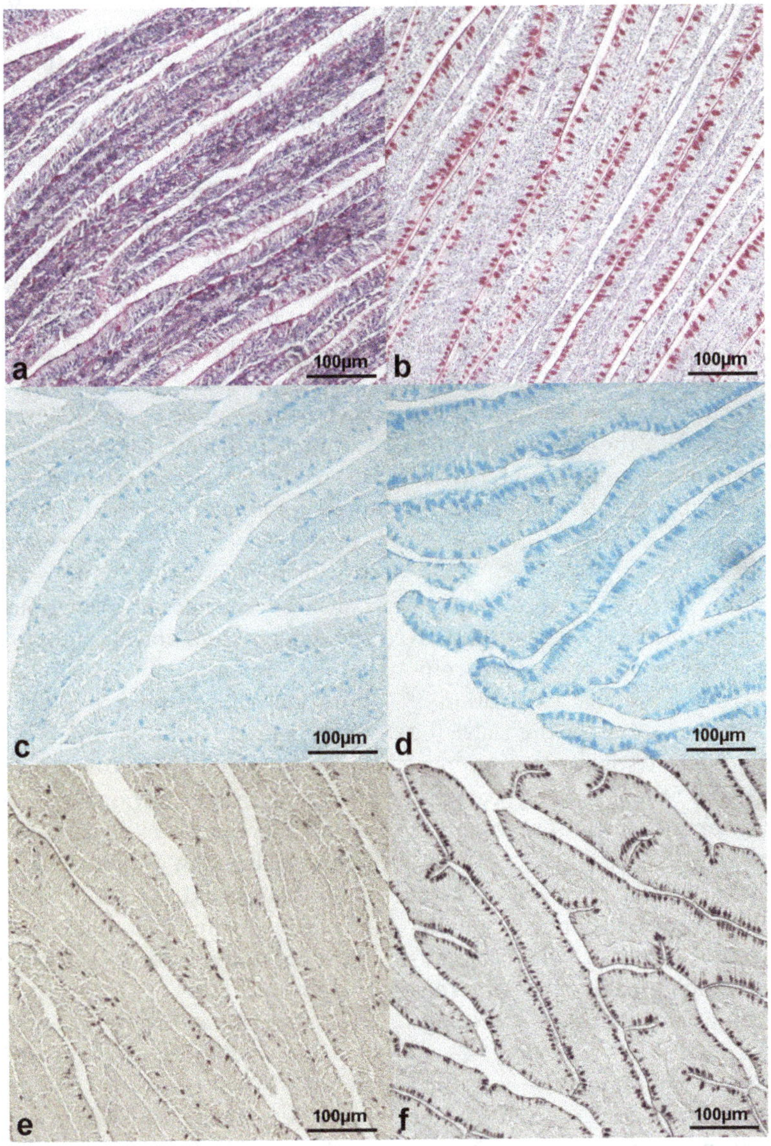

Fig. 7 Histological pictures of duodenal (**a**, **c**, **e**) and ileal (**b**, **d**, **f**) villi stained with (**a**, **b**) periodic-acid Schiff (10× magnification), (**c**, **d**) Alcian Blue pH 2.5 (10× magnification) and (**e**, **f**) high iron diamine (10× magnification). Ileal villi show higher mucin staining intensity than duodenal ones

Intestinal morphology

Dietary TM meal inclusion in free-range chickens of the current trial did not affect the gut morphometric indices, as already reported in details by Biasato et al. [21]. Features of the gastrointestinal tract have been reported to influence the efficiency of utilization of dietary protein [48], which is considered a crucial regulator of poultry growth and reproductive performance [7]. In particular, the microscopic structure of small intestine in terms of villus height and crypt depth is considered the main indicator of intestinal development, health and functionality, such influencing nutrient digestion and absorption [5]. Particular attention is given to crypt and villus since the former is the region in which new intestinal cells are formed and the latter has a fundamental role in the digestion and absorption of nutrients [48]. The ideal gut morphological asset appears to be characterized by long villi and shallow crypts. Indeed, longer villi are generally associated with increased total luminal absorptive area and subsequent satisfactory digestive enzyme action and higher transport of nutrients [7]. In parallel, shallower crypts reflect the prolonged survival of villi without the need for renewal [49], with reduced energy expenditure for this process and consequent enabled growth of other tissues [50]. Since no standardized measurements ranges referring to "long villi" and "shallow crypts" have been determined till now, the identification of unaffected gut morphometric indices and growth performance in the birds fed with insects of the present study is enough to suggest that TM meal does not alter the physiological intestinal morphology and subsequently the digestion efficiency.

Independently of TM meal utilization, the duodenum of the free-range chickens of the current trial showed greater development in relation to the other gut segments. It is well known that the intestine possesses an inherent ability to create and maintain regional differences with regard to mucosal structure [51], thus determining different absorption processes depending on the segment considered [52]. The identification of a proximodistal decreasing gradient of the morphometric indices from the duodenum to the ileum, which is in agreement with the previous studies [4, 52, 53], is another important aspect in terms of preservation of the physiological gut development and absorption processes.

Intestinal mucin composition

The present study is the first one to investigate the effects of dietary TM meal inclusion on gut mucin composition in poultry. Furthermore, the characterization of the three types of mucins (i.e., neutral, acidic sialylated and acidic sulfated) in the two intestinal mucosal compartments (i.e., crypts and villi) of four defined segments (i.e., duodenum, jejunum, ileum and cecum) represents a unique histochemical approach.

Dietary TM meal inclusion in free-range chickens of the current trial did not influence the mucin staining intensity either in the intestinal crypts or villi. This finding could reflect an innate feeding habit, since chickens naturally consume insects when reared in free-range systems [54].

Independently of TM meal utilization, the intestinal crypts of the birds of the current trial showed greater acidic sialylated mucins staining intensity. This represents an interesting finding, since sialic acid groups have protective properties [55] and increase in acidic sialylated mucins production has been hypothesized to represent a defense strategy against mucus degradation by bacterial colonization [4]. On the contrary, similar staining intensity of the three types of mucins was observed in the intestinal villi. The variations in the proportion of the goblet cells containing the mucin types observed in the intestinal crypts instead of villi could be related to the higher presence of mucin-producing goblet cells in the first compartment, as reported by Uni et al. [56].

The intestinal crypts of both chickens fed with control and TM diets of the present research showed lower mucin staining intensity in the cecum compared with the other gut segments. This finding is in agreement with what observed by Tsirtsikos et al. [9], even if the authors provided no explanation for that result. The avian cecum has been found to be a site for fermentation and further digestion of feed (especially for breakdown of cellulose), for utilization and absorption of water and nitrogenous components, for microbial action of both beneficial and disease-causing organism, and as a site for production of immunoglobulins and antibodies [57]. As previously reported mucins enhance the propulsion of chyme, modulate nutrient absorption and protect the intestinal mucosa from enteric pathogens [6]. Since the small intestine represents the major site of nutrient absorption [58] and the ileum has been suggested to be more predisposed to bacterial colonization [4], it is possible to speculate that the mucin synthesis for nutrient metabolism and gut protection could be reduced in the cecum. Furthermore, because the cecum is blind-ended, its contents can be retained for long periods, with no need for mucin secretion to facilitate the propulsion of chyme. Therefore, the decreased mucin staining intensity in the cecal crypts observed in the present study could be considered a physiological feature related to the different anatomy and physiology of the cecum.

Independently of dietary TM meal inclusion, the intestinal villi of the free-range chickens of the present study showed greater mucin staining intensity in the ileum compared with the other gut segments. This is consistent with previous findings in chickens [4, 59], which demonstrated a distal increase in the density of goblet cells along the duodenal-ileal axis [59]. Forder et al. [4] suggested that the distal ileum could be a preferred region for bacterial colonization, thus explaining the need for greater protection and subsequent higher mucin synthesis.

The intestinal crypts of both birds fed with control and TM diets of the current trial showed greater mucin staining intensity in the base compared with the other crypt fragments. This is in agreement with previous studies, in which the decreased stain in the crypt tip has been suggested to be related to the process of proliferation and maturation of goblet cells [9, 56]. Similarly, the intestinal villi showed higher mucin staining intensity in the base compared with the other villus fragments. This is in contrast with what observed by Tsirtsikos et al. [9, 10], which found higher staining intensity at the villus tip and explained this increased accumulation as a confirm for the key role of mucin in the protection of gut epithelium against luminal threats [9, 10]. However, the process of cell proliferation in chicken intestinal epithelium has also been reported to occur along the entire length of the villus, with proliferation activity decreasing from crypt to the upper half of the villus [56]. Therefore, the higher mucin staining intensity observed at the villus base in the birds of the present study can be related to the physiological goblet cells proliferation process occurring in villus compartment.

Conclusions

In conclusion, dietary TM meal inclusion may positively modulate the gut microbiota of the free-range chickens without influencing the intestinal morphology and mucin composition. The identification of physiological cecal community, gut morphological development and mucin dynamics also suggests that insect meal utilization does not negatively affect the gut health of the birds. Furthermore, since the rapid growth of chickens directly depends on morphological and functional integrity of the digestive tract (as herein confirmed), the gut health assessment by a post mortem multidisciplinary approach appears to be fundamental.

Methods

Birds and experimental design

The present study is a part of ongoing research work that aims to investigate the effects of TM meal inclusion on growth performance, haematochemical profile, carcass traits, histological features and gut health of free-range chickens [21]. In order to avoid unnecessary repetition of already published data, a brief summary of the experimental trial is reported below. The experimental protocol was approved by the Ethical Committee of the Department of Veterinary Sciences of the University of Turin [Italy].

A total of 140 43-days-old female Label Hubbard hybrid chickens (female: JA 57 × male: S77CN, mean weight: 716.3 ± 23.2 g, purchased from "Aglietto Natura SRL" farm [Bianzè, VC – Italy]), a medium-growing genotype, were randomly allotted to two dietary treatments. Each of them consisted of five replicate floor pens, with 14 chicks per pen. The experimental unit was the pen. Two diets were formulated: a control corn-soybean-gluten meal-based diet (C), normally used by the breeder, and an experimental diet (TM7.5), in which full-fat TM larvae meal (Gaobeidian Shannong Biology Co. Ltd., Gaobeidian, Hebei province - China) was included at 75 g/kg in complete substitution of corn gluten meal. Details of the diets are shown in Additional file 3. Nutrients digestibility, apparent metabolisable energy and nitrogen-corrected apparent metabolisable energy were previously assessed [17]. All the birds were free-range reared under the same environmental conditions throughout the experimental trial. Feed (isonitrogenous and isoenergetic diets) and water were provided ad libitum for the whole experimental periods. The chickens (showing overall good health conditions at the beginning of the experiment) were regularly vaccinated and showed no signs of illness or mortality throughout the trial. Growth (initial and final live weight, average feed intake and feed conversion ratio) and slaughtering performance (carcass, breasts, thighs, deboned thighs, spleen, bursa of Fabricius, liver, gizzard and abdominal fat weights) were also unaffected by dietary treatments. Welfare-related assessments (footpad dermatitis score evaluation) were also performed after the experimental trial. The experiment lasted 54 days.

Intestinal sampling and processing

A total of ten chickens per treatment (two birds per pen) were randomly selected and slaughtered in a commercial abattoir at 97 days of age. Animals were euthanized by electrical stunning and bleeding. The remaining birds were slaughtered using the same euthanasia procedures and carcasses were submitted to proper disposal methods. Cecal content was collected into sterile plastic tubes that were promptly refrigerated (for a maximum of 2 h) and frozen at – 80 °C until DNA extraction. Intestinal segment samples (approximately 5 cm in length) of duodenum, jejunum, ileum and cecum were excised and flushed with 0.9% saline to remove all the content. The collected segments of intestine were the loop of the duodenum (duodenum), the tract before Meckel's diverticulum (jejunum), the tract before the ileocolic junction (ileum) and the apex of the caeca (cecum). Gut segments were fixed in Carnoy's and 10% buffered formalin solutions for morphometric analysis and histochemical staining, respectively. Tissues were routinely embedded in paraffin wax blocks, sectioned at 5 μm thickness and mounted on glass slides.

DNA extraction and sequencing

A pool of the cecal content from two chickens per pen (five pools per feeding group) was submitted to DNA extraction and sequencing. DNA was extracted with a commercial kit (DNAzol® Reagent, Thermo Fisher Scientific) according to the manufacturer's instructions. One μl of RNase (Illumina Inc., San Diego, CA) was added to digest RNA in the DNA samples, with an incubation of 1 h at 37 °C. DNA was quantified using the NanoDrop and standardized at 5 ng/μl. Microbial diversity was studied by sequencing the amplified V3-V4 region of the 16S rRNA gene by using primers and PCR conditions previously reported [60]. Samples multiplexing, library purification pooling and sequencing was carried out as described in the "16S Metagenomic Sequencing Library Preparation" guide by Illumina. Libraries were sequenced by BMR genomics (Padova, Italy) on a MiSeq platform (Illumina Italy s.r.l., Milan, Italy), leading to 250 bp, paired-end reads. After the first purification step following the Illumina sample preparation procedure, the library was combined with the sequencing adapters and dual indices using the Nextera XT Index Kit (Illumina, San Diego, USA), obtaining the multiplexed paired-end libraries. Individual libraries concentration in nM were calculated based on the size of amplicons by using a Bioanalyzer (Agilent) and diluted to 4 nM, denaturated with 0.2 N NaOH and spiked with 20% (*v*/v) of PhiX.

Histomorphological investigations

Carnoy-fixed and paraffin-embedded intestinal sections of ten chickens per feeding group (two birds per pen) were submitted to Haematoxylin & Eosin staining and one slide per each intestinal segment was examined by light microscopy. Each slide was captured with a Nikon DS-Fi1 digital camera coupled to a Zeiss Axiophot microscope using a 2.5× objective lens. NIS-Elements F software was used for image capturing. Morphometric analysis was performed by Image®-Pro Plus software. The evaluated morphometric indices were villus height (Vh, from the tip of the villus to the crypt), crypt depth (Cd, from the base of the villus to the submucosa) and the villus height to crypt depth ratio (Vh/Cd) [7]. Morphometric measurements were performed on 10 well-oriented and intact villi and 10 crypts chosen from duodenum, jejunum and ileum [8].

Histochemical staining

Formalin-fixed and paraffin-embedded intestinal sections of ten chickens per feeding group (two birds per pen) were submitted to three different histochemical staining, in order to characterize the three types of mucins [4].

Neutral mucins were identified by periodic-acid Schiff staining. Sections were brought to water, immersed in 0.5% periodic acid solution for 20 min, washed in running tap water for 5 min and immersed in Schiff's reagent for a further 30 min. Sections were successively rinsed in running tap water for 10 min, dehydrated and mounted. Neutral mucins stained magenta [61].

Acidic sialylated mucins were identified by Alcian Blue pH 2.5 staining. Sections were brought to water, immersed in 8 G X alcian blue in 3% acetic acid solution for 30 min, washed in running tap water for further 5 min and successively dehydrated and mounted. Acidic sialylated mucins stained blue [62].

Acidic sulfated mucins were identified by high iron diamine staining. Sections were brought to water, oxidized in 1% periodic acid solution for 10 min and washed in running tap water for further 5 min. Sections were successively immersed in the high iron diamine solution (120 mg metadiamine, 20 mg paradiamine and 1.4 ml 10% ferric chloride in 50 ml distilled water) for 18 h, rinsed very rapidly in tap water, dehydrated and mounted. Acidic sulfated mucins stained purple-black [63].

Mucin staining intensity

One slide per histochemical staining for each intestinal segment was examined by light microscopy. A total of 10 crypts and 10 villi for each slide were evaluated. Crypts and villi were divided into three fragments (base, midsection and tip), according to Tsirtsikos et al. [9, 10]. Mucin staining intensity of goblet cells was scored semiquantitatively for each fragment as follows: grade 0 for absent staining, grade 1 for mild staining, grade 2 for moderate staining and grade 3 for marked staining. The score was formulated according to that proposed by Tsirtsikos et al. [9, 10] for the mucus layer and depended on the number of positive goblet cells and the intensity of the staining. All the slides were assessed blinded by three observers (IB, EB and MTC) and the discordant cases were reviewed at a multi-head microscope until a consensus was reached.

Bioinformatics and statistical analysis

The experimental unit was the pen for 16S rRNA sequences and bird for morphometric and histochemical data.

Paired-end reads were first merged using FLASH software [64] with default parameters. Joint reads were further quality filtered (at Phred > Q20) using QIIME 1.9.0 software [65] and the pipeline recently described [66]. Briefly, operational taxonomic units (OTUs) were picked at 97% of similarity by means of UCLUST clustering methods [67] and representative sequences of each cluster were used to assign taxonomy using the Greengenes 16S rRNA gene database (version 2013). Alpha diversity indices were calculated using the *diversity* function of the vegan package [68]. A pairwise t-test was used to assess differences between the theses. The diversity indices were further analyzed by pairwise comparisons using Wilcoxon rank sum test to assess differences between the dietary treatments. Weighted UniFrac distance matrices were used to perform Adonis and ANOSIM statistical tests in R environment. A filtered OTU table was generated at 0.1% abundance in at least 2 samples through QIIME. The table was then used to produce the Principal Component Analysis (PCA) in R environment (https://www.r-project.org). OTU table showed the higher taxonomy resolution reached by the 16S data. When the taxonomy assignment was not able to reach the genus level, the phylum, class, order or family were displayed. Pairwise Kruskal-Wallis tests were used to find significant differences in microbial taxa abundance according to the dietary treatment. The experimental unit was the pen. *P*-values were adjusted for multiple testing and a false discovery rate (FDR) < 0.05 considered as significant.

The statistical analysis of morphometric and histochemical data was performed using IBM SPSS Statistics V20.0.0 software. The influence of diet on intestinal morphometric measurements was tested using Student's *t* test for independent samples. Morphometric data were also analyzed by means of One-way ANOVA (post hoc test: Duncan's multiple range test) to evaluate the influence of intestinal segment within each dietary treatment. Results were expressed as mean and pooled standard error of the mean (SEM). *P* values < 0.05 were considered statistically significant [21]. Histochemical data were analyzed by fitting a generalized linear model (GLM) similar to those proposed by Tsirtsikos et al. [9, 10]. The GLM allowed the mean mucin staining intensity scores to depend on linear predictors such as diet, mucin type, intestinal segment and fragment within crypts or villi through a negative binomial response probability distribution with a nonlinear link function (log). The bird within treatment effect was also included in the GLM as the repeated factor. A hybrid method for parameter estimation was used and a type III analysis with Wald chi-square test was applied to assess the model effects. Results were expressed as least squares means and SEM and the interactions between factor levels were evaluated by pairwise comparisons. P values < 0.05 were considered statistically significant. Statistical analysis was performed by procedure "Generalized Linear Models".

Additional files

Additional file 1: Good's coverage and α-diversity measures of cecal microbiota of free-range chickens fed with control (C) and 7.5% inclusion level of *Tenebrio molitor* meal (TM7.5) diets. Description column indicates the 5 replicate pens of control (C_1, C_2, C_3, C_4 and C_5) and 7.5% inclusion level of *Tenebrio molitor* meal (TM7.5_1, TM7.5_2, TM7.5_3, TM7.5_4 and TM7.5_5) dietary treatments.

Additional file 2: Relative abundance of the main bacterial phyla and genera of cecal microbiota of free-range chickens fed with control (C) and 7.5% inclusion level of *Tenebrio molitor* meal (TM7.5) diets.

Additional file 3: Ingredients and chemical composition of the experimental diets. The mineral-vitamin premix (Trevit Volatili 3.5 - Trei - Rio Saliceto (RE) Italy) given values are supplied per kg: 650.000 IU of vitamin A; 65.000 IU of vitamin D3; 650 IU of vitamin E; 80 mg of vitamin K; 80 mg of vitamin B1; 150 mg of vitamin B2; 770 mg of vitamin B3; 80 mg of vitamin B6; 0.5 mg of vitamin B12; 240 mg of pantothenic acid; 4700 mg of betaine; 1750 mg of Iron (II) carbonate; 1835 mg of Magnesium oxide; 1612 mg of Zinc oxide; 178 mg of Copper (II) oxide; 18.3 mg of Potassium iodide; 6.6 mg of Sodium selenite; 4100 mg of DL-methionine; 5500 mg of L-lysine; 120 g Calcium carbonate; 450 g Calcium phosphate; 11.5 g of Sodium chloride. SFA: saturated fatty acids; MUFA: monounsaturated fatty acids; PUFA: polyunsaturated fatty acids. Other fatty acids (all less than 0.40 g/kg DM): C12:0, C14:0, C14:1 *cis*9, C16:1 *cis*9, C18:1 *cis*11, C20:0, C18:3 n6, C20:1 *cis*9, C20:1 *cis*11. TM, *Tenebrio molitor*; AME, apparent metabolizable energy; DM, dry matter; CP, crude protein; C = control; TM7.5 = 7.5% inclusion level of *Tenebrio molitor*.

Abbreviations

C: Control; Cd: Crypt depth; FDR: False discovery rate; GLM: Generalized linear model; OTUs: Operational taxonomic units; PD: Phylogenetic Diversity; TM: *Tenebrio molitor*; TM7.5: 7.5% inclusion level of *Tenebrio molitor*; Vh: Villus height; Vh/Cd: Villus height to crypt depth ratio

Acknowledgments

The authors are grateful to Dr. Oreste Massimino, Mr. Roberto Borgogno and Mr. Aldo Pozzo - "O.R.A. Agricola Srl" (Cherasco, CN - Italy) and Mr. Alessandro Varesio and Mrs. Roberta Lacopo - "Fattoria La Fornace", Montechiaro d'Asti (At - Italy) - for their technical support. The authors are also grateful to Dr. Paolo Montersino, Mr. Dario Sola and Mr. Mario Colombano for bird care and technical support.

Funding

The research was supported by the University of Turin (2014–2015) and by the European Knowledge and Innovation Community (KIC), within the EIT Food program 'INNOPOULTRY. The poultry food chain: tackling old problems with innovative approaches (Project ID 18023). The funding bodies were involved in the animals purchase, collection of data and article processing charges.

Authors' contributions

IB conceived the study, collected the samples, performed the histomorphological and histochemical investigations, analyzed the data, wrote the manuscript and prepared the tables and figures. IF analyzed the data and prepared the tables and figures. EB collected the samples and performed the histomorphological and histochemical investigations. EG collected the samples and performed the DNA extraction. SD and FG collected the samples. AS2 processed and cut the gut samples. LG, AS1 and LC conceived the study. MTC conceived the study and performed the histomorphological and histochemical investigations. All authors read, edited and approved the final manuscript.

Competing interests

The authors declare that they have no competing interests.

Author details

[1]Department of Veterinary Sciences, University of Turin, Largo Paolo Braccini 2, 10095 Grugliasco, TO, Italy. [2]Department of Agricultural, Forest and Food Sciences, University of Turin, Largo Paolo Braccini 2, 10095 Grugliasco, TO, Italy. [3]Institute of Science of Food Production, National Research Council, Largo Paolo Braccini 2, 10095 Grugliasco, TO, Italy. [4]Institute of Multidisciplinary Research on Sustainability, University of Turin, Via Accademia Albertina 13, 10100 Turin, Italy.

References

1. Kogut MH, Arsenault RJ. Editorial: gut health: the new paradigm in food animal production. Front Vet Sci. 2016;3:71.
2. Bailey RA. Gut health in poultry - the world within. Huntsville: Aviagen; 2013.
3. Pan D, Yu Z. Intestinal microbiome of poultry and its interaction with host and diet. Gut Microbes. 2014;5:108–19.
4. Forder RE, Howarth GS, Tivey DR, Hughes RJ. Bacterial modulation of small intestinal goblet cells and mucin composition during early posthatch development of poultry. Poult Sci. 2007;86:2396–403.
5. Wang JX, Peng KM. Developmental morphology of the small intestine of African ostrich chicks. Poult Sci. 2008;87:2629–35.
6. Forstner G, Forstner JF. Gastrointestinal mucus. In: Johnson LR, editor. Physiology of the gastrointestinal tract. New York: Raven Press; 1994. p. 1255–84.
7. Laudadio V, Passantino L, Perillo A, Lopresti G, Passantino A, Khan RU, Tufarelli V. Productive performance and histological features of intestinal mucosa of broiler chickens fed different dietary protein levels. Poult Sci. 2012;91:265–70.
8. Qaisrani SN, Moquet PC, van Krimpen MM, Kwakkel RP, Verstegen MW, Hendriks WH. Protein source and dietary structure influence growth performance, gut morphology, and hindgut fermentation characteristics in broilers. Poult Sci. 2014;93:3053–64.
9. Tsirtsikos P, Fegeros K, Kominakis A, Balaskas C, Mountzouris KC. Modulation of intestinal mucin composition and mucosal morphology by dietary phytogenic inclusion level in broilers. Animal. 2012a;6:1049–57.
10. Tsirtsikos P, Fegeros K, Balaskas C, Kominakis A, Mountzouris KC. Dietary probiotic inclusion level modulates intestinal mucin composition and mucosal morphology in broilers. Poult Sci. 2012b;91:1860–8.
11. Reed S, Neuman H, Glahn RP, Koren O, Tako E. Characterizing the gut (Gallus gallus) microbiota following the consumption of an iron biofortified Rwandan cream seeded carioca (*Phaseolus Vulgaris* L.) bean-based diet. PLoS One. 2017;12:e0182431.
12. Yegani M, Korver DR. Factors affecting intestinal health in poultry. Poult Sci. 2008;87:2052–63.
13. FAO (Food and Agriculture Organization of the United Nations). Edible insects – future prospects for food and feed security, FAO Forestry Paper, vol. IX; 2013. p. 171.

14. van Huis A. Potential of insects as food and feed in assuring food security. Annu Rev Entomol. 2013;58:563–83.
15. Makkar HPS, Tran G, Heuzé V, Ankers P. State of the art on use of insects as animal feed. Anim Feed Sci Technol. 2014;197:1–33.
16. Bovera F, Piccolo G, Gasco L, Marono S, Loponte R, Vassalotti G, Mastellone V, Lombardi P, Attia YA, Nizza A. Yellow mealworm larvae (*Tenebrio molitor*, L.) as a possible alternative to soybean meal in broiler diets. Br Poult Sci. 2015;56:569–75.
17. De Marco M, Martínez S, Hernandez F, Madrid J, Gai F, Rotolo L, Belforti M, Bergero D, Katz H, Dabbou S, Kovitvadhi A, Zoccarato I, Gasco L, Schiavone A. Nutritional value of two insect meals (*Tenebrio molitor* and *Hermetia illucens*) for broiler chickens: apparent nutrient digestibility, apparent ileal amino acid digestibility and apparent metabolizable energy. Anim Feed Sci Technol. 2015;209:211–8.
18. Bovera F, Loponte R, Marono S, Piccolo G, Parisi G, Iaconisi V, Gasco L, Nizza A. Use of *Tenebrio molitor* larvae meal as protein source in broiler diet: effect on growth performance, nutrient digestibility and carcass and meat traits. J Anim Sci. 2016;94:639–47.
19. Biasato I, Gasco L, De Marco M, Renna M, Rotolo L, Dabbou S, Capucchio MT, Biasibetti E, Tarantola M, Bianchi C, Cavallarin L, Gai F, Pozzo L, Dezzutto D, Bergagna S, Schiavone A. Effects of yellow mealworm larvae (*Tenebrio molitor*) inclusion in diets for female broiler chickens: implications for animal health and gut histology. Anim Feed Sci Technol. 2017;234:253–63.
20. Biasato I, Gasco L, De Marco M, Renna M, Rotolo L, Dabbou S, Capucchio MT, Biasibetti E, Tarantola M, Sterpone L, Cavallarin L, Gai F, Pozzo L, Bergagna S, Dezzutto D, Zoccarato I, Schiavone A. Yellow mealworm larvae (*Tenebrio molitor*) inclusion in diets for male broiler chickens: effects on growth performance, gut morphology, and histological findings. Poult Sci. 2018;97:540–8.
21. Biasato I, De Marco M, Rotolo L, Renna M, Dabbou S, Capucchio MT, Biasibetti E, Tarantola M, Costa P, Gai F, Pozzo L, Dezzutto D, Bergagna S, Martinez S, Gasco L, Schiavone A. Effects of dietary *Tenebrio molitor* meal inclusion in free-range chickens. J Anim Physiol Anim Nutr. 2016; 100:1104–12.
22. Xu Y, Yang H, Zhang L, Su Y, Shi D, Xiao H, Tian Y. High-throughput sequencing technology to reveal the composition and function of cecal microbiota in Dagu chicken. BMC Microbiol. 2016;16:259.
23. Mancabelli L, Ferrario C, Milani C, Mangifesta M, Turroni F, Duranti S, Lugli GA, Viappiani A, Ossiprandi MC, van Sinderen D, Ventura M. Insights into the biodiversity of the gut microbiota of broiler chickens. Environ Microbiol. 2016;18:4727–38.
24. Ferrario C, Alessandri G, Mancabelli L, Gering E, Mangifesta M, Milani C, Lugli GA, Viappiani A, Duranti S, Turroni F, Ossiprandi MC, Hiyashi R, Mackie R, van Sinderen D, Ventura M. Untangling the cecal microbiota of feral chickens by culturomic and metagenomic analyses. Environ Microbiol. 2017;19:4771–83.
25. Oakley BB, Lillehoj HS, Kogut MH, Kim WK, Maurer JJ, Pedroso A, Lee MD, Collett SR, Johnson TJ, Cox NA. The chicken gastrointestinal microbiome. FEMS Microbiol Lett. 2014;360:100–12.
26. Wise MG, Siragusa GR. Quantitative analysis of the intestinal bacterial community in one- to three-week-old commercially reared broiler chickens fed conventional or antibiotic-free vegetable-based diets. J Appl Microbiol. 2007;102:1138–49.
27. Tillman GE, Haas GJ, Wise MG, Oakley B, Smith MA, Siragusa GR. Chicken intestine microbiota following the administration of lupulone, a hop-based antimicrobial. FEMS Microbiol Ecol. 2011;77:395–403.
28. Wei S, Morrison M, Yu Z. Bacterial census of poultry intestinal microbiome. Poult Sci. 2013;92:671–83.
29. Gong J, Si W, Forster RJ, Huang R, Yu H, Yin Y, Yang C, Han Y. 16S rRNA gene-based analysis of mucosa-associated bacterial community and phylogeny in the chicken gastrointestinal tracts: from crops to ceca. FEMS Microbiol Ecol. 2007;59:147–57.
30. Danzeisen JL, Kim HB, Isaacson RE, Tu ZJ, Johnson TJ. Modulations of the chicken cecal microbiome and metagenome in response to anticoccidial and growth promoter treatment. PLoS One. 2011;6:e27949.
31. Stanley D, Denman SE, Hughes RJ, Geier MS, Crowley TM, Chen H, Haring VR, Moore RJ. Intestinal microbiota associated with differential feed conversion efficiency in chickens. Appl Microbiol Biotechnol. 2012; 96:1361–9.
32. Costa MC, Bessegatto JA, Alfieri AA, Weese JS, Filho JA, Oba A. Different antibiotic growth promoters induce specific changes in the cecal microbiota membership of broiler chicken. PLoS One. 2017;12:e0171642.
33. Callaway TR, Dowd SE, Wolcott RD, Sun Y, McReynolds JL, Edrington TS, Byrd JA, Anderson RC, Krueger N, Nisbet DJ. Evaluation of the bacterial diversity in cecal contents of laying hens fed various molting diets by using bacterial tag-encoded FLX amplicon pyrosequencing. Poult Sci. 2009;88: 298–302.
34. Xue Z, Zhang W, Wang L, Hou R, Zhang M, Fei L, Zhang X, Huang H, Bridgewater LC, Jiang Y, Jiang C, Zhao L, Pang X, Zhang Z. The bamboo-eating giant panda harbors a carnivore-like gut microbiota, with excessive seasonal variations. MBio. 2015;6:e00022–15.
35. Kuhn I, Katouli M, Lund A, Wallgren P, Mollby R. Phenotype diversity and stability of intestinal coliform flora in piglets during the first three months of age. Microbial Ecol Health Dis. 1993;6:101–7.
36. Stanley D, Geier MS, Hughes RJ, Denman SE, Moore RJ. Highly variable microbiota development in the chicken gastrointestinal tract. PLoS One. 2013;8:e84290.
37. Sergeant MJ, Constantinidou C, Cogan TA, Bedford MR, Penn CW, Pallen MJ. Extensive microbial and functional diversity within the chicken cecal microbiome. PLoS One. 2014;9:e91941.
38. Li Y, Xu Q, Huang Z, Lv L, Liu X, Yin C, Yan H, Yuan J. Effect of Bacillus subtilis CGMCC 1.1086 on the growth performance and intestinal microbiota of broilers. J Appl Microbiol. 2016;120:195–204.
39. Singh KM, Shah T, Deshpande S, Jakhesara SJ, Koringa PG, Rank DN, Joshi CG. High through put 16S rRNA gene-based pyrosequencing analysis of the fecal microbiota of high FCR and low FCR broiler growers. Mol Biol Rep. 2012;39:10595–602.
40. Pryde SE, Duncan SH, Hold GL, Stewart CS, Flint HJ. The microbiology of butyrate formation in the human colon. FEMS Microbiol Lett. 2002; 217:133–9.
41. Yang L, Liu S, Ding J, Dai R, He C, Xu K, Honaker CF, Zhang Y, Siegel P, Meng H. Gut microbiota co-microevolution with selection for host humoral immunity. Front Microbiol. 2017;8:1243.
42. Onrust L, Ducatelle R, Van Driessche K, De Maesschalck C, Vermeulen K, Haesebrouck F, Eeckhaut V, Van Immerseel F. Steering endogenous butyrate production in the intestinal tract of broilers as a tool to improve gut health. Front Vet Sci. 2015;2:75.
43. Liu C, Finegold SM, Song Y, Lawson PA. Reclassification of *Clostridium coccoides, Ruminococcus hansenii, Ruminococcus hydrogenotrophicus, Ruminococcus luti, Ruminococcus productus* and *Ruminococcus schinkii* as *Blautia coccoides* gen. Nov., comb. nov., *Blautia hansenii* comb. nov., *Blautia hydrogenotrophica* comb. nov., *Blautia luti* comb. nov., *Blautia producta* comb. nov., *Blautia schinkii* comb. nov. and description of *Blautia wexlerae* sp. nov., isolated from human faeces. Int J Syst Evol Microbiol. 2008;58: 1896–902.
44. Sunkara LT, Jiang W, Zhang G. Modulation of antimicrobial host defense peptide gene expression by free fatty acids. PLoS One. 2012;7:e49558.
45. Gillespie MJ Stanley D, Chen H, Donald JA, Nicholas KR, Moore RJ, Crowley TM. Functional similarities between pigeon 'milk' and mammalian milk: induction of immune gene expression and modification of the microbiota. PLoS One. 2012;7:e48363.
46. Chang CL, Chung CY, Kuo CH, Kuo TF, Yang CW, Yang WC. Beneficial effect of Bidens pilosa on body weight gain, food conversion ratio, gut Bacteria and coccidiosis in chickens. PLoS One. 2016;11:e0146141.
47. Wrigley DM. Inhibition of *Clostridium perfringens* sporulation by *Bacteroides fragilis* and short-chain fatty acids. Anaerobe. 2004;10:295–300.
48. Swatson HK, Gous R, Iji PA, Zarrinkalam R. Effect of dietary protein level, amino acid balance, and feeding level on growth, gastrointestinal tract, and mucosal structure of the small intestine in broiler chickens. Anim Res. 2002; 51:501–15.
49. Oliveira MC, Rodrigues EA, Marques RH, Gravena RA, Guandolini GC, Moraes VMB. Performance and morphology of intestinal mucosa of broilers fed mannan-oligosaccharides and enzymes. Arq Bras Med Vet Zootec. 2008;60: 442–8.
50. Miles RD, Butcher GD, Henry PR, Littell RC. Effect of antibiotic growth promoters on broiler performance, intestinal growth parameters, and quantitative morphology. Poult Sci. 2006;85:476–85.
51. Ferraris RP, Villenas SA, Diamond J. Regulation of brush-border enzyme activities and enterocyte migration rates in mouse small intestine. Am J Phys. 1992;262:G1047–59.

52. de Verdal H, Mignon-Grasteau S, Jeulin C, Le Bihan-Duval E, Leconte M, Mallet S, Martin C, Narcy A. Digestive tract measurements and histological adaptation in broiler lines divergently selected for digestive efficiency. Poult Sci. 2010;89:1955–61.
53. Iji PA, Saki A, Tivey DR. Body and intestinal growth of broiler chicks on a commercial starter diet. 1. Intestinal weight and mucosal development. Br Poult Sci. 2001;42:505–13.
54. Zuidhof MJ, Molnar CL, Morley FM, Wray TL, Robinson FE, Khan BA, Al-Ani L, Goonewardene LA. Nutritive value of house fly (*Musca domestica*) larvae as a feed supplement for Turkey poults. Anim Feed Sci Technol. 2003;105:225–30.
55. Corfield AP, Wagner SA, Clamp JR, Kriaris MS, Hoskins LC. Mucin degradation in the human colon: production of sialidase, sialate O-acetylesterase, N-acetylneuraminate lyase, arylesterase, and glycosulfatase activities by strains of fecal bacteria. Infect Immun. 1992;60:3971–8.
56. Uni Z, Platin R, Sklan D. Cell proliferation in chicken intestinal epithelium occurs both in the crypts and along the villus. J Comp Physiol B. 1998;168: 241–7.
57. Clench MH, Mathias JR. The avian cecum – a review. Wilson Bull. 1995;107: 93–121.
58. Dingle JG. Module 3: nutrient procurement and processing, study book: poultry husbandry 1. Toowoomba: DEC, UCSQ; 1991.
59. Uni Z, Smirnov A, Sklan D. Pre- and posthatch development of goblet cells in the broiler small intestine: effect of delayed access to feed. Poult Sci. 2003;82:320–7.
60. Klindworth A, Pruesse E, Schweer T, Peplies J, Quast C, Horn M, Glöckner FO. Evaluation of general 16S ribosomal RNA gene PCR primers for classical and next-generation sequencing-based diversity studies. Nucleic Acids Res. 2013;41:e1.
61. McManus JFA. Histological demonstration of mucin after periodic acid. Nature. 1946;158:202.
62. Pearse AGE. Histochemistry: Theoretical and Applied. 2nd ed. Churchill, Livingstone: London; 1960.
63. Spicer SS. Diamine methods for differentiating mucosubstances histochemically. J Histochem Cytochem. 1965;13:211–34.
64. Magoč T, Salzberg SL. FLASH: fast length adjustment of short reads to improve genome assemblies. Bioinformatics. 2011;27:2957–63.
65. Caporaso JG, Kuczynski J, Stombaugh J, Bittinger K, Bushman FD, Costello EK, Fierer N, Peña AG, Goodrich JK, Gordon JI, Huttley GA, Kelley ST, Knights D, Koenig JE, Ley RE, Lozupone CA, McDonald D, Muegge BD, Pirrung M, Reeder J, Sevinsky JR, Turnbaugh PJ, Walters WA, Widmann J, Yatsunenko T, Zaneveld J, Knight R. QIIME allows analysis of high-throughput community sequencing data. Nat Methods. 2010;7:335–6.
66. Ferrocino I, Bellio A, Romano A, Macori G, Rantsiou K, Decastelli L, Cocolin L. RNA-based amplicon sequencing reveals microbiota development during ripening of artisanal versus industrial lard d'Arnad. Appl Environ Microbiol. 2017;83. https://doi.org/10.1128/AEM.00983-17.
67. Edgar RC. Search and clustering orders of magnitude faster than BLAST. Bioinformatics. 2010;26:2460–1.
68. Dixon P. VEGAN, a package of R functions for community ecology. J Veg Sci. 2003;14:927–30.

18

Prebiotics and synbiotics – *in ovo* delivery for improved lifespan condition in chicken

M. Siwek[1*], A. Slawinska[1], K. Stadnicka[1], J. Bogucka[2], A. Dunislawska[1] and M. Bednarczyk[1]

Abstract

Commercially produced chickens have become key food-producing animals in the global food system. The scale of production in industrial settings has changed management systems to a point now very far from traditional methods. During the perinatal period, newly hatched chicks undergo processing, vaccination and transportation, which introduces a gap in access to feed and water. This gap, referred to as the hatching window, dampens the potential for microflora inoculation and as such, prevents proper microbiome, gastrointestinal system and innate immunity development. As a consequence, the industrial production of chickens with a poor microbial profile leads to enteric microbial infestation and infectious disease outbreaks, which became even more prevalent after the withdrawal of antibiotic growth promoters on many world markets (e.g., the EU).

This review presents the rationale, methodology and life-long effects of *in ovo* stimulation of chicken microflora. *In ovo* stimulation provides efficient embryonic microbiome colonization with commensal microflora during the perinatal period. A carefully selected bioactive formulation (prebiotics, probiotics alone or combined into synbiotics) is delivered into the air cell of the egg on day 12 of egg incubation. The prebiotic penetrates the outer and inner egg membranes and stimulates development on the innate microflora in the embryonic guts. Probiotics are available after the mechanical breakage of the shell membranes by the chick's beak at the beginning of hatching (day 19). The intestinal microflora after *in ovo* stimulation is potent enough for competitive exclusion and programs the lifespan condition. We present the effects of different combinations of prebiotic and probiotic delivered *in ovo* on day 12 of egg incubation on microflora, growth traits, feed efficiency, intestinal morphology, meat microstructure and quality, immune system development, physiological characteristics and the transcriptome of the broiler chickens.

We discuss the differences between *in ovo* stimulation (day 12 of egg incubation) and *in ovo* feeding (days 17–18 of egg incubation) and speculate about possible future developments in this field. In summary, decades of research on *in ovo* stimulation and the lifelong effects support this method as efficient programming of lifespan conditions in commercially raised chickens.

Keywords: Broilers, Synbiotics, Hatching window, Embryonic development, Day 12 of egg incubation, Microflora, Performance, Intestinal morphology, Gene expression

Introduction

Commercially produced chickens, together with other poultry species, have become key food-producing animals within the global food system. Both egg and meat-type chickens provide easily achievable and affordable animal protein, which is particularly important for food security in developing countries. An increase in global food production is critical to meet the demand of the rapidly growing human population. The recent prognosis for overall growth in agriculture is expected to be between 25 and 70% by 2050 compared to current production rates [1]. Poultry production is predicted to double worldwide, especially in developing countries, which will require further intensification of the chicken food chain [2]. These increased production rates will be accompanied by serious quality adjustments, mainly in animal welfare and food safety. For this reason, anti-antibiotic strategies will continue, and the introduction of prebiotic and/or probiotic preparations will be in demand as safe and efficient methods to improve food production in a manner called

* Correspondence: siwek@utp.edu.pl
[1]Department of Animal Biotechnology and Genetics, UTP University of Science and Technology, Mazowiecka, 28 85-084 Bydgoszcz, Poland
Full list of author information is available at the end of the article

"sustainable intensification." Meeting all these challenges in the future will be possible with the further development of technologies of precision livestock farming for egg [3] and meat [4] poultry production.

One of the precision livestock farming tools in poultry production is *in ovo* technology for modulating the conditions inside the egg through nutrients, vaccines and other bioactives. It allows depositing a certain amount of a carefully selected substance into a specific site within an incubating egg. *In ovo* technology is focused on the most critical time in the bird's development, which is the perinatal period. The perinatal period lasts from the final days of the egg's incubation to the first days post-hatching. During that time, the embryo has to adjust to a change in diet (from fat-rich to carbohydrate-rich) and exposure to environmental microbes. In commercial settings, newly hatched chicks are first processed at the hatchery and then transported to the farm before having received any feed or water. Given the large numbers of industrially hatched chicks per batch, these procedures can take a significant amount of time (the so-called hatching window, discussed further in Chapter 4). The disadvantages of the hatching window are practically unavoidable given the constantly growing scale of poultry production. For this reason, *in ovo* technology has been developed to facilitate manipulation of the chicken embryo before hatching. In principal, it is based on the mechanical delivery of substances directly into the incubating egg. This technology was primarily established for the vaccination of 18-day-old embryos against multiple infectious agents, including the Marek's disease virus and infectious bursal disease [5, 6]. Apart from vaccination, *in ovo* technology has been applied to the delivery of prebiotics, probiotics, synbiotics, vitamins, hormones, carbohydrates and peptides [7, 8]. Such precise manipulation of the embryo improves the robustness and resilience of the hatched chicks, which contributes to their further post-hatching development.

The literature describes two major time points of chicken embryo development that have been successfully used for substance delivery through *in ovo* technology. The first time point is around day 12 of egg incubation and has been used solely for the delivery of prebiotics and synbiotics. The deposition site at this stage of embryonic development is an air cell lined by two layers of egg membranes, which are in contact with the highly vascularized chorioallantoic membrane. The route mechanisms of prebiotic and synbiotic penetration through egg membranes are discussed in detail in Chapter 4. In principle, delivering prebiotics at this time point is aimed to stimulate native egg microflora. Pedroso et al. [9] has already proven that neonatal chicks possess microbiota within their intestinal tract even before hatching [9]. However, their diversity and abundance are not effective enough for the competitive exclusion of undesirable microorganisms. Nevertheless, these microbiota are most likely stimulated by *in ovo*-delivered prebiotic. Two independent studies have demonstrated that prebiotics administered *in ovo* on day 17 [10] and on days 12 or 17 [11] increased the number of bifidobacteria in newly hatched chicks. The second study especially clearly shows that the most effective time point for prebiotic delivery, defined by the number of bifidobacteria, is day 12 of egg incubation. When the bioactive formulation for *in ovo* delivery on day 12 of egg incubation is a synbiotic, then prebiotic stimulates native microflora from day 12 to 18 of egg incubation and probiotic is ingested on day 18 after the chick has started pipping. For this reason, the delivery of prebiotics and synbiotics on day 12 of egg incubation in chickens is called *in ovo* stimulation (as opposed to *in ovo* feeding, discussed further). This method has been developed and patented by K. Gulewicz and M. Bednarczyk [12].

The second time point for the *in ovo* delivery of bioactives is around days 17/18 of egg incubation. Then, the chicken embryo is completely developed and, until hatching, it will only grow, using the yolk as a source of nutrients. The perinatal period is crucial for the chick because it prepares it for life post-hatching. *In ovo* technology used at this time point aims to mitigate the negative effects of starvation during the hatching window. For this reason, the late-term chicken embryo is supplemented with nutrients sufficient for securing the newly hatched chick through the hatching window. This method is referred to as *in ovo* feeding and has been patented by Uni and Ferket [13]. The differences between *in ovo* stimulation and *in ovo* feeding are presented in Chapter 4. There have been trials to extend *in ovo* feeding to microbiome stimulation and competitive exclusion by injecting probiotics on day 17/18 of egg incubation, reviewed by Roto et al. [8]. The question could be asked here, what difference it makes for intestinal microflora development as long as the inoculation is done before hatching? In fact, the effect of the time point is tremendous when it comes to *in ovo* delivery of prebiotics and synbiotics. First, *in ovo* stimulation on day 12 of egg incubation is primarily based on the enhancement of growth of native flora present in the embryo by using prebiotics or synbiotics. Second, the injection site differs between day 12 and day 18 of egg incubation in embryos due to the size and structural changes of the egg. On day 12 of egg incubation, the site of *in ovo* injection is the air cell, which is safe for the embryo and easy to automate. A late-term embryo is more likely to be traumatized by an improperly performed injection. The injection site can be the amnion or the embryo, depending on the chicken embryo position in the egg. Third, *in ovo* feeding on day 18 of egg incubation is likely to reduce hatchability, unlike *in ovo* stimulation on day 12 of egg incubation, which leaves hatchability rates practically intact [7, 8].

In light of these considerations, this review aims to present the paradigms of *in ovo* stimulation of the embryonic microbiome in chicks, performed on day 12 of egg incubation, and its long-term effects. We discuss the development of the chicken microbiome (Chapter 2), means of its modulation (Chapter 3), various issues regarding *in ovo* microbiome stimulation (Chapter 4) and how it influences lifespan condition (Chapter 5). We also outline what lies ahead in microbiome research in poultry and what possible steps can be taken to forward the research into and application of *in ovo* microbiome stimulation (Chapter 6).

Development of the chicken microbiome

The microbiome is defined as the total of the genetic information provided by a community of microorganisms (microbiota) inhabiting one environment, for example, intestinal mucosa and lumen in animals. Microbiota consist of commensal, symbiotic and pathogenic microorganisms. Recently, the most accurate analysis of the chicken gastrointestinal tract (GIT) microbiome was done by Wei et al. [14]. According to these authors, there are 915 species defined as operational taxonomic units (OTUs) in the chicken GIT. Among them, there is a high prevalence of Firmicutes (70%), represented by 495 OTU, Bacteroidetes (12.3%), represented by 139 OTU and Proteobacteria (9.3%), represented by 124 OTU. Most of the genera identified in the phyla *Firmicutes, Proteobacteria, Bacteroidetes* are common intestinal residents. However, some intriguing genera were also identified, e.g., *Ethanoligenens—an ethanol-producing bacteria (phylum Firmicutes).* The chicken microbiome depends on diet, location and age. Therefore, the taxonomic characteristic of each GIT section varies between the studies and also differs in the conditions of hatching and rearing [15]. The microbiome composition differs also between the morphological and physiological sections of the GIT, with its diversity and abundance increasing towards the hindgut. In birds, double ceca are the functional equivalent of the mammalian large intestine and an important site for fermentation. Detailed information on the most abundant taxa in individual sections of chicken GIT was reviewed by Stanley et al. [16]

An earlier opinion that the internal environment of the developing egg is sterile [17] was recently widely questioned, thanks to next-generation sequencing, which allows an investigation of whole microbial communities. For example, some bacteria can affect chicken embryos via infection of the female chicken's reproductive organs, resulting in incorporation of bacteria into the egg during oogenesis [18]. Furthermore, the profile of gut microflora becomes more differentiated and the population number of particular groups becomes higher with the age of embryos [19].

In comparison to mammals, poultry has shorter GIT and faster digesta transit. This feature significantly influences the diversity of the birds' microbiome compared to other livestock animals [20]. In the first day after hatching, the digestive system is the most intensively developing organ in birds. The microbiome of the newly hatched chick develops rapidly from days 1 to 3 post-hatching [21]. Such quick development is facilitated by the fast movement of the intestinal content, including bacteria. Therefore, the rapid development of the chicken microbiome is characterized by the quick multiplication of intestinal bacteria and their ability to adhere to the mucosa. In the ceca, the intestinal content slows, which promotes the development of microorganisms and the formation of the proper intestinal microbiome [22].

Colonization of the GIT with bacterial microflora is a key factor in the development and regulation of immunity, digestion, absorption of nutrients and their metabolism. The beneficial intestinal microflora helps not only in the digestion of food compounds but also reduces the potential of pathogen colonization in the guts. The entire process of microbiota development is stimulated by a huge number of different factors, including feed intake, antibiotics, supplements and other nutraceuticals, enzyme activities, age of the host, genetic modification and the whole environment [23]. Stimulation of commensal microbiota as early as possible (discussed in Chapter 4 of this review) is important to life-long programming and development in animals [24]. The interaction between the microbiome and its host leads to the proliferation of beneficial bacteria and reduces the development of enteric pathogens in the chicken gut [25, 26].

In poultry nutrition, gastrointestinal microorganisms and their influence on the host are of great significance. Intestinal microflora can ensure a complementary source of exogenous vitamins. Bacteria inhabiting GIT have the ability to synthesize vitamin K as well as the majority of water-soluble B vitamins [27]. Beneficial microorganisms promote the development of intestinal layers of mucosa and epithelia. They are also capable of distribution of polysaccharides and provide amino acids and short-chain fatty acids (SCFA), which are an important energy source and metabolic modulators [28].

Intestinal microbiota are in constant cross-talk with the host. This is possible due to recognition between receptors present on the lining of the GIT and gut-associated lymphoid tissue (GALT) and the microbial ligands, called microbe-associated molecular patterns (MAMPs) [29]. Such interaction affects the immunological status and physiology of the individual host [22]. The continuous cross-talk between the host and intestinal microbiome is long term and contributes to maintaining homeostasis in the guts. The major role of commensal gut microbiota is to trigger the maturation of both innate defense mechanisms and adaptive immune response [30]. The impact of the chicken microbiome on innate and adaptive immunity is described in the review by Clavijo & Florez [15]. The host immune system continuously adapts to the intestinal microbiome. Intestinal epithelia and GALT can tell commensal microflora from pathogenic strains and activate inflammatory mechanisms only in response to the latter. This subtle recognition between commensals and pathogens allows for the stable growth of beneficial flora with instant and local immune response to pathogenic flora [29]. In this manner, animals contribute from fermentation products of the microbiome, but also eradicate any environmental pathogen that enters the GIT.

Microbiome modulation

In nature, animals receive the first inoculation of the microbiota from the dam and as such, the microbiome can be passed to another generation. During hatching and directly post-hatching, chicks are exposed to hen's microbiota present on the egg shell or in the litter, which serves as the first source of microbial inoculation [9]. Current breeding and production systems involve automated hatcheries that eliminate the contact between chicks and their dams. During chicken embryo development in the egg, the main route of colonization with environmental bacteria is from the air, through the pores of the egg shell into the embryo [17]. After hatching, initial inoculation with the bacteria is obtained while handling during chickens' delivery from the hatchery to the farm. To avoid random composition of chicken microbiota, a planned microbiome stimulation might be introduced through direct supplementation of chicken embryo with proper bioactive substances (i.e., prebiotics and probiotics separately or combined into synbiotics).

Prebiotics are non-fermentable polysaccharides, which similarly to probiotics (discussed below) need to fulfil specific criteria to be qualified for GIT microbiome modulation. These compounds need to be resistant to digestion in the upper gastrointestinal tract, selectively stimulate beneficial microorganisms, and improve colonic microbiota composition. The most popular naturally occurring prebiotics are: fructans (e.g., fructooligosacchardies, short-chain fructooligosaccharides, oligofructose, inulin), followed by mannooligosaccharides (*Saccharomyces cerevisae*), soy oligosaccharides, and galacto- or transgalactooligosaccharides [31].

Most of the microorganisms used as probiotics for animals belong to various bacteria species, such as: *Lactobacillus*, *Bifidobacterium*, *Bacillus* or *Enterococcus*. Certain yeast (e.g., *Saccharomycces bourlardii* or *Saccharomyces cerevisae*) and fungal (e.g., *Aspergillus oryzae* or *Candida pintolepsii*) species can also be used for the same purpose. The variety of potential probiotics is tremendous. A document prepared by FAO lists 16 species and 57 strains of different microorganisms, which are used as separate strains or multispecies in a single product in animal nutrition [32]. Microorganism used as a probiotic need to fulfill particular requirements. It should be non-pathogenic to the future host organism and resistant to low pH and have a high concentration of bile acids. On top of that, from the manufacturing perspective, the potential probiotic should be easy to transport, store, and apply. The routine process of probiotic selection is based on several analyses covering: the in vitro challenge of low pH, toleration of an acid environment and ability to adhere to the intestinal epithelium [32]. Finally, when selecting potential probiotic for animal nutrition, one should make sure that it grows on an inexpensive medium. This last feature is of less importance when probiotic is applied using *in ovo* technology, where only a small amount of the bioactive substances is given (details are discussed in Chapter 4). Scientific reports show that probiotics applied in a standard way (as DFM, in-feed or in-water) improve the growth rate of the animals, increase efficiency and stimulate intestinal histomorphology (reviewed by FAO, [32]). However, the results obtained during feed trials are inconsistent and need to be verified on a case-by-case basis.

In ovo stimulation of embryonic chicken microbiome

Principles of in ovo stimulation

In poultry practice, birds are offered prebiotics and probiotics at the hatchery by spray application or in the form of DFM (in-feed or in-water) during the first days post-hatching, after they have arrived at the farm. In this case, the whole embryonic and perinatal period (21 days) is disregarded. If we consider the total lifespan of a regular broiler chicken (42 days), applying DFM post-hatching seems like late intervention. The main concept of *in ovo* technology is to apply substances long before the bird hatches, which helps to program lifelong phenotypes (e.g., immunity, gut microbiome, performance, adaptive) already during the embryonic phase. In this manner, the *in ovo* strategy is focused solely on stimulating the colonization of embryonic GIT with native microbiota that will facilitate establishing the optimal microbiome already during egg incubation. We demonstrated that by delivering the volume of 0.2 ml of dissolved bioactive stimulus exactly on day 12 of egg incubation, we can initiate a whole cascade of events on different phenotypic levels, from gene expression modulation to growth performance. The timing of *in ovo* stimulation of microbial growth in embryonic GIT is quite crucial. If the *in ovo* injection into the air cell is performed earlier, then the bioactive cannot pass through the underdeveloped allantochorion, which at this point is not vascularized enough. If *in ovo* injection is performed on late-term embryos (day 17/18 of egg incubation), then the air cell can no longer be used and the injection site is the amnion or more often, the embryo muscles (refer to the Introduction for more details on the time point of *in ovo* stimulation). The timing of *in ovo* stimulation of microbiota colonization in embryonic GIT on day 12 of egg incubation was established experimentally based on measurements of the proliferation rates of Bifidobacteria at hatching. High proliferation rates were shown after injecting different doses of oligosaccharide prebiotic on day 12 of egg incubation, compared to days 1, 8 and 17 of egg incubation [11]. The doses of the bioactives for *in ovo* injections are determined empirically, based on the criterion of egg hatchability and intestinal bacteria abundance in 1-day-old chicks. Proper *in ovo* stimulation should result in high bacteria abundance in chicken meconium without dampening the hatchability of the injected eggs [33].

Mechanisms of in ovo stimulation by prebiotics and probiotics

Before we performed *in ovo* stimulation of the embryonic microflora on day 12 of egg incubation, *in ovo* technology had been routinely used on day 18 of egg incubation for *in ovo* vaccination or (more recently) *in ovo* feeding. The poultry community was convinced to keep using late-term embryos for injecting probiotics as well. The use of prebiotics *in ovo* was scarce. Our experimental data indicated day 12 of egg incubation as the most efficient time point to provide stimulation of embryonic microflora with no harm to the incubating eggs [11]. There are two groups of natural growth promoters that can be used for *in ovo* stimulation, i.e., prebiotics and probiotics (or a synergistic combination of both). Applying prebiotic on day 12 of egg incubation gives the unique opportunity to stimulate endogenous microbiota before hatching. At this time, the chorioallantoic membrane is highly vascularized and allows the transfer of the prebiotic from the air cell into the circulatory system and further to the developing intestine. On the other hand, probiotics enter the GIT the first moment the chick breaks the inner membrane at the early stages of hatching. Therefore, they might act as pioneer colonizers, which augment the development of complex microbiota by modifying the intestinal environment [9]. The same authors claim that pioneer colonizers determine the composition of the climax microbiota by creating the microenvironment necessary for the development of complex microbiota.

To demonstrate the transfer of the prebiotic and probiotic after *in ovo* stimulation on day 12 of egg incubation, we performed a simple experiment [34]. The hatching eggs were subjected to *in ovo* delivery of (1) a solution of blue dye with a molecular weight similar to a GOS prebiotic and (2) 10^5 CFU of a *Lactococcus lactis* subsp. *cremoris* probiotic strain. Both substances were injected into the air cell on day 12 of egg incubation. This approach allowed us to monitor the migration of prebiotic (mimicked by the blue dye) and probiotic through the outer and inner shell membranes. To analyze the migration rate of the bioactives, the eggs were dissected and analyzed daily from day 13 to 18 of egg incubation. We could determine the presence of the blue dye in the embryonic blood vessels by simple observation (Fig. 1). The presence of the probiotic inside the embryo was determined with the plating method. We have demonstrated that that prebiotic/blue dye migrates through the shell membrane and enters the blood circulation on day 3 after injection (i.e., day 15 of egg incubation onwards). Unlike the prebiotic (blue dye), the probiotic bacteria stay in the air cell until the beginning of hatching (i.e., day 19 of egg

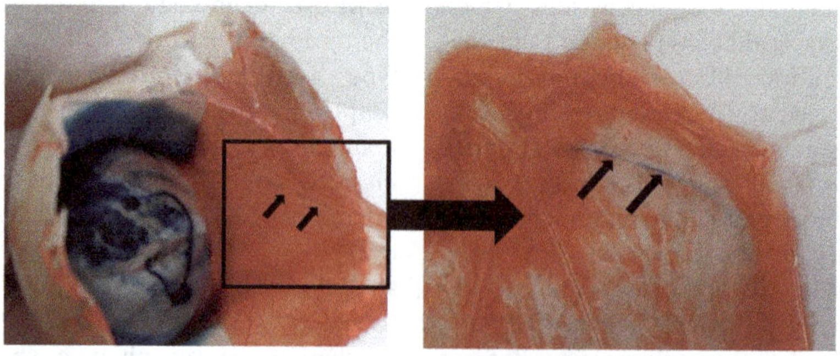

Fig. 1 Model of *in ovo* delivery and penetration of the bioactive solution through the chorioallantoic membrane into the circulatory system of the chicken embryo. For *in ovo* injection, 0.2 ml of a blue dye (E132, indigotine, artificial dye used for food coloring) at a concentration of 0.01 g/ml was injected *in ovo* on day 12 of egg incubation. Blue dye was deposited into the air cell. The eggs were sealed and incubation was continued until 19 of egg incubation. Penetration of the dye through chorioallantoic membrane was observed daily by dissecting permeable outer shell membrane from semi-permeable inner cell membrane. After 3 days (i.e., from 15 of egg incubation onwards) the dye was transferred to the blood vessel on the inner shell membrane. The stained vessel is marked with an arrow Steady influx of the dye through the chorioallantoic membrane was observed until 19 of egg incubation, when the experiment was terminated

incubation). Based on the above, we have demonstrated the way and the time point for both prebiotics and probiotics to be delivered *in ovo*. We argue for day 12 of egg incubation as the most optimal time point for the efficient delivery of prebiotic, or prebiotic combined with probiotic (i.e., synbiotic).

In ovo stimulation—A tool for early programming?

The concept of perinatal programming was first defined in human studies [35], providing the hypothesis that a stimulus (or adverse factor) provided to a fetus at the early stage of development leads to lifelong phenotypic changes. The animal organism is susceptible to programming at specific developmental stages when it is most receptive to the environment [35]. In poultry practice, two critical moments determine the life-long performance and health status of a bird and these are tightly related with microbiome status: (1) hatching window (the time from when the first chick hatches in the incubator until the last one, usually 48 h) and (2) fasting *post* hatching due to management in a hatchery, transportation [36] and medication treatments [37]. *In ovo* technology seems the only practical tool to provide direct stimuli to a developing embryo to help mitigate the stressors received around the hatch (perinatally) and later in life. Figure 2 presents the critical time points of chicken perinatal development and the life-long production losses that follow. The specific reproduction system of birds allows stimulation of the eggs during incubation in industrial settings. The delivery of prebiotic and probiotic *in ovo* on day 12 of egg incubation stimulates the development of GIT and GALT inside the embryo in response to microbial programming. A newly hatched chick that had been reinforced with *in ovo* stimulation can better handle the perinatal stresses resulting from the hatching window and post-hatching fasting.

In ovo stimulation vs. in ovo feeding

There are different approaches to using *in ovo* technology in poultry, i.e., *in ovo* stimulation (characterized in this paper) and *in ovo* feeding (developed by Uni and Ferket [13]). The differences between *in ovo* stimulation and *in ovo* feeding are fundamental when it comes to the strategy, biological mechanisms and technical tools. *In ovo* stimulation refers to the delivery of prebiotics or synbiotics in the early-stage embryo (day 12 of egg incubation). It aims to stimulate selective growth of indigenous microflora, which colonize embryonic guts. A causative agent that triggers this stimulation is a prebiotic, which is metabolized by the indigenous bacteria, stimulating their growth. Prebiotics and synbiotics can be delivered *in ovo* on day 12 of egg incubation to program the microbiota colonization and modulate gene expression related to the microbiome. Such modulation

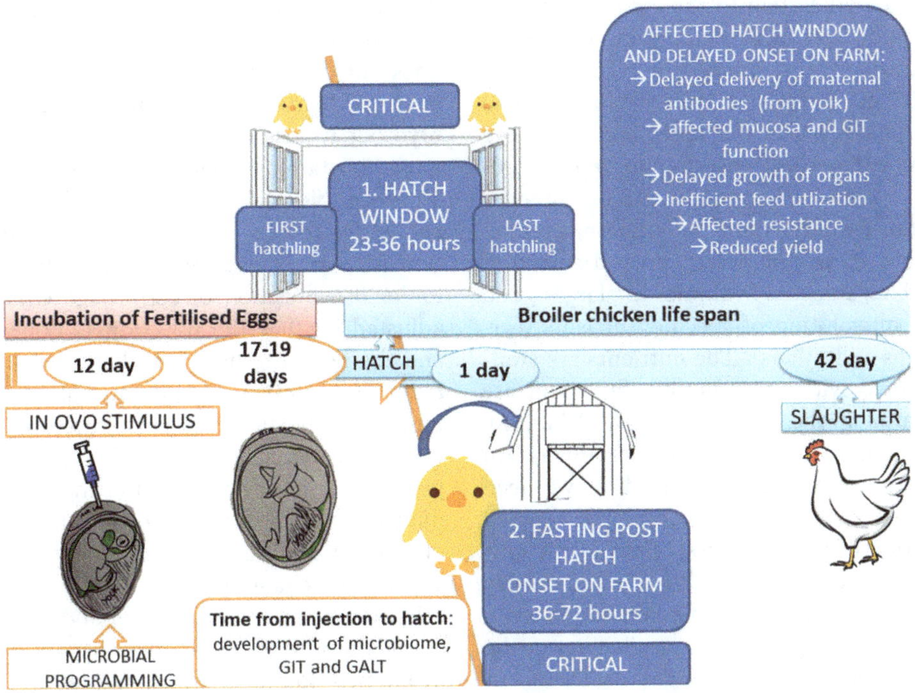

Fig. 2 Concept of early microbial programming *in ovo*. Prebiotic or probiotic given on day 12 of egg incubation influences embryonic factors (microbiome, GALT development and function, gene expression, nutrient absorption) which are critical for future phenotype of the broiler chicken. Two critical perinatal moments are shown (hatch window and fasting post-hatching), when the newly hatched chicken is the most receptive to environmental stressors

poses life-long beneficial effects in performance and mitigating life stressors. In turn, *in ovo* feeding is more of a nutritional strategy developed by aiming to enhance chicks' adaptation to different nutrients post-hatching. In particular, this technology was designed [13] for a late-stage poultry embryo (day 17/18 of egg incubation) to facilitate its adaptation to a switch from embryonic nutrition based on fat and proteins deposited in the yolk (during the embryonic stage) to autonomous nutrition, which is based on carbohydrates and proteins (during the post-hatching stage) [8]. The key parameters, which were measured to evaluate the efficacy of *in ovo* feeding (i.e., the transition from embryonic to independent nutrition), included the proliferation of enterocytes, the enzymatic activity of digestive organs and gut morphology [8]. In summary, *in ovo* stimulation implements a different strategy and underlying biological mechanisms in improving the perinatal development of a chick.

Embryonic development in a chicken is rapid, and the embryo differs from day to day. For this reason, there are differences between *in ovo* stimulation and *in ovo* feeding in the delivery of bioactive compounds. *In ovo* stimulation on day 12 of egg incubation allows for the deposition of the prebiotic/synbiotic solution onto the air cell and the consecutive transfer of small-weight oligosaccharides through the membrane into the blood vessels surrounding the embryo. The prebiotic is transferred through the blood system to the embryo. For this purpose, the prebiotic has to be soluble in water or it does not pass through the membrane. The volume injected is only 0.2 ml, because a larger volume would infiltrate the junction between the inner and outer egg membrane and immediately kill the embryo. *In ovo* feeding of late-term chicken embryos (day 17/18 of egg incubation) must be done directly into the amnion, because at this stage, the air cell is completely dry and devoid of vascularization. The injected volume is larger and amounts to 1–1.7 ml solution of specific nutrients (carbohydrates, proteins and others). The nutrients are swallowed by the embryo with the amniotic fluid and are made available through the intestinal tract. When we consider the distinction in the route of administration between *in ovo* stimulation and *in ovo* feeding, we need to acknowledge that only prebiotic compound is used for actual stimulation of the indigenous flora in the embryo. Probiotic, even though it might be injected on day 12 of egg incubation (in the form of synbiotic), should be considered *in ovo* feeding due to its availability to the late-stage embryo (after pipping).

Manipulation of the eggs at different stages of incubation requires the development of different technological strategies. In industrial settings, the most common and sensible time point for performing *in ovo* injection is day 18 of egg incubation. At this time, the eggs are candled and transferred from the incubator to the hatcher. The most common application of *in ovo* technology on day 18 of egg incubation is *in ovo* vaccination against Marek's disease virus. The attempts to combine *in ovo* vaccination with *in ovo* feeding has been scarce and it is hard to predict its effectiveness [38]. The injection machines that are available on the market are designed to perform intramuscular injection with the vaccine (e.g., www.egginject.com) on days referred to as the "*in ovo* injection window" (i.e., days 17–19 of egg incubation). These machines generally do not support amniotic delivery of nutrients (or other bioactives) due to the high probability of an imprecise injection site by the operating needle. The possible imprecision in targeting the amnion of late-term embryos is due to several factors, such as (1) the design of the needles in vaccination systems, (2) imperfectly controlled positioning of injected eggs, and (3) the fact that each incubated embryo uses and/or rearranges the compartments of its egg (air cell, allantoic sac, amniotic sac, yolk sac) in a very fast and individual manner in the later stage of incubation. The embryo is already filling almost the whole space of the egg on days 18–19.5 of egg incubation, which makes it harder to locate the amnion for automated *in ovo* feeding. Using early-stage embryos for *in ovo* injection is much more standardized and safer for the viability of the embryos. On day 12 of egg incubation, the air cell is large and easily available for the injection without having the embryo injected by a needle. For this reason, *in ovo* stimulation does not dampen hatchability results. *In ovo* stimulation provides beneficial and long-lasting effects to the embryo but it is still a new technology on the market. The major obstacle to using *in ovo* stimulation is the need for technological adjustment of production lines in hatcheries to add another time point for egg manipulation (day 12 of egg incubation). The injection machine for *in ovo* injection of early-stage embryos needs to be adjusted so that the injection is done into the air cell and the puncture hole is sealed to avoid embryo evaporation. A prototype of such an automated injection system for early-stage embryos exists and has been tested in commercial settings [39]. The pipeline includes three procedures: a hole is drilled in the eggshell, a solution is injected, and then the hole is sealed. This allows for *in ovo* injections of 30,000 eggs per hour while maintaining superior hatchability.

In summary, there are pros and cons to using *in ovo* stimulation with prebiotics or *in ovo* feeding with probiotics for embryonic microflora development in chicken. Both approaches require adjustment in the production line and additional handling of the eggs. However, *in ovo* stimulation is done on day 12 of egg incubation, and therefore it cannot be combined with other procedures (e.g., vaccination). On the other hand, *in ovo* stimulation assures stable and long-term effects compared with the more transient effects of *in ovo* feeding. Also, through *in ovo* stimulation we enhance the growth of indigenous, diverse microbial populations, which is biologically more relevant than delivering a single (or even multistrain) inoculation with arbitrarily selected microbiota. This is how we claim that the long-term effects on *in ovo* stimulation are based on natural mechanisms of early gut colonization and can provide a valuable addition to the perinatal treatment of chicks.

Life-long phenotypic effects

Life-long phenotypic effects that followed *in ovo* stimulation were determined on multiple levels, from industrially relevant traits, such as hatchability, mortality and performance among different genotypes of broiler chickens (e.g., Ross, Cobb and Hubbard) and native chickens (i.e., Green-legged Partridgelike), through the morphology and histology of the intestinal, muscular and immune tissue, to physiological and molecular modulation. In this chapter, we discuss the life-long phenotypic effects of microbiome stimulation with prebiotics or synbiotics administered *in ovo* on day 12 of egg incubation. The results presented here were derived from a series of experimental and industrial trials and give quite a good outlook on the effects of *in ovo* stimulation in chickens (Table 1). However, many trials are still in progress to complement the picture with data on the metagenome and challenge response of chickens stimulated *in ovo*.

Hatchability, performance and intestinal microbiota

One of the major concerns related to *in ovo* technology is its impact on egg hatchability and chick mortality. Hatchability depends on the incubation conditions, *in ovo* delivery technique, site and depth of injection (air cell vs. amnion), and the kind of bioactive and inoculated dose [8, 39, 40]. In some cases, especially when late-term embryos are handled or the dosage of the bioactive is exceeded, hatchability may drop rapidly, which decreases hatching success [40]. In general, properly optimized doses for *in ovo* stimulation will keep hatchability at a high level and improve bifidobacteria counts [11, 33]. In relation to the performance traits of broiler chickens, including body weight, feed intake and efficiency, carcass traits and meat quality, the results, though different between trials, showed very slight changes in *in ovo*-stimulated chickens in comparison to controls. But, our validation study conducted in industrial settings (as opposed to small-scale experimental trials that are usually reported) on 275,000 broilers delivered a proof of concept that *in ovo* stimulation with prebiotics in fact improves performance traits, including body weight (BW), carcass weight, carcass yield, and breast muscle weight [41]. But, the aforementioned data were collected under controlled conditions and the birds were not challenged with any environmental or immunological stresses. In our ongoing challenge study, we have detected a large improvement in *in ovo*-stimulated birds in mitigating the harmful effects of heat stress, expressed by significantly decreased mortality, faster recovery of feed intake and body weight gain and decreased number of myopathies in the meat [42]. We are convinced that further exploring the effects of *in ovo* stimulation on challenged chickens will prove them more resilient to handle stressful environments in a more robust way.

Intestinal morphology

The morphology of the intestinal mucosa is an important determinant of the digestive and absorptive intestinal functions, which in turn determine the growth performance of poultry. In our research, *in ovo* stimulation with prebiotics in combination with probiotics affected the histomorphology of the small intestine of chickens. The observed effects strictly depend on the type of bioactive substances used for *in ovo* stimulation and the age of the chicken. The effects of *Lactobacillus*-based synbiotics on histomorphology were assessed on days 1 and 42 post-hatching. On day 1, both synbiotics, *L. salivarius* (combined with prebiotic GOS - SYN1) and *L. plantarum* (with RFO - SYN2) increased villus height, width and surface in the duodenum of Cobb broiler chickens. Increasing the absorbent surface by increasing the surface of the intestinal villi as a result of SYN1 injection was also observed in the duodenum of chickens at 42 days of age. A similar effect of the synbiotic used was found in the jejunum in 1-day-old chicks. On day 42 post-hatching (end-point of the study), small intestinal length and weight, as well as histomorphology of the duodenum (increasing the width of villi, surface area and deepening of intestinal crypts), was improved after *in ovo* delivery of *L. salivarius* (with GOS). Additionally, it is worth mentioning that this synbiotic increased the number of neutral goblet cells in the jejunum and ileum. [43]. In intestinal crypts, epithelial cells undergo intensive mitotic divisions. Deeper crypts allow for faster regeneration of intestinal villi. The depth of the crypt is associated with faster turnover of epithelial cells and thus also goblet cells. Goblet cells produce mucus, which constitutes one of the components of the physical barrier in the guts [44, 45]. *Lactococcus*-based synbiotics had more transient effects. Their positive effects were found in the small intestine of Ross broiler chickens on the first days post-hatching but were different depending on the investigated section of the small intestine and age of chickens. On the first day of life, an increase in the height of the villi in the jejunum was found as a result of

Table 1 Overview of the life-long phenotypic effects of *in ovo* stimulation with prebiotics and probiotics on day12 of egg incubation

Prebiotic	Amount (mg/ml)	Probiotic	Amount CFU	Effects of *in ovo* stimulation	Reference
a. Hatchability, performance and intestinal microbiota					
RFO	0,1763; 0,8815; 1763	–	–	Confirmed effects of RFO injected *in ovo* on day 12 of egg incubation, increased number of bifidobacteria	Villaluenga et al. 2004 [11]
FOS, RFO	1763 2,1158	–	–	Decreased mortality, high level of bifidobacteria after hatching and before slaughter	Pilarski et al. 2005 [39]
RFO	1900	–	–	Increased body weight and FCR, increased and dose-dependent counts of bifidobacteria	Bednarczyk et al. 2011 [38]
RFO	1900	–	–	Improved performance traits of broiler chickens	Brudnicki et al., 2015 [40]
GOS, *Laminaria* spp. seaweed extract containing laminarin and fucoidan	3500 0,880	–	–	Increased number of lactobacilli and bifidobacteria in chicken feces, validation of *in ovo* vs. in-feed prebiotics administration	Bednarczyk et al., 2016 [33]
GOS, *Laminaria* spp. seaweed extract containing laminarin and fucoidan	3500 0,880	–	–	Improvement in a number of production parameters including carcass weight and yield	Maiorano et al., 2017 [41]
GOS, RFO	3500 1900	*Lactobacillus salivarius, Lactobacillus plantarum*		Modulation of the fecal microflora composition	Dunislawska et al.,2017 [42]
b. Intestinal morphology					
RFO	1500 3000 4500	–	–	Potential to enhance ileum mucosa morphology and improve immunity in the small intestine	Berrocoso et al., 2016 [43]
Inulin, GOS	1760 0,528	*Lactococcus lactis* subsp. *lactis, Lactococcus lactis* subsp. *cremoris*	10^3 10^3	Increase in the number of goblet cells in the duodenum and the jejunum on day 1 of life, followed by significant decrease on the day 4	Bogucka et al., 2016 [44]
Inulin, GOS	1760 0,528	*Lactococcus lactis* subsp. *lactis, Lactococcus lactis* subsp. *cremoris*	10^3 10^3	Significantly higher number of goblet cells	Bogucka et al., 2017 [45]
RFO	1900	–	–	Increased length of the small intestine and transient effect (days 1–3 post-hatching) on villi length and surface	Brudnicki et al., 2017 [46]
Laminaria spp. seaweed extract containing laminarin and fucoidan	0,880	–	–	Increase in villi width and crypt depth in duodenum on day 21	Sobolewska et al., 2017 [47]
GOS, RFO	2000 2000	*Lactobacillus salivarius, Lactobacillus plantarum*	10^5 10^5	Increased absorbent area of villi in different sections of the intestine on days 1 and 42, increase in the number of the goblet cells, reduced crypt depth	Sobolewska et al., 2017 [48]

Table 1 Overview of the life-long phenotypic effects of *in ovo* stimulation with prebiotics and probiotics on day12 of egg incubation *(Continued)*

Prebiotic	Amount (mg/ml)	Probiotic	Amount CFU	Effects of *in ovo* stimulation	Reference
c. Muscle histology and meat quality					
RFO	1900	–	–	Improved meat quality in terms of collagen content	Maiorano et al., 2012 [49]
d. Immune system development and function					
Oligosaccharide extract of palm kernel cake (PKC)	20	–	–	Increased IgG production and antioxidant capacity in serum and liver of prenatal chicks and had limited carrying-over effects on the post-hatched chicks	Jahromi et al., 2017 [50]
Inulin, GOS	1760 0,528	*Lactococcus lactis* subsp. *lactis, Lactococcus lactis* subsp. *cremoris*	10^3 10^3	Prebiotics and synbiotics can modulate the central and peripheral lymphatic organ development in broilers	Madej et al., 2015 [51]
Inulin, GOS	1760 0,528	*Lactococcus lactis* subsp. *lactis, Lactococcus lactis* subsp. *cremoris*	10^3 10^3	Stimulated GALT development after hatching and colonization with lymphocytes	Madej & Bednarczyk 2016 [52]
RFO	0,7765 0,8815 1763	–	–	Protective effect against *Salmonella enteritidis* infection	Ruiz Lopez et al., 2008 [53]
e. Gene expression modulation					
RFO	1900	*Lactococcus lactis* subsp. *lactis, Lactococcus lactis* subsp. *cremoris,*	10^3 10^3	Stimulation of an immune system by lymphatic organs development	Slawinska et al., 2014 [54]
Inulin	1760	*Lactococcus lactis* subsp. *lactis,*	10^3	Down-regulation of gene expression in cecal tonsils and spleen	Plowiec et al., 2015 [55]
Inulin GOS	1760 0,528	*Lactococcus lactis* subsp. *lactis, Lactococcus lactis* subsp. *cremoris*	10^3 10^3	Long-term transcriptomic effects	Slawinska et al., 2016 [56]
GOS, RFO	2000 2000	*Lactobacillus salivarius, Lactobacillus plantarum*	10^5 10^5	Significant impact on gene expression level in spleen and cecal tonsils	Dunislawska et al., 2017 [42]
GOS, RFO	2000 2000	*Lactobacillus salivarius, Lactobacillus plantarum*	10^5 10^5	Downregulated GLP-1 and GIP mRNA expression in the duodenum and GLP-1R in the pancreas	Kołodziejski et al., 2018 [57]
f. Avian physiology					
RFO	1900	–	–	Increased rate of yolk sac absorption	Brudnicki et al., 2015 [40]
Inulin GOS	1760 0,528	*Lactococcus lactis* subsp. *lactis, Lactococcus lactis* subsp. *cremoris*	10^3 10^3	Increased the activity of amylase, lipase, and trypsin in the pancreas	Pruszynska Oszmalek et al., 2015 [58]
Inulin GOS	1760 0,528	*Lactococcus lactis* subsp. *lactis, Lactococcus lactis* subsp. *cremoris*	10^3 10^3	Increased body weight, improved the short – chain, fatty acid cecal profile, increase the villus length	Miśta et al., 2017 [59]

the injection with prebiotics (inulin and GOS) and synbiotics *L. lactis* spp. *lactis* with inulin and *L. lactis* spp. *cremoris* with GOS. The stimulatory effect of both synbiotics on the villus surface area of the duodenum was also reported on days 1 and 4 post-hatching and of the jejunum on day 4 post-hatching. In the majority of the examined traits of the small intestine morphology, a positive effect of *in ovo* injections of synbiotics was demonstrated [46, 47]. At the end of rearing (day 35), a positive effect on the histomorphology of the small intestine was exerted by inulin as it increased the villus height in the duodenum as well as ileum (the main absorption organ). The absorptive area of the duodenum increased in response to inulin supplementation. In the jejunum and ileum, the villus surface area was negatively affected only by the synbiotic with the GOS preparation, but other injected substances had a positive effect on the absorptive area of the intestine. Chickens that received *in ovo* synbiotic with inulin (jejunum) and the synbiotic with GOS had deeper crypts in the jejunum and ileum, respectively. The injection of synbiotics also increased the number of goblet cells (synbiotic with GOS) in the jejunum and ileum [47], which suggests it has regulatory effects on intestinal morphology. In our other studies, the positive effect of GOS and RFO prebiotics on the width and area of the villi of Ross broiler chickens at day 21 post-hatching was found (Bogucka, personal communication). Moreover, it has also been demonstrated that the injection of an extract of *Laminaria* spp. containing laminarin and fucoidan significantly increased the width of the duodenal villi and the crypt depth of chickens on day 21 of rearing. This resulted in a greater surface area of villi in these birds; however, this was not confirmed statistically [43].

Muscle histology and meat quality

The effect of bioactive substances administrated *in ovo* on the meat quality and microstructure of birds' skeletal muscles is not homogeneous. However, an increasing number of studies suggest there is a relation between the gut microbiome, the metabolic pathway of substances absorbed in the intestine, and a later increase in muscle mass and the formation of the quality traits of meat, although the mechanisms of these processes are still difficult to identify. Maiorano et al. [48] studied the influence of prebiotic and three different synbiotics on the chicken pectoral muscle microstructure. They observed a decrease in the density of the muscle fiber per mm^2 in the synbiotic group (RFO with *L. lactis* spp. *cremoris*) in comparison to the control group. Moreover, intramuscular collagen was notably reduced in the prebiotic and synbiotic groups. There was no effect of synbiotics on abdominal fat, ultimate pH, and cholesterol in the pectoral muscle of chickens. In other studies that aimed to explore the effect of GOS and extract of *Laminaria* spp. injected *in ovo*, there was a considerable reduction in the diameter of the muscle fibers in the prebiotic groups; however, this was not confirmed by statistical analysis. Furthermore, meat from prebiotic-treated birds displayed higher lipid oxidation levels compared to those from the control during the entire storage time [41]. Intramuscular fat content has an effect on the flavor and juiciness of meat, as well as the visual aspect (marbling). Sometimes, the increased percentage of fat in the pectoral muscle of broiler chickens may be associated with the occurrence of meat defects. They are manifested by numerous pathological changes of muscle fibers leading to their necrosis, which may result in fibrosis and increased fat content with a modified fatty acid composition, reduced protein content, which leads to a decrease in the nutritional value of meat [49–51]. In the white meat of broiler chickens, lipids accumulate in the connective tissue surrounding the fibers and bundles of fibers. Dankowiakowska et al. [52] showed the effect of GOS prebiotic on increasing the intramuscular fat content in the breast of Ross 308 chickens. In addition, the *in ovo* injection of synbiotic (GOS with *L. salivarius*) has been shown to have a positive effect on the blood supply of broiler chickens' pectoral muscle. There were statistically more capillaries per unit area and per one muscle fiber in the synbiotic group. Better muscle capillarity influenced less occurrence of pathological changes, i.e., fiber necrosis and splitting, and hence, a higher percentage of normal fibers (Bogucka, personal communication). A similar effect of synbiotic dietary supplementation (RFO with *Lactococcus lactis*, *Carnobacte rium divergens*, *Lactobacillus casei*, *Lactobacillus plantarum*, and *Saccharomyces cerevisiae*), obtained in *pectoralis major* muscle of female broiler chickens [53]. As it turns out, not only the type of bioactive substances but also the way they are administered may affect the microstructure of the muscles and the quality of the meat. Tavaniello et al. [54] studied the effect of prebiotics administration on the meat-quality traits of broiler chicken by evaluating the different routes of their delivery (*in ovo* vs. in-water vs. *in ovo* and in-water combined). Irrespective of the delivery method, prebiotics showed a positive impact on breast muscle weight and yield, which was also associated with a greater thickness of muscle fibers. The redness (a*) of fillets was statistically decreased upon giving prebiotics, irrespective of the method used. In addition, the authors showed a higher content of PUFA (polyunsaturated fatty acids) and n-3 fatty acid in the meat from the prebiotic groups, displaying more favorable indexes for human health. Similar studies were conducted by Tavaniello et al. [55], which compared the effect of *in ovo* administration of two different synbiotic formulations (GOS with *L.*

salivarius - SYN1 and RFO with *L. plantarum* - SYN2) on carcass- and meat-quality traits in broiler chickens. The highest color lightness (L*) in 45 min after slaughter is characteristic of meat from *in ovo* SYN1-injected chickens. In contrast, in the SYN2 group, the highest content of MUFA (monounsaturated fatty acids), PUFA (polyunsaturated fatty acids) and n-6 fatty acid was recorded.

Immune system development and function

It was demonstrated that *in ovo* stimulation with prebiotics and probiotics delivered on day 12 of egg incubation influences immune system development, including the structure of the central (i.e., bursa of Fabricius and thymus) and peripheral (i.e., spleen) lymphatic organs [56–58]. *In ovo* application of synbiotics provide stimulus for the immune organs of the growing chickens, but the potency of the stimulation depends on the chicken genotype. The bursa and bursa-to-spleen index were significantly higher in broiler chickens after *in ovo* stimulation with RFO and *Lactococcus lactis* subsp. *cremoris.* In Green-legged Partidgelike chicken, the spleen index was higher, but only on day 21 post-hatching. The histological image of the thymus displayed the increased density of thymocytes in the cortex after *in ovo* stimulation with two different synbiotics [57]. More robust immune system development is beneficial for mounting immune responses later in life. At hatching, the GALT is immature and requires early stimulation, otherwise it will impair the health status and performance of the animal [59]. To address this issue, a bioactive stimulus provided on day 12 of egg incubation may influence, through modulation of gene expression [60], developmental events related to future immunological functions. The more so because the precursors of T and B lymphocytes, major players in cellular (T) and humoral (B) immune response, are established between the 9th and 15th days of egg incubation and begin to differentiate and proliferate continuously in immune organs until several days post-hatching [61].

In ovo stimulation with synbiotics influences the immune phenotype and cell distribution in the cecal tonsils, ileum and bursa of Fabricius of broiler chicken [58]. These tissues were immunohistochemically stained for the estimation of Bu-1+, CD3+, CD4+, CD8α + and TCRγδ+ properties. The results indicated that synbiotics stimulated post-hatching development of GALT in chickens. A temporary decrease in the B-cell number in the bursa of Fabricius on day 7 post-hatching was detected, which can suggest an increased colonization rate of the peripheral lymphoid organs as an effect of *in ovo* stimulation with *Lactococcus lactis* subsp. *cremoris* and GOS [58]. The chicken immune system undergoes rapid changes during its first weeks post-hatching; some effects, such as an increase in the spleen/bursa Fabricius ratio, were age-dependent [56]. In the cecal tonsils, a high colonization of GALT by T cells was observed after *in ovo* stimulation with two different synbiotics: *L. lactis* subsp. *lactis* combined with inulin and *L. lactis* subsp. *cremoris* with GOS [57]. These synbiotics stimulated also germinal center formation in the spleen, observed on days 21 and 35 post-hatching, which indicated enhanced B-cell proliferation in peripheral lymphatic organs.

Gene expression modulation

The injection of synbiotics into the air cell on day 12 of egg incubation effectively stimulates the host microbiome, which indirectly influences changes in gene expression in broiler chickens. A single bioactive dose applied *in ovo* allows for early intestinal contact with bacteria by the direct stimulation of the host microbiome and an indirect effect on the modulation of the host transcriptome. Probiotic bacteria *Lactococcus lactis* combined with RFO prebiotic had a significant effect on the gene expression of cytokines and chemokines in cecal tonsils (gene expression down-regulation) and spleen (gene expression up-regulation) of Green-legged Partridgelike chickens [62]. Further experiments on *Lactococcus lactis* combined with inulin prebiotic, performed in a time-course manner, proved that *in ovo* stimulation down-regulated immune-related gene expression in cecal tonsils and spleen of broiler chickens and those effects were more pronounced in older animals (day 42) [63]. Analysis of the spleen, cecal tonsils and large intestine transcriptome of chickens that received *Lactococcus lactis* combined with prebiotic (GOS or inulin) allowed a determination of the transcriptome modulation profile after the synbiotics administration [60]. The highest number of differentially expressed genes was detected after *in ovo* stimulation with GOS prebiotic. The activated molecular pathways took part in different immune processes, such as the Toll-like receptor signaling pathways, TCR signaling, BCR signaling, NF-KB signaling, hematopoietic cell lineage, TNF signaling, G-coupled protein signaling and cytokine signaling [60]. *Lactobacillus salivarius* combined with GOS and *Lactobacillus plantarum* combined with RFO showed involvement in incretin secretion and reception defined by analysis of the mRNA level for GLP-1 and GIP [64].

Avian physiology

Supplementation of probiotics aims to replace or reduce the number of potentially pathogenic bacteria in the intestine by enriching the populations of beneficial strains. It improves the health status of the intestine and may be reflected in the general metabolism, as well as organ-specific biochemical processes.

Current research shows that the functioning of the liver, fatty tissue, kidney, and pancreas can be affected by disorders of the intestinal microflora. The increase in enzyme activity may be related to the delivery of additional portions of the enzyme by bacteria living in the intestine, which causes an improvement in nutrient digestibility and an increase in body weight [65]. Pruszynska-Oszmalek et al. [66] demonstrated that the beneficial effects of *in ovo* stimulation with *Lactococcus lactis* probiotics combined with inulin or GOS prebiotic significantly increased the total activity of pancreatic enzymes—amylase, lipase, and trypsin—and, as a consequence, increased the chickens' body weight. Moreover, these authors show a positive effect of both synbiotics on the activities of the two enzymatic markers of the liver (aminotransferase: ALT and AST), which demonstrates the high health status of the liver.

Brudnicki et al. [67] demonstrated that *in ovo* stimulation of broiler chickens with RFO prebiotic significantly increased the yolk sac resorption rate in newly hatched chicks. By the end of day 14 post-hatching, the retention of the yolk sac in the population was 0% in the *in ovo*-stimulated group vs. 30% in the control group. Since the yolk sac is the major source of immunoglobulins contributing to the passive immunity in newly hatched chicks, faster yolk sac resorption results in the greater transfer of maternal antibodies into the chicks' blood stream. Also, the yolk sac is essential for the initiation of early growth post-hatching in broiler chickens [68]. In the aforementioned study, body weight and feed efficiency were indeed improved at the end of the study (day 42). Finally, we have demonstrated that *in ovo* stimulation improved the profile of short-chain fatty acids in the ceca of broiler chickens, expressed by a higher molar percentage of butyrate, propionate and valerate, especially in older birds (21 and 35 days post-hatching).

Future perspectives

In this paper, we gave insight on microbiome programming with the early delivery of prebiotics and probiotics during the neonatal period in chickens. We have demonstrated that *in ovo* stimulation is a powerful tool for the early colonization of the embryonic guts with beneficial microbes so that it results in improved productivity, health and welfare of the animals. There are still questions ahead that will drive the development of prebiotics and probiotics for animals even further. Being aware of the miscellaneous effects of different probiotic formulations, it is tempting to fit the bacteria strain to the specific needs of the individual animal, depending on environmental pressure, host genetics and production goals. Currently, a screening procedure for prebiotics and probiotics is carried out with different in vitro, in vivo and "-omics" approaches and includes response to stresses within the host, ability to adhere to the intestinal wall and colonize the guts, and beneficial functions such as antimicrobials production, metabolism stimulation and immunomodulation. In an excellent review paper, Papadimitriou et al. [69] presented a number of currently available methodologies applied to select safe and effective probiotic strains.

But even now that worldwide research is focused on handling microbiome composition and function, there is a knowledge gap in the precise prediction of new probiotics from known collections using conventional approaches [70]. These are also limited by the oxygen conditions used to co-culture aerobic strains with eukaryotic host cells in vitro, and as such omit the majority of gut microbiota, which are anaerobic. Technological development in this field resulted in modeling artificial guts for the simulation of the complete digestive process, from ingestion of bacteria, passing through different compartments of the GIT, to fermentation in the hindgut. With this, complex microbiome studies including interactions with the host can be carried out also in anaerobic conditions, which was reviewed elsewhere [71, 72]. One of the advanced gut models, referred to as TIM-2 (TNO computer-controlled, dynamic in vitro gastro-Intestinal Model of the colon), which was developed for human research [73] is now being adjusted to mimic chicken guts and will be used in poultry research [74]. Another line of technological advancements in microbiome research has been facilitated by nanotechnology and led to the development of gut-on-a-chip [75]. This microfluidic device allows for growing human Caco-2 intestinal epithelial cells under constant fluid flow and peristatic movements mimicked by vacuum chambers. Intestinal epithelia grown on gut-on-a-chip formed faster and were much taller than in other in vitro assays, the morphology typically found only in human epithelia in vivo.

In case advanced selection methods of potential probiotics do not suffice, bacteria can be engineered to contain a specific trait. First, probiotics were biotech nologically "designed" to express phenotype of interest, such as more efficient fermentation of prebiotic fibers, producing lactic acid, resistance to stress or improved adherence to intestinal mucosa. Some of those probiotics were modified to mimic a pathogen's survival strategy and they were all called "bioengineered bugs" [76]. More recently, the emergence of synthetic biology revolutionized this field, bringing newly designed and synthesized biological systems. This technology allows engineering microbes with novel therapeutic functions, such as a "sense-and-kill" strategy towards specific infectious agents or cancer cells [77]. Even though this approach is very sophisticated and currently used only in biomedical studies, antibiotic crisis in livestock followed by outbursts of infectious diseases and the need for new therapeutic approaches, may result in applying designer probiotics in animals.

Biotechnology is also used to produce next-generation non-digestible oligosaccharide prebiotics (NGOs), which was reviewed by Mano et al. [78]. A vast majority of oligosaccharide prebiotics is produced by hydrolysis of the naturally occurring polysaccharides. The specific enzymes that transform polysaccharides into NGOs are derived from microbes, which very often is insufficient. For this reason, oligosaccharide biotechnology is based on the development of the optimal bacterial or enzymatic system to process sugars into NGOs. Only properly balanced oligosaccharide prebiotics will pass the small intestine and undergo microbial fermentation in the hindgut. They can be extremely immunoactive and can be used to target literally any microbial community. To meet the increasing demand for prebiotics, new materials are being explored as sources for obtaining NGOs. They include biomass of industrial byproducts of different plants, such as apples, soy, carrot, beans and other malting produce (reviewed by Patel et al., 2012 [79]). Also, new super-plants from different cultures are being investigated, such as yacon, blueberry, pleuran mushrooms and green tea, just to mention a few. It appears that in nature there are many more novel sources of bioactive NGOs. Last but not least are resistant starches, even though they are not prebiotics per se but have become recognized as potent fermentation substrates of many microbes. Resistant starches are able to escape undigested from the small intestine only to be fermented in the large intestine and become a rich source of SCFA (especially butyrate) production [80]. In this way, resistant starches, which are inexpensive and ubiquitous, might be considered prebiotics of the future.

An alternative approach that would not require careful selection of the most beneficial microbes is to transplant the whole microbiome from individuals of preferred phenotype to recipient ones. This method, called fecal microbiota transplant (FMT), was recently tested on chicken lines that differed in feed efficiency [81]. The newly hatched chicks from the line of poor feed efficiency received FMT from a highly efficient line. In this study, FMT allowed colonization with specific feed efficiency associated bacteria, but did not affect the chicken phenotype as much. These data, though promising, have proven that microbiome transplantation is still an immature technique. Also, future issues associated with FMT include better control over microbial composition and developing a method (e.g., encapsulation) to preserve the transplants before use.

By definition, probiotics contain viable microorganisms because their efficiency is strongly correlated with the ability to colonize the guts. For this reason, stabilizing the probiotics so they can retain their viability after production processes and storage is a must. The core technology is to immobilize the live bacteria using an inert material. The protection strategy depends on the properties of the probiotic strain, such as the range of tolerance to pH, temperature and oxygen, and the stresses that may affect its viability in the intestine [82]. One of the emerging technologies for the immobilization of probiotics for livestock is probiotic encapsulation technology (PET), which uses a range of materials, from a common carbohydrate matrix to hydrogel-based delivery systems [83]. If the effort into selecting (or engineering) the proper probiotic is becoming more advanced and sophisticated, the PET should follow to protect the microbes during production, storage and ingestion. And the need for more advanced PET technologies will introduce a significant upgrade into feed mills.

When discussing the future of prebiotics and probiotics in animals, one cannot forget about postbiotics, a term that refers to the byproducts of bacterial metabolism, which have biological activity in the host. Postbiotics include bacteriocins, exopolysaccharides, vitamins, and short-chain fatty acids. They exert beneficial effects on the host without the risk of delivering live bacteria (e.g., horizontal transfer of antibiotic resistance genes) [84]. Killed probiotics are sometimes considered postbiotics, because they are still able to modulate the immune system. In a practical sense, postbiotics are much easier to standardize and deliver in encapsulated form. They also fulfill all safety standards, but the effects of using postbiotics may be transient compared to probiotics. Therefore, some authors report using postbiotics in combination with prebiotics to include long-term beneficial effects in the stimulation of indigenous flora [85].

Last but not least, when considering probiotics in animal production as a dietary supplement with health benefits, the cost of such supplementation is an important factor to consider. Therefore, there is an interest in a transgenerational effect of probiotics applied in livestock. Berghof et al. [86] published a very interesting review paper on transgenerational epigenetics effects on innate immunity in broiler chickens. The authors propose two possible scenarios for modulation of transgenerational effects in broiler chickens, including direct modulation of the chicken embryo or passing on transgenerational effects from the stimulated dams to their offspring. Both of those cases are interesting from an *in ovo* stimulation point of view. First, as we tried to document in this paper, early stimulation of the native microflora in chickens modifies the embryo environment, especially microbial population, which has an intrinsic role in embryo stimulation. Research has not been conducted so far to determine the molecular epigenetic effects at the DNA level. However, we have determined numerous effects of *in ovo* stimulation on different characteristics of the bird, including its transcriptome modulation. It is still unclear if we can successfully stimulate dams *in ovo* so that offspring can generate beneficial effects. It is quite crucial to emphasize that the transgenerational effects of *in ovo* stimulation can express themselves as the phenotype, for example, an increased level of maternal antibodies in the egg yolk or beneficial microbes stored at the surface of the eggshell. Hopefully, future research will shed some light on those questions.

Conclusions

In this review paper, we tried to synthesize the concepts and advancements of *in ovo* stimulation of chicken microflora by delivering prebiotics and probiotics into the air cell on day 12 of egg incubation. In a chicken embryo model at this stage, a bioactive substance deposited inside the air cell can enter the embryonic GIT from two different routes. Oligosaccharide prebiotics that are soluble in water can pass the outer and inner shell membranes and reach the embryo through the blood stream. This process takes place between day 12 and 18 of egg incubation following a single-dose injection with prebiotic. On the other hand, probiotic is engulfed by the embryo on day 19 of egg incubation, during the early phases of hatching. Carefully selected bioactives, delivered *in ovo* on day 12 of egg incubation is the earliest method to stimulate intestinal environment and trigger GALT maturation in chickens. Since it is delivered early in embryonic development, it exerts a strong modulatory effect on the embryo. The determined, so far life-long, phenotypic modulation includes performance traits, intestinal development and microflora abundance at hatching, immune system development, increased SCFA production, yolk-sac absorption and gene expression signatures regulation. The effects were stable and expressed up to six weeks post-hatching, which is the endpoint of the broiler trials. We are convinced that this method can be introduced into commercial settings, given the possible introduction of dedicated infrastructure compliant with production lines used in hatcheries. The ongoing studies in this area consider mitigating effects in chickens under environmental and immunological challenges as well as transgenerational effects that could increase the impact of *in ovo* stimulation.

Abbreviations

CFU: Colony-forming units; GALT: Gut-associated lymphoid tissue; GIT: Gastrointestinal tract; OUT: Operational taxonomic unit; SCFA: Short-chain fatty acids

Acknowledgments

Not applicable.

Funding

Funding - This review paper was supported by research projects: OVOBIOTIC (#RBSI14WZCL) from MIUR in Rome (Italy) (interpretation of the data) and UMO-2013/11/B/NZ9/00783 from National Science Centre in Cracow (Poland) (design of the study, data collection, analysis).

Authors' contributions

Authors' contributions: MS, AS: edited, supported and critically revised the manuscript; KS, JB, AD, MB: wrote the manuscript. All authors have given final approval to the current version to be published.

Competing interests

The authors declare that they have no competing interests.

Author details

[1]Department of Animal Biotechnology and Genetics, UTP University of Science and Technology, Mazowiecka, 28 85-084 Bydgoszcz, Poland.
[2]Department of Animal Physiology, Physiotherapy and Nutrition, UTP University of Science and Technology, Mazowiecka, 28 85-084 Bydgoszcz, Poland.

References

1. Hunter MC, Smith RG, Schipanski ME, Atwood LW, Mortensen DA. Agriculture in 2050: recalibrating targets for sustainable intensification. Bioscience. 2017;67:386–91.
2. Aleksandratos, Bruinsma. World Agriculture towards 2030/2050: The 2012 Revision. ESA Working Paper No. 12–03. Rome: FAO; 2012.
3. Xin H, Liu K. Precision livestock farming in egg production. Anim Front. 2017;7:24.
4. Mollo V. Okano. Precision livestock tools to improve products and processes in broiler production: a review. Braz J. Poult Sci. 2009;11:211–8.
5. Ricks CA, Avakian A, Bryan T, Gildersleeve R, Haddad E, Ilich R, et al. In ovo vaccination technology. Adv Vet Med. 1999;41:495–515.
6. Williams CJ, Zedek AS. Comparative field evaluations of in ovo applied technology. Poult Sci. 2010;89:189–93.
7. Kadam MM, Barekatain MR, Bhanja SK, Iji PA. Prospects of *in ovo* feeding and nutrient supplementation for poultry: the science and commercial applications-a review: *In ovo* feeding and nutrient supplementation for poultry. J Sci Food Agric. 2013;93:3654–61.
8. Roto SM, Kwon YM, Ricke SC. Applications of In Ovo Technique for the Optimal Development of the Gastrointestinal Tract and the Potential Influence on the Establishment of Its Microbiome in Poultry. Front Vet Sci. 2016 [cited 2018 Jan 31];3. Available from: http://journal.frontiersin.org/Article/10.3389/fvets.2016.00063/abstract
9. Pedroso AA, Batal AB, Lee MD. Effect of in ovo administration of an adult-derived microbiota on establishment of the intestinal microbiome in chickens. Am J Vet Res. 2016;77:514–26.
10. Tako E, Glahn RP, Knez M, Stangoulis JC. The effect of wheat prebiotics on the gut bacterial population and iron status of iron deficient broiler chickens. Nutr J. 2014;13:58.
11. Villaluenga CM, Wardeńska M, Pilarski R, Bednarczyk M, Gulewicz K. Utilization of the chicken embryo model for assessment of biological activity of different oligosaccharides. Folia Biol. 2004;52:135–42.
12. Gulewicz K, Bednarczyk M. Sposób stymulacji korzystnego profilu bakteryjnego wylężonych piskląt; 2008.
13. Uni Z, Ferket P. Enhancement of development of oviparous speciesbyinovofeeding.USRegularPatentUS6:B2. 2003.
14. Wei S, Morrison M, Yu Z. Bacterial census of poultry intestinal microbiome. Poult Sci. 2013;92:671–83.
15. Clavijo V, Flórez MJV. The gastrointestinal microbiome and its association with the control of pathogens in broiler chicken production: a review. Poult Sci. 2017;1;97(3):1006–21.
16. Stanley D, Hughes RJ, Moore RJ. Microbiota of the chicken gastrointestinal tract: influence on health, productivity and disease. Appl Microbiol Biotechnol. 2014;98:4301–10.
17. Wielen PWJJ, Keuzenkamp DA, Lipman LJA, Knapen F, Biesterveld S. Spatial and temporal variation of the intestinal bacterial Community in Commercially Raised Broiler Chickens during Growth. Microb Ecol. 2002;44:286–93.

18. Gantois I, Ducatelle R, Pasmans F, Haesebrouck F, Gast R, Humphrey TJ, et al. Mechanisms of egg contamination by *Salmonella* Enteritidis. FEMS Microbiol Rev. 2009;33:718–38.
19. Kizerwetter-Świda M, Binek M. Bacterial microflora of the chicken embryos and newly hatched chicken. J Anim Feed Sci. 2008;17:224–32.
20. Pan D, Yu Z. Intestinal microbiome of poultry and its interaction with host and diet. Gut Microbes. 2014;5:108–19.
21. Ballou AL, Ali RA, Mendoza MA, Ellis JC, Hassan HM, Croom WJ, et al. Development of the Chick Microbiome: How Early Exposure Influences Future Microbial Diversity. Front Vet Sci. 2016;3:2.
22. Cisek AA, Binek M. Chicken intestinal microbiota function with a special emphasis on the role of probiotic bacteria. Pol J Vet Sci. 2014;17:385–94.
23. ARI MM, IJI PA, BHUIYAN MM. Promoting the proliferation of beneficial microbial populations in chickens. Worlds Poult Sci J. Cambridge University Press on behalf of World's Poultry Science Association; 2016;72:785–792.
24. Patterson JA. The Commensal Microbiota. Direct-Fed Microb Prebiotics Anim. New York, NY: Springer New York; 2012. p. 3–11.
25. Patterson J A, Burkholder KM. Application of prebiotics and probiotics in poultry production. Poult Sci 2003;82:627–631.
26. Hajati H, Rezaei M. The application of prebiotics in poultry production. Int J Poult Sci. 2010;9:298–304.
27. Ichihashi T, Takagishi Y, Uchida K, Yamada H. Colonic absorption of menaquinone-4 and menaquinone-9 in rats. J Nutr. 1992;122:506–12.
28. Oakley BB, Lillehoj HS, Kogut MH, Kim WK, Maurer JJ, Pedroso A, et al. The chicken gastrointestinal microbiome. FEMS Microbiol Lett. 2014;360:100–12.
29. Kogut MH, Swaggerty CL. Effects of Prebiotics and Probiotics on the Host Immune Response. Direct-Fed Microb Prebiotics Anim. New York, NY: Springer New York; 2012. p. 61–72.
30. Brisbin JT, Gong J, Sharif S. Interactions between commensal bacteria and the gut-associated immune system of the chicken. Anim Health Res Rev Conf Res Work Anim Dis. 2008;9:101–10.
31. Charalampopoulos D, Rastall R, editors. Prebiotics and probiotics science and technology. New York: Springer Verlag; 2009.
32. Bajagai YS, Yadav S. Probiotics in animal nutrition: production, impact and regulation. Rome: Food and Agriculture Organization of the United Nations; 2016.
33. Bednarczyk M, Stadnicka K, Kozłowska I, Abiuso C, Tavaniello S, Dankowiakowska A, et al. Influence of different prebiotics and mode of their administration on broiler chicken performance. Animal. 2016;10:1271–9.
34. Plowiec A. Wpływ prebiotyków i synbiotyków podanych in ovo na zmianę ekspresji genomu kury. UTP Bydgoszcz; 2018.
35. Calkins K, Devaskar SU. Fetal origins of adult disease. Curr Probl Pediatr Adolesc Health Care. 2011;41:158–76.
36. Vieira SL, Moran ET. Effects of delayed placement and used litter on broiler Yields1. J Appl Poult Res. 1999;8:75–81.
37. Schokker D, Jansman AJM, Veninga G, de Bruin N, Vastenhouw SA, de Bree FM, et al. Perturbation of microbiota in one-day old broiler chickens with antibiotic for 24 hours negatively affects intestinal immune development. BMC Genomics [Internet]. 2017 [cited 2018 Feb 8];18. Available from: http://bmcgenomics.biomedcentral.com/articles/10.1186/s12864-017-3625-6
38. Teague KD, Graham LE, Dunn JR, Cheng HH, Anthony N, Latorre JD, et al. In ovo evaluation of FloraMax®-B11 on Marek's disease HVT vaccine protective efficacy, hatchability, microbiota composition, morphometric analysis, and Salmonella enteritidis infection in broiler chickens. Poult Sci. 2017;96:2074–82.
39. Bednarczyk M, Urbanowski M, Gulewicz P, Kasperczyk K, Maiorano G, Szwaczkowski T. Field and in vitro study on prebiotic effect of raffinose family oligosaccharides in chickens. Bull Vet Pulawy. 2011;55:465–9.
40. de Oliveira JE, van der Hoeven-Hangoor E, van de Linde IB, Montijn RC, van der Vossen JMBM. In ovo inoculation of chicken embryos with probiotic bacteria and its effect on posthatch Salmonella susceptibility. Poult Sci. 2014;93:818–29.
41. Maiorano G, Stadnicka K, Tavaniello S, Abiuso C, Bogucka J, Bednarczyk M. In ovo validation model to assess the efficacy of commercial prebiotics on broiler performance and oxidative stability of meat. Poult Sci. 2017;96:511–8.
42. Slawinska A, Zampiga F, Sirri S, Tavaniello S, Maiorano G. Performance parameters in response to heat stress in broiler chickens injected in ovo with GOS. In: XXIX Int Poult Sci Symp PB WPSA Sep 18–20 2017 Tarnowo Podgórne near Poznan Pol; 2018.
43. Sobolewska A, Bogucka J, Dankowiakowska A, Elminowska-Wenda G, Stadnicka K, Bednarczyk M. The impact of synbiotic administration through in ovo technology on the microstructure of a broiler chicken small intestine tissue on the 1st and 42nd day of rearing. J Anim Sci Biotechnol . 2017 [cited 2018 Jan 31];8. Available from: http://jasbsci.biomedcentral.com/articles/10.1186/s40104-017-0193-1
44. Rehman H, Rosenkranz C, Böhm J, Zentek J. Dietary inulin affects the morphology but not the sodium-dependent glucose and glutamine transport in the jejunum of broilers. Poult Sci. 2007;86:118–22.
45. Solis delos Santos F, Donoghue AM, Farnell MB, Huff GR, Huff WE, Donoghue DJ. Gastrointestinal maturation is accelerated in turkey poults supplemented with a mannan-oligosaccharide yeast extract (Alphamune). Poult Sci. 2007;86:921–30.
46. Bogucka J, Dankowiakowska A, Elminowska-Wenda G, Sobolewska A, Szczerba A, Bednarczyk M. Effects of prebiotics and Synbiotics delivered *in Ovo* on broiler small intestine Histomorphology during the first days after hatching. Folia Biol. 2016;64:131–43.
47. Bogucka J, Dankowiakowska A, Elminowska-Wenda G, Sobolewska A, Jankowski J, Szpinda M, et al. Performance and Small Intestine Morphology and Ultrastructure of Male Broilers Injected in Ovo with Bioactive Substances. Ann Anim Sci. 2017:17.
48. Maiorano G, Sobolewska A, Cianciullo D, Walasik K, Elminowska-Wenda G, Slawinska A, et al. Influence of in ovo prebiotic and synbiotic administration on meat quality of broiler chickens. Poult Sci. 2012;91:2963–9.
49. Kuttappan VA, Brewer VB, Apple JK, Waldroup PW, Owens CM. Influence of growth rate on the occurrence of white striping in broiler breast fillets. Poult Sci. 2012;91:2677–85.
50. Mudalal S, Lorenzi M, Soglia F, Cavani C. Petracci M. Implications of white striping and wooden breast abnormalities on quality traits of raw and marinated chicken meat animal. 2015;9:728–34.
51. Petracci M, Mudalal S, Babini E, Cavani C. Effect of white striping on chemical composition and nutritional value of chicken breast meat. Ital J Anim Sci. 2014;13:3138.
52. Dankowiakowska A, Bogucka J, Szczerba A, Sobolewska A, Kozlowska I, Maiorano G, et al. Assessment of intramuscular fat content and cho¬lesterol content in *M. pectoralis* superficialis of chicken injected in ovo with bioactive substances;Book of abstracts XXVI PO WPSA, 8-10.09.2014, Kazimierz Dolny near Wisla. 2014. p. 59.
53. Bogucka J, Ribeiro DM, da Costa R.P.R, Bednarczyk M. Effect of synbiotic dietary supplementation on histological and histopathological parameters of pectoralis major muscle of broiler chickens. Czech J Anim Sci 2018;63:263–217.
54. Tavaniello S, Maiorano G, Stadnicka K, Mucci R, Bogucka J, Bednarczyk M. Prebiotics offered to broiler chicken exert positive effect on meat quality traits irrespective of delivery route. Poult Sci. 2018;97:2979–87.
55. Tavaniello S, Mucci R, Stadnicka K, Ongwech A, Bednarczyk M, Maiorano G. Effect of in ovo administration of different synbiotics on carcass and meat quality traits in broiler chickens. Poult Sci. 2018.
56. Sławińska A, Siwek M, Zylińska J, Bardowski J, Brzezińska J, Gulewicz KA, et al. Influence of synbiotics delivered in ovo on immune organs development and structure. Folia Biol. 2014;62:277–85.
57. Madej JP, Stefaniak T, Bednarczyk M. Effect of *in ovo* -delivered prebiotics and synbiotics on lymphoid-organs' morphology in chickens. Poult Sci. 2015;94:1209–19.
58. Madej JP, Bednarczyk M. Effect of *in ovo*- delivered prebiotics and synbiotics on the morphology and specific immune cell composition in the gut-associated lymphoid tissue. Poult Sci. 2016;95:19–29.
59. Yegani M, Korver DR. Factors affecting intestinal health in poultry. Poult Sci. 2008;87:2052–63.
60. Slawinska A, Plowiec A, Siwek M, Jaroszewski M, Bednarczyk M. Long-Term Transcriptomic Effects of Prebiotics and Synbiotics Delivered In Ovo in Broiler Chickens. Wilson BA, editor. PLOS ONE. 2016;11:e0168899.
61. Brand A, Galton J, Gilmour DG. Committed precursors of B and T lymphocytes in chick embryo bursa of Fabricius, thymus, and bone marrow. Eur J Immunol. 1983;13:449–55.
62. Sławinska A, Siwek MZ, Bednarczyk MF. Effects of synbiotics injected in ovo on regulation of immune-related gene expression in adult chickens. Am J Vet Res. 2014;75:997–1003.
63. Płowiec A, Sławińska A, Siwek MZ, Bednarczyk MF. Effect of in ovo administration of inulin and *Lactococcus lactis* on immune-related gene expression in broiler chickens. Am J Vet Res. 2015;76:975–82.
64. Kolodziejski PA, Sassek M, Chalupka D, Leciejewska N, Nogowski L, Mackowiak P, et al. GLP1 and GIP are involved in the action of synbiotics in broiler chickens. J Anim Sci Biotechnol. 2018 [cited 2018 Feb 8];9. Available from: https://jasbsci.biomedcentral.com/articles/10.1186/s40104-017-0227-8
65. Xu ZR, Hu CH, Xia MS, Zhan XA, Wang MQ. Effects of dietary

fructooligosaccharide on digestive enzyme activities, intestinal microflora and morphology of male broilers. Poult Sci. 2003;82:1030–6.
66. Pruszynska-Oszmalek E, Kolodziejski PA, Stadnicka K, Sassek M, Chalupka D, Kuston B, et al. In ovo injection of prebiotics and synbiotics affects the digestive potency of the pancreas in growing chickens. Poult Sci. 2015;94:1909–16.
67. Brudnicki A, Szymeczko R, Bednarczyk M, Głowińska B, Pietruszynska D. Assesment of the effect of alpha-galactosides on yolk sac resorption rate in broiler chickens. J Anim Plant Sci. 2015;25:387–91.
68. Chamblee TN, Brake JD, Schultz CD, Thaxton JP. Yolk sac absorption and initiation of growth in broilers. Poult Sci. 1992;71:1811–6.
69. Papadimitriou K, Zoumpopoulou G, FolignÃ© B, Alexandraki V, Kazou M, Pot B, et al. Discovering probiotic microorganisms: in vitro, in vivo, genetic and omics approaches. Front Microbiol. 2015;6. [cited 2018 Jan 31] Available from: http://journal.frontiersin.org/Article/10.3389/fmicb.2015.00058/abstract
70. Vinderola G, Gueimonde M, Gomez-Gallego C, Delfederico L, Salminen S. Correlation between in vitro and in vivo assays in selection of probiotics from traditional species of bacteria. Trends Food Sci Technol. 2017;68:83–90.
71. Venema K, van den Abbeele P. Experimental models of the gut microbiome. Best Pract Res Clin Gastroenterol. 2013;27:115–26.
72. von Martels JZH, Sadaghian Sadabad M, Bourgonje AR, Blokzijl T, Dijkstra G, Faber KN, et al. The role of gut microbiota in health and disease: in vitro modeling of host-microbe interactions at the aerobe-anaerobe interphase of the human gut. Anaerobe. 2017;44:3–12.
73. Minekus M. The TNO Gastro-Intestinal Model (TIM). In: Verhoeckx K, Cotter P, López-Expósito I, Kleiveland C, Lea T, Mackie A, et al., editors. Impact Food Bioact Health. Cham: Springer International Publishing; 2015 [cited 2018 Feb 8]. p. 37–46. Available from: http://link.springer.com/10.1007/978-3-319-16104-4_5
74. Venema K. CALIMERO, an in vitro dynamic, computer-controlled chicken ALIMEntary tRact mOdel the 6th beneficial microbes conference, Oct 9–11, Amsterdam (Netherlands). 2017.
75. Kim HJ, Huh D, Hamilton G, Ingber DE. Human gut-on-a-chip inhabited by microbial flora that experiences intestinal peristalsis-like motions and flow. Lab Chip. 2012;12:2165.
76. Sleator RD, Hill C. 'Bioengineered bugs' – a patho-biotechnology approach to probiotic research and applications. Med Hypotheses. 2008;70:167–9.
77. Chua KJ, Kwok WC, Aggarwal N, Sun T, Chang MW. Designer probiotics for the prevention and treatment of human diseases. Curr Opin Chem Biol. 2017;40:8–16.
78. Mano MCR, Neri-Numa IA, da Silva JB, Paulino BN, Pessoa MG, Pastore GM. Oligosaccharide biotechnology: an approach of prebiotic revolution on the industry. Appl Microbiol Biotechnol. 2018;102:17–37.
79. Patel S, Goyal A. The current trends and future perspectives of prebiotics research: a review. 3. Biotech. 2012;2:115–25.
80. Bird A, Conlon M, Christophersen C, Topping D. Resistant starch, large bowel fermentation and a broader perspective of prebiotics and probiotics. Benef Microbes. 2010;1:423–31.
81. Siegerstetter S-C, Petri RM, Magowan E, Lawlor PG, Zebeli Q, O'Connell NE, et al. Fecal microbiota transplant from highly feed-efficient donors shows little effect on age-related changes in feed-efficiency-associated fecal microbiota from chickens. Dudley EG, editor. Appl Environ Microbiol. 2017; 84:e02330–17.
82. Šipailienė A, Petraitytė S. Encapsulation of Probiotics: Proper Selection of the Probiotic Strain and the Influence of Encapsulation Technology and Materials on the Viability of Encapsulated Microorganisms. Probiotics Antimicrob Proteins. 2017 [cited 2018 Jan 31]; Available from: http://link.springer.com/10.1007/s12602-017-9347-x
83. Gbassi GK, Vandamme T. Probiotic encapsulation technology: from microencapsulation to release into the gut. Pharmaceutics. 2012;4:149–63.
84. Imperial ICVJ, Ibana JA. Addressing the Antibiotic Resistance Problem with Probiotics: Reducing the Risk of Its Double-Edged Sword Effect. Front Microbiol. 2016 [cited 2018 Feb 7];07. Available from: http://journal.frontiersin.org/article/10.3389/fmicb.2016.01983/full
85. Kareem KY, Loh TC, Foo HL, Akit H, Samsudin AA. Effects of dietary postbiotic and inulin on growth performance, IGF1 and GHR mRNA expression, faecal microbiota and volatile fatty acids in broilers. BMC Vet Res. 2016 [cited 2018 Jan 31];12. Available from: http://bmcvetres.biomedcentral.com/articles/10.1186/s12917-016-0790-9
86. Berghof TVL, Parmentier HK, Lammers A. Transgenerational epigenetic effects on innate immunity in broilers: an underestimated field to be explored? Poult Sci. 2013;92:2904–13.

19

Comparative features of infections of two Massachusetts (Mass) infectious bronchitis virus (IBV) variants isolated from Western Canadian layer flocks

Aruna Amarasinghe[1], Upasama De Silva Senapathi[1], Mohamed Sarjoon Abdul-Cader[1], Shelly Popowich[2], Frank Marshall[3], Susan C. Cork[1], Frank van der Meer[1], Susantha Gomis[2] and Mohamed Faizal Abdul-Careem[1*]

Abstract

Background: Infectious bronchitis virus (IBV) is one of the leading causes of mortality and morbidity in chickens. There are numerous serotypes and variants, which do not confer cross protection resulting in failure of currently used IBV vaccines. Although variant IBV isolates with major genetic differences have been subjected to comparative studies, it is unknown whether minor genetic differences in IBV variants within a serotype are different in terms of pathogenesis and eliciting host responses. Two Massachusetts (Mass) variant IBV isolates recovered from commercial layer flocks in the Western Canadian provinces of Alberta (AB) and Saskatchewan (SK) were compared genetically and evaluated for their pathogenicity, tissue distribution and ability to recruit and replicate in macrophages.

Results: Although whole genome sequencing of these two Mass IBV isolates showed low similarity with the M41 vaccinal strain, they had an identical nucleotide sequence at open reading frames (ORFs) 3a, 3b, envelop (E), matrix (M), 5a and 5b. The rest of the ORFs of these 2 IBV isolates showed 99.9% nucleotide similarity. However, upon experimental infection, we found that the IBV isolate originating from AB was different to the one that originated in SK due to higher tracheal lesion scores and lower lung viral replication and lower genome loads in cecal tonsils. Nevertheless, both IBV isolates elicited host responses characterized by significant macrophage recruitment to the respiratory tract and there was evidence that both IBV isolates replicated within tracheal and lung macrophages.

Conclusions: Overall, this study shows that Mass variant IBV isolates, although possessing minor genetic variations, can lead to significant differences in pathogenicity in young chickens. Further studies are required to investigate the pathogenicity of these two Mass variant IBV isolates in laying hens.

Keywords: Infectious bronchitis virus, Whole genome sequencing, Tissue distribution, Pathogenicity, Macrophage response

Background

Infectious bronchitis virus (IBV) belongs to the family *Coronaviridae*. Traditionally, IBV is considered to be a host-specific respiratory pathogen in chickens and IBV initially replicates at the route of entry, the tracheal mucosa [1, 2]. However, identification of new variants and/or serotypes of IBV have shown a wide variation of tissue tropism including urinary [3–10], gastrointestinal [6, 9, 11, 12], oviduct [9, 13] and bursa of Fabricius [11, 14]. IBV is known to replicate in the reproductive tract epithelium in layers leading to reduced egg production and shell defective eggs [15, 16]. False layer syndrome, which is associated with cystic oviduct formation occurs with IBV infection in early life [17, 18]. IBV can also replicate in the testes of cockerels [19].

An array of serotypes and strains of IBV infect chickens throughout the world [2]. Genetic events such as insertion(s) and deletions [20, 21], point mutations [22], and recombination [23–27] contribute to genomic variations of IBV [28]. The spike 1 (S1) gene is highly

* Correspondence: faizal.abdulcareem@ucalgary.ca
[1]Department of Ecosystem and Public Health, Faculty of Veterinary Medicine, University of Calgary, Health Research Innovation Center 2C53, 3330 Hospital Drive NW, Calgary, AB T2N 4N1, Canada

variable among IBV strains and it encodes epitopes, which bind to neutralizing antibodies [29]. A change in the amino acid sequence as small as 2 to 3% in the S1 subunit can result in changes in the antigenicity of the virus [30]. Based on these changes of the S1 protein, numerous IBV strains have been characterized throughout the world [1]. Therefore, either the partial [31–33] or the full-length [34–39] of the S1 glycoprotein gene has been used in the molecular characterization of IBV isolates. However, using whole genome sequencing it has been observed that genes other than S1 may play a role in the pathogenicity of IBV infection [40, 41]. Currently, there are no IBV reference full genome sequences available for Canadian IBV isolates but partial [33] or complete [37] S1 sequences are available.

Previously, comparative studies have been conducted to elucidate differential pathogenicity and host responses between variant IBV isolates with larger genome variability such as nephropathogenic and Massachusetts (Mass) IBV isolates recovered from various geographical areas [18, 42–46]. Since it is well known that very minor changes in the genome of viruses [47] including IBV [48, 49] can lead to difference in pathogenicity, comparative studies involving variant IBV isolates are required.

A wide array of IBV variants are affecting commercial broiler, layer and breeder flocks in Canada [33, 37, 50, 51]. For example, Mass type IBV variants are impacting the commercial egg production in Western Canada [50]. This study showed that Mass type IBV infection of 27-week old layers at peak production, lead to loss of egg production for about 2 weeks followed by production of shell less, small and defective eggs for another 2 days before the egg production bounced back to normal production. The duration of the loss of marketable eggs appears to be 16 days. In an 8000 bird layer flock of Western Canada, we observed 47.6% drop in egg production for 10 days and we isolated a Mass type variant IBV from this flock. Based on Egg Farmers of Alberta's price of $2.15/dozen of large eggs, the loss for 10 days is calculated to be $6823 for this particular outbreak and the loss will be higher if the infection persists beyond 10 days or reoccur. In addition to this direct loss of egg production, egg grading stations need to outsource supply of eggs during these production drops and also some of the affected birds may succumb to secondary bacterial infections without a recovery, hence the indirect losses are substantially higher than the direct costs of IBV infection.

The present study was conducted with the aim to elucidate the genetic differences of two Mass variant IBV isolates (15AB-01 and 15SK-02) that originated from layer flocks in Western Canada, Alberta (AB) and Saskatchewan (SK). Additionally, we aimed to describe whether these genetic changes in Mass isolates 15AB-01 and 15SK-02 lead to differential pathogenicity, tissue distribution and macrophage responses in the respiratory tract.

Results

Molecular characterization of the mass IBV isolates, 15AB-01 and 15SK-02

For the isolate 15AB-01, total of 525,998 and 908,860 read sets were obtained from the direct allantoic fluid sample and layered virus sample, respectively from the Illumina MiSeq workflow. From these, a total of 418,446 and 825,147 reads were aligned with the reference IBV genomes. For isolate 15SK-02, a total of 4,355,902 and 3,731,578 reads were obtained from the direct allantoic fluid sample and layered virus sample, respectively, of which, a total of 3,146,584 and 3,152,622 reads were aligned with the IBV reference genomes. The number of reads obtained from Illumina MiSeq for 15SK-02 was 10 times higher than for 15AB-01 and the viral genome concentration of the submitted complimentary (c)DNA was also much higher for 15SK-02 than for 15AB-01 (real-time PCR cycle threshold or Ct value difference of 3.54). Entirely based on the NGS output, the complete genomes obtained for the Mass IBV isolates, 15AB-01 and 15SK-02 were 27.6 kb in length. As determined by the open reading frames (ORF) predictor, both isolates had similar genome organization with 11 ORFs and they were 5′-1a-1b-S-3a-3b-E-M-5a-5b-N-3′ (Table 1). Between the isolates, 15AB-01 and 15SK-02, there was 99.9% nucleotide similarity. With the M41 vaccine strain, 15AB-01 and 15SK-02 shared a genetic similarity of 92.8 and 92.7%, respectively. 15AB-01 and 15SK-02 Mass IBV isolates shared 92.8 and 92.7% nucleotide similarity with the M41 vaccine strain (GQ504725), respectively. Nucleotide similarities between Mass IBV isolates, 15AB-01 and 15SK-02 across the ORFs 3a, 3b, E, M, 5a and 5b were 100%. The remainder of genes were over 99% identical (Table 2). Nucleotide similarities

Table 1 Genes, coding regions, and deduced proteins of the current Mass IBV isolates, 15AB-01 and 15SK-02

Gene	Translation Frame	Gene location	Nucleotide length (bp)	Number of amino acids
1a	1	529–12,390	11,862	3953
1b	3	12,465–20,423	7959	2652
S	1	20,374–23,862	3489	1162
3a	3	23,862–24,035	174	57
3b	2	24,035–24,229	195	64
E	3	24,210–24,539	330	109
M	1	24,511–25,188	678	225
5a	3	25,539–25,736	198	65
5b	2	25,733–25,981	249	82
N	1	25,924–27,153	1230	409

Table 2 Nucleotide similarity of 15AB-01 and 15SK-02 across various genes with M41 vaccine strain GQ504725

Strain	ORF1a	ORF 1b	S1	S2	ORF3a	ORF3b	E	M	ORF5a	ORF5b	N
15AB-01	90.7	96.8	95.5	97.6	91.2	96.9	94.5	100	86.4	89.2	94.1
15SK-02	90.6	96.8	95.5	97.9	91.2	96.9	94.5	100	86.4	89.2	94.5
[a]15AB-01 Vs 15SK-02	99.9	99.9	99.6	99.8	100	100	100	100	100	100	99.5

[a]Third raw depicts the nucleotide similarities between current Mass IBV isolates 15AB-01 and 15SK-02 across various genes

between Mass IBV isolates, 15AB-01 and 15SK-02 across the hypervariable S gene regions, S1 and S2, respectively were 99.6 and 99.8%. The lowest nucleotide similarity these two isolates shared with the vaccine M41 strain was 86.4% in the ORF 5a and the highest similarity was 100% in the ORF M. The S1 and S2 identities of these two IBV isolates with the M41 vaccine strain were 95.5 and 97.6%, respectively (Table 2). Additionally, our IBV isolates, 15AB-01 and 15SK-02 were compared with important reference strains in Mass type cluster (Table 3) and they are most variable from the Beaudette IBV strain and very closely related to other Mass type IBV strains.

Full genome nucleotide distance matrix and heat maps of these isolates with 23 selected reference sequences are shown in the Fig. 1. The highest similarity of 15AB-01 and 15SK-02 shared with M41 strain (USA) was 99.9%. The lowest similarity was 87.5% with the United Kingdom (UK) strain 1148 (Fig. 1). Phylogenetic reconstruction with reference IBV strains based on the full genome sequence revealed a close relation with two IBV isolates to the USA Mass type virus (Fig. 2). The isolate 15AB-01 appeared to be more recently evolved than the 15SK-02 isolate. However, clear separation of these two IBV isolates (cluster M1) from the vaccine M41 strain (Cluster M11) is evident (Fig. 2).

Clinical observations and body weight changes following IBV infection of chickens

From 2 to 8 days post-infection (dpi), there were two birds with mildly increased respiration (more labored and higher frequency) from the IBV (15AB-01) infected group. On 9 and 10 dpi, there were 3 birds with mildly increased breathing. Only one bird continued to show a mild breathing increase by 11 dpi and thereafter all birds were bright and alert. In contrast, 2 birds from the 15SK-02 infected group started showing increased breathing as late as 7 dpi and continued until 11 dpi. The control birds were alert and active throughout. The highest body weight gain was achieved by the uninfected control group followed by the group infected with the isolate 15AB-01 and 15SK-02 (Fig. 3). However, the difference in bodyweights was not significant ($p > 0.05$).

Histopathological changes following IBV infection of chickens

Histopathological examination of both trachea and lung at 4 dpi showed lesions suggestive of infectious bronchitis. In the trachea, these histological changes included de-ciliation of tracheal mucosal cells, lymphoid infiltration into mucosa and underlying submucosa, and mucosal detachment from the lamina propria (Fig. 4a). Squamous metaplasia of the tracheal mucosa was prominent in trachea of both IBV infected groups (Fig. 4a). Average lesion score of the trachea infected with the isolate 15AB-01 ($p < 0.001$) and 15SK-02 ($p < 0.01$) were significantly higher than that of trachea from uninfected controls (Fig. 4b). Also lesion scores of the trachea from 15AB-01 infected chicken were significantly higher than that of trachea from chicken infected with isolate 15SK-02 ($p < 0.01$) (Fig. 4b). In the lungs, mononuclear cell infiltration and hemorrhage into the parabronchial lumen were observed in the infected lungs compared to

Table 3 Nucleotide identities of 15AB-01 and 15SK-02 across various genes with Mass-type reference strains

Strain	ORF 1a	ORF 1b	S1	S2	ORF 3a	ORF 3b	E	M	ORF 5a	ORF 5b	N
Beaudette	91.5/91.5	99.9/93.9	97.8/97.8	98.1/98.2	97.7/97.1	98.2/98.2	98.7/98.7	98.8/98.8	98.9/98.9	90.8/90.8	92.6/92.7
TN92–03 (India)	93.2/93.2	94.0/93.9	97.2/97.0	98.3/98.4	97.7/98.3	97.0/97.0	98.4/98.4	99.0/99.0	100/100	95.6/95.6	97.6/99.8
Mass41/1972 (USA)	99.7/99.7	99.5/99.5	99.8/99.8	99.2/99.3	98.8/99.4	100/100	99.7/99.7	99.0/99.0	100/100	100/100	99.6/99.8
Mass/USA (AY851295)	100/99.9	99.9/99.9	99.9/99.9	99.9/99.9	100/99.4	100/100	99.7/99.7	100/100	100/100	100/100	99.8/99.6
ck/CH/LHLJ/091205 (China)	99.9/99.9	99.9/99.9	99.9/99.9	99.9/99.9	100/99.4	100/100	100/100	100/100	100/100	100/100	100/99.8
M41/USA (DQ834384)	100/100	100/99.9	99.9/99.9	99.7/99.7	100/99.4	100/100	100/100	99.9/99.9	100/100	100/100	99.7/99.8
Mass/USA (GQ504724)	100/100	99.9/99.9	99.9/99.9	99.8/99.8	100/99.4	100/100	100/100	100/100	100/100	99.6/99.6	99.8/99.8
M41/1965 (USA)	99.9/99.9	100/100	99.9/99.9	99.8/99.8	98.8/99.4	100/100	100/100	100/100	100/100	100/100	99.6/99.8
M41/1985 (USA)	100/100	99.9/100	99.9/100	99.8/99.8	98.8/99.4	100/100	100/100	100/100	100/100	100/100	99.7/99.8

In each cell of the table, the first value is the nucleotide identity shared with the IBV isolate 15AB-01 and the second value is the nucleotide identity shared with the IBV isolate 15SK-02

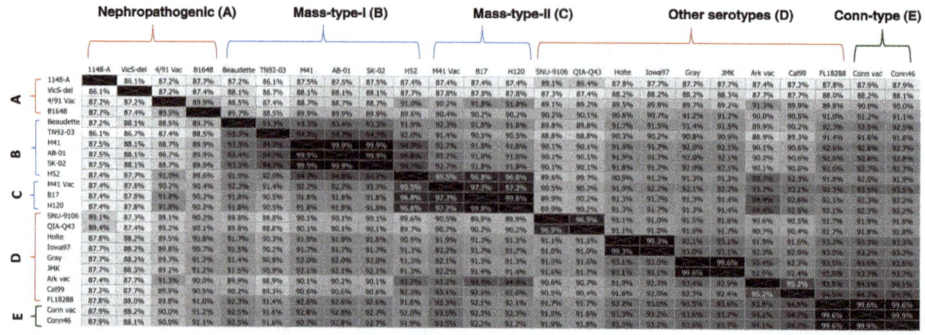

Fig. 1 Nucleotide similarity matrix and heat maps of 23-selected reference strains and 2 current Mass IBV isolates. Full genome sequences of Western Canadian isolates (15AB-01 and 15SK-02) were aligned with 23 reference sequences retrieved from the GenBank. The resulting genetic distance matrix along with heat maps based on the similarity index is shown

the uninfected controls (Fig. 4a). We observed variation in lung lesion scores within groups, yet significant outliers were not detected when the data were subjected to Grubbs test. The lung lesion scores of the lung from 15AB-01 infected chicken were significantly higher than that of the lungs from uninfected controls. However, the lung lesion score between the lung from 15SK-02 infected chicken and the uninfected controls was not statistically significant ($p > 0.05$) (Fig. 4c).

IBV genome loads in oropharyngeal and cloacal swabs and tissue samples

The average genome loads on oropharyngeal swabs were higher than the genome loads on cloacal swabs for both Mass IBV isolates at all time points (Fig. 5a and b). In the trachea, a genome load of 1×10^6 up to 3 dpi could be detected in chicken infected with either isolate. By 12 dpi, the genome loads in the trachea declined by 2-folds. The genome load in the cloaca was consistent for the 15AB-01 throughout the observation period. However, in the cloaca of birds infected with 15SK-02 no IBV genome at 12 dpi could be detected (Fig. 5b). No IBV genome load was detected in the uninfected controls (results not shown).

The highest genome loads in tissue were found in trachea and lung tissues of chicken infected with either isolate at 4 dpi, however, the genome load in lung tissue was significantly lower for IBV isolate 15AB-01 (Fig. 5c). At 4 and 14 dpi, the genome loads in trachea were similar for IBV isolates (15AB-01 and 15SK-02). By 14 dpi, tracheal and lung genome loads fell considerably for both isolates. Cecal tonsil genome load was declined by 14 dpi only in birds infected with the isolate 15AB-01

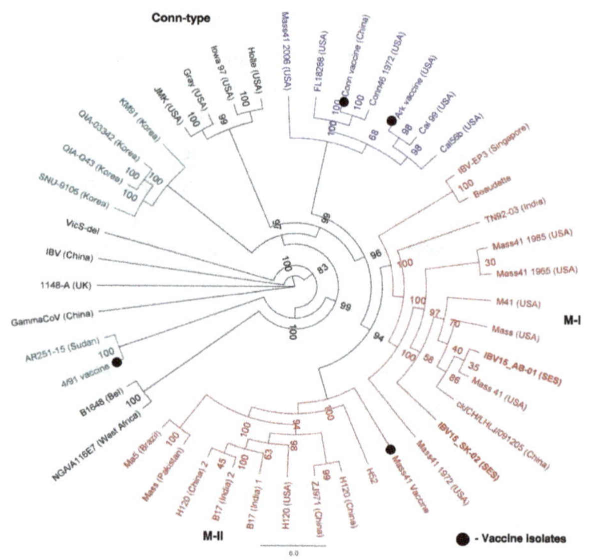

Fig. 2 Phylogenetic tree of 45 IBV whole genome sequences including Western Canadian Mass IBV isolates. A maximum likely hood phylogenetic tree was inferred based on the whole genomes of 15AB-01 and 15SK-02 isolates of IBV and 43 IBV genome sequences retrieved from GenBank using RAxML with GTR GAMMA nucleotide substitution model with 1000 bootstrap iterations

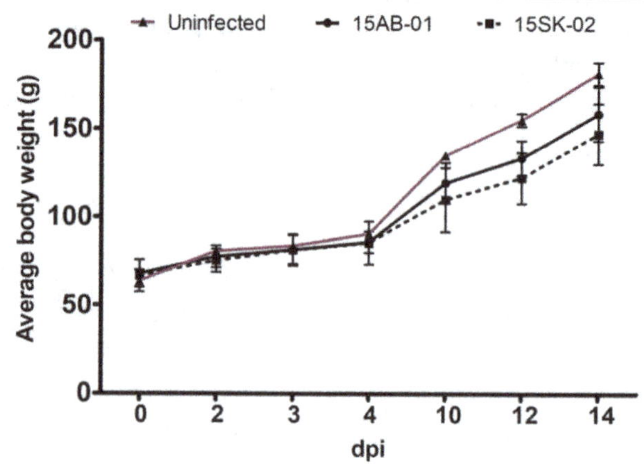

Fig. 3 Bodyweight changes following infection with Western Canadian Mass-type IBV isolates. The average body weight of each group (IBV 15AB-01, IBV 15SK-02 and uninfected control) of chickens at different time points are depicted in the graph. The error bars represent the ±standard error of mean (SEM)

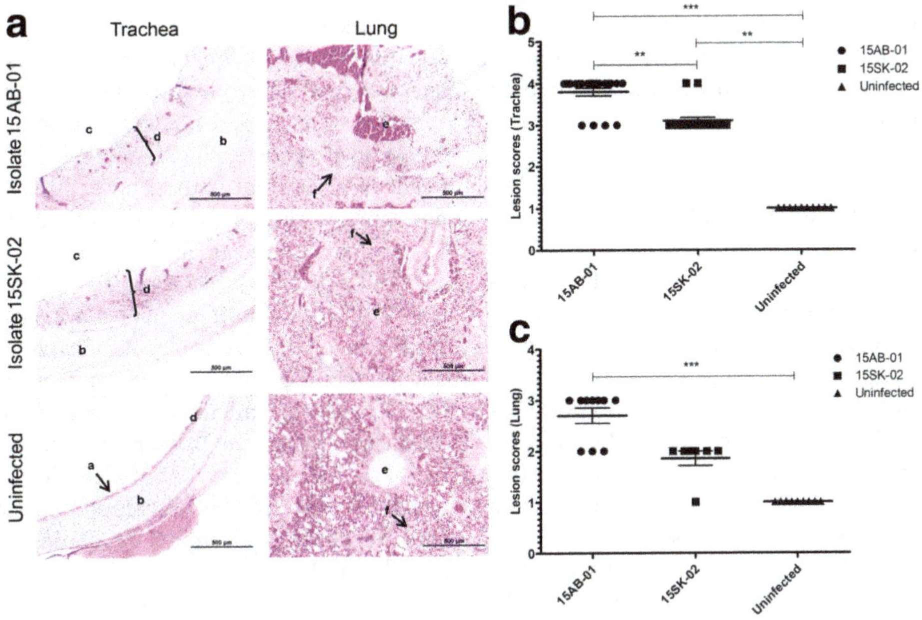

Fig. 4 Histological changes in trachea and lungs induced by Western Canadian Mass-type IBV isolates. The representative images of trachea and lungs (**a**) and average lesion scores of trachea (**b**) and lung (**c**) for each group of chickens (IBV 15AB-01, IBV 15SK-02 and uninfected control) at 4 dpi are depicted. a = cilia, b = tracheal cartilage, c = tracheal lumen, d = tracheal mucosa, e = parabronchial lumen and f = inter-parabronchial septum. The error bars represent the ±SEM. Statistical significance: $*p < 0.05$, $**p < 0.01$, $***p < 0.001$

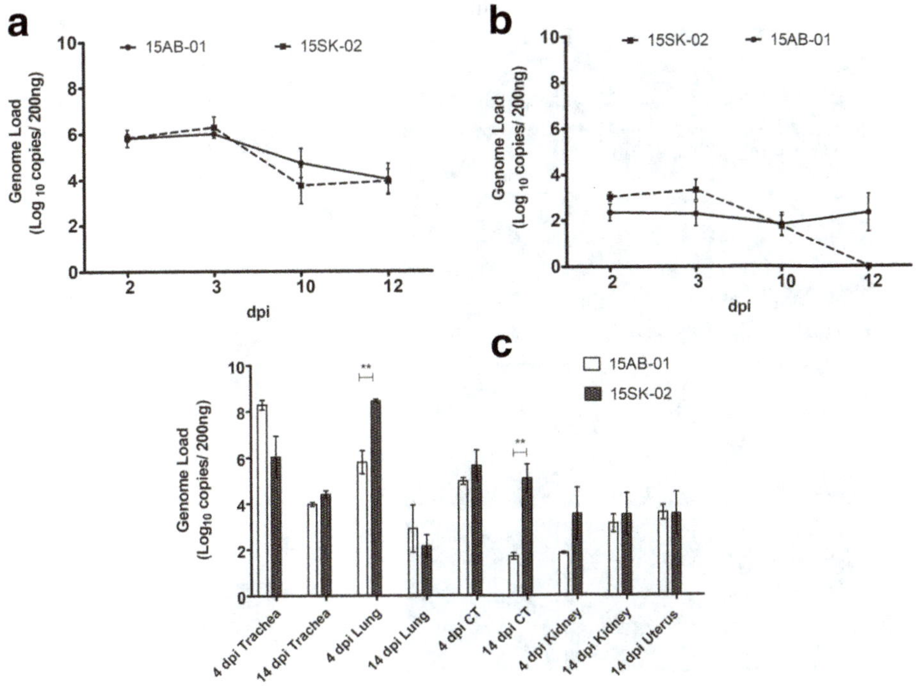

Fig. 5 IBV genome loads in swabs and tissues following infection with Western Canadian Mass-type IBV isolates. For each time point, the average $\log_{10}$ starting IBV copies per 200 ng of extracted RNA from the oropharyngeal (**a**) and cloacal (**b**) swabs and various tissues (**c**) at 4 and 14 dpi are shown with the SEM. Statistical significance: $*p < 0.05$, $**p < 0.01$, $***p < 0.001$

and the difference was statistically significant ($p < 0.01$). There was no difference in genome loads in the kidney tissue between two time points in both strains. The uterine genome loads were available only at 14 dpi for both strains. None of the uninfected control birds had detectable IBV genome loads (results not shown).

Immunofluorescence data for IBV antigens and macrophages at 4 dpi

Similar to IBV genome loads, we did not see a difference in IBV antigen expression in trachea of 15AB-01 and 15SK-02 infected chickens ($p > 0.05$, Fig. 6b). Similar to the IBV genome load, the amount of IBV antigen was significantly lower in 15AB-01 infected lung than that of 15SK-02 infected lung ($p < 0.01$, Fig. 7b). The lung and tracheal macrophage numbers were not significantly different in the two groups of experimentally infected chicken, indicating a similar response to the two strains of Mass IBV used in the study ($p > 0.05$). Macrophage infiltration was significantly higher ($p < 0.05$) in tracheas infected with both Mass IBV isolates compared to the tracheas of uninfected controls (Fig. 6a and c). Similarly, lungs infected with either Mass IBV isolates had significantly higher number of macrophages ($p < 0.01$) than lungs from uninfected controls (Fig. 7a and c). In both the trachea and lung, there was no significant difference in IBV positive macrophages between two Mass IBV isolates (Figs. 6d and 7d).

Discussion

The work described in this manuscript focused, first, to see the genetic differences of two Mass variant IBV isolates originating from layer flocks in two provinces of Western Canada. Second, we evaluated if infection of SPF chicken with genetically different Mass IBV variants would result in a different pathogenicity, tissue distribution and macrophage responses in the respiratory tract. Since the S1 protein is the main inducer of virus-neutralizing antibodies [52–54], either partial [31–33] or full-length [34–39] S1 glycoprotein gene have been used in many studies involving IBV molecular characterization throughout the world including previous Canadian studies. Using whole genome sequencing it has been shown that genes other than S1 may be playing roles in the pathogenicity of IBV [41, 55]. To our knowledge this is the first study, which investigated the full genome of Canadian IBV isolates. Although, IBV strains circulating in other countries have been characterized and changes in the IBV genome due to mutations and recombination have been shown [40, 55–59], our whole genome sequencing data shows that mutations are more common than the recombination and genes other than S1 have been affected by mutations. These changes in the genome of these two IBV isolates reflected in the differences in the pathogenicity although host response characterized by macrophage recruitment into the respiratory tract were similar.

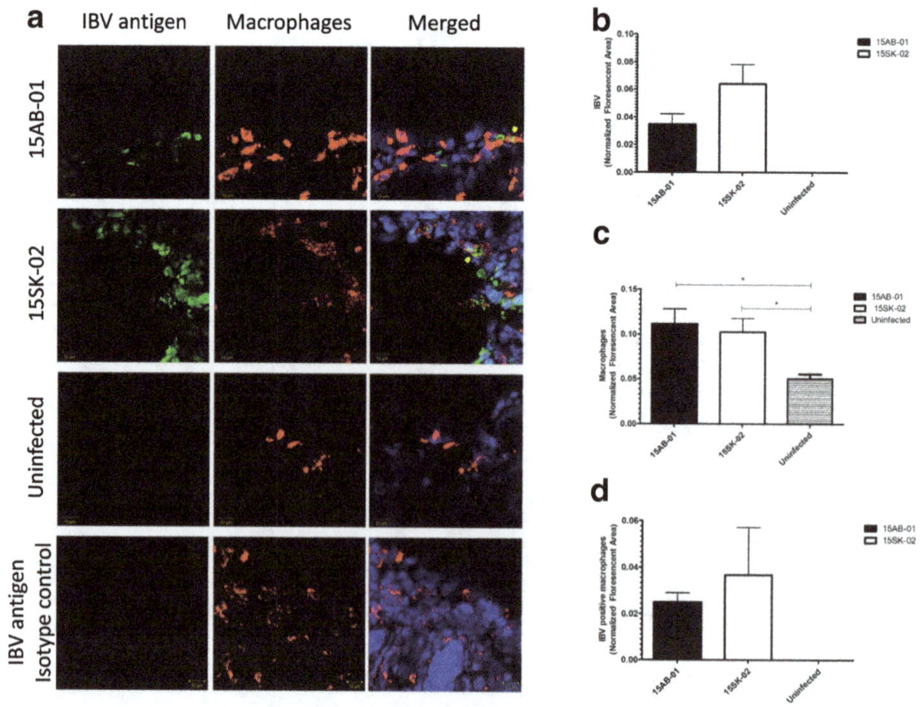

Fig. 6 Mass-type IBV isolates recovered from Western Canadian layer flocks increase macrophage numbers in trachea. The representative images of trachea stained for IBV antigens and macrophages (**a**) and quantitative data for IBV antigens (**b**), macrophages (**c**) and IBV antigen positive macrophages (**d**) at 4 dpi are depicted. The error bars represent the ± SEM. Statistical significance: $^{*}p < 0.05$, $^{**}p < 0.01$, $^{***}p < 0.001$

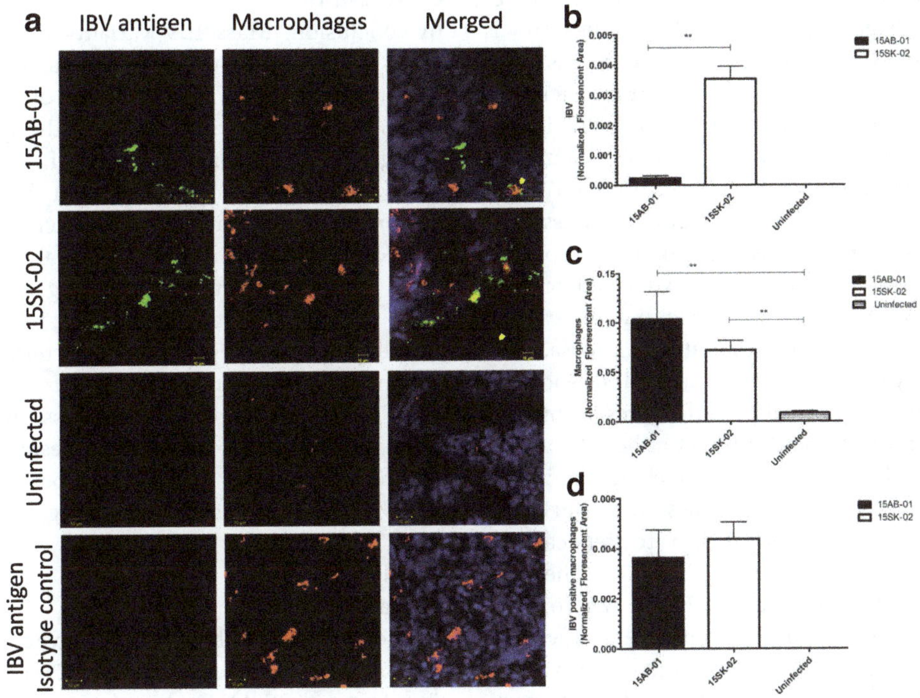

Fig. 7 Mass-type IBV isolates recovered from Western Canadian layer flocks increase macrophage numbers in lungs. The representative images of lungs stained for IBV antigens and macrophages (**a**) and quantitative data for IBV antigens (**b**), macrophages (**c**) and IBV antigen positive macrophages (**d**) at 4 dpi are depicted. The error bars represent the ± SEM. Statistical significance: $*p < 0.05$, $**p < 0.01$, $***p < 0.001$

Previously, studies have been conducted to compare pathogenicity and immune responses between nephropathogenic (i.e. QX like IBV strains) and Mass type IBV isolates recovered from various geographical areas [18, 42, 43, 45, 46]. Nephropathogenic QX-type strains are distantly related to the M41 strain and their genetic similarity is 78.6% [60]. Based on the partial S1 sequence analysis of three nephropathogenic IBV isolates, which belong to the Mass genotype revealed that they are more than 99.1 and 97.8% identical to the classical M41 strain in nucleotide and deduced amino acid sequences, respectively [61]. Since it is well known that very minor changes in the genome of viruses [47] including IBV [49, 62] lead to differences in pathogenicity and host responses, we chose two Mass type IBV isolates expected to have minor genome differences for evaluation of pathogenicity, tissue distribution and host responses. Mass type IBV isolates have been shown to replicate in the respiratory tissues, kidney, cecal tonsils and gastrointestinal tissues [18, 42, 43]. In agreement with these studies, we also observed that Mass IBV isolates originating from Western Canadian layer flocks replicate in tissues other than respiratory tissues. In our study, Mass IBV isolate originating in SK had significantly higher genome load in the lungs than the one originating in AB, though there was no difference of lung lesion score between these two IBV isolates. On the other hand, in trachea, the genome loads of these two Mass IBV isolates were not different, although the Mass IBV isolate originated in AB had significantly higher lesion scores. It is also noteworthy that Mass IBV isolate originating in SK was not detectable at 12 dpi in cloacal swabs. Since the major source of IBV genome in cloaca is due to IBV replication in gastrointestinal tract [63], it is possible that 15SK-02 IBV infection in gastrointestinal tract may have been cleared by 12 dpi. Overall, these two Mass IBV variants have similar pathogenicity with only slight differences which also agree with the minor genetic differences of these 2 Mass type IBV isolates.

Literature that describe pathogenicity using Canadian IBV isolates are scarce [8] Although this study evaluated the pathogenicity of 5 IBV isolates originated in Ontario poultry flocks, IBV shedding, tissue distribution and macrophage response have not been studied. Molecular characterization of IBV strains circulating in Canada has been done previously and it appears that there is a difference in distribution of IBV genotypes according to the geographic area within Canada [33, 37, 50, 51]. In Western Canada, mostly Mass and Conn IBV genotypes are common and in Eastern Canada, in addition to Mass and Conn genotypes nepharopathogenic strains, 4/91, CA1737 and DMV genotypes are common among poultry flocks.

It appears that host responses, characterized by respiratory macrophages following IBV infection is dependent on the IBV strain. For example, it has been shown that M41 but not Australian T strain increases macrophages in respiratory tract lavage [64]. Previously, it has also been shown that IBV ConnA5968 strain elicit higher macrophage response in tracheal mucosa when compared IBV M41 strain [65]. Although Mass IBV (M41) strain is known to increase macrophage responses in lungs and trachea [65], it is unknown whether variants of Mass IBV are capable of eliciting differential macrophage response in the respiratory mucosa. The current study shows that minor genetic differences in Mass IBV variants do not lead to differences in macrophage responses in the respiratory tract although each of the tested IBV isolates were able to elicit significant macrophage responses in these tissues. IBV M41 and ConnA5968 strains have been shown to replicate in macrophages in vivo and in vitro [65–67]. The IBV Beaudette strain is able to infect the HD11 macrophage cell line [68]. Although we did not observe a difference in macrophage infection in vivo by both Mass variant IBV isolates used in our experiments, it will be important to establish if these field IBV strains also target macrophages located in the respiratory tract for its replication. We observed a higher quantity of IBV antigens in 15SK-02 IBV infected lungs when compared to 15AB-01 infected lungs and that difference was not reflected in the number of IBV infected macrophages. This discrepancy may be due to the fact that IBV infects cells other than macrophages. Previously, we observed that IBV infection in macrophage of the respiratory tract was about 3–4% [65].

In spite of the novel aspect of our work comparing two Mass type IBV isolates with minor genetic variations for pathogenicity and host responses, there were some limitations to this study. First, we did not include a control group infected with the standard M41 IBV strain in our experiment and this was due to the limited infrastructure availability at our animal facility. However, in a separate study, we investigated the M41 IBV induced macrophage recruitment to respiratory tract [65]. Second, for the quantification of IBV, we employed real-time PCR technique targeting the most conserved N gene of IBV [69–71]. Since qPCR quantify both live and dead virus [72], our IBV genome load data do not reflect IBV replication. Alternatively, we would have determined viral titers using 9–11 day old SPF embryonated chicken eggs and expressed as EID_{50} [73]. Although this later method could be employed to determine IBV infectivity in our swabs and tissues, the use of this technique was precluded for quantifying IBV titer due to the requirement of large numbre of eggs.

Conclusion

In conclusion, Mass IBV variants recovered from layer flocks had identical nucleotide sequence at ORFs 3a, 3b, E, M, 5a and 5b. The remainder of the ORFs showed 99.9% nucleotide similarity. Although, the genetic changes were minor, the pathogenicity in vivo of the Mass IBV variant originating from AB was different to the one that originated in SK, whereby the AB variant caused significantly higher tracheal lesions and replicated to a lower titer in the lung and lower genome loads in cecal tonsils. Additionally, both Mass IBV variants elicited host responses characterized by significant macrophage recruitment to the respiratory tract and there was evidence that both Mass IBV variants replicated within macrophages in the respiratory tract. Studies are underway to evaluate the pathogenicity of these two Mass type IBV isolates in laying hens.

Methods

Animals

Day old white leghorn (SPF) layer chickens (White Leghorn) were obtained from the Canadian Food Inspection Agency (CFIA), Ottawa, ON, Canada and housed in high containment poultry isolators at the Prion- Virology Animal Facility of the University of Calgary's Foothills Campus with ad libitum access to feed and water.

Virus

Mass IBV isolates originating from layer flocks in AB (15AB-01) and SK (15SK-02) were used in the study. Viral titers were determined using 9-day old SPF embryonated chicken eggs and expressed as 50% embryo infectious dose (EID_{50}) according to the method described by Reed and Muench [73].

Experimental design

Whole genome sequencing of mass IBV isolates that originated from Western Canada

The Mass-type IBV isolates from the current study were sequentially passaged in 9 to 11-day old embryonated eggs and allantoic fluid was harvested and concentrated using ultracentrifugation at 50,000 g for 3 h. The resulting pellet was fractionated using the Optiprep™ kit (Millipore-Sigma, Missouri, USA). Ribonucleic acid (RNA) was extracted from the sixth fraction and allantoic fluid directly. cDNA was synthesized, and quantitative polymerase chain reaction (PCR) assay was performed to quantify the IBV genome load.

Infection of chickens with mass IBV isolates originated from Western Canada

Two groups of 6 day-old unsexed-SPF chickens ($n = 7$ per group) were infected with 50 μL inoculum containing 2.75×10^4 PFUs of IBV isolates 15AB-01 and 15SK-02 intra-tracheally under isoflurane anesthesia. Another 5 of 6-day old SPF chickens were kept as uninfected controls. The chickens were weighed at 0, 2, 3, 4, 10, 12 and 14 dpi. At 2, 3, 10 and 12 dpi, oropharyngeal and cloacal swabs were obtained using Puritan® Unitranz-RT transport system (Puritan Medical Products LLC, Maine, USA) and stored at −80 °C until further processing. At 4 dpi, 4 chickens from each infected and 3 chickens from the uninfected group were euthanized and trachea, lung, kidneys, cecal tonsils and reproductive tract samples were collected into RNASave® (Biological Industries, Beit Haemek, Israel) and stored at -20 °C. In addition, a portion of trachea, lung, kidneys, cecal tonsils and reproductive tract samples were collected into 10% neutral buffered formalin for histopathological examination and in Optimum Cutting Medium (OCT, Leica Biosystems, Wetzlar, Germany) for immunostaining. The remainder of chickens were continued to be observed and tissue samples were collected in RNASave® for genome load quantification at 14 dpi.

For the sampling of tissues, the chickens were euthanized using overdose of isoflurane anesthesia followed by cervical dislocation.

Techniques

Sample preparation for whole genome sequencing

The virus fractions, which gave the highest genome load or the lowest Ct were selected from both isolates and cDNA concentration was determined using the Qubit 4 Fluorometer (Thermo Fisher Scientific, MA, USA). A total of 10–12 ng of cDNA from each isolate was sent for whole genome sequencing to the Swine and Poultry Infectious Diseases Research Center (CRIPA), University of Montreal, Canada where library preparation and sequencing performed on Illumina® Miseq (San Diego, California, United States) next generation sequencing (NGS) platform generating 150 bp paired-end reads.

Histology

Formalin-fixed terminal tissue samples were processed at the Diagnostic Services Unit of the University of Calgary's Faculty of Veterinary Medicine including staining of sections with stained hematoxylin and eosin (H&E). Sections were examined under the light microscope and photomicrographs were taken under 40x magnification. The histological changes observed in the trachea [8] and lungs [69] were scored as described previously.

RNA extraction and cDNA synthesis

Tracheal and cloacal swab samples, which were in the transport media were vortexed and medium was transferred to 1.5 mL centrifuge tubes and centrifuged at 1000 g for 20 mins at −4 °C. A total of 250 µL of supernatant was used for RNA extraction using the Trizol LS® reagent (Ambion, Invitrogen Canada Inc., Burlington, ON, Canada) following manufacturer's instructions. Approximately 50 mg piece of tissue preserved in RNA Save® was homogenized in 1 mL of Trizol® reagent (Ambion, Invitrogen Canada Inc., Burlington, ON, Canada) using a Pro200 Power homogenizer (Diamed, Mississauga, ON, Canada) on ice and RNA was extracted following the manufacturer's protocol. The RNA pellet was re-suspended in 20 µL of RNase-free water and quantified using the Nanodrop 1000 spectrophotometer at 260 nm wavelength (Thermo Scientific, Wilmington, DE, USA). A total of 2000 ng of RNA was transcribed using 10X RT random primers (High Capacity cDNA Reverse Transcription Kit, Invitrogen Life Technologies, Carlsbad, CA, USA) as has been instructed by the manufacturer.

IBV genome load quantification

The IBV genome load quantification was carried out following the products and protocol as described previously [65]. Briefly, real-time PCR method was used for IBV genome load quantification using forward primer, IBV N Fw: 5'GACGGAGGACCTGATGGTAA3' and reverse primer, IBV-N Re: 5'CCCTTCTTCTGCTG ATCCTG3'. Thermal cycling conditions were 95 °C for 20 s; 40 cycles of amplification/extension at 95 °C for 3 s, and 60 °C for 30 s followed by a melting curve analysis was done between 65 °C to 95 °C with an increment of 0.5 °C at every 5 s. Fluorescent acquisition was done at 60 °C for 30 s.

Immunofluorescent assay

Frozen tissues in OCT blocks were sectioned at 5 µm thickness using a cryotome (Leica Biosystems, Richmond, Illinois, USA) and adhered on to positively charged slides. The sections were air dried on the bench for 20 min, fixed by immersing in ice-cold acetone for 5 min and stained for macrophage and IBV antigens sequentially as has been described previously [65]. Briefly, after protein blocking with 2.5% horse serum, the cryosections were incubated for 30 min with mouse monoclonal anti-chicken macrophage (KUL01) antibodies (Southern Biotech, Birmingham, Alabama, USA) (1:200 in PBS containing 2.5% horse serum). The staining specificity of KUL01 anti-chicken macrophage antibody was confirmed previously using an appropriate isotype control (Additional file 1: Figure 1). The secondary antibody, DyLight® 550 conjugated goat anti-mouse IgG (H + L) (Vector Laboratories, Inc., Burlingame, California, USA) (1:500 in PBS containing 2.5% horse serum) was used. Before being stained for IBV antigen using anti-IBV rabbit polyclonal serum (Federal Research Institute for Animal Health, Greifswald-Insel Riems, Germany; 1:3000 diluted in 2.5% horse serum), the sections were blocked again with 2.5% of horse serum. For subsequent staining,

Vectaflour RTU Antibody kit containing Dylight 488 conjugate (DK-1488, Vector Laboratories, Burlingame, California, USA). The sections were mounted on Vectashield® antifade mounting medium with 4′,6-Diamidine-2′-phenylindole dihydrochloride (DAPI) nuclear stain (Vector Laboratories, Burlingame, California, USA).

Data analyses

Whole genome sequencing data analysis

All reads were assembled to get a single contig. The resulting full genome sequences were aligned, and identity matrix was performed with 43 published sequences of IBV reference serotypes retrieved from the National Center for Biotechnology Information (NCBI) GenBank database using the BLAST® research tool and the Custal Omega 1.2.3 package with automated settings (Geneious® 10.1.3). The phylogenetic tree inferred based on the full-length IBV genomes using Randomized Axelerated Maximum Likelihood (RAxML 8.2.7) method with GTR GAMMA nucleotide substitution model with 1000 bootstrap iterations (Geneious software, version 10.1.3, Biomatters Ltd., Auckland, New Zealand). The resulting proportional tree of the two Mass IBV isolates of this study with 43 reference sequences (Additional file 2: Table S1) is shown with the bootstrap support. Nucleotide sequences of the two Mass IBV isolates, 15AB-01 and 15SK-02 were deposited in the NCBI GenBank public domain under the accession numbers MH539771 and MH539772, respectively. ORFs of the two new sequences were determined using the NCBI ORF finder (https://www.ncbi.nlm.nih.gov/orffinder/) and the genome annotation was carried out using Geneious software, version 10.1.3 (Biomatters Ltd., Auckland, New Zealand).

IBV genome load quantification

IBV genome copies per 200 ng of cDNA were calculated using a standard curve based on a serial dilution of plasmids as described previously [69].

Quantification of the number of macrophages, IBV antigen and IBV positive macrophages

For the quantification of macrophages and the IBV antigen in the trachea and lungs, five areas with the highest DyLight® 550 (macrophages) and DyLight® 488 (IBV antigen) fluorescent signals and corresponding nuclear stained (DAPI) areas were captured under 40x magnification using an epifluorescent microscope (Olympus IX51, Center Valley, Pensylvania, USA). The obtained images were transferred to the Image-J® software (National Institute of Health, Bethesda, Maryland, USA) and the areas of each micrograph covered with DyLight® 550 (macrophages), DyLight® 488 (IBV) signals, combined DyLight® 550 and DyLight® 488 signals (IBV positive macrophages) and the combined DyLight® 550 and DyLight® 488 signals with DAPI (total area of the section) were quantified. The areas covered with macrophage and IBV antigens and the combined signals were expressed as a percentage of the total area covered by the section.

Statistical analyses

Two-way analysis of variance (ANOVA) followed by Benferroni post-hoc test was used to determine the significant time points of body weight among the infected and control groups. Tracheal and lung histopathological lesion score data was analyzed using Kruskal-Wallis test followed by Dunn's multiple comparison test. The genome load data was analyzed by the two-way ANOVA followed by Bonferroni post-test. One-way ANOVA test followed by Tukey's posttest was used to determine the significance of immunofluorescence data. Grubbs' outlier test was performed in order to identify outliers before the data was analyzed. All the statistical procedures were performed using in-built statistical analysis feature of the GraphPad™ Prism 5 (GraphPad Software, La Jolla, California, USA). Differences were considered significant at $p < 0.05$. The significance on graphs are labelled as the following: $^{*}p < 0.05$, $^{**}p < 0.01$, $^{***}p < 0.001$, and $^{****}p < 0.0001$.

Additional files

Additional file 1: Figure S1. Specificity of avian macrophage staining. The power point slide contain lung sections stained with anti-chicken macrophage antibody (KUL-01) and isotype control. The sections were also stained with nuclear staining (DAPI).

Additional file 2: Table S1. List of reference IBV sequences used in this study. This table contain names and accession numbers of IBV whole genome reference sequences retrieved from the GenBank repository.

Acknowledgements

We acknowledge the staff involved in field poultry veterinary services in SK and AB, Canada. We acknowledge the staff of Prion-Virology Animal Facility, University of Calgary for routine monitoring of experimental animals.

Funding

This research including graduate student stipend was funded by the SK Agricultural Development Fund and SK Egg Producers (Project number: 20140185). Graduate student, AA was also supported partly by University of Calgary's Eyes-High International Doctoral Scholarship and Bridge Funding from the Department of Ecosystem and Public Health. These funding bodies were not involved in the design of the study, collection, analysis, interpretation of data and in writing the manuscript.

Authors' contributions

Conceptualization, funding acquisition and project administration was done by SG and MFA. Investigations, data collection and formal analysis were undertaken by AA, UDSS and MSA. SP, FM and Saskatchewan Poultry Extension (Tennille Knezacek, Jenny Fricke and Stephanie Derbawka) helped in sample collection leading to the isolation of Mass IBV isolates. Original draft was prepared by AA. Review of the manuscript including editing and student supervision and mentoring were done by SG, SCC, FvdM and MFA. All authors read and approved the final manuscript.

Authors' information
AA: PhD student, University of Calgary.
UDSS: MSc student, University of Calgary.
MSA: PhD student, University of Calgary.
SP: Research Technician-Poultry, University of Saskatchewan.
FM: Veterinarian, Marshall Swine and Poultry Health Services.
SCC: Professor and Head, Ecosystem and Public Health, University of Calgary.
FvdM: Associate Professor (Global Health), University of Calgary.
SG: Professor and Head, Department of Veterinary Pathology, University of Saskatchewan.
MFA: Associate Professor (Virology), University of Calgary.

Competing interests
The authors declare that they have no competing interests.

Author details
[1]Department of Ecosystem and Public Health, Faculty of Veterinary Medicine, University of Calgary, Health Research Innovation Center 2C53, 3330 Hospital Drive NW, Calgary, AB T2N 4N1, Canada. [2]Department of Veterinary Pathology, Western College of Veterinary Medicine, University of Saskatchewan, Saskatoon, SK S7N 5B5, Canada. [3]Marshall Swine and Poultry Health Services, 3831- Bay G- 44 Ave, Camrose, AB T4V 3T1, Canada.

References

1. Cavanagh D. Coronaviruses in poultry and other birds. Avian Pathology. 2005;34(6):439–48.
2. Cavanagh D. Coronavirus avian infectious bronchitis virus. Vet Res. 2007; 38(2):281–97.
3. Cumming RB. The control of avian infectious bronchitis/nephrosis in Australia. Aust Vet J. 1969;45(4):200–3.
4. Kingham BF, Keeler CL, Nix WA, Ladman BS, Gelb J. Identification of avian infectious bronchitis virus by direct automated cycle sequencing of the S-1 gene. Avian Dis. 2000;44(2):325–35.
5. Song C-S, Lee Y-J, Lee C-W, Sung H-W, Kim J-H, Mo I-P, Izumiya Y, Jang H-K, Mikami T. Induction of protective immunity in chickens vaccinated with infectious bronchitis virus S1 glycoprotein expressed by a recombinant baculovirus. J Gen Virol. 1998;79(4):719–23.
6. Liu S, Chen J, Han Z, Zhang Q, Shao Y, Kong X, Tong G. Infectious bronchitis virus: S1 gene characteristics of vaccines used in China and efficacy of vaccination against heterologous strains from China. Avian Pathology. 2006; 35(5):394–9.
7. Reddy VRAP, Theuns S, Roukaerts IDM, Zeller M, Matthijnssens J, Nauwynck HJ. Genetic characterization of the Belgian Nephropathogenic infectious bronchitis virus (NIBV) reference strain B1648. Viruses. 2015;7(8):4488–506.
8. Grgic H, Hunter DB, Hunton P, Nagy E. Pathogenicity of infectious bronchitis virus isolates from Ontario chickens. Can J Vet Res. 2008;72(5):403–10.
9. Boroomand Z, Asasi K, Mohammadi A. Pathogenesis and tissue distribution of avian infectious bronchitis virus isolate IRFIBV32 (793/B serotype) in experimentally infected broiler chickens. Sci World J. 2012;2012:402537.
10. Franca M, Woolcock PR, Yu M, Jackwood MW, Shivaprasad HL. Nephritis associated with infectious bronchitis virus Cal99 variant in game chickens. Avian Dis. 2011;55(3):422–8.
11. Ambali AG, Jones RC. Early pathogenesis in chicks of infection with an Enterotropic strain of infectious bronchitis virus. Avian Dis. 1990;34(4):809–17.
12. El-Houadfi M, Jones RC, Cook JKA, Ambali AG. The isolation and characterisation of six avian infectious bronchitis viruses isolated in Morocco. Avian Pathology. 1986;15(1):93–105.
13. Sevoian M, Levine P. Effects of infectious bronchitis on the reproductive tracts, egg production, and egg quality of laying chickens. Avian Dis. 1957;1(2):136–64.
14. MacDonald JW, McMartin DA. Observations on the effects of the h52 and h120 vaccine strains of infectious bronchitis virus in the domestic fowl. Avian Pathology. 1976;5(3):157–73.
15. Raj GD, Jones RC. Infectious bronchitis virus: Immunopathogenesis of infection in the chicken. Avian Pathology. 1997;26:677–706.
16. Cook JK, Jackwood M, Jones RC. The long view: 40 years of infectious bronchitis research. Avian Pathology. 2012;41(3):239–50.
17. Crinion RAP, Ball RA, Hofstad MS. Abnormalities in laying chickens following exposure to infectious bronchitis virus at one day old. Avian Dis. 1971;15(1):42–8.
18. Benyeda Z, Mato T, Suveges T, Szabo E, Kardi V, Abonyi-Toth Z, Rusvai M, Palya V. Comparison of the pathogenicity of QX-like, M41 and 793/B infectious bronchitis strains from different pathological conditions. Avian Pathology. 2009;38(6):449–56.
19. Boltz DA, Nakai M, Bahra JM. Avian infectious bronchitis virus: a possible cause of reduced fertility in the rooster. Avian Dis. 2004;48(4):909–15.
20. Liu S, Zhang X, Wang Y, Li C, Han Z, Shao Y, Li H, Kong X. Molecular characterization and pathogenicity of infectious bronchitis coronaviruses: complicated evolution and epidemiology in China caused by Cocirculation of multiple types of infectious bronchitis coronaviruses. Intervirology. 2009;52(4):223–34.
21. Alvarado IR, Villegas P, Mossos N, Jackwood MW. Molecular characterization of avian infectious bronchitis virus strains isolated in Colombia during 2003. Avian Dis. 2005;49(4):494–9.
22. Kusters JG, Niesters HGM, Bleumink-Pluym NMC, Davelaar FG, Horzinek MC, Van der Zeijst BAM. Molecular epidemiology of infectious bronchitis virus in the Netherlands. J Gen Virol. 1987;68(2):343–52.
23. Lee C-W, Jackwood MW. Evidence of genetic diversity generated by recombination among avian coronavirus IBV. Arch Virol. 2000;145(10):2135–48.
24. Kottier SA, Cavanagh D, Britton P. Experimental evidence of recombination in coronavirus infectious bronchitis virus. Virology. 1995;213(2):569–80.
25. Wang L, Xu Y, Collisson EW. Experimental confirmation of recombination upstream of the S1 hypervariable region of infectious bronchitis virus. Virus Res. 1997;49(2):139–45.
26. Mardani K, Noormohammadi A, Ignjatovic J, Browning G. Naturally occurring recombination between distant strains of infectious bronchitis virus. Arch Virol. 2010;155(10):1581–6.
27. Jia W, Karaca K, Parrish CR, Naqi SA. A novel variant of avian infectious bronchitis virus resulting from recombination among three different strains. Arch Virol. 1995;140(2):259–71.
28. Montassier HJ. Molecular epidemiology and evolution of avian infectious bronchitis virus. Revista Brasileira de Ciência Avícola. 2010;12:87–96.
29. Cavanagh D, Davis PJ, Darbyshire JH, Peters RW. Coronavirus IBV: virus retaining spike glycopolypeptide S2 but not S1 is unable to induce virus-neutralizing or haemagglutination-inhibiting antibody, or induce chicken tracheal protection. J Gen Virol. 1986;67(7):1435–42.
30. Cavanagh D, Davis PJ, Cook JKA. Infectious bronchitis virus: evidence for recombination within the Massachusetts serotype. Avian Pathology. 1992; 21(3):401–8.
31. Selim K, Arafa AS, Hussein HA, El-Sanousi AA. Molecular characterization of infectious bronchitis viruses isolated from broiler and layer chicken farms in Egypt during 2012. International Journal of Veterinary Science and Medicine. 2013;1(2):102–8.
32. Awad F, Baylis M, Ganapathy K. Detection of variant infectious bronchitis viruses in broiler flocks in Libya. International Journal of Veterinary Science and Medicine. 2014;2(1):78–82.
33. Martin EA, Brash ML, Hoyland SK, Coventry JM, Sandrock C, Guerin MT, Ojkic D. Genotyping of infectious bronchitis viruses identified in Canada between 2000 and 2013. Avian Pathology. 2014;43(3):264–8.
34. Jackwood MW, Lee DH. Different evolutionary trajectories of vaccine-controlled and non-controlled avian infectious bronchitis viruses in commercial poultry. PLoS One. 2017;12(5):e0176709.
35. Hong S-M, Kwon H-J, Kim I-H, Mo M-L, Kim J-H. Comparative genomics of Korean infectious bronchitis viruses (IBVs) and an animal model to evaluate pathogenicity of IBVs to the reproductive organs. Viruses. 2012;4(11):2670.
36. Marandino A, Pereda A, Tomas G, Hernandez M, Iraola G, Craig MI, Hernandez D, Banda A, Villegas P, Panzera Y, et al. Phylodynamic analysis of avian infectious bronchitis virus in South America. The Journal of General Virology. 2015;96(Pt 6):1340–6.
37. Smati R, Silim A, Guertin C, Henrichon M, Marandi M, Arella M, Merzouki A. Molecular characterization of three new avian infectious bronchitis virus (IBV) strains isolated in Quebec. Virus Genes. 2002;25(1):85–93.
38. Feng K, Wang F, Xue Y, Zhou Q, Chen F, Bi Y, Xie Q. Epidemiology and characterization of avian infectious bronchitis virus strains circulating in southern China during the period from 2013-2015. Sci Rep. 2017;7(1):6576.
39. Mo ML, Li M, Huang BC, Fan WS, Wei P, Wei TC, Cheng QY, Wei ZJ, Lang YH. Molecular characterization of major structural protein genes of avian coronavirus infectious bronchitis virus isolates in southern China. Viruses. 2013;5(12):3007–20.
40. Quinteros JA, Markham PF, Lee SW, Hewson KA, Hartley CA, Legione AR, Coppo MJ, Vaz PK, Browning GF. Analysis of the complete genomic

sequences of two virus subpopulations of the Australian infectious bronchitis virus vaccine VicS. Avian Pathology. 44(3):182–91.
41. van Beurden SJ, Berends AJ, Kramer-Kuhl A, Spekreijse D, Chenard G, Philipp HC, Mundt E, Rottier PJM, Verheije MH. Recombinant live attenuated avian coronavirus vaccines with deletions in the accessory genes 3ab and/or 5ab protect against infectious bronchitis in chickens. Vaccine. 2018;36(8):1085–92.
42. Benyeda Z, Szeredi L, Mató T, Süveges T, Balka G, Abonyi-Tóth Z, Rusvai M, Palya V. Comparative histopathology and immunohistochemistry of QX-like, Massachusetts and 793/B serotypes of infectious bronchitis virus infection in chickens. J Comp Pathol. 2010;143(4):276–83.
43. W-q F, H-n W, Zhang Y, Z-b G, Wang T, C-w X, A-y Z, Yang X. Comparative dynamic distribution of avian infectious bronchitis virus M41, H120, and SAIBK strains by quantitative real-time RT-PCR in SPF chickens. Biosci Biotechnol Biochem. 2012;76(12):2255–60.
44. Cheng J, Huo C, Zhao J, Liu T, Li X, Yan S, Wang Z, Hu Y, Zhang G. Pathogenicity differences between QX-like and mass-type infectious bronchitis viruses. Vet Microbiol. 2018;213:129–35.
45. Chhabra R, Kuchipudi SV, Chantrey J, Ganapathy K. Pathogenicity and tissue tropism of infectious bronchitis virus is associated with elevated apoptosis and innate immune responses. Virology. 2016;488:232–41.
46. Chhabra R, Ball C, Chantrey J, Ganapathy K. Differential innate immune responses induced by classical and variant infectious bronchitis viruses in specific pathogen free chicks. Developmental & Comparative Immunology. 2018;87:16–23.
47. Bae JY, Lee I, Kim JI, Park S, Yoo K, Park M, Kim G, Park MS, Lee JY, Kang C, et al. A single amino acid in the polymerase acidic protein determines the pathogenicity of influenza B viruses. J Virol. 2018;92(13):e00259-18.
48. Cavanagh D, Davis PJ, Cook JK, Li D, Kant A, Koch G. Location of the amino acid differences in the S1 spike glycoprotein subunit of closely related serotypes of infectious bronchitis virus. Avian Pathology. 1992;21(1):33–43.
49. Binns MM, Boursnell ME, Tomley FM, Brown DK. Comparison of the spike precursor sequences of coronavirus IBV strains M41 and 6/82 with that of IBV Beaudette. J Gen Virol. 1986;67(Pt 12):2825–31.
50. Amarasinghe A, Popowich S, De Silva Senapathi U, Abdul-Cader MS, Marshall F, van der Meer F, Cork SC, Gomis S, Abdul-Careem MF. Shell-less egg syndrome (SES) widespread in Western Canadian layer operations is linked to a Massachusetts (mass) type infectious bronchitis virus (IBV) isolate. Viruses. 2018;10(8). https://doi.org/10.3390/v10080437.
51. Stachowiak BKD, Hunton P, Gillingham S, Nagy É. Infectious bronchitis virus surveillance in Ontario commercial layer flocks. The Journal of Applied Poultry Research. 2005;14(1):141–6.
52. Ignjatovic J, McWaters PG. Monoclonal antibodies to three structural proteins of avian infectious bronchitis virus: characterization of epitopes and antigenic differentiation of Australian strains. J Gen Virol. 1991;72(12):2915–22.
53. Kant A, Koch G, van Roozelaar DJ, Kusters JG, Poelwijk FAJ, van der Zeijst BAM. Location of antigenic sites defined by neutralizing monoclonal antibodies on the S1 avian infectious bronchitis virus glycopolypeptide. J Gen Virol. 1992;73(3):591–6.
54. Cavanagh D, Davis PJ, Mockett APA. Amino acids within hypervariable region 1 of avian coronavirus IBV (Massachusetts serotype) spike glycoprotein are associated with neutralization epitopes. Virus Res. 1988;11(2):141–50.
55. Quinteros JA, Lee S-W, Markham PF, Noormohammadi AH, Hartley CA, Legione AR, Coppo MJC, Vaz PK, Browning GF. Full genome analysis of Australian infectious bronchitis viruses suggests frequent recombination events between vaccine strains and multiple phylogenetically distant avian coronaviruses of unknown origin. Vet Microbiol. 2016;197:27–38.
56. Abozeid HH, Paldurai A, Khattar SK, Afifi MA, El-Kady MF, El-Deeb AH, Samal SK. Complete genome sequences of two avian infectious bronchitis viruses isolated in Egypt: evidence for genetic drift and genetic recombination in the circulating viruses. Infect Genet Evol. 2017;53:7–14.
57. Naguib MM, Höper D, Arafa A-S, Setta AM, Abed M, Monne I, Beer M, Harder TC. Full genome sequence analysis of a newly emerged QX-like infectious bronchitis virus from Sudan reveals distinct spots of recombination. Infect Genet Evol. 2016;46:42–9.
58. Wu X, Yang X, Xu P, Zhou L, Zhang Z, Wang H. Genome sequence and origin analyses of the recombinant novel IBV virulent isolate SAIBK2. Virus Genes. 2016;52(4):509–20.
59. X-l L, J-l S, J-x Z, G-z Z. Complete genome sequence analysis of a predominant infectious bronchitis virus (IBV) strain in China. Virus Genes. 2009;38(1):56–65.
60. Khataby K, Souiri A, Kasmi Y, Loutfi C, Ennaji MM. Current situation, genetic relationship and control measures of infectious bronchitis virus variants circulating in African regions. The Journal of Basic & Applied Zoology. 2016; 76:20–30.
61. Li H, Yang H. Sequence analysis of nephropathogenic infectious bronchitis virus strains of the Massachusetts genotype in Beijing. Avian Pathology. 2001;30(5):535–41.
62. Yan S, Liu X, Zhao J, Xu G, Zhao Y, Zhang G. Analysis of antigenicity and pathogenicity reveals major differences among QX-like infectious bronchitis viruses and other serotypes. Vet Microbiol. 2017;203:167–73.
63. Khanh NP, Tan SW, Yeap SK, Lee HJ, Choi KS, Hair-Bejo M, Bich TN, Omar AR. Comparative pathogenicity of Malaysian QX-like and variant infectious bronchitis virus strains in chickens at different age of exposure to the viruses. J Comp Pathol. 2018;161:43–54.
64. Fulton RM, Reed WM, Thacker HL. Cellular response of the respiratory tract of chickens to infection with Massachusetts 41 and Australian T infectious bronchitis viruses. Avian Dis. 1993;37(4):951–60.
65. Amarasinghe A, Abdul-Cader MS, Nazir S, De Silva Senapathi U, van der Meer F, Cork SC, Gomis S, Abdul-Careem MF. Infectious bronchitis corona virus establishes productive infection in avian macrophages interfering with selected antimicrobial functions. PLoS One. 2017;12(8):e0181801.
66. Amarasinghe A, Abdul-Cader MS, Almatrouk Z, van der Meer F, Cork SC, Gomis S, Abdul-Careem MF. Induction of innate host responses characterized by production of interleukin (IL)-1beta and recruitment of macrophages to the respiratory tract of chickens following infection with infectious bronchitis virus (IBV). Vet Microbiol. 2018;215:1–10.
67. da Silva SEL, Ferreira HL, Garcia AF, Silva FES, Gameiro R, Fabri CUF, Vieira DS, Cardoso TC. Mitochondrial bioenergy alterations in avian HD11 macrophages infected with infectious bronchitis virus. Arch Virol. 2018;163(4):1043–9.
68. Han X, Tian Y, Guan R, Gao W, Yang X, Zhou L, Wang H. Infectious bronchitis virus infection induces apoptosis during replication in chicken macrophage HD11 cells. Viruses. 2017;9(8):198.
69. Kameka AM, Haddadi S, Kim DS, Cork SC, Abdul-Careem MF. Induction of innate immune response following infectious bronchitis corona virus infection in the respiratory tract of chickens. Virology. 2014;450–451:114–21.
70. Meir R, Maharat O, Farnushi Y, Simanov L. Development of a real-time TaqMan® RT-PCR assay for the detection of infectious bronchitis virus in chickens, and comparison of RT-PCR and virus isolation. J Virol Methods. 2010;163(2):190–4.
71. Breslin JJ, Smith LG, Fuller FJ, Guy JS. Sequence analysis of the Turkey coronavirus nucleocapsid protein gene and 3' untranslated region identifies the virus as a close relative of infectious bronchitis virus. Virus Res. 1999; 65(2):187–93.
72. Klein D. Quantification using real-time PCR technology: applications and limitations. Trends Mol Med. 2002;8(6):257–60.
73. Reed LJ, Muench H. A simple method of estimating fifty per cent endpoint. Am J Epidemiol. 1938;27(3):493–7.

Development of a lateral flow test for the rapid detection of *Avibacterium paragallinarum* in chickens suspected of having infectious coryza

Sandra Morales Ruiz[1*], Jorge Bendezu[1*], Ricardo Choque Guevara[1], Ricardo Montesinos[1], David Requena[1,2], Luz Choque Moreau[1], Ángela Montalván Ávalos[1] and Manolo Fernández-Díaz[1,3]

Abstract

Background: Infectious coryza (IC) is an acute respiratory disease of growing chickens and layers caused by *Avibacterium paragallinarum*. The development of tools that allow rapid pathogen detection is necessary in order to avoid disease dissemination and economic losses in poultry. An *Av. paragallinarum*-specific Ma-4 epitope of the TonB-dependent transporter (TBDT) was selected using bioinformatic tools in order to immunize a BalbC mouse and to produce monoclonal antibodies to be used in a lateral flow test (LFT) developed for *Av. paragallinarum* detection in chicken nasal mucus samples.

Results: The 1G7G8 monoclonal antibody was able to detect TBDT in *Av. paragallinarum* cultures (serogroups: A, B and C) by Western blot and indirect ELISA assay. Consequently, we developed a self-pairing prototype LFT. The limit of detection of the prototype LFT using *Av. paragallinarum* cultures was 1×10^4 colony-forming units (CFU)/mL. Thirty-five nasal mucus samples from chickens suspected of having infectious coryza were evaluated for the LFT detection capacity and compared with bacterial isolation (B.I) and polymerase chain reaction (PCR). Comparative indicators such as sensitivity (Se), specificity (Sp), positive predictive value (PPV), negative predictive values (NPV) and the kappa index (K) were obtained. The values were 100.0% Se, 50% Sp, 65.4% PPV, 100% NPV, and 0.49 K and 83.9% Se, 100% Sp, 100% PPV, 44.4% NPV, and 0.54 K for the comparison of the LFT with B.I and PCR, respectively. Additionally, the LFT allowed the detection of *Av. paragallinarum* from coinfection cases of *Av. paragallinarum* with *Gallibacterium anatis*.

Conclusions: The results indicate that the self-pairing prototype LFT is suitable for the detection of TBDT in *Av. paragallinarum* cultures as well as in field samples such as nasal mucus from *Av. paragallinarum*-infected chickens. Therefore, this prototype LFT could be considered a rapid and promising tool to be used in farm conditions for *Av. paragallinarum* diagnosis.

Keywords: *Avibacterium paragallinarum*, Infectious coryza, Lateral flow test (LFT), Monoclonal antibody, TonB-dependent transporter (TBDT)

* Correspondence: sandra.morales@farvet.com; smr_sandra@yahoo.es; jorge.bendezu@farvet.com; eduar156@yahoo.es
[1]Laboratorios de Investigación y Desarrollo, FARVET SAC, Carretera Panamericana Sur N°766 Km 198.5, Chincha Alta 11702, Ica, Peru
Full list of author information is available at the end of the article

Background

Infectious coryza (IC) is caused by *Avibacterium paragallinarum* (*Av. paragallinarum*), a gram-negative bacteria previously called *Haemophilus paragallinarum* [1]. IC is an acute respiratory disease of growing chickens and layers and is associated with reduced egg production in laying flocks and delay in growth due to decreased food and water consumption in young chickens [2]. The most common clinical signs are serous or mucous nasal exudates, sneezing, swelling of infraorbital sinuses, facial edema and conjunctivitis [2]. All these clinical signs caused by *Av. paragallinarum* have been associated with economic losses in the poultry industry [2] that highlight the necessity of developing reliable tools for *Av. paragallinarum* detection.

Previous studies emphasized the use of PCR methods in comparison with bacterial isolation for the detection of *Av. paragallinarum*. The difficulties and challenges of *Av. paragallinarum* identification are well known using the latter method, which is a time-consuming process [3–6]. Moreover, the use of PCR-based methods is less time-consuming, but the methods require well-trained personnel and sophisticated infrastructure in laboratories. On the other hand, the use of a lateral flow test for detection of *Av. paragallinarum* is an alternative method that is not demanding and is an easily performable task. This method requires the identification of a specific epitope in the target protein that is detected through the use of a monoclonal antibody.

TonB-dependent transporters (TBDTs) are membrane proteins that have high affinity for iron, vitamin B_{12}, siderophores and carbohydrates, which are important for bacteria. Iron is the main substrate for TBDTs, and it participates in many bacterial metabolic processes [7].

These TBDT proteins have been identified inside outer membrane vesicles (OMVs) in culture supernatants of members of Pasteurellaceae, such as *Pasteurella multocida* [8]. The content of these OMVs has been previously associated with extracellular virulence factors released from bacteria that can damage host tissue [9]. These OMVs were also identified in *Av. paragallinarum* cultures, but TBDT presence inside these vesicles is still uncertain in culture supernatants [9].

The potential extracellular presence of TBDT makes it a promising candidate for *Av. paragallinarum* identification due to its simple detection by monoclonal antibodies. The application of monoclonal antibodies in previous works with *Av. paragallinarum* has been limited to serotyping [10, 11], strain-vaccine differentiation [12], inhibition of hemagglutination and vaccine development [13, 14]. This study will take advantage of the specificity that a monoclonal antibody provides for the identification of *Av. paragallinarum* through the development of a lateral flow assay.

Results

Monoclonal antibody characterization

Two hybridoma clones, 1G7G8 and 3A3D8, were selected based on their reactivity against the peptide (Ma-4); 1G7G8 and 3A3D8 produced monoclonal antibodies, which were characterized as IgG1 (k) and IgG2b (k) isotypes, respectively. The antibody titer was 4.1×10^{-4} for each hybridoma culture.

Identification of recombinant TBDT using the 1G7G8 and 3A3D8 monoclonal antibodies in a Western blot

Five hundred nanograms of recombinant protein were used for Western blot using purified monoclonal antibodies from 1G7G8 and 3A3D8 hybridoma clones. We obtained a unique reactive band of 87 kDa corresponding to recombinant TBDT, demonstrating that the recombinant protein was recognized by 1G7G8 and 3A3D8 (Fig. 1).

Identification of TBDT in *Av. paragallinarum* cultures (serogroups a, B and C) using the 1G7G8 and 3A3D8 monoclonal antibodies in a Western blot

The 1G7G8 and 3D3A8 monoclonal antibody reactivity was evaluated using bacterial lysates, pellets and supernatants from *Av. paragallinarum* cultures. Using the 1G7G8 antibody, in transfer protein membranes from bacterial lysates, we detected specific bands (~90 kDa) for serogroup B at 500 and 50 µg/mL, whereas for serogroup A, we detected specific stronger bands (~90 kDa) at the same concentrations. However, for serogroup C, multiple bands were observed at 500 µg/mL, and only two bands had similar molecular weight (~90 kDa) at 50 µg/mL. Regarding transfer protein membranes from the bacterial supernatant, we detected a specific band slightly above 90 kDa for serogroups B and A until dilutions of 1:4 and 1:16 were achieved, respectively. However, nonspecific weak bands were also observed in serogroup A at the lowest dilutions. For serogroup C, we observed two stronger bands of different molecular weights of ~90 and ~48.5 kDa until a dilution of 1:8 was reached (Fig. 2). For pellet samples, no bands were detected for serogroups A and B. However, we observed non-specific bands including a band of ~90 kDa at 500 µg/mL and a weak band at 50 µg/mL for serogroup C. On the other hand, no bands were detected for any samples using the 3A3D8 antibody in Western blot (data not shown). Therefore, this antibody was not used for further evaluations.

Evaluation of the 1G7G8 monoclonal antibody reactivity against related and non-related bird bacterial cultures in Western blot

1G7G8 reactivity was evaluated against other bacteria, such as *Gallibacterium anatis, Pasteurella multocida, Ornithobacterium rhinotracheale* and *Bortedella bronchiseptica,*

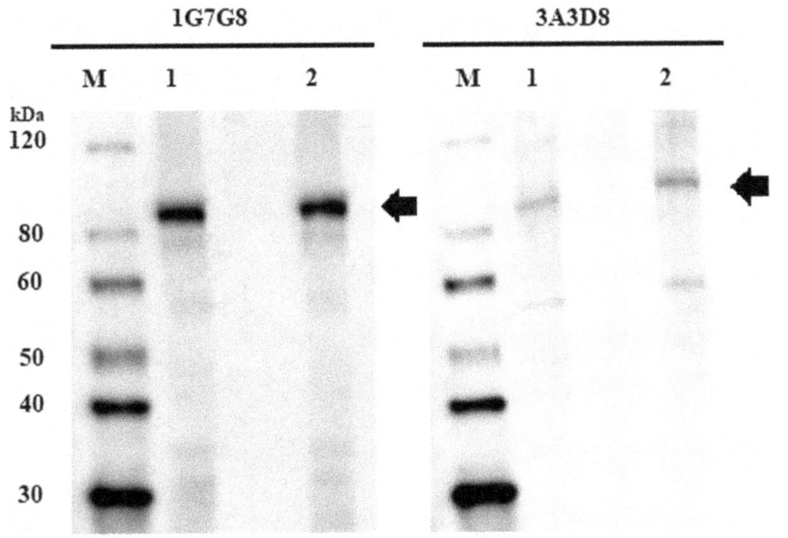

Fig. 1 Identification of recombinant TBDT protein by Western blot assay using 1G7G8 and 3A3D8 antibodies. M: Molecular weight marker (30–120 kDa); Lane 1 and 2: recombinant TBDT (500 ng). Black arrows indicate the reactive band in each lane

using bacterial pellets, lysates and supernatants in Western blot.

For supernatant samples, no bands were detected for *G. anatis*, *P. multocida*, *O. rhinotracheale* or *B. bronchiseptica*. Furthermore, for lysate samples, a weak, nonspecific band of low molecular weight was detected for *B. bronchiseptica* at 500 and 50 μg/mL, whereas no bands were detected for other bacteria.

For pellet samples, nonspecific bands were detected at 500 μg/mL for all bacteria, which included a band of ~90 kDa only for *G. anatis*, *P. multocida* and *O. rhinotracheale*. We also detected a specific band of ~90 kDa at 50 μg/mL for *P. multocida*. In addition to a ~90 kDa band, other bands of different molecular weights were also detected for *O. rhinotracheale* at 50 μg/mL. No additional bands were detected for other concentrations (Fig. 3).

Reactivity of the 1G7G8 monoclonal antibody against *Av. paragallinarum* serogroups in indirect ELISA assay

Reactivity differences between lysates (CFU/mL) and supernatant dilutions from the three *Av. paragallinarum* serogroups were detected. Positive reactivity was observed at 1:10 and 1:100 dilutions of the culture supernatants, and a higher reactivity was observed for serogroups A and C than for serogroup B ($p = 0.001$). Moreover, positive reactivity slightly above the cutoff (0.05) was also detected at a 1:1000 dilution for serogroups A and C (Fig. 4). Regarding lysates, we observed positive reactivity at 10^7 and 10^6 CFU/mL for serogroups A, B and C with higher reactivity observed for serogroup C than for the other serogroups ($p = 0.001$). Furthermore, positive reactivity slightly above the cutoff (0.086) was also detected at 1×10^5 CFU/mL for serogroup A (Fig. 5).

Characterization of the prototype LFT

The assembly of the prototype was performed by Creative Diagnostics (Shirley, NY, USA) using 1G7G8 as a detection and capture antibody (self-pairing). The self-pairing LFT was initially tested using recombinant TBDT protein. The results showed a red line in the T and C line regions, which was considered a positive result (data not shown). All technical details related to prototype assembly are of commercial interest and therefore these cannot be provided.

The assembly prototypes were sent to FARVET Laboratories in Peru and used to evaluate the performance of the LTF using a variety of culture samples. A positive red line was observed in the T line region as well as in the C line region for all three serogroups (serogroup A, B and C) obtained from culture supernatants and lysates of *Av. paragallinarum*. On the other hand, no red line in the T line region was observed for lysates and supernatants from other bacterial cultures such as *G. anatis*, *P. multocida*, *O. rhinotracheale* and *B. bronchiseptica*. In addition, no positive T lines were observed for supernatants obtained from cell cultures infected with infectious laryngotracheitis virus, avian metapneumovirus or Newcastle disease virus (Table 1).

The limit of detection was only determined for FARPER-140 (serogroup C) because it had the highest reactivity observed in indirect ELISA assay. The limit of detection was 1×10^4 CFU/mL (Fig. 6). On the other hand, a total of 35 chicken nasal mucus samples were collected from two farms located in the south of Peru in October and

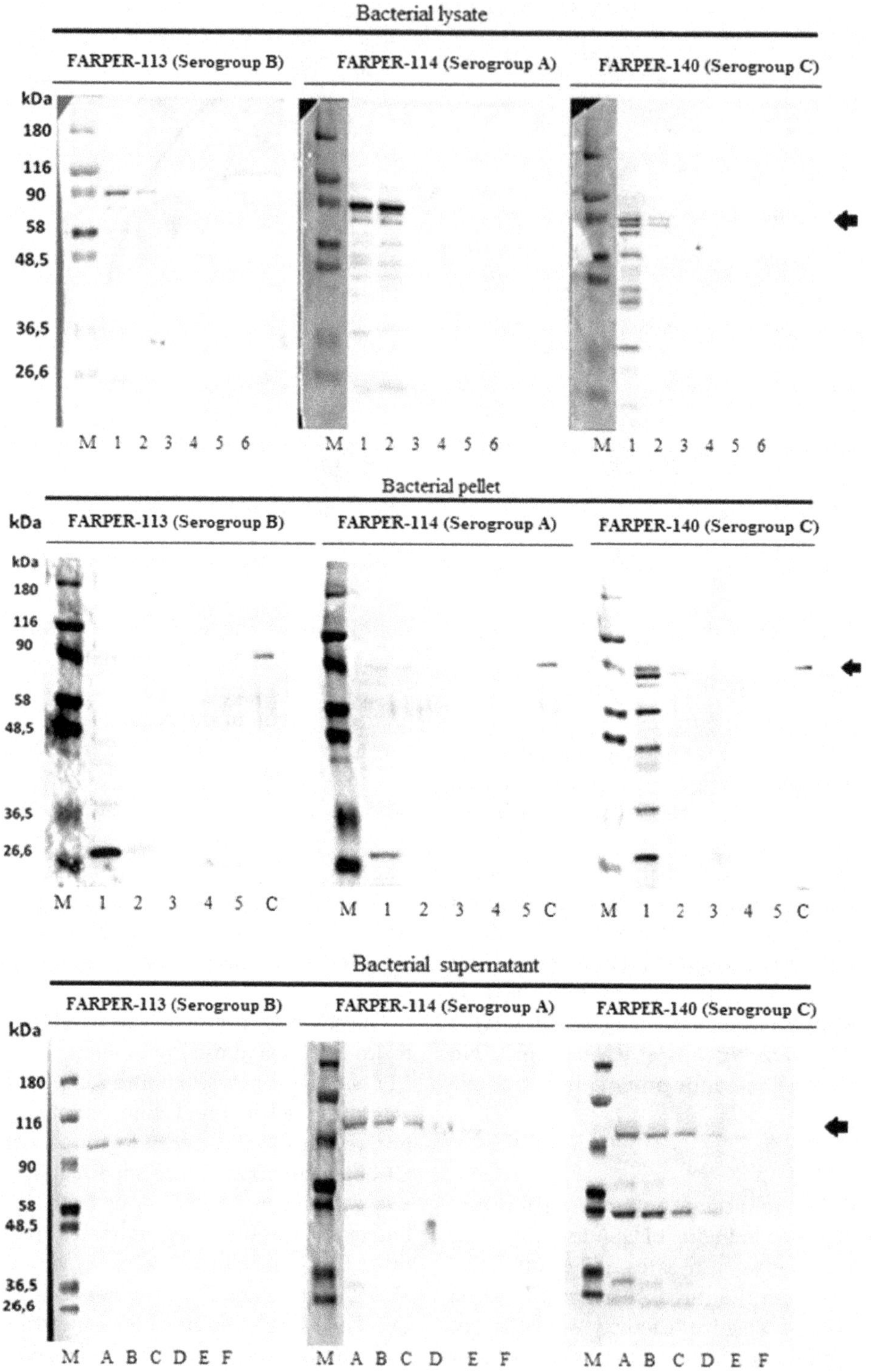

Fig. 2 Identification of TBDT in *Avibacterium paragallinarum* cultures using the 1G7G8 monoclonal antibody in Western blot. *Avibacterium paragallinarum* cultures: FARPER-113 (serogroup B), FARPER-114 (serogroup A) and FARPER-140 (serogroup C). Lane M: Molecular weight marker (26.6–180 kDa). For bacterial lysate concentrations: lane 1 (500 µg/mL), lane 2 (50 µg/mL), lane 3 (5 µg/mL), lane 4 (0.5 µg mL), lane 5 (0.05 µg/mL), and lane 6 (0.005 µg/mL). For bacterial pellet: lane 1 (500 µg/mL), lane 2 (50 µg/mL), lane 3 (5 µg/mL), lane 4 (0.5 µg mL), lane 5 (0.05 µg/mL), and lane C (recombinant TBDT). For bacterial supernatant concentrations: (**a**) (undiluted supernatant), (**b**) (diluted 1:2), (**c**) (diluted 1:4), (**d**) (diluted 1:8), (**e**) (diluted 1:16), and (**f**) (BHI broth). Black arrows indicate the specific reactive band in each lane

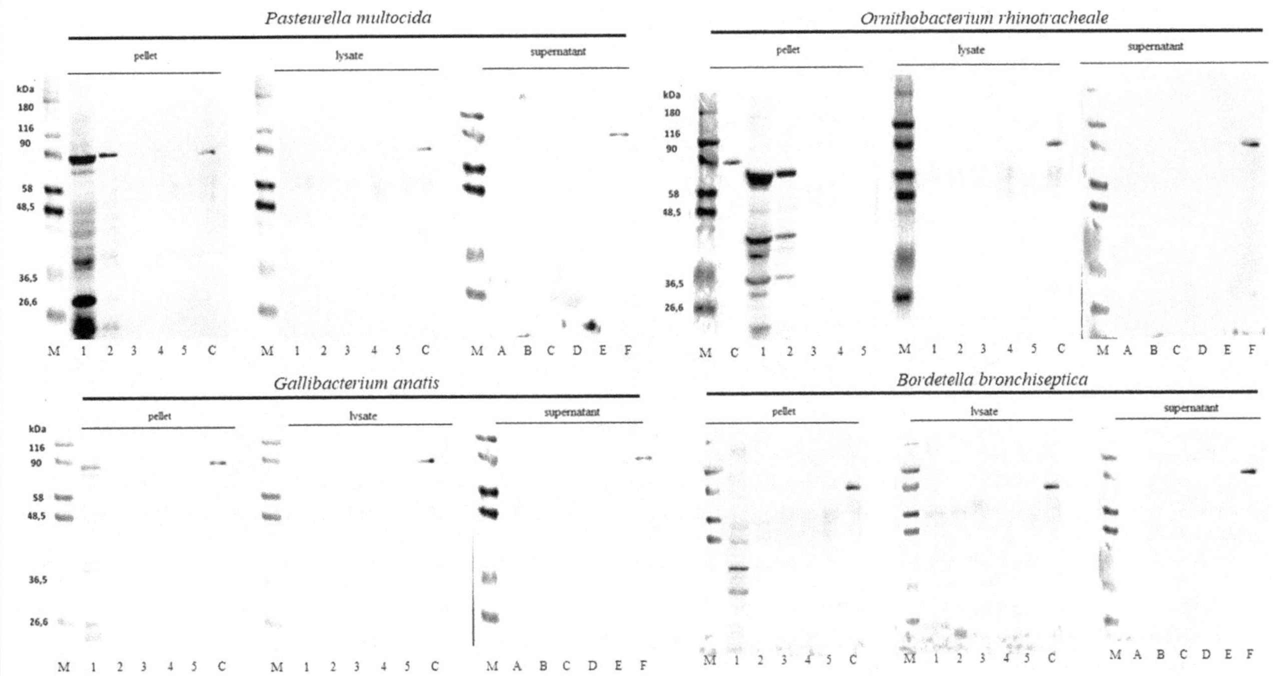

Fig. 3 Evaluation of 1G7G8 antibody reactivity against related and non-related bird bacterial cultures in Western blot. Related bird bacteria cultures: *Pasteurella multocida* (FARPER-069), *Ornithobacterium rhinotracheale* (FARPER-172), and *Gallibacterium anatis* (FARPER-173). Non-related bird bacteria culture: *Bordetella bronchiseptica* (FARPER-118). Lane M: Molecular weight marker (26.6–180 kDa). For bacterial lysate and pellet concentrations: lane 1 (500 μg/mL), lane 2 (50 μg/mL), lane 3 (5 μg/mL), lane 4 (0.5 μg mL), lane 5 (0.05 μg/mL), and lane C (recombinant TBDT). For bacterial supernatant concentrations: (**a**) (undiluted supernatant), (**b**) (diluted 1:2), (**c**) (diluted 1:4), (**d**) (diluted 1:8), (**e**) (BHI broth), and (**f**) (recombinant TBDT)

December 2017 and in January 2018 (Table 2). To evaluate the specificity and diagnostic sensitivity of the self-pairing prototype LFT, three detection methods, bacterial isolation (B.I), polymerase chain reaction (PCR) and self-pairing LFT, were compared (Fig. 7). The results from the three detection methods are shown in Table 3.

We identified 17 out of 35 clinical samples (51.4%) as positive *Av. paragallinarum* samples by B.I. Additionally, 11 out of 35 samples were positive for *G. anatis* using B.I. However, 88.57% (31/35) of the chicken samples were positive for *Av. paragallinarum* by PCR. From these 31 samples, 28 samples were typed by PCR as serogroup B and 3 samples as serogroup A. The self-pairing prototype LFT gave a positive result for *Av. paragallinarum* in 74.28% (26/35) of the chicken samples. All positive samples from B.I (17/35) were detected by PCR and LFT. Only 11.4% (4/35) samples were negatives from all tests. Nine samples were only positive from LFT and

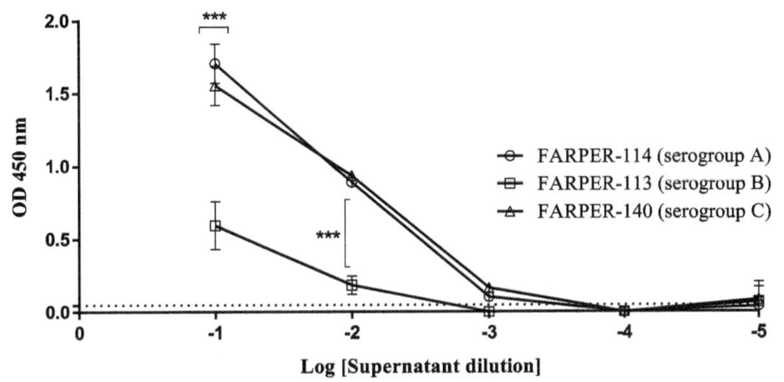

Fig. 4 Reactivity evaluation of the 1G7G8 antibody using *Avibacterium paragallinarum* culture supernatants. *Avibacterium paragallinarum* culture: FARPER-113 (serogroup B), FARPER-114 (serogroup A) and FARPER-140 (serogroup C). Error bars represent the standard deviation (SD). The dotted line represents the cutoff that was calculated as the mean + 2SD. ***: $p = 0.001$. A two-way ANOVA and Tukey's post hoc test were used

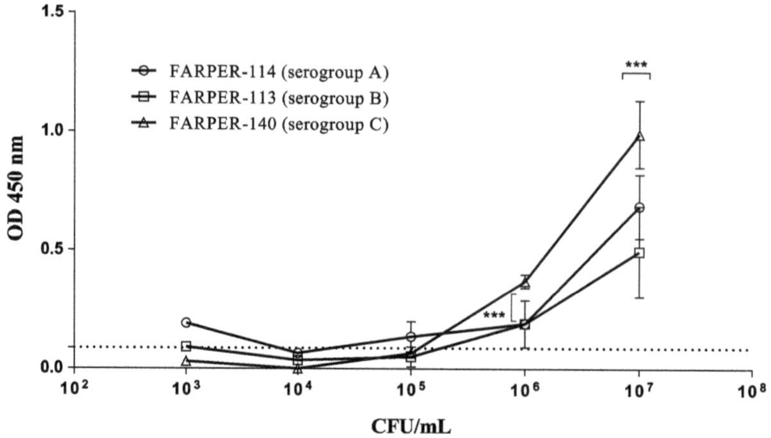

Fig. 5 Reactivity evaluation of the 1G7G8 antibody using *Avibacterium paragallinarum* culture lysates. *Avibacterium paragallinarum* culture: FARPER-113 (serogroup B), FARPER-114 (serogroup A) and FARPER-140 (serogroup C). Error bars represent the standard deviation (SD). The dotted line represents the cutoff that was calculated as the mean + 2SD. CFU: colony-forming units. ***: p = 0.001. A two-way ANOVA and Tukey's post hoc test were used

PCR. Additionally, 5 samples were exclusively positive from PCR and negative from B.I and LFT (Fig. 8).

For indicator comparison, self-pairing LFT was compared with B.I and PCR, which were used as reference methods. For comparison between B.I and self-pairing LFT, we obtained 100% Se, 50% Sp, 65.4% PPV, 100% NPV and K of 0.49. This index was considered a moderate agreement between LFT with B.I. For comparison between PCR and self-pairing LFT, we obtained 83.9% Se, 100% Sp, 100% PPV, 44.4% NPV and K of 0.54 (Table 4). This index was considered a moderate agreement between LFT with PCR. Finally, a higher sensitivity was observed when the self-pairing prototype LFT was compared with B.I than when it was compared with PCR (100 and 83.9%, respectively). However, greater specificity was observed when the self-pairing prototype LFT was compared with PCR than when it was compared with B.I (100 and 50.0%, respectively).

Discussion

IC is caused by *Av. paragallinarum*. This disease is diagnosed by clinical signs and confirmed by bacterial isolation [6]. Several factors such the requirement of specific growth media make *Av. paragallinarum* diagnosis a difficult and tedious task [5]. Some molecular techniques such as PCR have been used in order to achieve highly sensitive results [5, 6]. Despite this technique being faster than bacterial isolation, it is still expensive to implement in local laboratories and includes costs transporting samples from farms.

In the present study, we were able to develop a lateral flow test based on the detection of the TonB-dependent transporter (TBDT) through the use of a monoclonal

Table 1 Specificity evaluation for the prototype LFT using bacterial and viral cultures

Organism	host/cell culture	LFT results	
		Lysates	Supernatants
Bacteria			
Avibacterium paragallinarum (A) (FARPER-114)	Chicken	+	+
Avibacterium paragallinarum (B) (FARPER-113)	Chicken	+	+
Avibacterium paragallinarum (C) (FARPER-140)	Chicken	+	+
Ornithobacterium rhinotracheale (FARPER-172) (isolated from Chincha, Ica, Peru)	Chicken	–	–
Gallibacterium anatis (FARPER-173) (isolated from Pucallpa, Ucayali, Peru)	Chicken	–	–
Pasteurella multocida (FARPER-069) (isolated from Pucallpa, Ucayali, Peru)	Duck	–	–
Bordetella bronchiseptica (FARPER-118) (isolated from Cañete, Lima, Peru)	Guinea pig	–	–
Viruses			
Infectious laryngotracheitis virus (VFAR-043)	LMH	N.A	–
Avian metapneumovirus (SHS-FAR)	VERO	N.A	–
Newcastle disease virus (LaSota)	DF-1	N.A	–

+: positive result (T and C red lines); –: negative results (C red line). N.A (Not applicable)

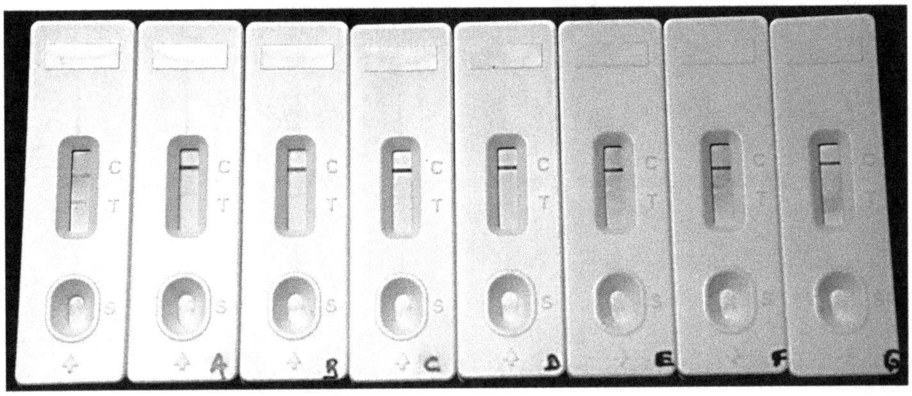

Fig. 6 Limit of detection for the prototype LFT. Control (C) line; Test (T) line. *Avibacterium paragallinarum* lysate FARPER-140 (serogroup C) at 10-fold dilution. From left to right: Not labeled test, 10^7 CFU/mL; (**a**), 10^6 CFU/mL; (**b**), 10^5 CFU/mL; (**c**), 10^4 CFU/mL; (**d**), 10^3 CFU/mL; (**e**), 10^2 CFU/mL; (**f**), 10^1 CFU/mL; (**g**), BHI broth. The detection threshold for the LFT was 10^4 CFU/mL

antibody (1G7G8). This antibody was able to detect the expected molecular weight of TBDT (~ 90 kDa) from bacterial cell cultures in Western blot (Fig. 2). Likewise, 1G7G8 allowed us to determine the reactivity differences between *Av. paragallinarum* serogroups using an indirect ELISA assay. Despite all serogroups showing positive reactivity, serogroups A and C had the highest reactivity for supernatants and lysates (Figs. 4 and 5). Therefore, all these results highlighted the ability to use 1G7G8 to detect all three serogroups, which emphasized its use for the development of the lateral flow test (LFT).

The use of the same antibody (self-pairing) as a detection and capture reagent for prototype LFT development has been previously reported for the detection of foot-and-mouth disease virus (FMDV) in cloven-hoofed animals [15]. In this study, the self-pairing prototype LFT allowed us to detect the TonB-dependent transporter from nasal mucus samples from chickens suspected of having infectious coryza.

Nasal mucus discharge is one of the main clinical signs in chickens infected with *Av. paragallinarum* [16]. *Av. paragallinarum* presence in nasal mucus discharge has been detected using molecular techniques such as the TaqMan Real-Time PCR assay [17]. Consequently, we chose to use nasal mucus as a suitable type of sample for lateral flow test evaluation.

The evaluation of thirty-five nasal mucus samples using LFT showed high sensitivity (100%) when it was compared with bacterial isolation (B.I), which indicated that LFT is capable of detecting *Av. paragallinarum*. However, the LFT evaluation showed low specificity (50%) and low positive predictive value (PPV) (65.4%) (Table 4), which could be related either to B.I. limitations in detecting low bacterial load in nasal mucus samples or to overgrowth of fast-growing bacteria [5, 6, 17].

In the same way, we observed high sensitivity (83.9%) and high specificity (100%) when LFA were compared with PCR, but a low negative predictive value (NPV) (44.4%) was observed (Table 4). This low NPV is due to discrepancies when both methods classify samples as negative. For instance, five samples were negative by LFA and positive by PCR. These discrepancies could be explained by either low bacterial load beneath the limit of detection of LFT or bacterial DNA remnants after antibiotic treatment.

According to the Venn diagram (Fig. 8), we observed that 17 out of 35 evaluated samples were positive using all three methods of detection, and nine samples were positive using only LFT and PCR. Despite 8 out of these 9 positive samples from LFT and PCR showing clinical signs, B.I was not able to detect *Av. paragallinarum* in them. This result indicated that LFT is a more reliable and sensitive method of detection than B.I, which could explain the low specificity (50%) observed when LFT was compared with B.I (Table 4).

On the other hand, despite 4 out of 35 samples being negative for all three methods of detection, PCR was the only method that was able to detect a high portion of positive samples (31/35). This difference with LFT (26/35 positive samples) could explain the obtained sensitivity (83.9%) (Fig. 8 and Table 4). The high proportion of positive samples detected by PCR could have arisen from free DNA remnants of non-viable bacteria resulting from antibiotic treatment, which could be a reason for unnecessary and expensive prevention strategy implementation. Therefore, all these results indicated that LFT is a reliable and trustworthy method of detection that can be used in field conditions.

Previous studies reported coinfection cases of *Av. paragallinarum* with *G. anatis* or *O. rhinotracheale* in chickens [16, 18]. In the present study, we identified *G. anatis* in nasal mucus samples (11/35) by B.I., but only 8 out of these 11 samples were identified as coinfection cases by B.I. These 8 samples were also positive for *Av.*

Table 2 Detailed description of chickens involved in the study

Farm localization/date	Chicken	Type of chicken	Age of chicken (weeks)	*Coryza vaccine*[a]	Antibiotic treatment[b]	Clinical signs[c]
City of Arequipa						
Suspected farm 1 Oct-2017	A1	Laying hens	23	Yes	No	+
	A2	Laying hens	23	Yes	No	–
	A3	Laying hens	23	Yes	No	+
	A4	Laying hens	23	Yes	No	–
	A5	Laying hens	23	Yes	No	+
	A6	Laying hens	23	Yes	No	+
City of Ica						
Suspected farm 2 Dic-2017	T1	Laying hens	4	Yes	Yes	+
	T2	Laying hens	4	Yes	Yes	+
	T3	Laying hens	4	Yes	Yes	–
	T4	Laying hens	4	Yes	Yes	+
	T5	Laying hens	4	Yes	Yes	–
	T6	Laying hens	4	Yes	Yes	–
	T7	Laying hens	4	Yes	Yes	–
	T8	Laying hens	4	Yes	Yes	–
Suspected farm 3 Jan-2018	I1	Laying hens	7	Yes	No	+
	I2	Laying hens	7	Yes	No	+
	I3	Laying hens	7	Yes	No	+
	I4	Laying hens	7	Yes	No	+
	I5	Laying hens	7	Yes	No	+
	I6	Laying hens	7	Yes	No	+
	I7	Laying hens	7	Yes	No	+
	I8	Laying hens	7	Yes	No	–
	I9	Laying hens	7	Yes	No	–
	I10	Laying hens	7	Yes	No	–
	I11	Laying hens	7	Yes	No	+
	I12	Laying hens	7	Yes	No	+
	I13	Laying hens	7	Yes	No	+
	I14	Laying hens	7	Yes	No	+
	I15	Laying hens	7	Yes	No	+
	I16	Laying hens	7	Yes	No	+
	I17	Laying hens	7	Yes	No	–
	I18	Laying hens	7	Yes	No	+
	I19	Laying hens	7	Yes	No	–
	I20	Laying hens	7	Yes	No	–
	I21	Laying hens	7	Yes	No	–

[a]The vaccination program varies between farms
[b]The antibiotic treatment varies between farms
[c]Clinical signs was judged by veterinary experts. +: positive for clinical signs and -: negative for clinical signs

paragallinarum using LFT (Table 3), which supported the capability of LFT to target the protein of interest.

Particular attention should be paid to the fact that nasal mucus samples, which were typed as serogroup B (Table 3), could be detected by LFT despite the fact that bacterial culture samples from serogroup B were the least reactive by indirect ELISA assay (Figs. 4 and 5). This result emphasizes the ability of LFT to detect all three serogroups. On the other hand, recent studies indicated the poor performance of mPCR [19, 20], which was used in this study

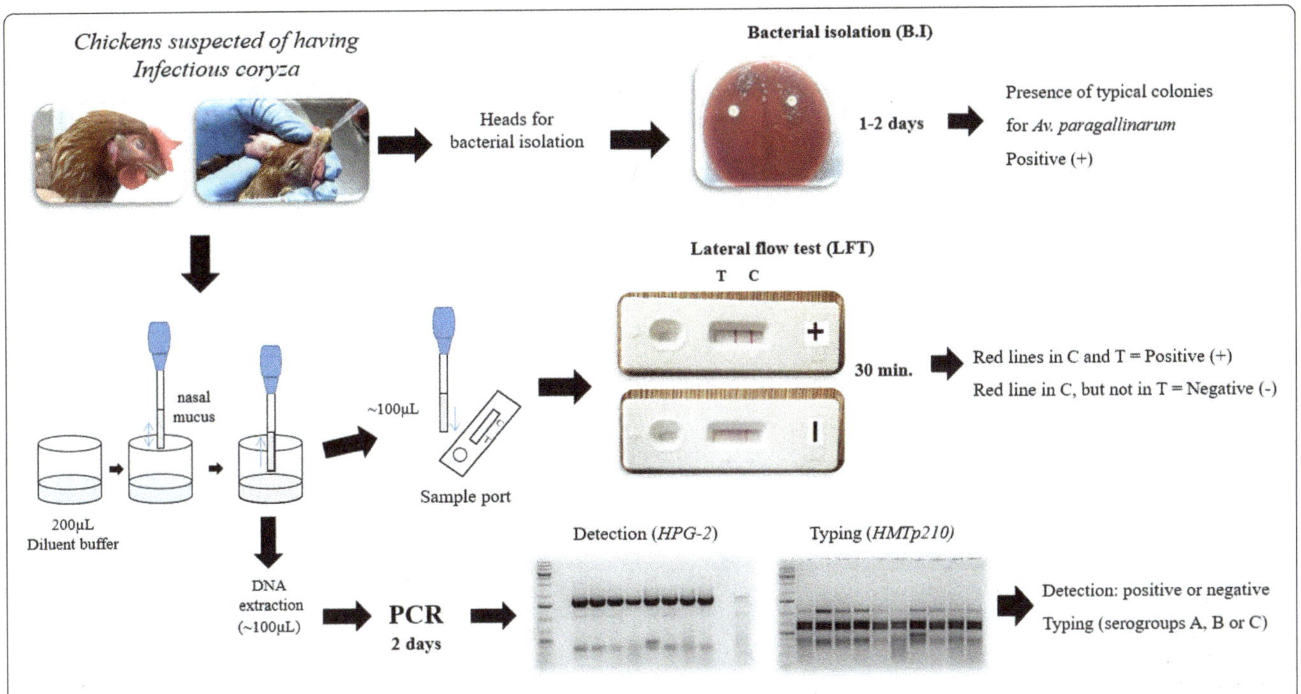

Fig. 7 Workflow for evaluation of chickens suspected of having infectious coryza. The evaluation was carried out using bacterial isolation (B.I), polymerase chain reaction (PCR) and lateral flow test (LFT)

for sample typing. Nonetheless, the conventional serotyping technique of hemagglutination inhibition (HI) has also been reported to have sensitivity limitations and to be a laborious technique [21]. Moreover, HI is a technique that depends on bacterial isolation, and in the present study, only 51.4% of all samples were successfully isolated using B.I. Therefore, in order to type the largest number of samples, we used an alternative and sensitive technique based on PCR assay, which is still being used to type field samples [6]. Interestingly, we could detect the serogroups A and B in the same farm (Table 3). This event is supported by a coinfection case of two serogroups A and B reported previously in other geographic region [22]. These two independent cases indicate the need for future assessments in order to understand the dynamic of *Av. paragallinarum* serogroups in a given farm.

Likewise, the limit of detection of the present LFT was 1×10^4 CFU/mL, which is similar to other lateral flow tests previously reported. Thus, a limit of detection of 1×10^4 CFU/mL and 1.8×10^5 CFU/mL was found for *Salmonella sp.* in chickens [23] and for *Escherichia coli* O157:H7 in bovine and swine animals [24], respectively. This comparative description underscores that the LFT developed in the present study has a similar limit of detection to that reported in other studies.

Finally, one of the two limitations of this study was the lack of specificity evaluation using other *Avibacterium spp.* (*Av. avium*, *Av. endocarditis*, *Av. gallinarum* and *Av. volantium*), and the other one was related to the number of cultures of positive isolates from *Av. paragallinarum*. Therefore, these two limitations should be taken into consideration for further validation assays of this prototype.

Conclusion

A prototype self-pairing LFT for *Av. paragallinarum* detection was successfully developed in this study. The use of this prototype in the field does not require specially trained personnel, and this test evaluation can be performed in 30 min. In light of our results, the prototype self-pairing LFT has potential applications for rapid detection of *Av. paragallinarum* in chickens suspected of having infectious coryza in field. However, further validation assays are required using reference and field strains from different laboratories in other geographical regions.

Methods

Bioinformatics prediction

Using as reference reported sequences of TBDT in *Av. paragallinarum* (GenBank: WP_052716793.1, KKA99715.1, WP_051185305.1 and WP_017806794.1), we annotated the TBDT gene (GenBank: MG242130.1) on the previously reported genome of *Av. paragallinarum* strain 72 [25]. The following regions were flagged on the protein sequence: the signal peptide (predicted using SignalP 4.1) and regions with an identity of 50% or higher with any protein of other avian pathogens (identified using BLASTp). Then, peptides were predicted on the unflagged regions using Bebipred

Table 3 Comparative description of *Avibacterium paragallinarum* detection using bacterial isolation (B.I), PCR and the prototype LFT

Chicken	Bacterial isolation			PCR		LFT
	Avibacterium paragallinarum	*Gallibacterium anatis*	*Ornithobacterium rinotracheale*	Detection[a]	Typing[b]	
A1	+	+	−	+	B	+
A2	+	−	−	+	B	+
A3	+	−	−	+	B	+
A4	−	−	−	−	−	−
A5	+	−	−	+	B	+
A6	+	−	−	+	B	+
T1	−	−	−	+	B	+
T2	−	−	−	+	A	+
T3	−	−	−	−	−	−
T4	−	−	−	+	A	+
T5	−	−	−	+	A	−
T6	−	−	−	−	−	−
T7	−	−	−	+	B	−
T8	−	−	−	−	−	−
I1	+	+	−	+	B	+
I2	−	+	−	+	B	+
I3	+	+	−	+	B	+
I4	+	+	−	+	B	+
I5	+	+	−	+	B	+
I6	−	−	−	+	B	+
I7	+	+	−	+	B	+
I8	−	−	−	+	B	+
I9	−	+	−	+	B	−
I10	+	+	−	+	B	+
I11	+	+	−	+	B	+
I12	+	−	−	+	B	+
I13	−	−	−	+	B	+
I14	−	−	−	+	B	+
I15	+	−	−	+	B	+
I16	−	−	−	+	B	+
I17	−	−	−	+	B	−
I18	−	+	−	+	B	−
I19	+	−	−	+	B	+
I20	+	−	−	+	B	+
I21	+	−	−	+	B	+

+: positive results; −: negative results. *Avibacterium paragallinarum* serogroups: A, B and C
[a]Target gene *HPG-2*
[b]Target gene *HMTp210*

2.0 and Optimum Antigen (GenScript Laboratories, Piscataway, NJ, USA), selecting those peptides with the highest conservation and located in the outer part of the protein structure (by analyzing the 3D model). The best candidate obtained (named Ma-4) was used in the downstream analysis.

Monoclonal antibody production using a peptide

For monoclonal antibody production, the peptide (Ma-4) was synthesized and used to immunize a BalbC mouse which was provided by GenScript Laboratories. After 5 weeks, spleen cells were isolated for hybridoma production by fusion with the SP2/0 mouse myeloma line. Peptide-

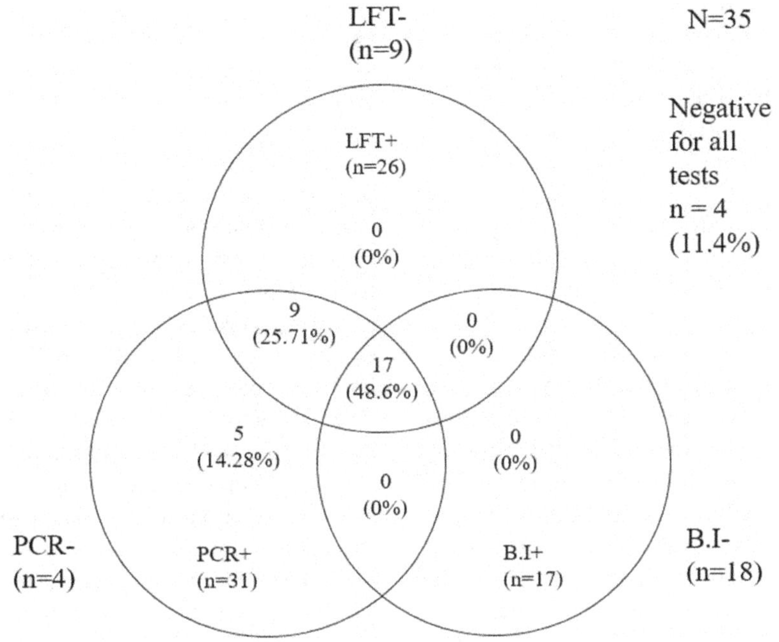

Fig. 8 Venn diagrams showing results from the evaluation of chicken nasal mucus samples ($N = 35$). These results were obtained using bacterial isolation (B.I), PCR and the prototype LFT. Samples positive (+) for *Avibacterium paragallinarum* are represented as n (%) inside the diagram, and negative samples (−) are represented as n outside diagram. n = numbers. Areas of overlap indicate positive results in common. Adapted from Stewart et al. BMC Veterinary Research (2017) 13:131

reactive clones were finally selected by ELISA assay and then isotyped. Antibody titers in hybridoma supernatants were calculated as the highest positive dilution against the peptide by indirect ELISA assay. The peptide synthesis, hybridoma production, clone selection, isotyping and antibody titers were performed by GenScript Laboratories. The positive clones were sent to FARVET Laboratories for monoclonal antibody production according to a previously published methodology [26]. The hybridoma culture supernatants were used for the antibody purification steps.

The mouse was euthanized using cervical dislocation without anesthesia following the American Veterinary Medical Association (AVMA) guidelines and the terminated animal was disposed by contracted biological waste disposal company [27].

Antibody purification

Chromatographic runs were conducted on a commercial prepacked affinity HiTrap rProtein A FF column (0.7 cm × 2.5 cm) (GE Healthcare, Uppsala, Sweden), which occurred at room temperature, and an AKTA Pure system (GE Healthcare). The column was equilibrated with 8 column volumes (CV) of 200 mM sodium phosphate buffer and 3 M NaCl at pH 7. The hybridoma culture supernatants were previously centrifuged for 10 min at 4 °C at 13000 x g, filtered using a 0.22 μm filter and then loaded into the column at 156 cm/h linear flow velocity using a 150 mL superloop (GE Healthcare). After washing unbound compounds with 8 CV of the equilibration buffer, the antibodies were eluted stepwise with 8 CV of 0.1 M sodium citrate at pH 3.5. Column eluates were continuously collected as 1 mL fractions in a F9-R fraction collector (GE Healthcare).

Table 4 Comparative indicators of the prototype LFT using bacterial isolation (B.I) and PCR

Prototype LFT					
Comparative techniques	Sensitivity % (95% CI)	Specificity % (95% CI)	PPV[a] % (95% CI)	NPV[b] % (95% CI)	Kappa index (95% CI)
B.I	100.0	50.0	65.4	100	0.49
	(81.6–100)	(29.0–71.0)	(46.2–80.6)	(70.1–100.0)	(0.21–0.78)
PCR	83.9	100.0	100.0	44.4	0.54
	(66.4–92.9)	(51.0–100.0)	(87.1–100.0)	(18.9–49.0)	(0.17–0.91)

95% CI Confidence interval at 95%
[a]Positive predictive value
[b]Negative predictive value

Preparation of recombinant protein

The sequence of the gene encoding the TonB-dependent transporter (TBDT) was obtained from NCBI reference sequence MG242130. The protein sequence consists of 756 amino acids with a theoretical molecular weight of 86.5 kDa. The expression of this recombinant protein was carried out in an *E. coli* BL21 strain, and a 6x-histidine tag was added at the amino terminus (Creative Diagnostics, Shirley, NY, USA). This recombinant protein was produced to be used as a positive control in Western blots.

Bacterial pellet, lysate and culture supernatant

The *Av. paragallinarum* isolates FARPER-113 (serovar B-1), FARPER-114 (serovar A-2) and FARPER-140 (serovar C-1) from an outbreak in Peru were serotyped (Page and Kume scheme), characterized and grown using methods that were previously reported [2]. The bacteria were grown in brain-heart infusion (BHI) broth (Sigma Aldrich Co., St Louis, Missouri, USA) supplemented with 5% equine serum and 0.01% NAD (TM/SN) for 18 h under agitation at 37 °C [2]. Later, the *Av. paragallinarum* cultures were centrifuged at 11000 x g for 5 min at 4 °C. The supernatants were filtered using a microfilter (0.45 μm) and stored at − 80 °C until use. The bacterial pellet was resuspended with 10 mL of phosphate-buffered saline, pH 7.4, plus Tween 20 5% (*v*/v) (Sigma Aldrich Co.), and incubated on ice for 30 min. Subsequently, the resuspended bacterial pellet was sonicated at an amplitude of 100 (On/Off time, 30 s/10 s) for 2 min on ice using a sonicator S-4000 (Qsonica, LLC, Newton, CT, USA) and centrifuged at 11000 x g for 30 min at 4 °C. Finally, the bacterial lysates obtained from sonication were stored at − 80 °C, the pellet was resuspended using 1 mL of 8 M urea (Sigma Aldrich Co.) and was stored at − 20 °C. Culture lysates and pellets were quantified using a Bradford assay (Merck KGaA, Darmstadt, Germany). Additionally, the pellets, lysates and supernatants of related bird bacteria cultures, such as *Ornithobacterium rhinotracheale* (FARPER-172), *Gallibacterium anatis* (FARPER-173), and *Pasteurella multocida* (FARPER-069) and non-related bird bacteria *Bortedella bronchiseptica* (FARPER-118), were locally isolated and cultivated using standard bacteriological techniques [16, 28–30]. For a better understanding, the serovars B-1, A-2 and C-1 obtained using Page and Kume scheme were referred as serogroups B, A and C, respectively.

Identification of TBDT protein using Western blot

Bacterial lysate, pellet and culture supernatant samples were separated by 4–20% sodium dodecyl sulfate-polyacrylamide gel electrophoresis (SDS-PAGE) and transferred to a nitrocellulose membrane using an e-blot device (GenScript Laboratories). The membranes were blocked with Azure Chemi Blot clocking buffer (Azure Biosystems, Dublin, CA, USA) with shaking for 1 h at room temperature. Membranes were washed three times for 10 min with phosphate-buffered saline (PBS)-Tween 20 and were incubated with purified monoclonal antibodies (0.4 μg/mL). Membranes were washed three times with phosphate-buffered saline-Tween 20. The immune reaction was performed with peroxidase-labelled goat IgG anti-mouse antibody (GenScript Laboratories) using radiance as a substrate and revealed by a CCD camera (Azure Biosystems). For bacterial lysates and pellets, 10-fold dilutions were prepared with the previously quantified samples (500, 50, 5, 0.5, 0.05 and 0.005 μg/mL). For bacterial culture supernatants, 2-fold dilutions were prepared with the supernatant (1:2, 1:4, 1:8 and 1:16) in PBS at pH 7.0. For recombinant TBDT protein, 500 ng was used. For molecular weight estimation of the studied proteins, we used a pre-stained molecular weight marker that ranged from 26,600 to 180,000 Da (Sigma Aldrich Co.) and a molecular weight marker that ranged from 30,000 to 120,000 Da (GenScript Laboratories).

Reactivity of monoclonal antibodies using indirect enzyme-linked immunosorbent assay (ELISA)

The 96-well microplate (Greiner Bio-One GmbH, Kremsmünster, Austria) was coated with 100 μL of antigen. The antigens used were bacterial lysates (10^3–10^7 CFU/mL) that were diluted with buffer carbonate/bicarbonate, pH 9.5, and supernatants from bacterial culture (dilutions from 1:10 to 1:100,000). All antigens except bacterial lysates were diluted in PBS at pH 7.4. Microplates were incubated at 4 °C overnight. Microplates were washed once with 150 μL of wash buffer (PBS + 0.05% Tween 20), blocked for 1 h at room temperature with 100 μL of blocking buffer (PBS + 0.5% BSA) and washed three times with 150 μL of washing buffer. One-hundred microliters purified monoclonal antibodies (5 μg/mL) were added to the plates and incubated for 1 h at 37 °C. Microplates were washed three times with 150 μL of washing buffer. Then, 100 μL of horseradish peroxidase conjugated goat anti-mouse IgG H&L (IgG-HRP) (Abcam, Cambridge, MA, USA) was added, and the microplates were incubated for 1 h at 37 °C. The microplates were washed four times using 150 μL of washing buffer; then, 100 μL of 3,3′,5,5′-tetramethylbenzidine (TMB) ELISA substrate (high-sensitivity) (Abcam) was added, and the microplates were incubated at room temperature for 30 min. Finally, the reaction was stopped using 100 μL of 2 N H_2SO_4, and the microplates were read at 450 nm using an EON microplate reader instrument (Biotek, Winooski, VT, USA).

Assembly of a prototype lateral flow test (LFT)

All possible combinations of antibodies were evaluated in order to determine which antibody pair was the most suitable for test development. All antibodies were tested as capture and detection reagents. Thus, all antibodies were conjugated to gold nanoparticles and used for the test. The detection antibody was immobilized at the Test (T) line region, and an anti-mouse IgG antibody was immobilized at the Control (C) line region of a nitrocellulose membrane. Briefly, two red lines in C and T regions meant a positive result, and only one red line in the C region meant a negative result for *Av. paragallinarum*. No band in the C region meant an invalid LFT.

Specificity evaluation of the prototype LFT

Evaluation of specificity was carried out using culture lysates and supernatants of related and non-related bird bacterial cultures that were prepared as described above. All culture lysates were diluted until reaching 1×10^7 CFU/mL, from which 100 µL was mixed with 100 µL of diluent buffer. Finally, 100 µL was added to the sample port of the LFT.

Additionally, cell cultures were infected with avian respiratory viruses in order to evaluate LFT specificity. Cell lines such as chicken hepatocellular carcinoma cell line (LMH), chicken fibroblast cell line (DF-1) and African green monkey kidney cell line (VERO) were seeded at 1×10^6 cells/mL with Dulbecco's modified Eagle medium (DMEM) F12 (HyClone, Logan, UT, USA) supplemented with 5% heat-inactivated fetal bovine serum (FBS) in a 75 cm^2 flask and cultivated in 5% CO_2 at 37 °C until reaching 70% confluency. These cell lines were infected with infectious laryngotracheitis virus (VFAR-043 strain) [31], which was initially adapted to the LMH cell culture following a previously reported method [32], Newcastle disease virus (LaSota strain) [33] or avian metapneumovirus (SHS-FAR strain) [34] at 0.02 multiplicity of infection (MOI). The infected cells were cultivated until a cytopathic effect was observed. Then, the supernatants were collected and centrifuged at 3000 g for 5 min at room temperature. Finally, the supernatants with viruses were quantified using plate assay and expressed in plaque-forming units (PFU)/mL. The quantified supernatants were diluted until reaching 1×10^6 PFU/mL, and 100 µL was mixed with 100 µL of diluent buffer and added to the sample port.

Limit of detection of the prototype LFT

The limit of detection was performed using *Av. paragallinarum* culture lysates from FARPER-140 (serogroup C). Colony count was performed following a previously reported method [17]. A bacterial culture (50 mL) was quantified (1×10^6 CFU/mL) and centrifuged at 11000 x g for 5 min at 4 °C. The bacterial pellet was resuspended with 5 mL of phosphate-buffered saline, pH 7.4, plus 5% Tween 20 (*v*/*v*) (Sigma Aldrich Co.), incubated on ice for 30 min and sonicated based on factors mentioned above in order to obtain 1×10^7 CFU/mL. The lysates were serially diluted at 10-fold from 1×10^7 to 1×10^1 CFU/mL in diluent buffer. The lowest positive dilution was considered the limit of detection of the test.

Chicken sample evaluation using the prototype LFT, PCR and bacterial isolation

Thirty-five chickens suspected of having infectious coryza were sent from farms located in the south of Peru (cities: Arequipa and Ica) to FARVET laboratories, after 2 days of clinical sign onset, for routine evaluations. Chicken nasal mucus was obtained by exerting pressure on the nasal sinuses. The nasal mucus was taken using a 2-mL transfer pipette and mixed with 200 µL of diluent buffer in a 1.5-mL microtube until reaching a homogeneous mixture. From this mixture, 100 µL was taken to perform the LFT, and 100 µL was used for the PCR assays. The homogeneous mixture was added to the sample port of the LFT, and the results were assessed visually after 30 min. PCR assays based on the *HPG-2* gene [35] were used for detection, and a multiplex PCR (mPCR) based on the *HMTp210* gene [21] was only used for chicken nasal mucus typing. Bacterial isolation was performed using bird heads according to a previous report [35].

All chickens were euthanized using cervical dislocation without anesthesia following the American Veterinary Medical Association (AVMA) guidelines [36] and the disposal of dead animals was performed according to the Peruvian Regulation [37].

Statistical analysis

All quantitative data were analyzed using GraphPad Prism 6.0 (GraphPad Software, La Jolla, CA, USA). Indirect ELISA assays performed for duplicate and optical density (O.D.) values were represented as the mean and standard deviation (SD). The cutoff was calculated as the mean O.D. value +2SD. The O.D. value used for cutoff calculation was obtained from the 1:10,000 dilution for supernatants and 1×10^4 CFU/mL for lysates. The O.D. values obtained above the cutoff were considered positive, and those below the cutoff were negative. Two-way ANOVAs with Tukey's post hoc test were used for serogroup comparison analysis ($p < 0.05$).

Two-by-two contingency tables were created to analyze associated values among the three methods of detection. The following indicators for the prototype LFT were calculated using bacterial isolation and PCR as reference methods: sensitivity (Se), specificity (Sp), positive predictive value (PPV), negative predictive value (NPV) and kappa index (K). Each indicator was calculated using a 95% confidence interval. Kappa index (agreement level) was denoted using Landis and Koch values [38], where 0 is poor, 0.01–0.2 is slight, 0.21–0.4 is fair, 0.41–0.6 is moderate, 0.61–0.8 is substantial and 0.81–1.0 is almost perfect.

Abbreviations

Av. paragallinarum: *Avibacterium paragallinarum*; *B. bronchiseptica*: *Bortedella bronchiseptica*; B.I: Bacterial isolation; BHI: Brain-heart infusión; BSA: Bovine serum albumin; CFU: Colony-forming unit; CV: Column volumes; DF-1: Chicken fibroblast cell line; DMEN: Dulbecco's modified Eagle medium;
ELISA: Enzyme-Linked ImmunoSorbent Assay; FBS: Fetal bovine serum; FMDV: Foot-and-mouth disease virus; *G. anatis*: *Gallibacterium anatis*; HI: Hemagglutination inhibition; IC: Infectious coryza; IgG-HRP: Horseradish peroxidase conjugated goat anti-mouse IgG H&L; K: Kappa index; kDa: Kilo dalton; LFT: Lateral flow test; LMH: Chicken hepatocellular carcinoma cell line; mPCR: Multiplex Polymerase Chain Reaction; NAD (TM/SN): Complete growth medium that contains both chicken serum and nicotinamide adenine dinucleotide; NCBI: National Center for Biotechnology Information;
NPV: Negative predictive value; *O. rhinotracheale*: *Ornithobacterium rhinotracheale*; O.D: Optical density; OMVs: Outer membrane vesicles; *P. multocida*: *Pasteurella multocida*; PBS: Phosphate-buffered saline;
PCR: Polymerase chain Reaction; PFU: Plaque-forming unit; PPV: Positive predictive value; SD: Standart desviation; SDS-PAGE: Sodium dodecyl sulfate–polyacrylamide gel electrophoresis; Se: Sensitivity; Sp: Specificity; TBDTs: TonB-dependent transporters; TMB: 3,3′,5,5′-tetramethylbenzidine; VERO: African green monkey kidney cell line

Acknowledgments

We would like to thank to Dr. Adriana Paredes and Dr. Miguel Marzal for their support with the hybridoma culture. We thank to MSc. Aldo Rojas and MSc. Vladimir Longa for their technical support. Additionally, we thank to Dr. Vladimir Morales for his valuable recommendations for the manuscript and his constructive discussions. We thank to DVM. Julio Ticona for organizing farm sample transportation.

Funding

This work was financially supported by the Programa Nacional de Innovación para la Competitividad y Productividad (Innóvate Perú) under the contract 184-FIDECOM-FINCyT-PIPEA-2014.

Authors' contributions

SMR, JB and MF-D authors conceived and designed the study. DR performed bioinformatic analyses. SMR, JB, RCG, RM, LCM and AMA acquired the data of the study. SMR and JB performed data analyses and wrote the manuscript. All authors read and approved the final manuscript.

Competing interests

The authors declare that they have no competing interests.

Author details

[1]Laboratorios de Investigación y Desarrollo, FARVET SAC, Carretera Panamericana Sur N°766 Km 198.5, Chincha Alta 11702, Ica, Peru. [2]Laboratorio de Bioinformática y Biología Molecular, Laboratorio de Investigación y Desarrollo, Universidad Peruana Cayetano Heredia, Av. Honorio Delgado 430 San Martín de Porres, Lima, Lima, Peru. [3]FARVET SPF SAC, Carretera Panamerica Sur N°766 Km 198.5, Chincha Alta 11702, Ica, Peru.

References

1. Blackall PJ, Christensen H, Beckenham T, Blackall LL, Bisgaard M. Reclassification of Pasteurella gallinarum, [Haemophilus] paragallinarum, Pasteurella avium and Pasteurella volantium as Avibacterium gallinarum gen. Nov., comb. nov., Avibacterium paragallinarum comb. nov., Avibacterium avium comb. nov. and Avibacterium volantium comb. nov. Int J Syst Evol Microbiol. 2005;55(Pt 1):353–62.
2. Blackall PJ, Soriano-Vargas E. Infectious Coryza and related bacterial infections. In: *Diseases of poultry*. 13th edn. Edited by Swayne DE, Glisson JR, McDougald LR, Nolan LK, Suarez DL, Nair V. Ames, IA: Wiley; 2013. p. 859–68.
3. Chen X, Chen Q, Zhang P, Feng W, Blackall PJ. Evaluation of a PCR test for the detection of Haemophilus paragallinarum in China. Avian Pathol. 1998; 27(3):296–300.
4. Chen X, Miflin JK, Zhang P, Blackall PJ. Development and application of DNA probes and PCR tests for Haemophilus paragallinarum. Avian Dis. 1996; 40(2):398–407.
5. Muhammad TM, Sreedevi B. Detection of Avibacterium paragallinarum by polymerase chain reaction from outbreaks of infectious coryza of poultry in Andhra Pradesh. Vet World. 2015;8(1):103–8.
6. Patil W, Mishra D, Mane DV. 16S ribosomal RNA sequencing and molecular serotyping of Avibacterium paragallinarum isolated from Indian field conditions. Vet World. 2017;10(8):1004–7.
7. Abascal EN, Guerra AC, Vazquez AS, Tenorio VR, Cruz CV, Zenteno E, Contreras GP, Pacheco SV. Identification of iron-acquisition proteins of Avibacterium paragallinarum. Avian Pathol. 2009;38(3):209–13.
8. Fernandez-Rojas MA, Vaca S, Reyes-Lopez M, de la Garza M, Aguilar-Romero F, Zenteno E, Soriano-Vargas E, Negrete-Abascal E. Outer membrane vesicles of Pasteurella multocida contain virulence factors. Microbiologyopen. 2014;3(5): 711–7.
9. Ramon Rocha MO, Garcia-Gonzalez O, Perez-Mendez A, Ibarra-Caballero J, Perez-Marquez VM, Vaca S, Negrete-Abascal E. Membrane vesicles released by Avibacterium paragallinarum contain putative virulence factors. FEMS Microbiol Lett. 2006;257(1):63–8.
10. Bragg RR, Gunter NJ, Coetzee L, Verschoor JA. Monoclonal antibody characterization of reference isolates of different serogroups of Haemophilus paragallinarum. Avian Pathol. 1997;26(4):749–64.
11. Yamaguchi T, Takigami S, Iritani Y, Hayashi Y. Characterization and use of monoclonal antibodies to identify Haemophilus paragallinarum serovars. Avian Dis. 1990;34(1):52–7.
12. Verschoor JA, Coetzee L, Visser L. Monoclonal antibody characterization of two field strains of Haemophilus paragallinarum isolated from vaccinated layer hens. Avian Dis. 1989;33(2):219–25.
13. Noro T, Oishi E, Kaneshige T, Yaguchi K, Amimoto K, Shimizu M. Identification and characterization of haemagglutinin epitopes of Avibacterium paragallinarum serovar C. Vet Microbiol. 2008;131(3–4):406–13.
14. Noro T, Yaguchi K, Amimoto K, Oishi E. Identification and expression of a gene encoding an epitope that induces hemagglutination inhibition antibody to Avibacterium paragallinarum serovar a. Avian Dis. 2007;51(1):84–9.
15. Oem JK, Ferris NP, Lee KN, Joo YS, Hyun BH, Park JH. Simple and rapid lateral-flow assay for the detection of foot-and-mouth disease virus. Clin Vaccine Immunol. 2009;16(11):1660–4.
16. Morales-Erasto V, Falconi-Agapito F, Luna-Galaz GA, Saravia LE, Montalvan-Avalos A, Soriano-Vargas EE, Fernandez-Diaz M. Coinfection of Avibacterium paragallinarum and Ornithobacterium rhinotracheale in chickens from Peru. Avian Dis. 2016;60(1):75–8.
17. Corney BG, Diallo IS, Wright L, Hewitson G, De Jong A, Tolosa X, Burrell P, Duffy P, Rodwell B, Boyle DB, et al. Rapid and sensitive detection of Avibacterium paragallinarum in the presence of other bacteria using a 5′ Taq nuclease assay: a new tool for diagnosing infectious coryza. Avian Pathol. 2008;37(6):599–604.
18. Paudel S, Ruhnau D, Wernsdorf P, Liebhart D, Hess M, Hess C. Presence of Avibacterium paragallinarum and histopathologic lesions corresponds with clinical signs in a co-infection model with Gallibacterium anatis. Avian Dis. 2017;61(3):335–40.
19. Morales-Erasto V, Posadas-Quintana Jde J, Fernandez-Diaz M, Saravia LE, Martinez-Castaneda JS, Blackall PJ, Soriano-Vargas E. An evaluation of serotyping of Avibacterium paragallinarum by use of a multiplex polymerase chain reaction. J Vet Diagn Investig. 2014;26(2):272–6.
20. Wang H, Sun H, Blackall PJ, Zhang Z, Zhou H, Xu F, Chen X. Evaluation of a proposed molecular methodology for the serotyping of Avibacterium paragallinarum. J Vet Diagn Investig. 2016;28(5):555–60.
21. Sakamoto R, Kino Y, Sakaguchi M. Development of a multiplex PCR and PCR-RFLP method for serotyping of Avibacterium paragallinarum. J Vet Med Sci. 2012;74(2):271–3.
22. Christy N, Carrasco S, Soriano E, Morales V. A severe outbreak of Avibacterium paragallinarum serovars A-2 and B-1 in coinfection with Gallibacterium anatis biovar haemolytica and the non-hemolytic biovar anatis in commercial layers. In: 2018 International Poultry Scientific Forum: 29–30 January, 2018 2018. Atlanta, Georgia: The International Production and Processing Expo; 2018. p. 42.
23. Bautista DA, Elankumaran S, Arking JA, Heckert RA. Evaluation of an immunochromatography strip assay for the detection of Salmonella sp. from poultry. J Vet Diagn Invest. 2002;14(5):427–30.
24. Jung B, Jung S, Kweon C. Development of a rapid immunochromatographic strip for detection of Escherichia coli O157. J Food Prot. 2005;68(10):3.

25. Requena D, Chumbe A, Torres M, Alzamora O, Ramirez M, Valdivia-Olarte H, Gutierrez AH, Izquierdo-Lara R, Saravia LE, Zavaleta M, et al. Genome sequence and comparative analysis of Avibacterium paragallinarum. Bioinformation. 2013;9(10):528–36.
26. Paredes A, Saenz P, Marzal MW, Orrego MA, Castillo Y, Rivera A, Mahanty S, Guerra-Giraldez C, Garcia HH, Nash TE, et al. Anti-Taenia solium monoclonal antibodies for the detection of parasite antigens in body fluids from patients with neurocysticercosis. Exp Parasitol. 2016;166:37–43.
27. Vertebrate Animals Section [https://www.genscript.com/NIH-grant-application-vertebrate-animal-section.html].
28. Glisson JR. Pasteurellosis and other respiratory bacterial infections. In: *Diseases of poultry*. 13th edn. Edited by Swayne DE, Glisson JR, McDougald LR, Nolan LK, Suarez DL, Nair V. Ames, IA: Wiley; 2013. p. 807–58.
29. Logue CM. Other Bacterial Diseases. In: *Diseases of poultry*. 13th edn. Edited by Swayne DE, Glisson JR, McDougald LR, Nolan LK, Suarez DL, Nair V. Ames, IA: Wiley; 2013. p. 1020–1.
30. Nakagawa M, Muto T, Nakano T, Yoda H, Ando K, Isobe Y, Im Aizumi K. Some observations on diagnosis of Bordetella bronchiseptica infection in Guinea pigs. Exp Animals. 1968;18(3):11.
31. Morales Ruiz S, Bendezu Eguis J, Montesinos R, Tataje-Lavanda L, Fernandez-Diaz M. Full-genome sequence of infectious Laryngotracheitis virus (Gallid Alphaherpesvirus 1) strain VFAR-043, Isolated in Peru. Genome Announc. 2018;6(10). https://doi.org/10.1128/genomeA.00078-18.
32. Devlin JM, Viejo-Borbolla A, Browning GF, Noormohammadi AH, Gilkerson JR, Alcami A, Hartley CA. Evaluation of immunological responses to a glycoprotein G deficient candidate vaccine strain of infectious laryngotracheitis virus. Vaccine. 2010;28(5):1325–32.
33. Chumbe A, Izquierdo-Lara R, Calderon K, Fernandez-Diaz M, Vakharia VN. Development of a novel Newcastle disease virus (NDV) neutralization test based on recombinant NDV expressing enhanced green fluorescent protein. Virol J. 2017;14(1):232.
34. Jones RC, Rautenschlein S. Newcastle disease, other Avian Paramyxoviruses, and Avian Metapneumovirus infections: Avian Metapneumovirus. In: *Diseases of poultry*. 13th edn. Edited by Swayne DE, Glisson JR, McDougald LR, Nolan LK, Suarez DL, Nair V. Ames, IA: Wiley; 2013. p. 112–9.
35. Falconi-Agapito F, Saravia LE, Flores-Perez A, Fernandez-Diaz M. Naturally occurring beta-nicotinamide adenine dinucleotide-independent Avibacterium paragallinarum isolate in Peru. Avian Dis. 2015;59(2):341–3.
36. Euthanasia AP. AVMA guidelines for the Euthanasia of animals (2013 edition): American Veterinary Medical Association; 2013.
37. SENASA. Normas Legales. In: Decreto Supremo N° 029–2007-AG "Aprueban Reglamento del Sistema Sanitario Avícola". Lima: El Peruano; 2007. p. 356402–12.
38. Landis JR, Koch GG. The measurement of observer agreement for categorical data. Biometrics. 1977;33(1):159–74.

21

Pathogenicity and molecular characterization of a fowl adenovirus 4 isolated from chicken associated with IBH and HPS in China

Limin Li[1,2,3†], Jianchang Wang[4†], Ping Chen[1†], Shan Zhang[1], Jiguo Sun[1,2,3] and Wanzhe Yuan[1,2,3*]

Abstract

Background: Since July in 2015, an emerging infectious disease, Fowl adenovirus (FAdV) species C infection with Hepatitis-Hydropericardium syndrome was prevalent in chicken flocks in China. In our study, one FAdV strain was isolated from commercial broiler chickens and was designated as SDSX1.The phylogenetic information, genetic mutations and pathogenicity of SDSX1 were evaluated.

Results: The phylogenetic analysis indicated that SDSX1 is a strain of serotype 4, FAdV-C. The amino acid analysis of fiber-2 showed that there were more than 20 mutations compared with the non-virulent FAdV-C strains. The pathogenic evaluation of SDSX1 showed that the mortality of one-day-old chickens inoculated SDSX1 was 100%. The typical histopathological changes of SDSX1 were characterized by the presence of basophilic intranuclear inclusion bodies in hepatocytes. The virus copies in different tissues varied from10^7 to 10^{11} per 100 mg tissue and liver had the highest virus genome copies.

Conclusion: In conclusion, the isolate SDSX1, identified as FAdV-4, could cause one-day-old chicks' typical inclusion body hepatitis (IBH) and hepatitis-hydropericardium syndrome (HHS) with 100% mortality. The virus genome loads were the highest in the liver. Molecular analysis indicated that substitutions in fiber-2 proteins may contribute to the pathogenicity of SDSX1.

Keywords: Fowl adenovirus 4, Hydropericardium syndrome, Inclusion body hepatitis, Virulent, Pathogenicity, Phylogenetic analysis, Chickens

Background

Fowl adenoviruses (FAdVs) are non-enveloped, double-stranded DNA linear viruses, belonging to the genus *Aviadenovirus*of the family *Adenoviridae* [1, 2]. The viral genome is approximately 43–46 kb, encoding a number of structural and nonstructural proteins [3–5]. The virion capsid consists of three major structural proteins, hexon, fiber and penton base proteins. Hexon is the major protein of the adenovirus capsid known to have a region related to virus neutralization and serotype specificity.

FAdVs are grouped into 5 species (FAdV-A to FAdV-E) based on genomic restriction enzyme digest patterns with 12 serotypes identified so far: FAdV-A (serotype FAdV-1); FAdV-B (serotype FAdV-5);FAdV-C (serotypes FAdV-4 and -10); FAdV-D (serotypes FAdV-2, – 3, – 9, and – 11); and FAdV-E (serotypes FAdV-6, – 7, –8a, and -8b) [6]. FAdVs have a worldwide distribution, but different serotypes or genotypes are discovered in different geographic regions [7–11]. FAdVs infection (the strain was closely related to serotype 11 and 8a) was first reported in Morocco recently [12]. All 12 serotypes of FAdVs have been associated with the outbreak of inclusion body hepatitis (IBH) with a ~ 10% mortality rate

* Correspondence: yuanwanzhe@126.com
†Limin Li, JianchangWang and Ping Chen contributed equally to this work.
[1]College of Veterinary Medicine, Hebei Agricultural University, Baoding 071001, Hebei, China
[2]Hebei Engineering and Technology Research Center of Veterinary Biotechnology, Baoding 071001, Hebei, China

[7]. Chickens between 3 and 4 weeks old are especially vulnerable to IBH that is characterized by congested and enlarged liver with hepatocyte necrosis and petechial hemorrhage. As the pathognomonic lesion of FAdVs infection, basophilic intranuclear inclusion bodies are usually observed in degenerated hepatocytes surrounded by a clear halo or filling the entire nucleus [13]. FAdV-4 has been implicated in hepatitis-hydropericardium syndrome (HHS) characterized by accumulation of transparent or straw-colored fluid in the pericardial sac, nephritis, and hepatitis with a 30–100% mortality rate [14–16]. Also, pathogenicity assays confirmed that FAdV-4 could induce HHS [8].

From June to November in 2015, a disease emerged in large-scale broiler and egg-laying chickens in farms in Shandong, Henan, Hebei, Liaoning, Anhui and Jilin provinces in China. The disease inflicted mostly the 20-to-30-day-old broilers with a mortality rate of 20–30%, and occasionally affected layers aged 20–70 days or 200–300 days but with a lower mortality rate. Usually, there were no obvious clinical signs with a sudden death. The death peak was 4–8 days after onset and the course of disease could last 8–15 days. The diseased animals had flabby hearts with straw-colored fluid in the pericardial sac and the swollen liver with foci of hemorrhage and/or necrosis. The aim of this study was to molecular characterization and pathogenicity assessment of FAdV strain responsible for disease outbreak in chickens.

Results

Sequence analysis

The virus was isolated from the SPF chicken's embryos by 5 blind passages and it was designated as SDSX1. The whole genome nucleotide sequence and hexon gene sequence of SDSX1 are available in the GenBank (accession number is KY636400 andKT932640, respectively) and the complete genome for SDSX1 was found to be 43,630 bp in length. The percent sequence identity for available whole genomes showed that SDSX1 has 100% sequence homology with three isolates from China (CH/SDDZ/2015, CH/AHBZ/2015, and CH/JSXZ/2015, [17] and 98.6% sequence identity with a KR5 isolate from Austria. Nevertheless, SDSX1 showed a low sequence identity (< 57.0%) with the strains of other FAdV species. According to the phylogenetic analysis based on the full genome (Fig. 1a), the SDSX1 isolate was classified into serotype 4 of FAdV-C. The similar evolutionary relationships were observed from the phylogenetic tree constructed based on the hexon gene (Fig. 1b).

The size of fiber-2 in SDSX1 isolate is 479 amino acid residues, the same as the strain KR5 (HE608152, FAdV-C), CH/SDDZ/2015(KU558761, FAdV-C), CH/AHBZ/2015 (KU569295, FAdV-C), and CH/JSXZ/2015(KU569296, FAdV-C) (Table 1) and 5 amino acid residues more than ON1(GU188428, FAdV-C) (Fig. 2). The basic residues rich sequence KRPK/KRAK (site 27–30) and VYPF (site 41–45) in fiber proteins of SDSX1were identified. Particularly, there were more than 20 substitutions compared with the non-pathogenic strains, ON1 and KR5 (Table 2 and Fig. 2) and shared only 94.1%~ 94.2% with ON1 and KR5, respectively. The fiber-2 protein identity of SDSX1 is 100% compared with CH/SDDZ/2015, CH/AHBZ/2015, and CH/JSXZ/2015.

Pathogenicity analysis of SDSX isolate

All the chickens inoculated with SDSX1 died within 5 days post inoculation and the survival curve was shown in Fig. 3. The median survival of inoculated group was 4 and the statistical differences between groups in survival was obvious significant ($p < 0.0001$). There were no obvious clinical manifestations for the inoculated chickens and the chickens had normal intake before death. Typical gross pathological changes in chickens were the flabby heart with straw-colored fluid in the pericardial sac and the swollen liver with foci of hemorrhages (Fig. 4a). The histopathology revealed that fatty degeneration of hepatocytes, basophilic intranuclear inclusion bodies in hepatocytes and necrosis of hepatocytes (Fig. 4b). All control chickens, the instead of that did not show any symptoms at the end of the observation.

Virus load analysis

Tissues including heart, liver, spleen, lung, kidney, and bursa of Fabricius were collected and the virus copies in tissues were determined using Real-time PCR. The results showed that virus can be detected in these tissues and the number of virus copies was between 10^7 and 10^{11} copies per 100 mg tissue with the highest viral genome copies in liver (Fig. 5).

Discussion

Since 2015, FAdVs infection with high mortality was endemic in the poultry farms in China, resulting in considerable economic losses to the poultry industry [14, 18–21]. It is reported that most cases have been observed in broilers of 3 to 5 weeks of age, and occasionally in layers and breeder pullets aged 10 to 20 weeks and the infection was characterized by hepatitis hydropericardium syndrome (HHS) [10, 22]. In addition to the cases mentioned above, the isolation of FAdV-C from ducks with HHS was also reported [23, 24]. In our study, one strain was isolated from the chickens with HHS infection, and the whole genome was sequenced and analyzed. In addition, pathogenicity of the isolate SDSX1 was evaluated.

Phylogenetic analysis based on the FAdV isolates from different geographic areas worldwide showed that SDSX1 was clustered into serotype-4, FAdV-C species. Nucleotide alignment of SDSX1isolate with theidentity

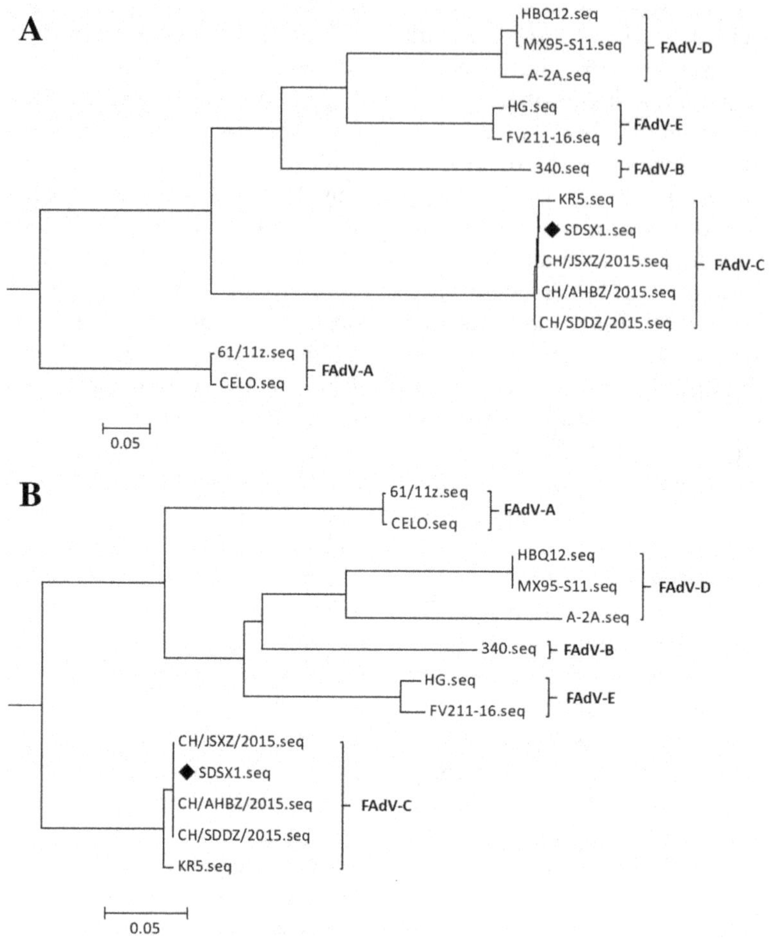

Fig. 1 The phylogenetic analysis of the isolate SDSX1. Phylogenetic trees of SDSX1 were constructed based on the whole genome sequence (**a**) and hexon gene sequence of FAdVs strains (**b**). FAdV isolate evaluated in the study was labeled with "♦"

of reference strains in 5 species showed thehigher identity of 98.6–100% with FAdV-C using the complete genome sequence. The alignment of the hexon of SDSX1 had the similar result in the percentage identity compared to the full genome sequence. Especially, SDSX1 had 100% identity with the three isolates in China (CH/SDDZ/2015, CH/AHBZ/2015 and CH/JSXZ/2015).

The data indicated that SDSX1 was a highly pathogenic isolate. Regarding the pathogenicity of the SDSX1 isolate, the survival curve showed that all of 10 one-day-old SPF chickens inoculated with the SDSX1 died within 14 days. The results were consistent with several previous studies in which parts of serotype 4 were highly pathogenic strains [19, 25, 26]. The 100% of mortalityshowed in our study indicated that SDSX1 isolate was a virulent strain, and these were also accounted for the pathogenicity of the virus, serotypes, and dosage of infected strain, especially the susceptibility of the chickens. Different pathogenicity, dynamic distribution and replication pattern in chickens and ducks were observed in the recent report [25].

It is reported that fiber-2 played a significant role in virulence of FAdV [27]. Several scientists' studies showed that some amino acid residues, positions located at 219 and 380 in the fiber-2 protein of KR5, were related to the virulence based on the amino acid residues comparison between strains induced HHS and non-virulent strains [20, 28]. The fiber-2 protein not only plays a critical role in mediating infection by FAdV-4 but also has a potential capacity to be the candidate vaccine. Immunogenicity of the recombinant fiber-2 protein was evaluated and the results showed that both the level of IgY and the $CD4^+$ T cell proliferative response of the chickens immunized with fiber-2 protein were significantly higher than that of the chickens immunized with inactivated vaccine, and fiber-2 could provide 100% protection [29]. The neutralizing epitope recognized by a mAb 3C2 was identified to be located between aa 416–448 in the fiber-2 protein of FAdV-4 and the fiber-2 protein could efficiently block infection by FAdV-4 [30]. In our study, there were more than 20 amino acids residues compared with non-virulent strains (ON1 and KR5), including the

Table 1 Reference isolates of FAdVs in the study

FAdV isolates	GenBank accession number	Geographic origin	Viral species
61/11z	KX247012	Poland	FAdV-A
CELO	U46933	Austria	FAdV-A
340	KC493646	Northern Ireland	FAdV-B
ON1	GU188428	Canada	FAdV-C
KR5	HE608152	Austria	FAdV-C
CH/SDDZ/2015	KU558761	China	FAdV-C
CH/AHBZ/2015	KU569295	China	FAdV-C
CH/JSXZ/2015	KU569296	China	FAdV-C
A-2A	AF083975	Canada	FAdV-D
HBQ12	KM096545	China	FAdV-D
MX95-S11	KU746335	Mexico	FAdV-D
HG	GU734104	Canada	FAdV-E
FV211–16	KX258422	Peru	FAdV-E

position 219(Glycine-Aspartic acid) and position 380(Alanine-Threonine) which had been identified. Whether the remaining amino acid resides substitutions of fiber-2 is responsible for the virulence of SDSX1 is unknown and need to be further studied in future.

Conclusions

In conclusion, the isolate SDSX1, identified as FAdV-4, could cause in one-day-old chicks' typical inclusion body hepatitis (IBH) and hepatitis-hydropericardium syndrome (HHS) with 100% mortality. The virus genome loads were highest in the liver. Molecular analysis indicated that substitutions in fiber-2 proteins may contribute to the pathogenicity of SDSX1.

Methods

Samples used in this study

Twenty 10~20-day-olddead "817" broilers were collected from various farms in Liaocheng in Shandong province. Gross lesions were examined, and liver tissues were fixed in 10% formalin for 48 h at room temperature and then HE staining was performed following routine histological procedures. Animal work performed in this study was approved by Research Animal Use and Care Committee, the Agricultural University of Hebei, in compliance with the Guidance for the Care and Use of Laboratory Animals issued by the Ministry of Science and Technology of China.

Viral DNA extraction and PCR

A TaqMan-based real-time PCR assay developed in-house for specific detection of FAdV-4 was used to determine the presence of FAdV-4 in collected liver samples as described previously [31]. The primers sequences were: sense, 5'-TTACGCTTACGGTGCCTACGT-3′;and antisense, 5'-CCGCGTTATTCATGATCCAGTA-3′. The TaqMan probe was as follows, FAM-5'CGACGGTTCCCAGTCCCTCACG-3′-Eclipse. The product size was 89 bp. Briefly, 10 g liver tissue was homogenized in 5 ml 0.01 M PBS. After freeze thawing for three times, the samples were centrifuged in 10000 rpm for 10 min. The supernatants of each sample were collected for viral DNA extraction. Viral DNA extraction was done TIANamp Virus DNA kit (Tiangen Biotech Co., Ltd., Beijing, China) according to the manufacturer's instructions. The viral DNA from each sample was eluted in a final volume of 100 μl of nuclease-free water.

Next generation sequencing

The whole sequence of the isolated virus was sequenced by Next Generation Sequencing (NGS). Briefly, the DNA samples were ultrasonicated to generate fragments less than 500 bp. The DNA fragments were polished using T4 polynucleotide kinase. The adaptors were then ligated and loaded on the HiSeq 2000 for sequencing.

Virus isolation

Virus isolation was performed on the samples with positive PCR results. The liver tissue was homogenized in 0.01 M PBS containing penicillin (100 IU/μL) and streptomycin (100 μg/μL) with the ration of 1:5(*W*/*V*), filtered and centrifuged. The supernatant from each sample was then collected and used for inoculation of 9-day-old specific pathogen free (SPF) chicken embryos (Beijing Vital River Laboratory Animal Technology Co., Ltd.) via allantoic cavity. The dose of inoculation was 0.1 ml per embryo. The allantoic fluid from infected embryos was harvested and used for repeated inoculation of chicken embryos. 5 embryos were used for each inoculum. The inoculation dose of allantoic fluid was 0.1 ml per embryo for each time. After 5 passages, allantoic fluid was harvested in the biohazard safety equipment and virus nucleic acid

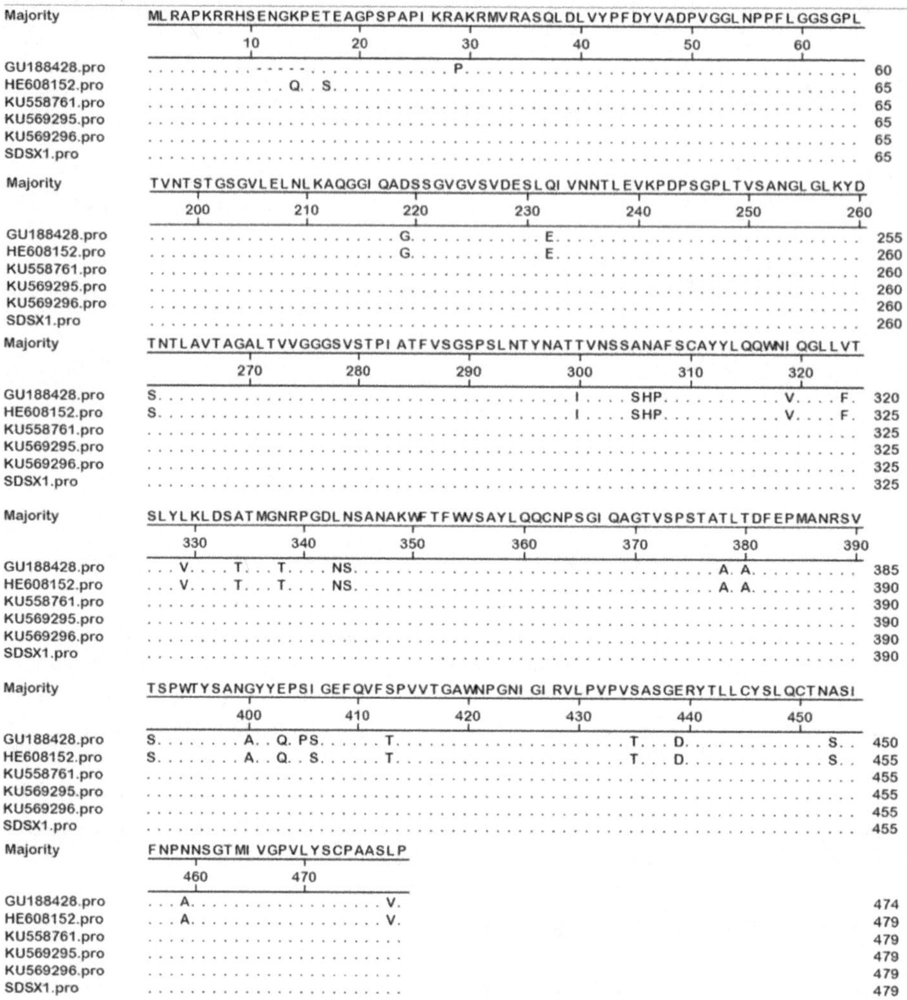

Fig. 2 Alignment of amino acids residues of fiber-2 of SDSX1 isolate with other FAdV-C strains. Dots indicate conserved residues; dashes indicate deleted amino acids

was extracted for DNA sequencing analysis. A specific fragment of 724 bp from the hexon gene was amplified by PCR for detecting the virus and the following primers were used: sense, 5'GACCTTCGCGGACTACTTGG3'; and antisense, 5'GACGGTTTGGTTGGAACTGAT3'.

Sequence analysis

To compare the complete genomic sequence and partial hexon gene sequence of the isolate determined in this study, 11 published genomic sequences available in GenBank (Table 1) were aligned by the MegAlign program of the DNAStar software version 7.0 with the Clustal W method (DNASTAR, Madison, Wisconsin, USA). The phylogenetic tree was constructed using the Maximum-Likelihood (ML) approach and MEGA6.06 [32] on the aligned data set. The analysis of amino acids residues of fiber-2 of SDSX1 isolate was done using DNAStar software version 7.0.

Determination of EID_{50}

Virus titer in chicken embryos for the allantoic fluid of five passage viruses was determined by chicken embryos infectivity assay. The result was recorded as EID_{50} per milliliter by using the Reed-Muench method. Briefly, 9-day-old chicken embryos were inoculated the allantoic fluid of the fifth passage viruses (0.1 ml/embryo), which were prepared by serial 10-fold dilution. The embryos were incubated at 37 °C for 6 days and observed every day. Embryos that died within the first 24 h were discarded, those that died than 24 h were selected for the allantoic fluid collection. PCR was used for detecting virus in allantoic fluid. Results were recorded.

Pathogenicity analysis of SDSX1 isolate

Twenty one-day SPF chickens were randomly allotted to two groups (10 chickens per group). For the inoculated group, each chicken was intraperitoneally inoculated with

Table 2 Amino acid substitutions of fiber-2 in FAdV-C strains with different pathogenicity

Sites of amino acid	ON1 (GU188428)	KR5 (HE608152)	CH/SDDZ/2015 (KU558761)	CH/AHBZ/2015 (KU569295)	CH/JSXZ/2015 (KU569296)	SDSX1 (KY636400)
11–15	–	ENGQP	ENGKP	ENGKP	ENGKP	ENGKP
29	P	A	A	A	A	A
219	G	G	D	D	D	D
232	E	E	Q	Q	Q	Q
261	S	S	T	T	T	T
300	I	I	T	T	T	T
305–307	SNP	SNP	ANA	ANA	ANA	ANA
319	V	V	I	I	I	I
324	F	F	V	V	V	V
329	V	V	L	L	L	L
334	T	T	A	A	A	A
338	T	T	N	N	N	N
343–344	NS	NS	LN	LN	LN	LN
378	A	A	T	T	T	T
380	A	A	T	T	T	T
391	S	S	T	T	T	T
400	A	A	G	G	G	G
403	Q	Q	E	E	E	E
405	P	S	S	S	S	S
406	S	S	I	I	I	I
413	T	T	S	S	S	S
435	T	T	S	S	S	S
439	D	D	E	E	E	E
453	S	S	A	A	A	A
459	A	A	N	N	N	N
478	V	V	L	L	L	L

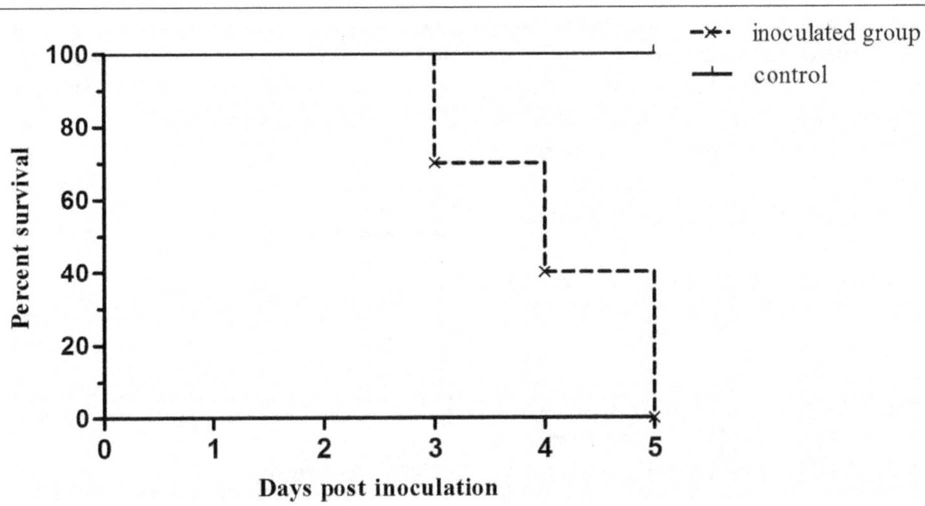

Fig. 3 The survival curve of the inoculated and control chickens. "×" indicates the censored observation. Three, three and four chickens were censored at 3, 4 and 5 days post inoculation, respectively

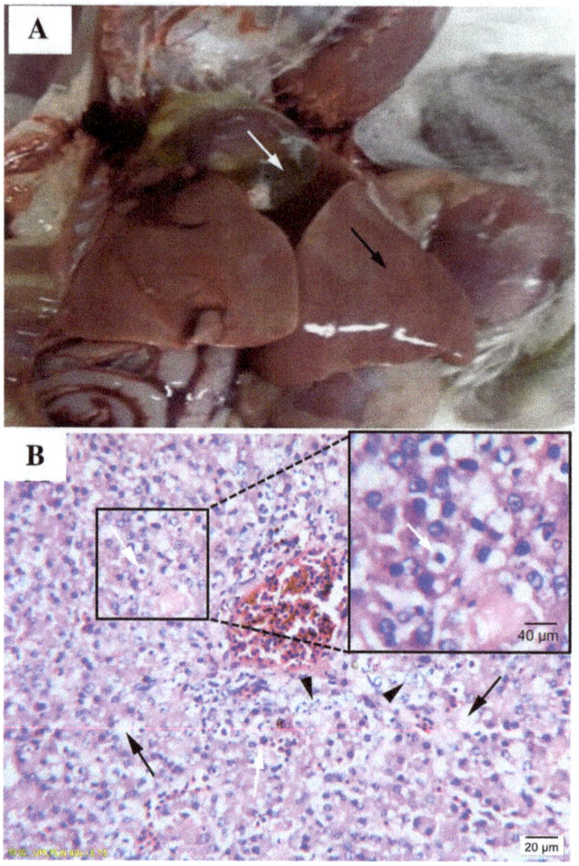

Fig. 4 Gross pathological changes and histopathological analysis. **a** showed a representative image showing the enlarged liver with fatty degeneration (black arrow) and the heart with straw-colored fluid in the pericardial sac (white arrow); **b** showed the histopathological analysis (HE staining) of the liver tissue. Hepatocyte fatty degeneration (black arrow); basophilic inclusion bodies in the hepatocyte (white arrow); hepatocyte necrosis (arrowhead) were observed

1 ml SDSX1 virus containing 10^6 EID_{50} per chicken. Virus used for inoculation was prepared in the biohazard safety equipment. Chickens in the control group were mock inoculated with the same dosage of PBS. The dosage and pathway of inoculation was selected based on the pre-experiment (data not shown). All the chickens were observed daily for clinical signs up to all chickens are dead. The date and time of the death of each animal were recorded. When the chickens died, the liver samples were collected for the histopathological examination and virus detection.

Virus load analysis

Quantification of the SDSX1 load in each of the analyzed organs was performed by Real-time PCR method reported in the previous study [31]. Briefly, an equal amount of heart, liver, spleen, lung, lung, kidney and bursa of Fabricicus (100 mg each) from each bird was homogenized in 2 ml PBS. Viral DNA extraction was done according to the method mentioned above. The viral DNA from each sample was eluted in a final volume of 100 μl of nuclease-free water. The TaqMan real-time PCR was performed in a final volume of 25 μl containing 2 × premix Ex Taq Master Mix12.5 μl, 1 μl of 10 μM primers, 0.8 μl of 10 μM probes, 1 μl DNA template and 8.7 μl RNase-free water. The reaction was carried out in ABI 7500 Real-Time PCR System and the optimal reaction procedure was as follows: pre-denaturing for 30s at 95 °C, followed by 40 cycles of denaturation at 95 °C for 10s, annealing at 65 °C for 35 s. A standard curve was used to calculate the virus copies: $Y = -3.56LogX + 46.08$. "Y" indicates cycles and "X" indicates copy number.

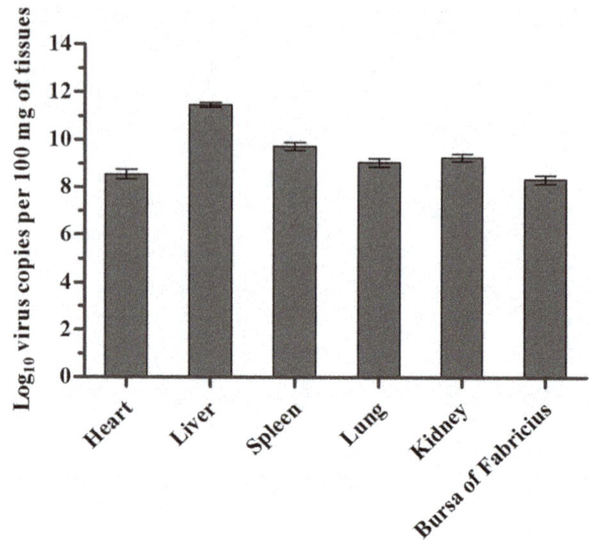

Fig. 5 The virus distribution of SDSX1 in experimented SPF chickens

Statistical analysis

Data were expressed as mean ± standard deviations. The survival curve of the inoculated and control chickens and the photograph of SDSX1 virus distribution in experimented SPF chickens were made using GraphPad Prism (version 5.0). One-way analysis of variance test was used.

Abbreviations

bp: Base pairs; EID_{50}: Egg 50% infectious dose; HE: Hematoxylin &Eosin;; PBS: Phosphate buffer solution; PCR: Polymerase chain reaction

Acknowledgements

Not applicable.

Funding

This work was partly supported by the Earmarked Fund for Hebei Lays Innovation Team of Modern Agro-industry Technology Research System (HBCT2018150210), Science and Technology Project Foundation of Hebei Province (18227517D), and by grants for one-hundred outstanding innovative talents from Hebei institution of higher learning (SLRC2017039).

Authors' contributions

LL and WY wrote the main manuscript text. LL and PC prepared figures. JW, PC, SZ and JS conducted the experiments and generate data. LL and WY analyzed the data. All authors reviewed the manuscripts. All authors read and approved the final manuscript.

Competing interests
The authors declare that they have no competing interests.

Author details
[1]College of Veterinary Medicine, Hebei Agricultural University, Baoding 071001, Hebei, China. [2]Hebei Engineering and Technology Research Center of Veterinary Biotechnology, Baoding 071001, Hebei, China. [3]North China Research Center of Animal Epidemic Pathogen Biology, China Agriculture Ministry, Baoding 071001, Hebei, China. [4]Inspection and Quarantine Technical Center of Hebei Entry-Exit Inspection and Quarantine Bureau, 318 Heping West Rd, Xinhua District, Shijiazhuang 050051, Hebei, China.

References

1. Adair B, Fitzgerald S. Group I adenovirus infections. In: Saif Y, Fadly A, Glisson J, McDougald L, Nolan L, Swayne D, editors. Diseases of poultry. 12th ed. Hoboken: Wiley-Blackwell; 2008. p. 252–66.
2. Benko M, Harrach B, Both G, Russell W, Adair B, A' dam E', de Jong J, Hess M, Johnson M, et al. In: Fauquet C, Mayo M, Maniloff J, Desselberger U, Ball L, editors. FamilyAdenoviridae. In Virus Taxonomy Eighth Report of the International Committee on the Taxonomy of Viruses. San Diego: Elsevier Academic Press; 2005. p. 213–28.
3. Griffin BD, Nagy E. Coding potential and transcript analysis of fowl adenovirus 4: insight into upstream ORFs as common sequence features in adenoviral transcripts. J Gen Virol. 2011;92(6):1260–72.
4. Shah MS, Ashraf A, Khan MI, Rahman M, Habib M, Chughtai MI, Qureshi JA. Fowl adenovirus: history, emergence, biology and development of a vaccine against hydropericardium syndrome. Arch Virol. 2017;162(7):1833–43.
5. Shah MA, Ullah R, March M, Shah MS, Ismat F, Habib M, Iqbal M, Onesti S, Rahman M. Overexpression and characterization of the 100K protein of fowl adenovirus-4 as an antiviral target. Virus Res. 2017;238:218–25.
6. Meulemans G, Couvreur B, Decaesstecker M, Boschmans M, van den Berg TP. Phylogenetic analysis of fowl adenoviruses. Avian pathol. 2004;33(2):164–70.
7. Mase M, Nakamura K, Minami F. Fowl adenoviruses isolated from chickens with inclusion body hepatitis in Japan, 2009-2010. J Vet Med Sci. 2012;74(8):1087–9.
8. Kim JN, Byun SH, Kim MJ, Kim JJ, Sung HW, Mo IP. Outbreaks of hydropericardium syndrome and molecular characterization of Korean fowl adenoviral isolates. Avian Dis. 2008;52(3):526–30.
9. Hess M, Raue R, Prusas C. Epidemiological studies on fowl adenoviruses isolated from cases of infectious hydropericardium. Avian pathol. 1999;28(5): 433–9.
10. Zhao J, Zhong Q, Zhao Y, Hu YX, Zhang GZ. Pathogenicity and complete genome characterization of fowl adenoviruses isolated from chickens associated with inclusion body hepatitis and Hydropericardium syndrome in China. PLoS One. 2015;10(7):e0133073.
11. Alemnesh W, Hair-Bejo M, Aini I, Omar AR. Pathogenicity of fowl adenovirus in specific pathogen free chicken embryos. J Comp Pathol. 2012;146:223–9.
12. Redondo H, Fragoso JS, Tahala MA, Y B, Gil I, Elbachir E, Rodríguez MJ, Abad Moreno JC. Characterization of strain of fowl adenoviruses circulating in Morocco. Poult Sci. 2018;97(11):4057–62.
13. Nakamura K, Mase M, Yamaguchi S, Shibahara T, Yuasa N. Pathologic study of specific-pathogen-free chicks and hens inoculated with adenovirus isolated from hydropericardium syndrome. Avian Dis. 1999;43(3):414–23.
14. Lim TH, et al. Outbreak of gizzard erosion associated with fowl adenovirus infection in Korea. Poult Sci. 2012;91(5):1113–7.
15. Schachner A, Marek A, Jaskulska B, Bilic I, Hess M. Recombinant FAdV-4 fiber-2 protein protects chickens against hepatitis-hydropericardium syndrome (HHS). Vaccine. 2014;32(9):1086–92.
16. Ganesh K, Raghavan R. Hydropericardium hepatitis syndrome of broiler poultry: current status of research. Res Vet Sci. 2000;68(3):201–6.
17. Liu Y, Wan W, Gao D, Li Y, Yang X, Liu H, Yao H, Chen L, Wang C, Zhao J. Genetic characterization of novel fowl aviadenovirus 4 isolates from outbreaks of hepatitis-hydropericardium syndrome in broiler chickens in China. Emerg Microbes Infect. 2016;5(11):e117.
18. Niu YJ, Sun W, Zhang GH, Qu YJ, Wang PF, Sun HL, Wang L, Xiao YH, Liu SD. Corrigendum to Hydropericardium syndrome outbreak caused by fowl adenovirus serotype 4 in China in 2015. J Gen Virol. 2016;97(10):2797.
19. Li H, Wang J, Qiu L, Han Z, Liu S. Fowl adenovirus species C serotype 4 is attributed to the emergence of hepatitis-hydropericardium syndrome in chickens in China. Infect Genet Evol. 2016;45:230–41.
20. Vera-Hernández PF, Morales-Garzón A, Cortés-Espinosa DV, Galiote-Flores A, García-Barrera LJ, Rodríguez-Galindo ET, Toscano-Contreras A, Lucio-Decanini E, Absalón AE. Clinicopathological characterization and genomic sequence differences observed in a highly virulent fowl Aviadenovirus serotype 4. Avian Pathol. 2016;45(1):73–81.
21. Slaine PD, Ackford JG, Kropinski AM, Kozak RA, Krell PJ, Nagy É. Molecular characterization of pathogenic and nonpathogenic fowl aviadenovirus serotype 11 isolates. Can J Microbiol. 2016;62(12):993–1002.
22. Morshed R, Hosseini H, Langeroudi AG, MHB F, Charkhkar S. Fowl adenoviruses D and E cause inclusion body hepatitis outbreaks in broiler and broiler breeder pullet flocks. Avian Dis. 2017;61(2):205–10.
23. Chen H, Dou Y, Zheng X, Tang Y, Zhang M, Zhang Y, Wang Z, Diao Y. Hydropericardium hepatitis syndrome emerged in Cherry Valley ducks in China. Transbound Emerg Dis. 2017;64(4):1262–7.
24. Pan Q, Liu L, Wang Y, Zhang Y, Qi X, Liu C, Gao Y, Wang X, Cui H. The first whole genome sequence and pathogenicity characterization of a fowl adenovirus 4 isolated from ducks associated with inclusion body hepatitis and hydropericardium syndrome. Avian Pathol. 2017;46(5):571–8.
25. Pan Q, Yang Y, Shi Z, Liu L, Gao Y, Qi X, Liu C, Zhang Y, Cui H, Wang X. Different dynamic distribution in chickens and ducks of the Hypervirulent, novel genotype fowl adenovirus serotype 4 recently emerged in China. Front Microbiol. 2017;8:1005.
26. Niu YJ, Sun W, Zhang GH, Qu YJ, Wang PF, Sun HL, Xiao YH, Liu SD. Hydropericardium syndrome outbreak caused by fowl adenovirus serotype 4 in China in 2015. J Gen Virol. 2016;97(3):2684–90.
27. Pallister J, Wright PJ. SheppardM. A single gene encoding the fiber is responsible for variations in virulence in the fowl adenoviruses. J Virol. 1996;70(8):5115–22.
28. Marek A, Nolte V, Schachner A, Berger E, Schlötterer C, Hess M. Two fiber genes of nearly equal lengths are a common and distinctive feature of fowl adenovirus C members. Vet Microbiol. 2012;156(3–4):411–7.
29. Chen L, Yin L, Zhou Q, Li Q, Luo Y, Xu Z, Zhang Y, Xue C, Cao Y. Immunogenicity and protective efficacy of recombinant fiber-2 protein in protecting spf chickens against fowl adenovirus 4. Vaccine. 2018;36(9):1203–8.
30. Ping W, Zhang J, Wang W, Li T, Liang G, Shao H, Wei G, Qin A, Ye J. A novel monoclonal antibody efficiently blocks the infection of serotype 4 fowl adenovirus by targeting fiber-2. Vet Res. 2018;49(1):29.
31. Wang J, Wang J, Chen P, LiuL YW. Development of a TaqMan-based real-time PCR assay for rapid and specific detection of fowl aviadenovirus serotype 4. Avian Pathol. 2017;46(3):338–43.
32. Tamura K, Stecher G, Peterson D, Filipski A, Kumar S. MEGA6: molecular evolutionary genetics analysis version 6.0. Mol Biol Evol. 2013;30(12):2725–9.

22

In vivo pharmacokinetic/Pharmacodynamic modeling of Enrofloxacin against *Escherichia coli* in broiler chickens

Xia Xiao[1,2,3], Lijie Jiang[1], Weixuan Lan[1], Yongjia Jiang[1] and Zhiqiang Wang[1,2,3*]

Abstract

Background: Systemic *Escherichia coli* infections cause early mortality of commercial broiler chickens. Although enrofloxacin has long been used in poultry, the in vivo pharmacokinetic/pharmacodynamic (PK/PD) relationship of enrofloxacin against *E. coli* is unclear. The present study aimed to establish an in vivo PK/PD model of enrofloxacin against *E. coli* in seven-day-old chicks and to ascertain whether the selection of target organ for PD determination is critical for parameter magnitude calculation in enrofloxacin PK/PD modeling.

Results: The in vivo effectiveness of enrofloxacin against *E. coli* in different organs varied, with the E_{max} ranging from − 4.4 to − 5.8 Log_{10} colony forming units (cfu)/mL or cfu/g. Both the surrogate AUC_{0-24}/MIC of enrofloxacin or AUC_{0-24}/MIC of the combination of enrofloxacin and ciprofloxacin correlated well with effectiveness in each organ. The AUC_{0-24}/MIC ratio of the combination of enrofloxacin and ciprofloxacin producing bactericidal and elimination effects were 21.29 and 32.13 in blood; 41.68, and 58.52 in the liver; and 27.65 and 46.22 in the lung, respectively.

Conclusions: The in vivo effectiveness of enrofloxacin against *E. coli* in different organs was not identical after administration of the same dosage. To describe the magnitude of PK/PD parameter exactly, bacterial loading reduction in different organs as PD endpoints should be evaluated and compared in PK/PD modeling. The selection of a target organ to evaluate PDs is critical for rational dosage recommendation.

Keywords: *Escherichia coli*, Enrofloxacin, In vivo, PK/PD modeling, Chicken

Background

It is estimated that 50% of total poultry loses could be attributed to first week mortalities [1]. Among them, over 50% of mortalities are caused by bacterial infections, primarily *Escherichia coli* [2]. Systemic *E. coli* infections contribute significantly to the early mortality of commercial broiler chickens [3]. However, because of the high diversity in virulence-associated genes and serotypes, effective vaccines against *E. coli* challenge are rare [4]. Until now, using antimicrobials has been the main strategy to control *E. coli* infections in the poultry industry.

Enrofloxacin, a second-generation fluoroquinolone, is commonly used in chickens because of its favorable pharmacokinetic (PK) profile and its excellent activity against gram-negative aerobic bacteria and some gram-positive bacteria [5, 6]. However, with the extensive use of enrofloxacin, resistance has emerged [7]. Enrofloxacin is metabolized to ciprofloxacin, which is used clinically in humans, and there are reports showing that resistance genes for fluoroquinolones could transfer to other organisms under antimicrobial pressure [8, 9]. Thus, the non-rational usage of enrofloxacin runs the risk of leading to bacterial resistance and potential health hazards in humans [10–12]. Thus, there is a growing need to optimize the use of enrofloxacin.

Optimizing the use of an antimicrobial should be based on a good understanding of its PK and pharmacodynamic (PD) relationship in target animals against specific bacterial species [13]. Although ex vivo PK/PD modeling of enrofloxacin has been evaluated in buffalo

* Correspondence: zqwang@yzu.edu.cn
[1]College of Veterinary Medicine, Yangzhou University, Yangzhou 225009, Jiangsu, People's Republic of China
[2]Jiangsu Co-innovation Center for Prevention and Control of Important Animal Infectious Diseases and Zoonoses, Yangzhou, Jiangsu 225009, People's Republic of China
Full list of author information is available at the end of the article

calves, swine, and chickens against *E. coli*, *Pasteurella multocida*, and *Salmonella typhimurium* [14–20], to the best of our knowledge, there are no in vivo PK/PD modeling studies of enrofloxacin against *E. coli* in chicks. In vivo PK/PD modeling has great advantages over ex vivo modeling in describing the PK/PD relationship [21], especially for enrofloxacin, whose metabolite, ciprofloxacin, (another fluoroquinolone) has almost the same potency as enrofloxacin. Therefore, for PK/PD modeling of enrofloxacin, it is important to involve its metabolism to ciprofloxacin in the modeling. According to our previous study, the selection of the target organ for PD determination is critical for parameter magnitude calculation in antimicrobial PK/PD modeling [22]. Whether this is true for enrofloxacin requires further investigation.

In the present study, to further understand the PK/PD relationship of enrofloxacin, especially whether the selection of a target organ for PD determination is critical for parameter magnitude calculation in PK/PD modeling, broilers were used as an animal model. The following aspects were investigated: (1) The pharmacokinetics of enrofloxacin and its metabolism to ciprofloxacin were determined at three different dosage administrations; (2) the in vivo PK/PD modeling of enrofloxacin against *E. coli* was developed using the reductions in the bacterial burden in the blood, liver, and lung as the PD endpoints; (3) whether the concentration of ciprofloxacin influence the in vivo PK/PD modeling of enrofloxacin was evaluated; and (4) the corresponding magnitude of PK/PD parameters for a certain efficacy were determined.

Results

In vitro susceptibility

The MICs of enrofloxacin and ciprofloxacin against *E. coli* O78 were the same (0.5 μg/mL). The corresponding MBC values were 0.5 and 1 μg/mL respectively. The MIC and MBC of enrofloxacin in serum were identical (0.5 μg/mL). The MPC of enrofloxacin was 3.2 μg/mL (Table 1).

Table 1 MIC and MPC values (μg/mL) of enrofloxacin and ciprofloxacin against *E. coli* O78 in Mueller–Hinton broth and serum

Parameters (μg/mL)	ENR (broth)	ENR (serum)	CIP (broth)
MIC	0.5	0.5	0.5
MBC	0.5	0.5	1
MPC	3.2	—	—

MIC minimum inhibitory concentration, *MPC* mutant prevention concentration, *MBC* minimum bactericidal concentration, *ENR* enrofloxacin, *CIP* ciprofloxacin

E. coli infection model

Clinical signs of colibacillosis, such as depression, decreased feeding, diarrhea, and fever were observed 24 h after challenge with *E. coli* O78. After dissection, perihepatitis and pericarditis were obvious. The bacteria load in the blood, liver, and lung were 7.2 ± 0.92 Log_{10} cfu/mL, 6.4 ± 0.14 Log_{10} cfu/g, and 6.1 ± 0.17 Log_{10} cfu/g respectively. The bacterial load in the three organs was similar among different chicks. The death rate was 10%.

Pharmacokinetics

The serum drug concentration-time profiles of enrofloxacin and ciprofloxacin after enrofloxacin administration at three dosages are illustrated in Fig. 1 and Fig. 2. The PK parameters of enrofloxacin and ciprofloxacin are shown in Table 2 and Table 3. The time of peak concentration (T_{max}) for enrofloxacin and ciprofloxacin were about 3.3~ 3.4 h and 4.3~ 5 h, respectively, with peak concentrations (C_{max}) of 0.16, 1.76, and 2.86 μg/mL for enrofloxacin at 1, 10, and 20 mg/kg, respectively; and of 0.03, 0.10, and 0.37 μg/mL at corresponding doses for ciprofloxacin. The C_{max} of ciprofloxacin was much lower and emerged later than enrofloxacin. The AUC_{0-24} of enrofloxacin at 1, 10, and 20 mg/kg b.w. was 1.64, 17.95, and 30.07 h, respectively, and the corresponding values for ciprofloxacin were 0.61, 1.42, and 4.93 h, respectively. The AUC_{0-24} values of enrofloxacin were 2.6, 12.6, and 6.09 times higher than those of ciprofloxacin at doses of 1, 10, and 20 mg/kg b.w. respectively. Dose proportionality was observed for the AUC_{0-24} of enrofloxacin and ciprofloxacin in the range of 1–20 mg/kg with r^2 of 0.9868 and 0.9035, respectively. Thus, the AUC_{0-24} of other doses between 1 and 20 mg/kg could be calculated.

In vivo PK/PD analysis

The AUC_{0-24}/MIC ratios of enrofloxacin against *E. coli* O78 for doses of 1, 2, 5, 7.5, 10, 12.5, 15, and 20 mg/kg were 5.27, 8.26, 17.20, 24.66, 32.11, 39.57, 47.02, and 61.93 h respectively. The corresponding AUC_{0-24}/MIC values of ciprofloxacin were 0.36, 0.82, 2.2, 3.35, 4.50, 5.65, 6.80, and 13.24 h, respectively.

The relationships between the effectiveness (bacteria loading reduction) of enrofloxacin in different organs and PK/PD indices of enrofloxacin, or the combination of enrofloxacin and ciprofloxacin, are shown in Fig. 3 and Fig. 4. The surrogate AUC_{0-24}/MIC correlated well with effectiveness in each organ, with r^2 values greater than 0.85. The in vivo effectiveness of enrofloxacin against *E. coli* in different organs varied, with E_{max} ranging from − 4.4 to − 5.8 Log_{10} cfu/mL. Using the PK of enrofloxacin for simulation, then the AUC_{0-24}/MIC values of enrofloxacin for the bactericidal effect in the blood, liver, and lung were 19.32, 32.15, and 23.41,

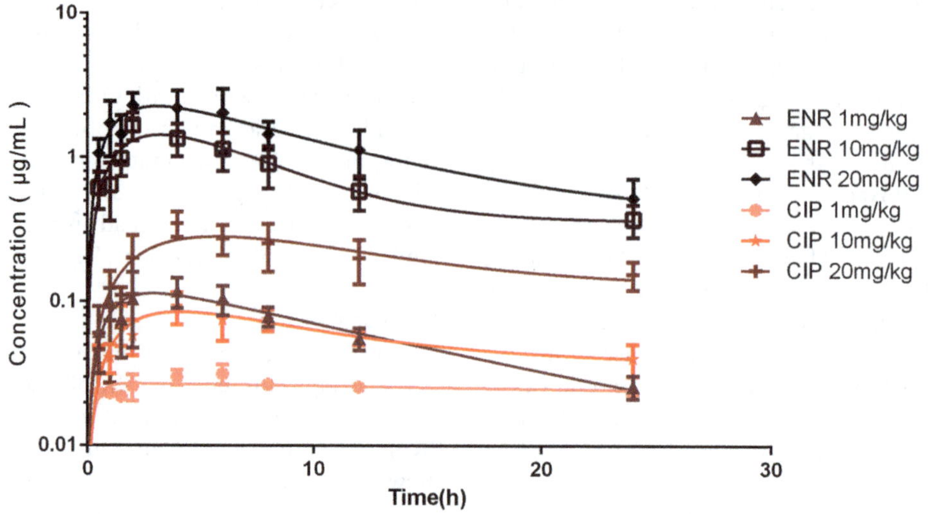

Fig. 1 The time-concentration profiles of enrofloxacin or ciprofloxacin in serum after oral administration of enrofloxacin at doses of 1, 10, and 20 mg/kg in *E. coli* O78-infected chicks. [enrofloxacin (ENR); ciprofloxacin (CIP)]

respectively. The slopes were 2.58, 2.36, and 3.01 for the blood, liver, and lung, respectively (Table 4). Using the combined PK of enrofloxacin and ciprofloxacin for simulation, then the AUC_{0-24}/MIC values for the bactericidal effect in the blood, liver, and lung were 21.29, 41.68, and 27.65, respectively (Table 5). The slopes were 4.19, 2.67, and 4.86 for the blood, liver, and lung, respectively.

Discussion

There are several infection routes for *E. coli* when simulating colibacillosis, such as intramuscular (IM) and oral administration [16, 23]. The complicated nature of the gastrointestinal tract could result in the bacterial load in different organs after oral administration being unstable. To obtain a stable bacterial concentration in the different organs, the IM infection route was chosen. In the present study, colibacillosis was achieved through inoculation of $\sim 10^7$ cfu/mL *E. coli* in chickens. However, the infection dose was lower than that used in a previous report [16], where colibacillosis was induced by oral gavage with 8 mL of *E. coli* culture containing 1.2×10^9 cfu/mL. The differences in infection dose might be explained in two ways: The broilers used by Sang were 39 days old, whereas ours were 7 days old, and the inoculation method was different, oral gavage in Sang's report vs. IM injection in this study.

The PKs of enrofloxacin have been investigated in goats, pigs, calves, horses, and sheep [24–28]. It has also

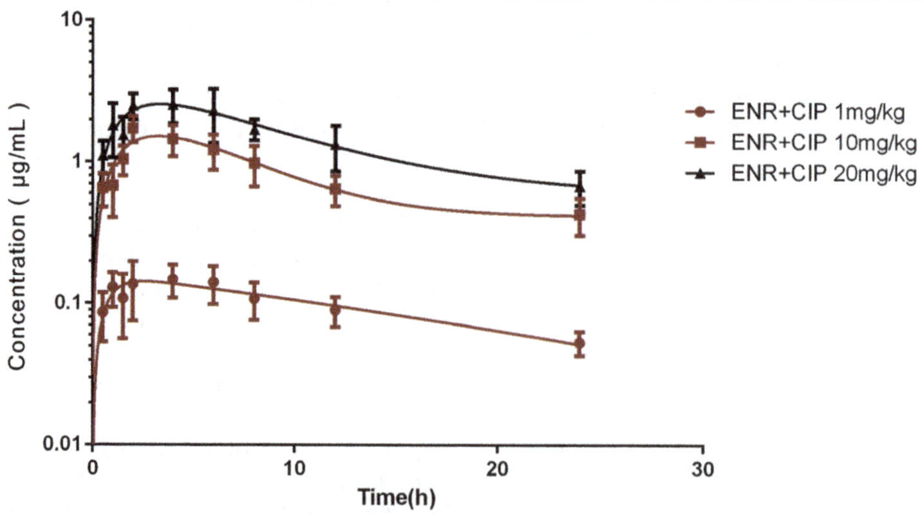

Fig. 2 The time-concentration profiles of the combination of enrofloxacin and ciprofloxacin in serum after oral administration of enrofloxacin at doses of 1, 10, and 20 mg/kg in *E. coli* O78-infected chicks. [enrofloxacin (ENR); ciprofloxacin (CIP)]

Table 2 Pharmacokinetic parameters of enrofloxacin in serum after oral administration of 1, 10, and 20 mg/kg body weight enrofloxacin in *E. coli* O78-infected chicks (mean and SD, $n = 10$)

Parameters	1 mg/kg	10 mg/kg	20 mg/kg
C_{max} (μg/mL)	0.16 ± 0.041	1.76 ± 0.32	2.86 ± 0.52
T_{max} (h)	3.35 ± 1.8	3.4 ± 2.12	3.3 ± 1.77
$T_{1/2\beta}$ (h)	10.75 ± 1.8	11.44 ± 1.41	9.78 ± 1.26
AUC_{0-24} (h·μg/mL)	1.64 ± 0.28	17.95 ± 2.45	30.07 ± 4.64
CL/F (mL/h/kg)	488 ± 114	420 ± 85	548 ± 114

C_{max} peak concentration, T_{max} time of peak concentration, $T_{1/2\beta}$ elimination half-life, AUC_{0-24} area under the concentration-time curve from 0 to 24 h, *CL/F* oral clearance

been studied in chickens [29–31]. The elimination half-life ($T_{1/2\beta}$) values in this study (9.78–11.4 h) were similar to those in previous reports [30]. After oral administration of 10 mg/kg of enrofloxacin, the AUC_{0-24} value (17.95 h) in this study was much lower than that reported in previously (35 h in Da Silva's report and 25.35 in Mekala's) [30, 32]. The difference illustrates that the pathological state affects the total amount of the drug in the blood. Although ciprofloxacin is the main active metabolite of enrofloxacin, few studies have reported the concentration of ciprofloxacin. Da Silva reported that the concentration of ciprofloxacin was lower than their limit of quantification (LOQ) (0.2 μg/mL) in healthy chickens [30]. With an LOQ of 0.02 μg/mL, the concentration of ciprofloxacin in the present study was detected even after enrofloxacin administration at dose of 1 mg/kg b.w. The biotransformation of enrofloxacin to ciprofloxacin at doses of 10 and 20 mg/kg b.w. were 7.9 and 15.3%, which was in accordance with a previous study [16]. However, the biotransformation rate was as 37% higher for the low dose (1 mg/kg b.w.). The moderate concentration of ciprofloxacin indicated that the role of ciprofloxacin should be considered in pharmacodynamic studies of enrofloxacin. A good linear relationship between dosage and AUC_{0-24} was observed for enrofloxacin and ciprofloxacin. This phenomenon was similar to that reported in previous studies [21, 33].

Table 3 Pharmacokinetic parameters of ciprofloxacin in serum after oral administration of 1, 10, and 20 mg/kg body weight enrofloxacin in *E. coli* O78-infected chicks (mean and SD, $n = 10$)

Parameters	1 mg/kg	10 mg/kg	20 mg/kg
C_{max} (μg/mL)	0.03 ± 0.001	0.10 ± 0.03	0.37 ± 0.05
T_{max} (h)	5 ± 1.05	4.3 ± 1.98	4.4 ± 1.58
$T_{1/2\beta}$ (h)	27.25 ± 7.76	19.51 ± 2.60	18.77 ± 4.11
AUC_{0-24} (h·μg/mL)	0.61 ± 0.03	1.42 ± 0.15	4.93 ± 0.76
CL/F (mL/h/kg)	0.67 ± 0.11	3.96 ± 0.78	2.26 ± 0.46

C_{max} peak concentration, T_{max} time of peak concentration, $T_{1/2\beta}$ elimination half-life, AUC_{0-24} area under the concentration-time curve from 0 to 24 h, *CL/F* oral clearance

To the best of our knowledge, there has been no in vivo PK/PD modeling study of enrofloxacin against *E. coli* in chicks. One of the best PK/PD parameter for fluoroquinolones is AUC_{0-24}/MIC [34–36], and this study further confirmed this conclusion. Both the surrogate AUC_{0-24}/MIC for enrofloxacin or the combination of enrofloxacin and ciprofloxacin correlated well with effectiveness in each organ. It seems that the metabolism of ciprofloxacin has little influence on the PK/PD modeling of enrofloxacin. However, whether the active metabolite plays a role in emerging resistance or has an impact on dosing optimization needs further investigation [37], because optimization of the dosing regimen involves not only maximizing therapeutic outcome, but also minimizing the risk of developing resistance [38–40]. The values of AUC_{0-24}/MIC for the bactericidal effect were 19.3–32.15 in the different organs using only the concentration of enrofloxacin for simulation, and 21.29–41.68 when using the combined concentrations of enrofloxacin and ciprofloxacin for simulation. The values were much lower than those of enrofloxacin against *E. coli* or *Salmonella typhimurium* in the intestinal content of infected chicken calculated from ex vivo PK/PD modeling (1065.93 and 719.33, respectively) [14, 16]. Several reasons may explain this significant discrepancy. First, the components of the intestinal content are very complex, a large proportion of drugs may exist in a bound form and show no antimicrobial effect; however, in our PK/PD modeling, the whole amount of the drug was involved in the AUC_{0-24}/MIC calculation; therefore, the value of AUC_{0-24}/MIC in the intestines may be higher than that in serum to achieve the same effect. Second, as reported previously, the discrepancy between ex vivo PK/PD modeling and in vivo PK/PD modeling was obvious [21].

Using the bacterial burden reduction in each organ as PD endpoints, the value of the PK/PD parameter, AUC_{0-24}/MIC, to attain the same effect was different. The AUC_{0-24}/MIC value for the bactericidal effect in the liver was higher than that in lung, and twice than that in blood. This phenomenon was also observed in our previous study. The AUC_{0-24}/MIC values of danofloxacin against *Salmonella typhimurium* for the same effect in different organs also showed marked differences [22]. Similar results were also reported in other studies [20, 41]. The possible reason for these differences may lie in the differences in the initial bacterial load, the concentration diversity of drugs in each organ, and the complicated structures among different organs. The precise explanation requires further study. Usually, bacterial load reduction in a single organ is used for PD evaluation in most in vivo PK/PD studies, and for ex vivo PK/PD studies, the antibacterial effect of drugs in serum or other body fluid is used for PD calculation [12, 40, 42–47].

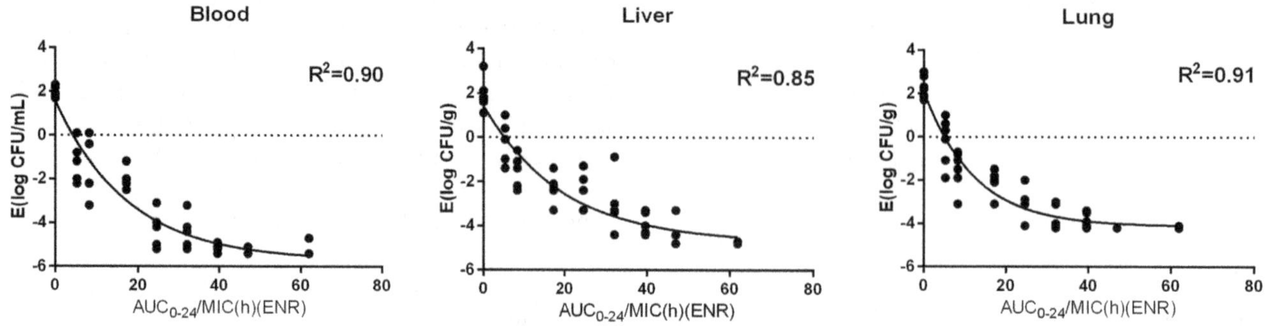

Fig. 3 Sigmoid maximum effect (E_{max}) relationship between the in vivo antimicrobial effect against *E. coli* O78 in three tissues of chicken and the area under the concentration-time curve from 0 to 24 h (AUC_{0-24})/minimum inhibitory concentration (MIC) ratio of enrofloxacin. [enrofloxacin (ENR)]

However, according to our results, for a systemic infection by bacteria, to describe the relationship of PK and PD exactly, bacterial loading reduction in different organs, as PD endpoints, should be compared in PK/PD modeling and the selection of a target organ for PD evaluation is critical in rational dosage recommendation.

The results obtained using this model need to be validated by clinical trials in relevant animal species. However, it is still a critical step to increase our understanding of PK/PD relationships for antimicrobials. To simulate the clinical use of enrofloxacin, enrofloxacin was administrated via gavage in this study; however, the most recent basic principles of the prudent use of antimicrobials do not support the group oral administration of fluoroquinolones in animals.

Conclusions

In conclusion, an in vivo PK/PD model of enrofloxacin against *E. coli* in seven-day-old chicks was established using bacteria loading reduction in several organs as PD endpoints. The in vivo effectiveness of enrofloxacin against *E. coli* in different organs varied with E_{max} ranging from −4.4 to −5.8 Log_{10} cfu/mL. Both the surrogate AUC_{0-24}/MIC of enrofloxacin or the combination AUC_{0-24}/MIC of enrofloxacin and ciprofloxacin correlated well with effectiveness in each organ. The combined AUC_{0-24}/MIC ratios producing bactericidal and elimination effects were 21.29 and 32.13 in blood; 41.68 and 58.52 in liver; and 27.65 and 46.22 in lung, respectively. The magnitude of PK/PD index was lowest for the same effect in blood, but highest in the liver, indicating that at the same dosage, the in vivo effectiveness of enrofloxacin against *E. coli* in different organs was not identical. This study emphasized the importance of target organ selection for PD evaluation in PK/PD modeling.

Methods

Organisms, chemicals, and animals

The clinical *E. coli* O78 strain, which was used in our previous study, was isolated from a broiler showing colibacillosis [23, 48]. The quality control standard *E. coli* strain ATCC 25922 was purchased from the Chinese Veterinary Culture Collection. The enrofloxacin reference standard and ciprofloxacin (98% purity) were purchased from Solarbio Life Sciences Co. Ltd. (Beijing, China). Acetonitrile (ACN) and methanol (MeOH) were purchased from TEDIA (Fairfield, CT, USA). All reagents used in this experiment were of high

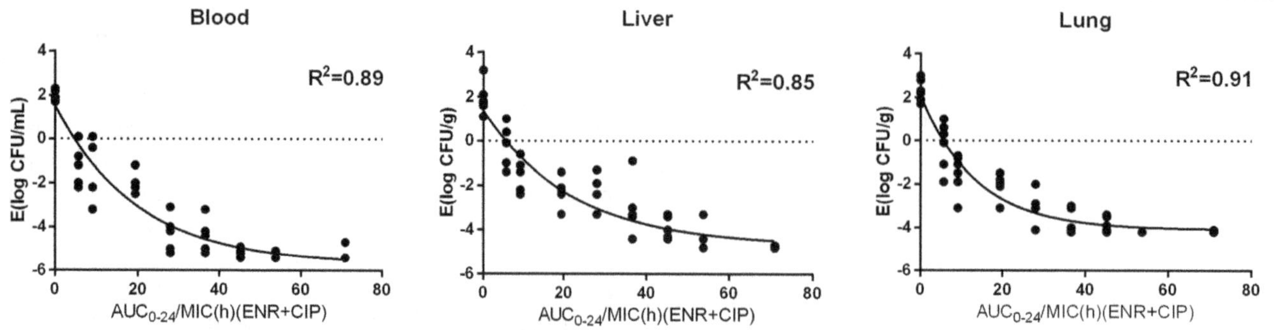

Fig. 4 Sigmoid maximum effect (E_{max}) relationship between the in vivo antimicrobial effect against *E. coli* O78 in three tissues of chicken and the area under the concentration-time curve from 0 to 24 h (AUC_{0-24})/minimum inhibitory concentration (MIC) ratio of enrofloxacin and ciprofloxacin. [enrofloxacin (ENR); ciprofloxacin (CIP)]

Table 4 In vivo PK/PD parameters of enrofloxacin in *E. coli* O78-infected chicks using the concentration of enrofloxacin for simulation

Parameters	Blood	Liver	Lung
E_{max} (Log $_{10}$ cfu/mL)	−5.8	− 4.7	−4.4
EC_{50}	16.61	16.93	10.33
E_0 (Log $_{10}$ cfu/mL)	1.78	1.68	2.29
AUC_{0-24}/MIC for 1 log_{10} cfu/mL killing	9.29	12.42	9.1
AUC_{0-24}/MIC for 3 log_{10} cfu/mL killing	19.32	32.15	23.41
AUC_{0-24}/MIC for 4 log_{10} cfu/mL killing	28.17	52.52	56.36
Slope (N)	2.58	2.36	3.01

E_{max} maximum effect, EC_{50} concentration of a drug that gives half-maximal response, E_0 concentration at baseline, AUC_{0-24} area under the concentration-time curve from 0 to 24 h, *cfu* colony forming units

performance liquid chromatography (HPLC) grade. The culture medium used in this experiment was purchased from Hope Biol-Technology Co. Ltd. (Qingdao, China). One-day-old healthy broilers ($n = 324$), obtained from the Jiangsu Institute of Poultry Sciences (Yangzhou, China), were used in this study. These broilers were mixed sex (50:50), weighting from 35.2 to 45.6 g. A six-day acclimation period was set for animals before the study. The animals were maintained according to the National Standards for Laboratory Animals of China (GB 14925–2010). This study was approved by the Animal Experiments Ethics Committee at Yangzhou University (SYXK (Su) IACUC 2017–0045).

Susceptibility testing

The MICs of enrofloxacin and ciprofloxacin against *E. coli* O78 and ATCC 25922 were determined using the micro-dilution method, according to the Clinical and Laboratory Standards Institute (CLSI) reference method [49]. The MIC of enrofloxacin in serum was also determined using the micro-dilution method according to a previous report [50]. To determine the minimum bactericidal concentration (MBC), 100 μL aliquots from the MIC determination procedure were diluted with Mueller-Hinton (MH) broth. The colony forming units (cfu) of each dilution were counted by spreading 100 μL dilutions on MH agar plates after 24 h of incubation at 37 °C. The lowest concentration of enrofloxacin that killed 99.9% of the bacteria was defined as the MBC. The mutant prevention concentration (MPC) determination was conducted according to a previous report [51]. Briefly, a series MH agar plates containing different drug concentrations (1 MIC to 64 MIC) were inoculated with more than 10^{10} cfu of *E. coli* and then incubated at 37 °C for 72 h. The MPC was determined as the lowest drug concentration that prevented bacterial growth.

Table 5 In vivo PK/PD parameters of enrofloxacin in *E. coli* O78-infected chicks using the combined concentration of enrofloxacin and ciprofloxacin for simulation

Parameters	Blood	Liver	Lung
E_{max} (Log $_{10}$ cfu/mL)	−5.8	−4.7	−4.4
EC_{50}	18.28	17.66	13.59
E_0 (Log $_{10}$ cfu/mL)	1.78	1.68	2.29
AUC_{0-24}/MIC for 1 log_{10} cfu/mL killing	9.74	14.01	9.49
AUC_{0-24}/MIC for 3 log_{10} cfu/mL killing	21.29	41.68	27.65
AUC_{0-24}/MIC for 4 log_{10} cfu/mL killing	32.13	60.52	57.24
Slope (N)	4.19	2.67	4.86

E_{max} maximum effect, EC_{50} concentration of a drug that gives half-maximal response, E_0 concentration at baseline, AUC_{0-24} area under the concentration-time curve from 0 to 24 h, *cfu* colony forming units

E. coli infection model

Preliminary experiments were conducted to confirm the inoculation amount. After a 6-day acclimation period, broilers were inoculated with 0.5 mL of *E. coli* culture containing ~ 10^7 cfu/mL through intramuscular injection (IM) in the chest muscle. The clinical symptoms and pathological changes were observed. Then, animals were sacrificed by a lethal intravenous injection of beuthanasia (0.3 mL/kg) after anesthesia with ketamine-Xylazine. The bacterial load in blood, liver, and lung were determined by the agar plate dilution method at 24 h post inoculation.

Pharmacokinetics

Enrofloxacin was administered orally to 280 infected broilers at doses of 1, 10, and 20 mg/kg body weight (b.w.). Ten blood samples were collected at each time point (0, 0.5, 1, 1.5, 2, 4, 6, 8, 12, and 24 h after administration). After sampling, animals were narcotized with ketamine-Xylazine and sacrificed by a lethal intravenous injection of beuthanasia (0.3 mL/kg). After incubation at room temperature, samples were centrifuged for 10 min at 3000×*g* to obtain serum. The serum was stored at − 20 °C until analysis. The concentrations of the drug in the serum were determined using HPLC with a fluorescence detector, as described previously, with some modifications [17, 52]. Briefly, 0.1 mL of serum was added to 1 mL of ACN containing 0.1% acetic acid, vortexed for 3 min, and then centrifuged at 12000×*g* for 10 min. The supernatant was transferred to a clean tube, dried under nitrogen, and re-dissolved with 0.1 mL 17% ACN. The sample was filtered through a 0.22-μm membrane before injecting into the HPLC apparatus. The recovery rate was between 80.2 and 91.3%, and the intra and inter coefficient of variation was less than 7%. The serum concentration–time data of enrofloxacin and ciprofloxacin for each animal was used to calculate the PK parameters in broilers using the WinNonlin software (version 6.1, Pharsight Corporation, Mountain View, CA, USA).

Pharmacodynamics determination

Infected broilers ($n = 54$) were randomly divided into nine groups ($n = 6$ in each group) and treated with enrofloxacin for 3 successive days at doses ranging from 0 to 20 mg/kg b.w. (0, 1, 2, 5, 7.5, 10, 12.5, 15, and 20 mg/kg per day). At 24 h after the last dose, the broilers were humanely killed through a lethal intravenous injection of beuthanasia (0.3 mL/kg) after anesthesia with ketamine-Xylazine to collect blood, liver, and lung samples. The bacterial loading in each organ was determined via plating dilutions onto MH agar and counting the colonies after incubation at 37 °C for 24 h. The effectiveness of enrofloxacin was expressed as the bacterial reduction after treatment compared with that before treatment in each organ.

Pharmacokinetics and Pharamcodynamics integration and modeling

The best PK/PD parameter for fluoroquinolones is AUC/MIC or C_{max}/MIC; therefore, in the present study, we chose the AUC_{0-24}/MIC method to model the PK data and in vitro PD data for enrofloxacin and its active metabolite ciprofloxacin. The sigmoid E_{max} model in the WinNonlin software (version 6.1; Pharsight) was used to simulate the relationship between AUC_{0-24}/MIC of enrofloxacin, or the combination AUC_{0-24}/MIC of enrofloxacin and ciprofloxacin and in vivo effectiveness. The equation for this model was as follows:

$$E = E_0 + \frac{E_{\max} \times C_e^N}{EC_{50}^N + C_e^N}$$

In the above formula, E_0 is the change in $\log_{10}$ cfu/mL or $\log_{10}$ cfu/g in the control sample (absence of drug). E_{max} is the difference in effect between the greatest amount of growth (as seen for the growth control, E_0) and the greatest amount of killing. C_e is the tested AUC_{0-24}/MIC ratio; EC_{50} is the AUC_{0-24}/MIC value that reached 50% of the E_{max}; and N is the Hill coefficient that describes the steepness of the AUC_{0-24}/MIC-effect curve [52]. The in vivo antibacterial effects of enrofloxacin were quantified into three levels including: (1) 1 log10 cfu/mL killing (E = – 1), (2) bactericidal action (99.9% reduction, E = – 3), and (3) bacterial elimination (99.99% reduction, E = – 4).

Data analysis

The PK data, PK/PD data, and PK/PD curve fitting were analyzed using the WinNonlin software (version 6.1; Pharsight). T-tests were conducted for other data using SPSS software (IBM, Armonk, NY, USA). $P < 0.05$ was considered statistically significant.

Abbreviations

ACN: Acetonitrile; AUC_{0-24}: Area under the concentrations-time curve from 0 to 24 h; CLSI: Clinical and Laboratory Standards Institute; CL_{β}/F: Clearance divided by bioavailability; C_{max}: Maximum serum concentration; F: Bioavailability; HPLC: High performance liquid chromatography; MBC: Minimum bactericidal concentration; MeOH: Methanol; MIC: Minimum inhibitory concnetration; MPC: Mutant prevention cocnetration; PK/PD: Pharmacokinetic/pharmacodynamic; $T_{1/2\beta}$: Elimination half-life; T_{max}: Time of maximum serum concentration; V_z/F: Volume of distribution scaled by bioavailability

Acknowledgements

Not applicable.

Funding

This work was funded by the National Key Research and Development Program of China (grant number 2016YFD0501310) in the design of the study; it was also supported by the National Natural Science Foundation of China (grant numbers 31702291, 31872526) in collection, analysis and interpretation of data and writing the manuscript; The Priority Academic Program Development of Jiangsu Higher Education Institutions, P. R. China, also paid publication charges for this article.

Authors' contributions

XX and ZW designed this study and revised and guided the experiment. XX wrote this manuscript and participated in the whole experiment process; LJ managed the whole experiment and analyzed the data; LJ and WL participated all the experiments; WL and YJ helped for the sampling process and concentration detection. All authors read and approved the final manuscript.

Competing interests

The authors declare that they have no competing interests.

Author details

[1]College of Veterinary Medicine, Yangzhou University, Yangzhou 225009, Jiangsu, People's Republic of China. [2]Jiangsu Co-innovation Center for Prevention and Control of Important Animal Infectious Diseases and Zoonoses, Yangzhou, Jiangsu 225009, People's Republic of China. [3]Institutes of Agricultural Science and Technology Development, 48 East Wenhui Road, Yangzhou, Jiangsu 225009, People's Republic of China.

References

1. Yassin H, Velthuis AG, Boerjan M, van Riel J. Field study on broilers' first-week mortality. Poult Sci. 2009;88(4):798–804.
2. Olsen RH, Frantzen C, Christensen H, Bisgaard M. An investigation on first-week mortality in layers. Avian Dis. 2012;56(1):51–7.
3. Kemmett K, Williams NJ, Chaloner G, Humphrey S, Wigley P, Humphrey T. The contribution of systemic Escherichia coli infection to the early mortalities of commercial broiler chickens. Avian Pathol. 2014;43(1):37–42.
4. Dziva F, Hauser H, Connor TR, van Diemen PM, Prescott G, Langridge GC, Eckert S, Chaudhuri RR, Ewers C, Mellata M, et al. Sequencing and functional annotation of avian pathogenic Escherichia coli serogroup O78 strains reveal the evolution of E. coli lineages pathogenic for poultry via distinct mechanisms. Infect Immun. 2013;81(3):838–49.
5. Salvadori M, Vercelli C, De Vito V, Dezzutto D, Bergagna S, Re G, Giorgi M. Pharmacokinetic and pharmacodynamic evaluations of a 10 mg/kg enrofloxacin intramuscular administration in bearded dragons (Pogona vitticeps): a preliminary assessment. J Vet Pharmacol Ther. 2017;40(1):62–9.

6. Pallo-Zimmerman LM, Byron JK, Graves TK. Fluoroquinolones: then and now. Compendium. 2010;32(7):E1–9 quiz E9.
7. Vanni M, Meucci V, Tognetti R, Cagnardi P, Montesissa C, Piccirillo A, Rossi AM, Di Bello D, Intorre L. Fluoroquinolone resistance and molecular characterization of gyrA and parC quinolone resistance-determining regions in Escherichia coli isolated from poultry. Poult Sci. 2014;93(4):856–63.
8. Dheilly A, Le Devendec L, Mourand G, Jouy E, Kempf I. Antimicrobial resistance selection in avian pathogenic E. coli during treatment. Vet Microbiol. 2013;166(3–4):655–8.
9. Dheilly A, Le Devendec L, Mourand G, Bouder A, Jouy E, Kempf I. Resistance gene transfer during treatments for experimental avian colibacillosis. Antimicrob Agents Chemother. 2012;56(1):189–96.
10. Ma R, Yang L, Ren T, Dong Y, Liu T, Zhuang P, Fang W, Yang X, Hu K. Enrofloxacin pharmacokinetics in Takifugu flavidusafter oral administration at three salinity levels. Aquac Res. 2017;48(8):4545–53.
11. Phu TM, Phuong NT, Scippo ML, Dalsgaard A. Quality of antimicrobial products used in striped catfish (Pangasianodon hypophthalmus) aquaculture in Vietnam. PLoS One. 2015;10(4):e0124267.
12. Koc F, Uney K, Ozturk M, Kadioglu Y, Atila A. Pharmacokinetics of florfenicol in the plasma of Japanese quail. N Z Vet J. 2009;57(6):388–91.
13. Aliabadi FS, Ali BH, Landoni MF, Lees P. Pharmacokinetics and PK-PD modelling of danofloxacin in camel serum and tissue cage fluids. Vet J. 2003;165(2):104–18.
14. Li J, Hao H, Cheng G, Wang X, Ahmed S, Shabbir MAB, Liu Z, Dai M, Yuan Z. The effects of different enrofloxacin dosages on clinical efficacy and resistance development in chickens experimentally infected with Salmonella typhimurium. Sci Rep. 2017;7(1):11676.
15. Wang J, Hao H, Huang L, Liu Z, Chen D, Yuan Z. Pharmacokinetic and Pharmacodynamic integration and modeling of Enrofloxacin in swine for Escherichia coli. Front Microbiol. 2016;7:36.
16. Sang K, Hao H, Huang L, Wang X, Yuan Z. Pharmacokinetic-Pharmacodynamic modeling of Enrofloxacin against Escherichia coli in broilers. Frontiers in veterinary science. 2015;2:80.
17. Ramalingam B, Sidhu PK, Kaur G, Venkatachalam D, Rampal S. Mutant prevention concentration, pharmacokinetic-pharmacodynamic integration, and modeling of enrofloxacin data established in diseased buffalo calves. J Vet Pharmacol Ther. 2015;38(6):529–36.
18. Daundkar PS, Vemu B, Dumka VK, Sharma SK. Pharmacokinetic-pharmacodynamic integration of enrofloxacin and its metabolite ciprofloxacin in buffalo calves. Vet Med Sci. 2015;1(2):63–71.
19. Balaje RM, Sidhu PK, Kaur G, Rampal S. Mutant prevention concentration and PK-PD relationships of enrofloxacin for Pasteurella multocida in buffalo calves. Res Vet Sci. 2013;95(3):1114–24.
20. Haritova A, Urumova V, Lutckanov M, Petrov V, Lashev L. Pharmacokinetic-pharmacodynamic indices of enrofloxacin in Escherichia coli O78/H12 infected chickens. Food Chem Toxicol. 2011;49(7):1530–6.
21. Xiao X, Sun J, Yang T, Fang X, Wu D, Xiong YQ, Cheng J, Chen Y, Shi W, Liu YH. In vivo pharmacokinetic/pharmacodynamic profiles of valnemulin in an experimental intratracheal mycoplasma gallisepticum infection model. Antimicrob Agents Chemother. 2015;59(7):3754–60.
22. Xiao X, Pei L, Jiang LJ, Lan WX, Xiao JY, Jiang YJ, Wang ZQ. In Vivo Pharmacokinetic/Pharmacodynamic Profiles of Danofloxacin in Rabbits Infected With Salmonella typhimurium After Oral Administration. Front Pharmacol. 2018;9:391.
23. Xiao X, Sun J, Chen Y, Zou M, Zhao DH, Liu YH. Ex vivo pharmacokinetic and pharmacodynamic analysis of valnemulin against mycoplasma gallisepticum S6 in mycoplasma gallisepticum and Escherichia coli co-infected chickens. Vet J. 2015;204(1):54–9.
24. Elmas M, Tras B, Kaya S, Bas AL, Yazar E, Yarsan E. Pharmacokinetics of enrofloxacin after intravenous and intramuscular administration in angora goats. Can J Vet Res. 2001;65(1):64–7.
25. McKellar Q, Gibson I, Monteiro A, Bregante M. Pharmacokinetics of enrofloxacin and danofloxacin in plasma, inflammatory exudate, and bronchial secretions of calves following subcutaneous administration. Antimicrob Agents Chemother. 1999;43(8):1988–92.
26. Anadon A, Martinez-Larranaga MR, Diaz MJ, Fernandez-Cruz ML, Martinez MA, Frejo MT, Martinez M, Iturbe J, Tafur M. Pharmacokinetic variables and tissue residues of enrofloxacin and ciprofloxacin in healthy pigs. Am J Vet Res. 1999;60(11):1377–82.
27. Giguere S, Sweeney RW, Belanger M. Pharmacokinetics of enrofloxacin in adult horses and concentration of the drug in serum, body fluids, and endometrial tissues after repeated intragastrically administered doses. Am J Vet Res. 1996;57(7):1025–30.
28. Kaartinen L, Salonen M, Alli L, Pyorala S. Pharmacokinetics of enrofloxacin after single intravenous, intramuscular and subcutaneous injections in lactating cows. J Vet Pharmacol Ther. 1995;18(5):357–62.
29. Guo QJ, Huang LL, Fang K, Wang YL, Chen DM, Tao YF, Dai MH, Liu ZL, Peng DP, Yuan ZH. Population pharmacokinetics of enrofloxacin and its metabolite ciprofloxacin in chicken based on retrospective data, incorporating first-pass metabolism. J Vet Pharmacol Ther. 2010;33(1):84–94.
30. da Silva RG, Reyes FG, Sartori JR, Rath S. Enrofloxacin assay validation and pharmacokinetics following a single oral dose in chickens. J Vet Pharmacol Ther. 2006;29(5):365–72.
31. Garcia Ovando H, Gorla N, Luders C, Poloni G, Errecalde C, Prieto G, Puelles I. Comparative pharmacokinetics of enrofloxacin and ciprofloxacin in chickens. J Vet Pharmacol Ther. 1999;22(3):209–12.
32. Mekala P, Jagadeeswaran A, Arivuchelvan A, Sethilkumar P, Nanjappan K, Krishnamurthy TRG. Comparison of pharmacokinetics of Enrofloxacin after single Oral bolus and pulse dose Administration in Broiler Chicken. Proc Nat Acad Sci India Sect B. 2015;87(3):679–83.
33. Schneider M, Paulin A, Dron F, Woehrle F. Pharmacokinetics of marbofloxacin in pigs after intravenous and intramuscular administration of a single dose of 8 mg/kg: dose proportionality, influence of the age of the animals and urinary elimination. J Vet Pharmacol Ther. 2014;37(6):523–30.
34. Zeng QL, Mei X, Su J, Li XH, Xiong WG, Lu Y, Zeng ZL. Integrated pharmacokinetic-Pharmacodynamic (PK/PD) model to evaluate the in vivo antimicrobial activity of Marbofloxacin against Pasteurella multocida in piglets. BMC Vet Res. 2017;13(1):178.
35. Dorey L, Pelligand L, Lees P. Prediction of marbofloxacin dosage for the pig pneumonia pathogens Actinobacillus pleuropneumoniae and Pasteurella multocida by pharmacokinetic/pharmacodynamic modelling. BMC Vet Res. 2017;13(1):209.
36. Xu L, Wang H, Yang X, Lu L. Integrated pharmacokinetics/pharmacodynamics parameters-based dosing guidelines of enrofloxacin in grass carp Ctenopharyngodon idella to minimize selection of drug resistance. BMC Vet Res. 2013;9:126.
37. Zhang B, Gu X, Li Y, Li X, Gu M, Zhang N, Shen X, Ding H. In vivo evaluation of mutant selection window of cefquinome against Escherichia coli in piglet tissue-cage model. BMC Vet Res. 2014;10:297.
38. Mouton JW, Ambrose PG, Canton R, Drusano GL, Harbarth S, MacGowan A, Theuretzbacher U, Turnidge J. Conserving antibiotics for the future: new ways to use old and new drugs from a pharmacokinetic and pharmacodynamic perspective. Drug Resista Updat. 2011;14(2):107–17.
39. Ambrose PG, Bhavnani SM, Rubino CM, Louie A, Gumbo T, Forrest A, Drusano GL. Pharmacokinetics-pharmacodynamics of antimicrobial therapy: it's not just for mice anymore. Clin Infect Dis. 2007;44(1):79–86.
40. Toutain PL, Lees P. Integration and modelling of pharmacokinetic and pharmacodynamic data to optimize dosage regimens in veterinary medicine. J Vet Pharmacol Ther. 2004;27(6):467–77.
41. Swanson RV, Adamson J, Moodley C, Ngcobo B, Ammerman NC, Dorasamy A, Moodley S, Mgaga Z, Tapley A, Bester LA, et al. Pharmacokinetics and pharmacodynamics of clofazimine in a mouse model of tuberculosis. Antimicrob Agents Chemother. 2015;59(6):3042–51.
42. Ahmad I, Hao H, Huang L, Sanders P, Wang X, Chen D, Tao Y, Xie S, Xiuhua K, Li J, et al. Integration of PK/PD for dose optimization of Cefquinome against Staphylococcus aureus causing septicemia in cattle. Front Microbiol. 2015;6:588.
43. Khan DD, Friberg LE, Nielsen EI. A pharmacokinetic-pharmacodynamic (PKPD) model based on in vitro time-kill data predicts the in vivo PK/PD index of colistin. J Antimicrob Chemother. 2016;71(7):1881–4.
44. Hawkins MG, Taylor IT, Byrne BA, Armstrong RD, Tell LA. Pharmacokinetic-pharmacodynamic integration of orbifloxacin in Japanese quail (Coturnix japonica) following oral and intravenous administration. J Vet Pharmacol Ther. 2011;34(4):350–8.
45. Zhang B, Gu X, Li X, Gu M, Zhang N, Shen X, Li Y, Ding H. Pharmacokinetics and ex-vivo pharmacodynamics of cefquinome against Klebsiella pneumonia in healthy dogs. J Vet Pharmacol Ther. 2014;37(4):367–73.
46. Govendir M. Review of some pharmacokinetic and pharmacodynamic properties of anti-infective medicines administered to the koala (Phascolarctos cinereus). J Vet Pharmacol Ther. 2017.
47. Haritova AM, Rusenova NV, Parvanov PR, Lashev LD, Fink-Gremmels J.

Integration of pharmacokinetic and pharmacodynamic indices of marbofloxacin in turkeys. Antimicrob Agents Chemother. 2006;50(11): 3779–85.
48. Xiao X, Zhao DH, Yang X, Shi W, Deng H, Ma J, Zhang S, Liu YH. Mycoplasma gallisepticum and Escherichia coli mixed infection model in broiler chickens for studying valnemulin pharmacokinetics. J Vet Pharmacol Ther. 2014;37(1):99–102.
49. CLSI: Performance Standards for Antimicrobial Disk and DilutionSusceptibility Tests for Bacteria Isolated From Animals: Approved Standard. 3rd Edition 2008:M31–A33.28(38).
50. Zhao DH, Zhou YF, Yu Y, Shi W, Yang X, Xiao X, Deng H, Qiao GG, Fang BH, Liu YH. Integration of pharmacokinetic and pharmacodynamic indices of valnemulin in broiler chickens after a single intravenous and intramuscular administration. Vet J. 2014;201(1):109–15.
51. Cao C, Qu Y, Sun M, Qiu Z, Huang X, Huai B, Lu Y, Zeng Z. In vivo antimicrobial activity of marbofloxacin against Pasteurella multocida in a tissue cage model in calves. Front Microbiol. 2015;6:759.
52. Burton ME, Shaw LM, Schentag JJ, Evans WE. Applied Pharmacokinetics & Pharmacodynamics: principles of therapeutic drug monitoring. Philadelphia: Lippincott Williams and Wilkins Press; 2005. p. 61–5.

PERMISSIONS

All chapters in this book were first published in VR, by BioMed Central; hereby published with permission under the Creative Commons Attribution License or equivalent. Every chapter published in this book has been scrutinized by our experts. Their significance has been extensively debated. The topics covered herein carry significant findings which will fuel the growth of the discipline. They may even be implemented as practical applications or may be referred to as a beginning point for another development.

The contributors of this book come from diverse backgrounds, making this book a truly international effort. This book will bring forth new frontiers with its revolutionizing research information and detailed analysis of the nascent developments around the world.

We would like to thank all the contributing authors for lending their expertise to make the book truly unique. They have played a crucial role in the development of this book. Without their invaluable contributions this book wouldn't have been possible. They have made vital efforts to compile up to date information on the varied aspects of this subject to make this book a valuable addition to the collection of many professionals and students.

This book was conceptualized with the vision of imparting up-to-date information and advanced data in this field. To ensure the same, a matchless editorial board was set up. Every individual on the board went through rigorous rounds of assessment to prove their worth. After which they invested a large part of their time researching and compiling the most relevant data for our readers.

The editorial board has been involved in producing this book since its inception. They have spent rigorous hours researching and exploring the diverse topics which have resulted in the successful publishing of this book. They have passed on their knowledge of decades through this book. To expedite this challenging task, the publisher supported the team at every step. A small team of assistant editors was also appointed to further simplify the editing procedure and attain best results for the readers.

Apart from the editorial board, the designing team has also invested a significant amount of their time in understanding the subject and creating the most relevant covers. They scrutinized every image to scout for the most suitable representation of the subject and create an appropriate cover for the book.

The publishing team has been an ardent support to the editorial, designing and production team. Their endless efforts to recruit the best for this project, has resulted in the accomplishment of this book. They are a veteran in the field of academics and their pool of knowledge is as vast as their experience in printing. Their expertise and guidance has proved useful at every step. Their uncompromising quality standards have made this book an exceptional effort. Their encouragement from time to time has been an inspiration for everyone.

The publisher and the editorial board hope that this book will prove to be a valuable piece of knowledge for researchers, students, practitioners and scholars across the globe.

LIST OF CONTRIBUTORS

Roselene Ecco
Departamento de Clínica e Cirurgia Veterinárias, Escola de Veterinárias, Universidade Federal de Minas Gerais, Av. Antônio Carlos, 6627, Campus Pampulha, 30161-970 Minas Gerais, Brazil

Juliana Fortes Vilarinho Braga
Departamento de Clínica e Cirurgia Veterinárias, Escola de Veterinárias, Universidade Federal de Minas Gerais, Av. Antônio Carlos, 6627, Campus Pampulha, 30161-970 Minas Gerais, Brazil
ISP, INRA, Université François Rabelais de Tours, 37380 Nouzilly, France

Nathalie Katy Chanteloup, Angélina Trotereau, Sylvie Baucheron, Rodrigo Guabiraba and Catherine Schouler
ISP, INRA, Université François Rabelais de Tours, 37380 Nouzilly, France

Feng Cong, Yujun Zhu, Jing Wang, Li Xiao, Ren Huang, Yu Zhang, Meili Chen and Pengju Guo
Guangdong Laboratory Animals Monitoring Institute and Guangdong Provincial Key Laboratory of Laboratory Animals, Guangzhou 510633, China

Yuexiao Lian and Xiangnan Liu
Guangdong Laboratory Animals Monitoring Institute and Guangdong Provincial Key Laboratory of Laboratory Animals, Guangzhou 510633, China
Guangdong Provincial Key Laboratory of Zoonosis Prevention and Control, College of Veterinary Medicine, South China Agricultural University, Guangzhou 510640, China

Baoyuan Liu, Yani Sun, Yiyang Chen, Taofeng Du, Yuchen Nan, Xinjie Wang, Huixia Li, Baicheng Huang, En-Min Zhou and Qin Zhao
Department of Preventive Veterinary Medicine, College of Veterinary Medicine, Northwest A&F University, 22 Xinong Road, Yangling, Shaanxi, China
Scientific Observing and Experimental Station of Veterinary Pharmacology and Veterinary Biotechnology, Ministry of Agriculture, Yangling, Shaanxi, China

Gaiping Zhang
Department of Preventive Veterinary Medicine, College of Veterinary Medicine, Henan Agriculture University, Zhengzhou, Henan 450002, China

Sara Joller and Cord Drögemüller
Institute of Genetics, Vetsuisse Faculty, University of Bern, Bern, Switzerland

Christine Flury and Flurina Bertschinge
School of Agricultural, Forest and Food Sciences, Bern University of Applied Sciences, Zollikofen, Switzerland

Erwin Kump
ProSpecieRara, Basel, Switzerland

Astrid Spiri
Züchterverein für ursprüngliches Nutzgeflügel, Neukirch an der Thur, Switzerland

Alois von Rotz
Divison of Veterinary Anatomy, Vetsuisse Faculty, University of Bern, Bern, Switzerland

Daniela Schweizer-Gorgas
Division of Clinical Radiology, Vetsuisse Faculty, University of Bern, Bern, Switzerland

Ke Shang, Bai Wei and Min Kang
Department of Veterinary Infectious Diseases and Avian Diseases, College of Veterinary Medicine and Center for Poultry Diseases Control, Chonbuk National University, 79 Gobong-ro, Iksan 54596, South Korea

Ting-Rong Luo
State Key Laboratory for Conservation and Utilization of Subtropical Agro-Bioresources, Guangxi University, Nanning, Guangxi 530004, China

Shi-Kai Sun
State Key Laboratory for Conservation and Utilization of Subtropical Agro-Bioresources, Guangxi University, Nanning, Guangxi 530004, China
Guangdong Enterprise Key Laboratory for Animal Health and Environmental Control, WENS Group Academy, Yunfu, Guangdong 527439, China

Xin Lin, Ding-Ai Wang, Jun-Peng Lu and Jian-Ping Qin
Guangdong Enterprise Key Laboratory for Animal Health and Environmental Control, WENS Group Academy, Yunfu, Guangdong 527439, China

Feng Chen
College of Animal Science, South China Agricultural University, Guangzhou, Guangdong 510642, China

Letícia G. Pacheco, Nilva K. Sakomura, Rafael M. Suzuki and Gabriel S. Viana
Department of Animal Science, Unesp, Jaboticabal, SP, Brazil

Juliano C. P. Dorigam
Evonik Nutrition & Care GmbH, Hanau, Germany

Jaap Van Milgen
INRA, Agrocampus Ouest, Rennes, France

Juliana C. Denadai
Department of Animal Science, Unesp, Botucatu, SP, Brazil

Behnam Ahmadipour and Fariborz Khajali
Department of Animal Science, Shahrekord University, Shahrekord 88186-34141, Iran

Hossein Hassanpour
Department of Basic Science, Shahrekord University, Shahrekord 88186-34141, Iran

Sheng Niu, Ali Raza Jahejo, Fa-jie Jia, Xin Li, Guan-bao Ning, Ding Zhang, Hai-li Ma, Wen-wei Gao, Yu-jun Zhao, Shi-min Gao, Gui-lan Li, Jian-hui Li, Fang Yan, Rong-kun Gao and Wen-xia Tian
College of Animal Science and Veterinary Medicine, Shanxi Agricultural University, Taigu 030801, China

George F. Gao and Yu-hai Bi
College of Animal Science and Veterinary Medicine, Shanxi Agricultural University, Taigu 030801, China
CAS Key Laboratory of Pathogenic Microbiology and Immunology, Collaborative Innovation Center for Diagnosis and Treatment of Infectious Disease, Institute of Microbiology, Center for Influenza Research and Early-warning (CASCIRE), Chinese Academy of Sciences, Beijing 100101, China

Wei-fang Hao
Taiyuan Center for Disease Control and Prevention, Taiyuan 030024, China

Ling-xia Han
Department of Laboratory Animal and Comparative Medicine, State Key Laboratory of Veterinary Biotechnology, Harbin Veterinary Research Institute, Chinese Academy of Agricultural Sciences, Harbin 150001, China

Krishna Prasad Vadalasetty and Radhika Vadalasetty
Department of Veterinary and Animal Sciences, University of Copenhagen, 1870 Frederiksberg, Denmark

André Chwalibog
Department of Veterinary and Animal Sciences, University of Copenhagen, 1870 Frederiksberg, Denmark
Department of Animal Nutrition and Biotechnology, Warsaw University of Life Sciences, 02-786 Warsaw, Poland

Charlotte Lauridsen and Ricarda Margarete Engberg
Department of Animal Science, Aarhus University, 8830 Tjele, Denmark

Ewa Sawosz and Marta Kutwin
Department of Animal Nutrition and Biotechnology, Warsaw University of Life Sciences, 02-786 Warsaw, Poland

Henny Akit and Anjas Asmara Samsudin
Department of Animal Science, Faculty of Agriculture, Universiti Putra Malaysia, 43400 UPM, Serdang, Selangor, Malaysia

Teck Chwen Loh
Department of Animal Science, Faculty of Agriculture, Universiti Putra Malaysia, 43400 UPM, Serdang, Selangor, Malaysia
Institute of Tropical Agriculture, Universiti Putra Malaysia, 43400 UPM, Serdang, Selangor, Malaysia

Karwan Yaseen Kareem
Department of Animal Science, Faculty of Agriculture, Universiti Putra Malaysia, 43400 UPM, Serdang, Selangor, Malaysia
Department of Animal Resource, University of salah al-Din, Erbil, Iraq

Hooi Ling Foo
Department of Bioprocess Technology, Faculty of Biotechnology and Biomolecular Science, Universiti Putra Malaysia, 43400 UPM, Serdang, Selangor, Malaysia
Institute of Bioscience, Universiti Putra Malaysia, 43400 UPM, Serdang, Selangor, Malaysia

Hany F. Ellakany, Ahmed R. Gado, Ahmed R. Elbestawy and Hatem S. Abd El-Hamid
Department of Poultry and Fish Diseases, Faculty of Veterinary Medicine, Damanhour University, Elgomhoria st, 63, Damanhour, Elbehira 22511, Egypt

Hafez M. Hafez
Institute of Poultry Diseases, Free University Berlin, Königsweg, 63, 14163 Berlin, Germany

Mohamed E. Abd El-Hack
Department of Poultry, Faculty of Agriculture, Zagazig University, Zagazig 44511, Egypt

Abdullah Al-Owaimer
Department of Animal Production, College of Food and Agriculture Sciences, King Saud University, Riyadh 11451, Saudi Arabia

Ayman A. Swelum
Department of Animal Production, College of Food and Agriculture Sciences, King Saud University, Riyadh 11451, Saudi Arabia
Department of Theriogenology, Faculty of Veterinary Medicine, Zagazig University, Zagazig 44511, Egypt

Islam M. Saadeldin
Department of Animal Production, College of Food and Agriculture Sciences, King Saud University, Riyadh 11451, Saudi Arabia
Department of Physiology, Faculty of Veterinary Medicine, Zagazig University, Zagazig 44511, Egypt

Xiao Chuan Zheng and Tian Wang
College of Animal Science and Technology, Nanjing Agricultural University, No. 6, Tongwei Road, Xuanwu District, Nanjing 210095, Jiangsu, People's Republic of China

Qiu Jue Wu
College of Animal Science and Technology, Nanjing Agricultural University, No. 6, Tongwei Road, Xuanwu District, Nanjing 210095, Jiangsu, People's Republic of China
State Key Laboratory of Animal Nutrition, NO. 2, Yuan Ming Yuan West Road, HaiDian District, Beijing 100193, People's Republic of China
College of Animal Science and Technology, Henan University of Science and Technology, Luoyang 471003, Henan, People's Republic of China

Tie Ying Zhang
State Key Laboratory of Animal Nutrition, NO. 2, Yuan Ming Yuan West Road, HaiDian District, Beijing 100193, People's Republic of China

Y. W. Kristeen-Teo, S. K. Yeap, S. W. Tan and A. R. Omar
Institute of Bioscience, Universiti Putra Malaysia, 43400 Serdang, Selangor, Malaysia

N. B. Alitheen
Institute of Bioscience, Universiti Putra Malaysia, 43400 Serdang, Selangor, Malaysia
Faculty of Biotechnology and Biomolecular Sciences, Universiti Putra Malaysia, 43400 Serdang, Selangor, Malaysia

A. Ideris
Faculty of Veterinary Medicine, Universiti Putra Malaysia, 43400 Serdang, Selangor, Malaysia.

S. G. Tan
3Faculty of Biotechnology and Biomolecular Sciences, Universiti Putra Malaysia, 43400 Serdang, Selangor, Malaysia

Sasidhar Malladi, Peter J. Bonney, David A. Halvorson and Carol J. Cardona
Secure Food Systems Team, University of Minnesota, Veterinary and Biomedical Sciences, 301C Veterinary Science Building, 1971 Commonwealth Avenue, Saint Paul, MN 55108, USA

Amos Ssematimba
Secure Food Systems Team, University of Minnesota, Veterinary and Biomedical Sciences, 301C Veterinary Science Building, 1971 Commonwealth Avenue, Saint Paul, MN 55108, USA
Department of Mathematics, Faculty of Science, Gulu University, Gulu, Uganda

Cristian Flores-Figueroa and Jeannette Muñoz-Aguayo
Mid Central Research and Outreach Center, 1802 18th St NE, Willmar, MN 56201, USA

A. M. Dalia and T. C
Department of Animal Science, Faculty of Agriculture, Universiti Putra Malaysia, 43400 Serdang, Selangor, Malaysia
Department of Animal Nutrition, Faculty of Animal Production, University of Khartoum, Khartoum, Sudan

M. F. Jahromi, Loh, A. Q. Sazili and A. A. Samsudin
Institute of Tropical Agriculture, Universiti Putra Malaysia, 43400 Serdang, Selangor, Malaysia

Ilaria Biasato, Elena Biasibetti, Elena Grego, Sihem Dabbou, Alessandra Sereno and Maria Teresa Capucchio
Department of Veterinary Sciences, University of Turin, Largo Paolo Braccini

Achille Schiavone
Department of Veterinary Sciences, University of Turin, Largo Paolo Braccini
Institute of Multidisciplinary Research on Sustainability, University of Turin, Via Accademia Albertina 13, 10100 Turin, Italy

Luca Cocolin and Ilario Ferrocino
Department of Agricultural, Forest and Food Sciences, University of Turin, Largo Paolo Braccini 2, 10095 Grugliasco, TO, Italy

Laura Gasc
Department of Agricultural, Forest and Food Sciences, University of Turin, Largo Paolo Braccini 2, 10095 Grugliasco, TO, Italy Institute of Science of Food Production, National Research Council, Largo Paolo Braccini 2, 10095 Grugliasco, TO, Italy

Francesco Gai
Institute of Science of Food Production, National Research Council, Largo Paolo Braccini 2, 10095 Grugliasco, TO, Italy
Institute of Multidisciplinary Research on Sustainability, University of Turin, Via Accademia Albertina 13, 10100 Turin, Italy

M. Siwek, A. Slawinska, K. Stadnicka, A. Dunislawska and M. Bednarczyk
Department of Animal Biotechnology and Genetics, UTP University of Science and Technology, Mazowiecka, 28 85-084 Bydgoszcz, Poland

J. Bogucka
Department of Animal Physiology, Physiotherapy and Nutrition, UTP University of Science and Technology, Mazowiecka, 28 85-084 Bydgoszcz, Poland

Aruna Amarasinghe, Upasama De Silva Senapathi, Mohamed Sarjoon Abdul-Cader, Susan C. Cork, Frank van der Meer and Mohamed Faizal Abdul-Careem
Department of Ecosystem and Public Health, Faculty of Veterinary Medicine, University of Calgary, Health Research Innovation Center 2C53, 3330 Hospital Drive NW, Calgary, AB T2N 4N1, Canada

Susantha Gomis and Shelly Popowich
Department of Veterinary Pathology, Western College of Veterinary Medicine, University of Saskatchewan, Saskatoon, SK S7N 5B5, Canada

Frank Marshall
Marshall Swine and Poultry Health Services, 3831- Bay G- 44 Ave, Camrose, AB T4V 3T1, Canada

Sandra Morales Ruiz, Jorge Bendezu, Ricardo Choque Guevara, Ricardo Montesinos, Luz Choque Moreau and Ángela Montalván Ávalos
Laboratorios de Investigación y Desarrollo, FARVET SAC, Carretera Panamericana Sur N°766 Km 198.5, Chincha Alta 11702, Ica, Peru

David Requena
Laboratorios de Investigación y Desarrollo, FARVET SAC, Carretera Panamericana Sur N°766 Km 198.5, Chincha Alta 11702, Ica, Peru
Laboratorio de Bioinformática y Biología Molecular, Laboratorio de Investigación y Desarrollo, Universidad Peruana Cayetano Heredia, Av. Honorio Delgado 430 San Martín de Porres, Lima, Lima, Peru

Manolo Fernández-Díaz
Laboratorios de Investigación y Desarrollo, FARVET SAC, Carretera Panamericana Sur N°766 Km 198.5, Chincha Alta 11702, Ica, Peru
FARVET SPF SAC, Carretera Panamerica Sur N°766 Km 198.5, Chincha Alta 11702, Ica, Peru

Limin Li, Ping Chen, Shan Zhang, Jiguo Sun and Wanzhe Yuan
College of Veterinary Medicine, Hebei Agricultural University, Baoding 071001, Hebei, China
Hebei Engineering and Technology Research Center of Veterinary Biotechnology, Baoding 071001, Hebei, China
North China Research Center of Animal Epidemic Pathogen Biology, China Agriculture Ministry, Baoding 071001, Hebei, China

Jianchang Wang
Inspection and Quarantine Technical Center of Hebei Entry-Exit Inspection and Quarantine Bureau, 318 Heping West Rd, Xinhua District, Shijiazhuang 050051, Hebei, China

Lijie Jiang, Weixuan Lan and Yongjia Jiang
College of Veterinary Medicine, Yangzhou University, Yangzhou 225009, Jiangsu, People's Republic of China

Xia Xiao and Zhiqiang Wang
College of Veterinary Medicine, Yangzhou University, Yangzhou 225009, Jiangsu, People's Republic of China
Jiangsu Co-innovation Center for Prevention and Control of Important Animal Infectious Diseases and Zoonoses, Yangzhou, Jiangsu 225009, People's Republic of China
Institutes of Agricultural Science and Technology Development, 48 East Wenhui Road,
Yangzhou, Jiangsu 225009, People's Republic of China

Hua Zhang, Zhenlei Zhou, Jingwen Luo and Jiafa Hou
College of Veterinary Medicine, Nanjing Agricultural University, Nanjing, Jiangsu 210095, China

Pete Kaiser and David W. Burt
The Roslin Institute & R(D)SVS, University of Edinburgh, Easter Bush, Midlothian EH25 9RG, UK

Index

www.ingramcontent.com/pod-product-compliance
Lightning Source LLC
LaVergne TN
LVHW080918210926
848485LV00004B/680